AF882801

Signals and Communication Technology

More information about this series at http://www.springer.com/series/4748

Wei Gao

Energy and Bandwidth-Efficient Wireless Transmission

Springer

Wei Gao
Broadcom Limited
Sunnyvale, CA, USA

ISSN 1860-4862 ISSN 1860-4870 (electronic)
Signals and Communication Technology
ISBN 978-3-319-44220-4 ISBN 978-3-319-44222-8 (eBook)
DOI 10.1007/978-3-319-44222-8

Library of Congress Control Number: 2016963595

© Springer International Publishing Switzerland 2017
This work is subject to copyright. All rights are reserved by the Publisher, whether the whole or part of the material is concerned, specifically the rights of translation, reprinting, reuse of illustrations, recitation, broadcasting, reproduction on microfilms or in any other physical way, and transmission or information storage and retrieval, electronic adaptation, computer software, or by similar or dissimilar methodology now known or hereafter developed.
The use of general descriptive names, registered names, trademarks, service marks, etc. in this publication does not imply, even in the absence of a specific statement, that such names are exempt from the relevant protective laws and regulations and therefore free for general use.
The publisher, the authors and the editors are safe to assume that the advice and information in this book are believed to be true and accurate at the date of publication. Neither the publisher nor the authors or the editors give a warranty, express or implied, with respect to the material contained herein or for any errors or omissions that may have been made.

Printed on acid-free paper

This Springer imprint is published by Springer Nature
The registered company is Springer International Publishing AG
The registered company address is: Gewerbestrasse 11, 6330 Cham, Switzerland

Dedicated to my parents
for their endless love and constant encouragement

Preface

This book provides an overview of the designs of energy and bandwidth-efficient or spectrum-efficient modulation and transmission for wireless cellular and Wireless Local Area Network (WLAN) systems, and focuses especially on system-on-chip (SoC) architecture that integrates all-digital, analog/mixed-signal, and often radio-frequency functions on a single chip substrate. Because of the unique aspects of the SoC architecture, the properties of energy efficiency (EE) and spectrum efficiency (SE) are two high-priority targets in these system designs and highly depend on the modulation scheme that is chosen and the operation region where a power amplifier runs. Trade-offs between EE and SE due to their contradictory nature, however, should be made depending on the specific modulation scheme and actual transmission system.

As demand has grown for high-data-rate applications in wireless networks, energy consumption in wireless communications has recently drawn increasing scientific and industrial attention. Energy-efficient solutions to reducing energy consumption in order to achieve significant reductions in the amount of carbon dioxide emissions in wireless radio communications are classified under the rubric Green Radio. To enhance energy-efficient utilization in wireless communications, different Green Radio programs across the world aim at investigating and creating innovative techniques for the reduction of the total energy used to operate radio access networks and to identify appropriate radio architectures that enable such power reduction.

In the early years of digital satellite communication (i.e., the-1980s and 1990s), energy and bandwidth-efficient modulation techniques and schemes were studied and developed in applications for digital satellite earth station systems in which voice and data transmissions in digital form were supported. Overlapped pulse-shaping modulation formats, such as raised-cosine pulse shaping or other doubled-symbol interval pulse-shaping methods were proposed and developed for these earth station systems so that on-board power amplifiers could operate in saturated or close to saturated regions to achieve high energy efficiency. These overlapped pulse-shaping modulation schemes have an advantage over Nyquist-type-signaling–based modulation formats in achieving higher energy efficiency because of their smaller envelope fluctuations; meanwhile they also have satisfactory spectrum efficiency. When further modifications were applied to the existing overlapped pulse-shaping modulation schemes in 2000s, the most representative modulation format, called Feher-patented quadrature phase shift keying (FQPSK), showed a good trade-off between energy and spectrum efficiency and lead to its application to deep-space communications.

After the launch of the 802.11a WLAN standard using 52-subcarrier orthogonal frequency division multiplexing (OFDM) format signaling in 1999 and the adoption of OFDM technology as the core technology of 4G in 2010 because of its anti-multipath–fading characteristics and high spectrum efficiency, achieving energy efficiency (EE) becomes more difficult. Improving the EE usually involves running the power amplifier in a saturated or close to saturated region, which is a great challenge for most transmitters used in the 802.11 WLAN systems and the 4G cellular communication systems due to large Peak-to-Average Power Ratio (PAPR) of the OFDM signaling. The trade-off between energy efficiency and spectrum efficiency for OFDM signals becomes particularly important and can be achieved by using a number of techniques and methods described in this book.

In 2002, I changed my career focus from traditional digital communication fields, where I had worked for a decade on included satellite/earth station and microwave systems, to the newly developed personal digital communication field, which mainly focuses on 2G/3G cellular and 802.11b/g/a WLAN communication systems implemented with SoC architectures, where I have worked until now. Since then, I have been involved with many new system concepts and circuit block implementations that are different from what I had learned from textbooks or classes in colleges and what I had experienced with traditional digital communication systems. These new concepts, architectures, and hardware implementations covered in cellular and WLAN communication systems initially motivated me to write a technique book and to share my design experiences with readers who are interested in these practical system design concepts.

The book primarily covers some major energy and bandwidth-efficient modulation and transmission techniques with applications to cellular and the 802.11 WLAN systems. For either constant envelope modulation or nearly constant envelope modulation formats, the transmitter can operate in a saturated region to achieve high energy efficiency. For non-constant envelope modulation schemes with large peak-to-average power ratio (PAPR) values, on the other hand, the power amplifier employing a pre-distortion technique is capable of operating close to a saturated region to improve energy efficiency. A variety of fundamental and practical topics associated with the SoC architectures, from communication system design concepts to baseband signal-processing algorithms and trade-offs are introduced in the book. These topics and some design examples are drawn from projects and products I previously designed or was involved with.

This book is aimed at recently graduated college students with electrical-engineering related backgrounds and practicing engineers who may be new to digital modulations and digital transmission systems, but already have a general broad knowledge of some digital communication system concepts. Especially for those readers who wish to learn and understand the system concepts and circuit block designs of RF IC systems, the book provides a variety of design examples and practical topics with analysis graphics to assist them for easily understanding these contents. After reading this book, the reader should have gained a strong system-level overview of various wireless systems associated with SoC-based implementations, and should understand the advantages and disadvantages that particular modulation techniques and certain system architectures have.

I would like to acknowledge my previous colleagues at the Communication Division of Harris Corp. and Via Communication Corp. where we worked together to make great contributions to many microwave modem products, 2G/3G cellular and Wi-Fi transceiver chips. I would also like to thank Broadcom for giving me a chance to contribute to and support their cellular and Wi-Fi transceiver chips and their applications.

Finally, I wish to express my great gratitude to my parents for their lifelong love and constant encouragement; my wife, Lutong, for her endless love, tolerance, and support during the long journey of writing the book; and my son, Will, for his feedback and opinions as a first reader of the manuscript during his last year pursuing his master's degree in Electrical and Computer Engineering.

Milpitas, CA Wei Gao
June, 2016

Contents

About the Author

Wei Gao received his Ph.D. in electrical and computer engineering from the University of California, Davis, in 2001. He currently works at Broadcom Limited, first with the Cellular group and now with the Wi-Fi group. Before joining Broadcom, he specialized in designing and validating energy and spectrally efficient modulation/demodulation for satellite earth station communication systems. He also worked on digital modems with adaptive equalization and cross-polarization cancellation features for microwave communication systems. In addition, his background includes wireless architectures/systems and power amplifier linearization modules for 2G/3G Cellular and IEEE 802.11 Wi-Fi standard transceivers. He has over 30 years of industry experience in algorithm development and circuit/system designs at companies including the Harris Corporation and Via Communication Technology.

Abbreviations

2G	Second generation wireless communication networks
3G	Third generation wireless communication networks
3GPP	Third Generation Partnership Project
4G	Fourth generation wireless communication networks
5G	Fifth generation wireless communication networks
8DPSK	8-angle differential phase shift keying
8PSK	8-angle phase shift keying
π/4-DQPSK	π/4 shift differential quadrature phase shift keying
16-QAM	16-ary quadrature amplitude modulation
32-QAM	32-ary quadrature amplitude modulation
AAC	Alternative adjacent channel
AC	Alternating current or adjacent channel
ACI	Adjacent channel interference
ACK	Acknowledgement
ACLR	Adjacent channel leakage ratio
ACPR	Adjacent channel power ratio
ACR	Adjacent channel rejection
ACS	Adjacent channel selectivity
ADC	Analog-to-digital converter
ADPLL	All digital phase-locked loop
AGC	Automatic gain control
AM	Amplitude modulation
AM-AM	Amplitude modulation to amplitude modulation
AM-PM	Amplitude modulation to phase modulation
AMP	Amplitude-modulation pulse
AMPS	Advanced mobile phone system
AP	Access point
APD	Analog pre-distortion (or -distorter)
ASK	Amplitude shift keying
AWGN	Additive white Gaussian noise

BB	Baseband
BER	Bit error rate
Bi-CMOS	Bipolar and complementary metal-oxide-semiconductor
BO	Back off
BPF	Bandpass filter
BPSK	Binary phase shift keying
BTS	Base transceiver station
CCDF	Complementary cumulative distribution function
CCI	Co-channel interference
CCK	Complementary code keying
CCSDS	Consultative committee for space data systems
CDMA	Code division multiple access
CF	Crest factor
CFO	Carrier frequency offsets
CFR	Channel frequency response
CIR	Channel impulse response
C/I	Carrier-to-interference
CLK	Clock
CM	Cross-modulation
CMA	Constant modulus algorithm
CMOS	Complementary metal-oxide-semiconductor
CP	Cyclic prefix
CPFSK	Continuous phase frequency shift keying
CPM	Continuous phase modulation
CPW	Clipping and peak window
CSDF	Channel selection digital filter
CW	Continuous waveform
DA	Data-aided
DAC	Digital-to-analog converter
DBB	Digital baseband
DC	Direct current
DCF	Distributed coordination function
DCOC	Direct current offset cancellation (or correction)
DDC	Direct down-conversion
DECT	Digital enhanced cordless telecommunications
DFE	Decision-feedback equalizer
DFT	Discrete Fourier transform
DMA	Direct memory access
DoD	Department of defense
DPD	Digital pre-distortion (or -distorter)
DRSSI	Digital received signal strength indictor
DSB	Double side band
DSP	Digital signal processing (or processor)
DSSS	Direct sequence spread spectrum

DTV	Digital TV
DUC	Direct up-conversion
DUT	Device under test
DVB	Digital video broadcasting
E_b/N_o	Energy per bit to noise power spectral density ratio
EDGE	Enhanced data rates for GSM evolution
EE	Energy efficiency
ENB	Equivalent noise bandwidth
ENOB	Effective number of bits
ET	Envelope tracking
ETSI	European Telecommunication Standards Institute
EVM	Error vector magnitude
FCC	Federal Communications Commission
FDM	Frequency division multiplexing
FDMA	Frequency division multiple access
FEM	Front-end module
FER	Frame error rate
FFT	Fast Fourier transform
FHSS	Frequency-hopping spread spectrum
FIR	Finite impulse response
FPGA	Field-programmable gate array
FQPSK	Feher-patented quadrature phase shift keying
FQPSK-B	Filtered FQPSK
GI	Guard interval
GFSK	Gaussian frequency shift keying
GLPF	Gaussian lowpass filter
GMSK	Gaussian filtered minimum shift keying
GPRS	General packet radio service
GSM	Global system for mobile communications
HD2	Second-order harmonic distortion
HG	High gain
HIU	Host interface unit
HP	Highpass
HPF	Highpass filter
HQPSK	Hybrid quadrature phase shift keying
HSDPA	High speed down-link packet access
HSUPA	High speed up-link packet access
IC	Integrated circuit
I-CH	In-phase channel
ICI	Intercarrier interference
I&D	Integrate and dump
IDFT	Inverse discrete Fourier transform
IF	Intermediate frequency
IFFT	Inverse fast Fourier transform

IIP2	Input second-order intercept point
IIP3	Input third-order intercept point
IJF-OQPSK	Intersymbol interference and jitter free OQPSK
IL	Insertion loss
IM2	Second-order intermodulation
IM3	Third-order intermodulation
IMD	Intermodulation distortion
IP2	Second-order intercept point
IP3	Third-order intercept point
I-Q	In-phase and quadrature channels (or branches)
IQMPGA	I-Q modulation program gain amplifier
IRR	Image rejection ratio
ISI	Intersymbol interference
ISM	Industrial scientific and medical
ISSCC	International Solid-State Circuits Conference
ITU	International Telecommunication Union
JF	Jitter-free
LFC	Low-frequency component
LG	Low gain
LI	Linear interpolation
LMS	Least-mean square
LNA	Low noise amplifier
LO	Local oscillator
LOFT	LO feed-through
LP	Lowpass
LPF	Lowpass filter
LS	Least square
LSE	Least square error
LTE	Long-term evolution
LTI	Linear time-invariant
LUT	Look-up table
MAC	Media access control
MCS7	Modulation and coding scheme rate index 7
MCS9	Modulation and coding scheme rate index 9
MFM	Multipath fading margin
MG	Middle gain
MIMO	Multi-input and multi-output
ML	Maximum-likelihood
MMSE	Minimum mean square error
MP	Memory polynomial
M-QAM	M-ary quadrature amplitude modulation
MSE	Mean square error
MSK	Minimum shift keying
MU-MIMO	Multi-user MIMO

NADC	North American Digital Cellular
NASA	National Aeronautics and Space Administration
NCO	Numerically controlled oscillator
NDA	Non-data-aided
NF	Noise figure or noise factor
NMSE	Normalized mean square error
NRSSI	Narrowband received signal strength indictor
NRZ	Non-return-to-zero
OFDM	Orthogonal frequency division multiplexing
OIP3	Output third-order intercept point
OPLL	Offset phase-locked loop
OQPSK	Offset quadrature phase shift keying
ORFS	Output RF spectrum
P1dB	Power 1 dB compression point
P1dB/S	Power 1 dB compression point-to-signal
PA	Power amplifier
PAE	Power-added efficiency
PAM	Pulse amplitude modulation
PAPR	Peak-to-average power ratio
PC	Peak cancellation or personal computer
PCB	Printed circuit board
PCF	Point coordination function
PCU	Protocol control unit
PD	Pre-distortion
PE	Power efficiency
PER	Packet error rate
PFD	Phase and frequency detector
PGA	Program gain amplifier
PHY	Physical
PLL	Phase-locked loop
PN	Phase noise
PPA	Pre-power amplifier
PSD	Power spectral density
PSK	Phase-shift keying
Q-CH	Quadrature channel
QORC	Quadrature overlapped raised-cosine
QPSK	Quadrature phase-shift keying
RAM	Random-access memory
RBER	Residual bit error rate
RC	Raised-cosine
RF	Radio frequency
RFIC	RF integrated circuit
RFPAL	RF power amplifier linearizer
RMS	Root-mean-square

RSSI	Received signal strength indictor
RX	Receiver
RZ	Return-to-zero
SA	Spectrum analyzer
SAW	Surface-acoustic wave
SCMSK	Self-convolving minimum shift keying
SCPC	Single channel per carrier
SFDR	Spurious-free-dynamic range
SG	Signal generator
SIR	Signal-to-interferer ratio
SISO	Single-input/single-output
SNR	Signal-to-noise ratio
SoC	System on chip
S/P	Serial-to-parallel
SPSA	Simultaneous perturbation stochastic approximation
SQAM	Superposed quadrature amplitude modulation
SQORC	Staggered QORC
SQPSK	Staggered QPSK
SRRC	Square root of raised-cosine
SSB	Single side band
SU-MIMO	Single-user multiple-input multiple-output
TDD	Time division duplexing
TDM	Time division multiplexing
TDMA	Time division multiple access
TD-SCDMA	Time division synchronous code division multiple access
TFM	Tamed frequency modulation
TIA	Transimpedance amplifier
TSM	Transmit spectrum mask
TSMC	Taiwan semiconductor manufacturing company
TWIF	Three-wire interface
TX	Transmitter
TX/RX	Transmitter and receiver
UHF	Ultra-high frequency
UMTS	Universal Mobile Telecommunications System
US	User equipment
UWB	Ultra-wideband
VA	Viterbi algorithm
VCO	Voltage-controlled oscillator
VGA	Variable-gain amplifiers
VHT	Very high throughput
VLSI	Very large scale integrated
WCDMA	Wideband code division multiple access
WEP	Wireless encryption protocol
WGN	White Gaussian noise

W-H	Wiener-Hammerstein
Wi_Fi	Wireless fidelity
WiMAX	Worldwide interoperability for microwave access
WLAN	Wireless local area network
WRSSI	Wideband received signal strength indictor
XPSK	Cross-correlated PSK
ZF	Zero-forcing

Chapter 1
Introduction

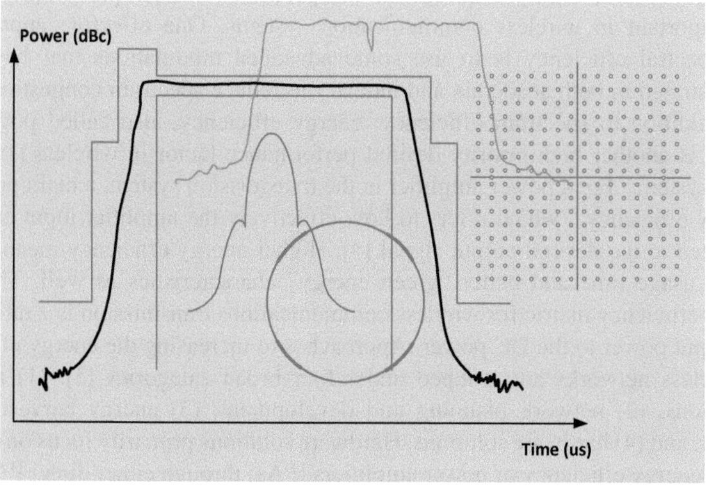

One of the primary and long-standing goals of the Federal Communications Commission (FCC) has been to promote more efficient use of spectrum for the radiofrequency (RF) wireless radio transmission. The FCC's 1999 Spectrum Policy Statement highlighted that "with increased demand of a finite supply of spectrum, the Commission's spectrum management activities must focus on allowing spectrum markets to become more efficient"; and the Strategic Plan for Fiscal Year 2003–2008 (published in 2002) indicated its general spectrum management goal was to "encourage the highest and best use of spectrum" [1]. Even though there are other, different metrics, one of the accepted common spectrum efficiency metrics for wireless communications is in terms of *bits/second/hertz*. To approach this goal, many government organizations and commercial communications companies have

© Springer International Publishing Switzerland 2017
W. Gao, *Energy and Bandwidth-Efficient Wireless Transmission*, Signals and Communication Technology, DOI 10.1007/978-3-319-44222-8_1

adopted advanced modulation and amplification transmission techniques to utilize spectrum more efficiently and as much as possible.

Wireless technology is transforming our society and people's lifestyles, with over six billion people communicating over cellular networks and over a billion people using Wi-Fi internet globally in 2014 [2, 3]. Total mobile subscriptions are expected to grow to over nine billion by the end of 2019. Cellular systems of 3G and 4G provide wide coverage areas, full mobility and roaming, but traditionally offer relatively low bandwidth connectivity. On the other hand, Wireless Local Area Networks (WLANs) provide high data rates at low cost, but only within a limited area. Most smart cellular phones can automatically be connected to Wi-Fi networks from cellular data connections when they are located within hot-spot areas. Therefore, WLANs have come to be relied on as a suitable complementary technology to the existing cellular radio access networks. The addition of more and more new mobile and Wi-Fi subscribers could result in congestion in the limited radio spectrum. Hence, highly efficient use of spectrum becomes particularly crucial and important in wireless communication systems. One effective approach to high spectral efficiency is to use some advanced modulations that have been demonstrated in both academia and industry to reduce spectrum congestion.

In addition to spectrum efficiency, energy efficiency, also called power efficiency, is another high-priority desired performance factor in wireless communication systems. For a power amplifier in the transmission system, a main priority is "energy efficiency", which refers to how effectively the amplifier input energy is converted to the desired output signal [4]. Higher energy efficiency means longer battery usage time and better "green energy" characteristics as well. The basic energy-efficiency metric for wireless communications transmission is a ratio of the RF output power to the DC power. Approaches to increasing the energy efficiency of wireless networks are grouped under four broad categories [5]: (1) resource allocations, (2) network planning and development, (3) energy harvesting and transfer, and (4) hardware solutions. Hardware solutions primarily focus on increasing the energy efficiency of power amplifiers (PAs) through either direct PA design architectures or modulation signal design techniques that aim at either constant envelope characteristics or pre-distortion (PD)-based linearization methods. In our discussion of hardware solutions, only modulation signal design techniques, however, will be covered and discussed in more detail.

There is no one modulation technique that possesses both the best spectrum efficiency and the best energy efficiency. Hence, a trade-off involving optimization between them should be made, depending on actual applications. In addition, in order to improve energy efficiency, polar transmitter [6–8] and envelope-tracking techniques [9, 10] have been developed around the last decade and some significant progress has been achieved in the recent years.

In classic polar transmitters, the input RF modulated signal to a power amplifier is first split into an envelope signal through an envelope detector and a phase-modulated signal via a limiter; then, the resulting low-frequency envelope signal is amplified by a low-frequency amplifier to control the supply voltage regulator of the final PA. The phase-modulated signal on the RF path to the input of the PA can be amplified by a highly efficient nonlinear PA. Finally, the amplified envelope signal and phase-modulated signal are combined inside the PA to restore the original RF-modulated signal as much as possible.

The envelope-tracking (ET) technique requires the supply voltage of the PA to be dynamically varied with the instantaneous value of the input RF signal envelope while the input to the PA is an original RF modulated signal. Compared with traditional fixed DC supply voltage, which has a significant amount of power loss as heat, in the envelope-tracking technique the RF signal envelope modulates the supply voltage of the PA to track the envelope of the input RF signal to reduce the amount of power dissipated as heat [9, 10]. As a result, high energy efficiency is achieved because of reduced DC power consumption. Any nonlinear distortion caused by modulating the supply voltage of the PA with the envelope signal can be compensated for using the pre-distortion method on the envelope path.

Compared with pre-distortion (PD)-based linearization techniques, which have been developed for several decades and have been widely used in wireless communication systems, the requirements of the envelope amplifier or modulator for polar transmitters are very stringent because they are responsible for amplifying the envelope signal and then precisely combining it with the phase-modulated signal inside the PA to regenerate the originally modulated signal at the output of the PA. With regard to the ET technique, even though some improvements in power supply modulation have been made experimentally in the past years, it still has a long way to go before this technique can actually be applied due to the limitations of accuracy and bandwidth in practical implementations. In addition, the pre-distortion method is still needed to compensate for nonlinear distortion in either the polar transmitters or ET-based transmitters. Hence, the PD technique is a fundamental and necessary approach to the linearization of the PA.

This book mainly focuses on the most recent advances in both energy and bandwidth-efficient modulation and transmission techniques, especially as they are applied to cellular and WLAN transceivers. Due to the portable properties of these products, besides the requirements of energy and spectrum efficiency, low cost is also highly desirable as a third priority in these applications. Starting with some fundamental modulation properties, and then moving to energy and bandwidth-efficient overlapped raised-cosine pulse-shaping modulations and the most bandwidth-efficient modulation formats with intersymbol interference (ISI)-free Nyquist pulse shaping—which are suitable to nonlinear and linear amplification channels, respectively, the discussion will cover the two most energy and bandwidth-efficient modulation types of techniques, or constant envelope modulations and nearly constant envelope modulations. In order to achieve highly efficient transmission strategies for non-constant envelope modulation signals, especially for those having larger peak-to-average power ratio (PAPR) values, a pre-distortion linearization technique of the power amplifier will be addressed in detail. In the linearization technique, a simple approximation to both memory effects and nonlinear behavior of a PA with the Volterra polynomial model will be first presented in the complex baseband domain, and then a pre-distorter as an approximate inverse of the PA will be introduced and analyzed by using the Volterra polynomial model as well. After that, modeling approximation analysis, performance simulation results, and pre-distortion implementations with both analog and digital realizations will be addressed. Finally, from a system-design point of view, RF transceivers for wireless communication systems will be discussed in detail, including most common design issues or challenges that RFIC designers may face and drawbacks or disadvantages that some certain architectures may have. Included in the discussion will be several commercial transceiver products that utilize highly efficient modulation schemes in cellular and WLAN applications so that reader can gain actual design strategies and ideas from these applications.

References

1. Oversight of the Federal Communications Commission, *Hearing before the Committee on Commerce, Science, and Transportation United States Senate,* one hundred twelfth congress, second session, May 16, 2012.
2. Rysavy, P. (2014). Challenges and considerations in defining spectrum efficiency. *Proceedings of the IEEE, 102*(9), 386–392.
3. Ericsson Mobility Report, June 2014.
4. McCune, E. (2015). A technical foundation for RF CMOS power amplifiers. *IEEE Solid-State Circuits Magazine, 7*(3), 81–85.
5. Bussi, S., Chih-Lin, I., Klein, T. E., Vincent Poor, H., Yang, C., & Zappone, A. (2016). A survey of energy-efficient techniques for 5G networks and challenges ahead. *IEEE Journal on Selected Areas in Communications, 34*(4), 697–709.
6. Groe, J. (2007). Polar transmitters for wireless communications. *IEEE Communications Magazine, 45*(9), 58–63.
7. Johnson, J. (2006, June). Power amplifier design for open loop EDGE large signal polar modulation systems. *RF Design*, 42–50.
8. Sowlati, T., Rozenblit, D., Pullela, R., Damgaard, M., McCarthy, E., Koh, D., et al. (2004). Quad-band GSM/GPRS/EDGE polar loop transmitter. *IEEE Journal of Solid-State Circuits, 39*(12), 2179–2189.
9. Application Report, "GC5325 envelope tracking," *Texas Instruments.*
10. Wimpenny, G. (2012, February 6). *Envelope tracking power amplifier characterization* (White Paper). Nujira limited. Retrieved from www.nujira.com

Chapter 2
Bandwidth-Efficient Modulation with Frequency Division Multiple Access (FDMA)

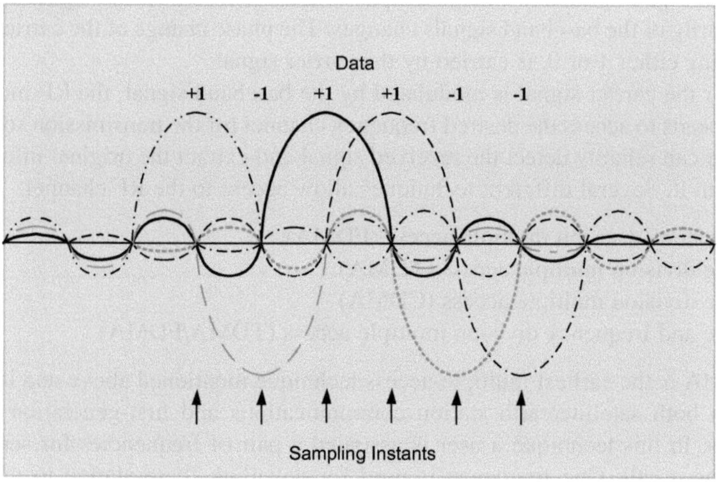

2.1 Introduction

In a wide variety of communication systems, modulation as a fundamental technique plays a very important role in data transmission through air within a specified spectral bandwidth. A modulation signal, which is usually represented by a low-frequency baseband signal and is commonly referred to as an information-bearing signal, varies or modulates one of three parameters: *amplitude*, *phase*, and *frequency* of the radio frequency (RF) carrier signal such that the baseband signal is carried by a varied parameter of the carrier signal through atmosphere propagation to the destination. Why does the information-bearing signal need to modulate a high-frequency carrier signal for this transmission? This is necessary because the

© Springer International Publishing Switzerland 2017

W. Gao, *Energy and Bandwidth-Efficient Wireless Transmission*, Signals and Communication Technology, DOI 10.1007/978-3-319-44222-8_2

size of the antenna used to radiate the signal to free space depends on the wavelength λ of the transmitted signal. The wavelength λ is equal to c/f, where c is the speed of light and equals 3×10^8 m/s, and f is the frequency of the transmitted signal. For cellular communication systems, antennas are typically $\lambda/4$ in size [1]. If a baseband signal with a frequency of 15 kHz were to be transmitted through an antenna without modulating a carrier signal, the size of the antenna would be $\lambda/4 = 5{,}000$ m. However, it is only 8 cm if a carrier signal with a frequency of 900 MHz is modulated by such a baseband signal. For this reason, a high-frequency signal usually called a carrier signal is needed for all wireless communication systems to carry the modulation baseband signal. Thus, modulation is a necessary process in all wireless communication systems. Through this book, we mainly consider either the phase modulation scheme or a combination of an amplitude and phase modulation scheme such as M-QAM for its simple implementation and robust performance. In the phase modulation scheme, the information-bearing baseband signal is used to change the phase of a sinusoidal carrier signal whenever the polarity of the baseband signals changes. The phase change of the carrier signal, indicating either 1 or 0, is carried by the carrier signal.

After the carrier signal is modulated by the baseband signal, the RF-modulated signal needs to access the desired frequency channel for the transmission so that the receiver can reliably detect the received signal and extract the original information bits from it. Several different techniques allow access to the RF channel:

- Frequency division multiple access (FDMA)
- Time division multiple access (TDMA)
- Code division multiple access (CDMA)
- Time and frequency division multiple access (TDMA/FDMA)

FDMA is the earliest multiple-access technique mentioned above and is widely used in both satellite/earth station communications and first-generation cellular systems. In this technique a user is assigned a pair of frequencies for sending or receiving a call. One frequency is used for downlink (base station to mobile in cellular systems or satellite to earth station in satellite systems) and one for uplink (mobile to base station or earth station to satellite). This process is called "frequency division multiplexing." Even though the user may not be talking, a pair of frequencies cannot be reassigned as long as a call is in place. Second-generation cellular systems, such as the global system for mobile communications (GSM), use an FDMA/TDMA technique for voice and data transmissions. The available spectrum band is divided into frequency slots, each with a 200-kHz sub-band. In addition, each frequency slot is also divided into time slots. Each user is assigned a pair of frequencies for uplink and downlink and a time slot during a frame. Therefore, the FDMA technique still plays an important role in modern digital communication systems.

In the early stages (during the late 1980s) satellite digital communication systems adopted an SCPC/FDMA (single channel per carrier) technique, where a total of 800 data/voice channels occupied a 36-MHz transponder bandwidth of the satellite. In this system each user can transmit and receive either data at a rate of 64 kbps with a Quadrature Phase Shift Keying (QPSK) format or voice at a rate of 32 kbps with a Binary Phase Shift Keying (BPSK) format. The channel spacing is 36 MHz/800 = 45 kHz. In the 2G GSM systems with a Gaussian Filtered Minimum Shift Keying (GMSK) modulation format, every symbol represents one bit, which means that symbol rate and bit rate are equal. In the 2.5G GSM Evolution Enhanced Data for Global Evolution (EDGE) systems with an 8PSK modulation format, every symbol represents three consecutive bits, which indicates that bit rate is three times the symbol rate. With the same symbol rate of 270.833 kbps for GMSK modulation, a bit rate up to 812.5 kbps for the EDGE system with an 8PSK modulation can be achieved within the same transmission bandwidth (200 kHz) as GSM systems.

FDMA systems have some common features: the relative low data transmission rate, narrow channel spacing, and restrictive transmission bandwidth. These features dictate that spectrally efficient modulation schemes are the best approach to achieving bandwidth-efficient transmission without significantly causing interference in adjacent channels. Meanwhile, the modulated signals in these applications prefer to have constant envelope or small envelope fluctuation in order to achieve energy efficiency when the power amplifiers operate in the saturation region or close to it. The energy-efficient operation of the power amplifier can extend the usage time of the direct current battery.

2.2 Definition of Energy and Spectral Efficiency

Energy- and spectrum-efficient modulation and transmission techniques are the first two high-priority requirements among three categories of most wireless communication systems: spectrum efficiency, energy efficiency, and cost efficiency. These three categories in the list order above are top priorities for most electrical system designers in designing or choosing a wireless communication system.

Recognizing the fact that a power amplifier (PA) is one of the most critical component and consumes most part of the total energy at the transmitter, we take some approaches to improving the energy efficiency of the PA from the perspective of the *signal design* without significantly causing the degradation of the spectral efficiency. With great demands for high-data-rate applications in the next wireless generations, such as 5G cellular networks and IEEE 802.11ax WLAN, the importance of the energy efficiency has been greatly recognized over the past several years. More and more research and development efforts in industry and academia focus on how to enhance and improve the energy efficiency of future wireless communication networks from perspectives of PA design, signal design and network design.

2.2.1 Bandwidth or Spectrum Efficiency

The bandwidth of a channel is the frequency range over which a modulated signal is transmitted and then reliably detected in the receiver. Since frequency spectrum resource is limited, it has to be utilized efficiently. The term *spectral efficiency* is used to describe the rate of maximum information being transmitted over a given bandwidth in a specific communication system. Hence, spectral efficiency may also be called "bandwidth efficiency."

In general, spectral efficiency refers to the information rate that can be transmitted over a given bandwidth in a specific communication system to achieve reliable performance. Spectral efficiency is viewed as bits per second per hertz (bits/s/Hz) and defined as

$$\eta_s = \frac{R}{B_w} (\text{bit/s/Hz}) \tag{2.1}$$

where R is the information rate or transmission rate in bit/s, and B_w is the *passband* (or double-side) transmission bandwidth in Hz. For a certain information rate, the spectral efficiency can be maximized by minimizing the transmission bandwidth. The minimum transmission bandwidth for intersymbol interference (ISI) free is determined by the Nyquist bandwidth (B_N) criterion, which states that the theoretical minimum bandwidth B_N of an ideal and linear phase brick-wall channel lowpass filter used for impulse transmission at a transmission rate of f_s symbols per second without ISI is equal to the Nyquist frequency f_N, or $B_N = f_N = f_s/2$. For a passband transmission system, the minimum *passband* bandwidth of B_w for ISI free is twice the Nyquist bandwidth, or $B_w = 2B_N = 2f_N$. If the transmission bandwidth of B_w is less than twice the Nyquist bandwidth B_N, the responses to these impulses at the output of the channel lowpass filter have ISI at the optimal sampling instants.

If rectangular pulses rather than impulses are used in the transmission channel, an inverse SINC function, or $x/\sin(x)$, shaped amplitude equalizer should be added before the ideal brick-wall channel filter so that the rectangular pulses can be transferred to the impulses before the ideal brick-wall channel filter.

Spectrum efficiency or bandwidth efficiency in (2.1) describes how efficiently the allocated bandwidth is utilized to accommodate the higher data transmission rate. For a given information rate R, the symbol rate f_s is associated with the information rate R or bit rate f_b through a modulation format such as M-order QAM. Therefore, the relationship between f_b and f_s is determined by the modulation format.

For a signaling alphabet with M alternative symbols, each symbol contains $N = \log_2 M$ bits, or N-bit represents the number of $M = 2^N$ symbols. The relationship between the bit rate f_b and symbol rate f_s is expressed as

$$f_b = Nf_s \tag{2.2}$$

or

$$f_s = \frac{f_b}{N} = \frac{f_b}{\log_2 M} \tag{2.3}$$

In the case of M-QAM, M represents M alternative symbols in (2.3). The theoretical spectral efficiency for the passband BPSK transmission, where the bit rate is equal to the symbol rate ($f_b = f_s$), is given by

$$\eta_s = \frac{f_b}{B_w} = \frac{f_s}{2f_N} = \frac{f_s}{2(f_s/2)} = 1\,\text{bit/s/Hz} \tag{2.4}$$

here the transmission bandwidth B_w for the passband signal is twice the Nyquist frequency f_N, or $B_w = 2f_N$. For a passband 16-QAM transmission, where M is equal to 16, the spectral efficiency is calculated by

$$\eta_s = \frac{f_b}{B_w} = \frac{\log_2 16 \times f_s}{2(f_s/2)} = 4\,\text{bit/s/Hz} \tag{2.5}$$

Table 2.1 illustrates the theoretical bandwidth efficiency limits for some main modulation formats. It should be noted that these figures cannot actually be achieved in practical radios since the ideal brick-wall channel filter is required, which is impossible to practically design.

In practice, a raised cosine filter with a roll-off factor of α is widely used to approximate the ideal brick-wall filter. The double-side bandwidth of the raised cosine filter is given by

$$B_w = 2(1 + \alpha)f_N, \quad 0 \le \alpha \le 1 \tag{2.6}$$

In (2.6), for $\alpha = 0$, the minimum double-side bandwidth is equal to twice the Nyquist frequency, or $B_w = 2f_N$, whereas for $\alpha = 1$, the maximum double-side bandwidth is four times the Nyquist frequency, or $B_w = 4f_N$. Theoretically, beyond the bandwidth expressed in (2.6) the attenuation has an infinite value. In practice, the raised cosine filter with a finite attenuation value can be realized, depending on

Table 2.1 Theoretically spectral efficiencies of main modulation formats

Modulation scheme	Theoretical bandwidth efficiency (bits/s/Hz)
BPSK	1
QPSK	2
8PSK	3
16-QAM	4
32-QAM	5
64-QAM	6
128-QAM	7
256-QAM	8

the allowed amount of adjacent channel interference. In the case of the channel raised cosine filter, spectral efficiency can be calculated using the bandwidth of the raised cosine filter in (2.6).

It has been shown so far that the minimally occupied bandwidth would be made equal to the symbol rate if a raised cosine filter with $\alpha = 0$ were implemented as the channel filter. From a system-design point of view, the occupied bandwidth is usually larger than the bandwidth of the channel filter given in (2.6) because the guard band should be included in the occupied bandwidth. The width of the guard band depends on the allowed amount of adjacent channel interference required by the specification of the system.

The TDMA version of the North American Digital Cellular (NADC) system adopts $\pi/4$ - DQPSK modulation with a root-raised cosine filter response with a roll-off factor of $\alpha = 0.35$. This system provides a 48.6-kbits/s data rate over a 30-kHz channel bandwidth. In this example, we can calculate spectral efficiency using three different bandwidths:

- Theoretical minimum bandwidth
- Actual filter bandwidth
- System channel bandwidth

Theoretical minimum bandwidth: Spectral efficiency with minimum bandwidth for the passband signal is calculated by setting $\alpha = 0$ in (2.6):

$$\eta_s = \frac{f_b}{B_w} = \frac{f_b}{2f_N} = \frac{f_b}{2(f_s/2)} = \frac{2f_s}{f_s} = 2 \text{ bit/s/Hz} \tag{2.7}$$

Actual filter bandwidth: Spectral efficiency with actual filter bandwidth is calculated by setting $\alpha = 0.35$ in (2.6):

$$\eta_s = \frac{f_b}{B_w} = \frac{f_b}{2(1+0.35)f_N} = \frac{f_b}{1.35f_s} = \frac{2f_s}{1.35f_s} = 1.48 \text{ bit/s/Hz} \tag{2.8}$$

System channel bandwidth: Spectral efficiency with system channel bandwidth is obtained by substituting the channel bandwidth of 30 kHz with $B_w = 30$ kHz into (2.4):

$$\eta_s = \frac{f_b}{B_w} = \frac{48.6 \text{ kbit/s}}{30 \text{ kHz}} = 1.62 \text{ bit/s/Hz} \tag{2.9}$$

It can be seen that the spectral efficiency of the system channel bandwidth is 1.62 bit/s/Hz in (2.9), where the channel bandwidth of 30 kHz used in the calculation is greater than the minimum bandwidth of 24.3 kHz in (2.7), but less than the actual filter bandwidth of 32.8 kHz in (2.8). Hence, the bandwidth efficiency with the system channel bandwidth is larger than that with the actual filter bandwidth because the system bandwidth in the former has a certain finite attenuation that meets the system requirement while the actual filter bandwidth in the latter is supposed to have an infinite attenuation calculated in (2.6).

2.2.2 Energy Efficiency

In highly integrated wireless systems, such as wireless system on chip (SoC) devices, the RF power amplifier is the subsystem that consumes most DC power in the whole transmission system. For example, it may consume up to more than 70% of DC power in the transmit path of certain RF mobile transceivers [2]. For this reason, the "energy efficiency" of the RF power amplifier, which is often described as "power efficiency" in the literature (and this is incorrect [3]), is directly proportional to and mainly represents the efficiency of the overall system, especially in the transmission path. In addition to power consumption, it is now apparent that energy consumption is an important metric for transmitter circuits. Energy consumption more accurately predicts the battery life, especially when a portable system operates with a wide range of the output power. Hence, energy efficiency tends to be a better metric of the performance than power efficiency in terms of the battery life. Actually, energy consumption depends on the power consumption and time spent in the power consumption duration.

Usually, in the published literature, the power efficiency of the power amplifier refers to how efficiently the input power to the power amplifier, including the input AC and DC powers, is converted to the output AC power to a load or an antenna, regardless of time. Energy efficiency of the power amplifier, however, is identical to its power efficiency as long as all powers, including the input AC power to the PA, power supply DC power, dissipated power as heat, and the output AC power from the PA are measured equivalently in time [3]. Therefore, we use the term *energy efficiency* instead of *power efficiency* in this book even though definitions are the same [3].

All power amplifiers can be represented by a four-port network: an input DC port, a RF input signal port, a RF output signal port, and a ground. The measured input DC power P_{DC} includes the power associated with all the bias lines of the power amplifier. It is assumed here that the PA is perfectly matched and has infinite reverse isolation. Therefore, the measured power P_{IN} at the input RF port corresponds to the input signal power at the fundamental frequency only. The measured power P_{OUT} at the output RF port corresponds to the output power at the fundamental frequency and all the spurious frequencies, which are generated by the PA itself.

A power amplifier is evaluated for efficiency using a conservation of power calculation based on the flows identified in Fig. 2.1:

$$P_{\mathrm{IN}} + P_{\mathrm{DC}} = P_{\mathrm{OUT}} + P_{\mathrm{D}} \tag{2.10}$$

The energy efficiency of the PA is a measure of its ability to convert the DC power of the power supply into the RF signal power delivered to the load. There are two definitions of power amplifier efficiency. One is basic PA efficiency and the other is power-added efficiency.

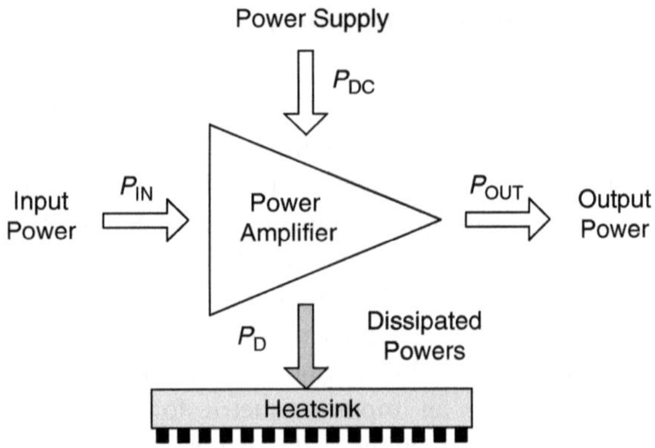

Fig. 2.1 Power flows in a power amplifier. Redrawn from [3]

Basic PA Efficiency: The basic PA efficiency is a ratio of the RF output power to the DC power and is derived from (2.10) as

$$\eta_B = \frac{P_{OUT}}{P_{DC}} = 1 - \frac{P_D - P_{IN}}{P_{DC}} \tag{2.11}$$

More precisely, PA efficiency is also called the overall efficiency [4]. If the power amplifier has relatively high power gain, the direct contribution of the RF input signal power to the RF output signal power is insignificant, and therefore it can be neglected. The basic efficiency can be approximated by

$$\eta_B \approx 1 - \frac{P_D}{P_{DC}} \tag{2.12}$$

Power-Added Efficiency: When the gain of the power amplifier is not significantly high, the RF input power needs to be subtracted from the RF output power in the efficiency expression, and then the power efficiency is referred to as the power-added efficiency (PAE):

$$\eta_{PAE} = \frac{P_{OUT} - P_{IN}}{P_{DC}} = 1 - \frac{P_D}{P_{DC}} \tag{2.13}$$

If the PA has a relatively large power gain, then, $\eta_B \approx \eta_{PAE}$, as expressed in (2.12) and (2.13). The power-added efficiency can be interpreted as the efficiency of the network to convert the input DC power into the amount of the RF net output power. PAE definition in (2.13) is widely used as a useful metric for evaluating the efficiency of the RF power amplifier. PAE becomes zero when the power gain of G_{PA} is unit or $P_{OUT} = G_{PA} \times P_{IN} = P_{IN}$. This means the power amplifier does not convert any DC power to the RF output power.

It can be seen from (2.11) and (2.13) that the PA efficiency increases as DC power of P_{DC} is reduced. The DC power P_{DC} is given by

Table 2.2 Classical modes of power amplifier operation

Classical mode	Conduction angle (°)	Operation range
A	360	Linear
AB	180–360	Either linear or nonlinear[a]
B	180	Nonlinear
C	0–180	Nonlinear

[a]Class-AB is not a complete linear amplifier; a RF signal with non-constant envelope will be distorted significantly at its peak power level

$$P_{\mathrm{DC}} = V_{DC} I_{\mathrm{DC}} \tag{2.14}$$

The DC current I_{DC} can decrease monotonically as the conduction angle of the power amplifier is reduced [5]. The conduction angle of the power amplifier determines the classical modes of the power amplifier operation as listed in Table 2.2. In 3G and 4G cellular and 802.11 WLAN communication systems, power amplifiers usually operate in a class AB mode to achieve high efficiency by reducing the DC current I_{DC} and to avoid severe nonlinear distortion on the RF output signal as well. In general, DC supply voltage of V_{DC} is fixed during the power amplifier operation except the envelope tracking technique based power amplifier [6, 7], where the supply voltage dynamically and synchronously tracks or follows the envelope of the RF input signal. Furthermore, even in the class AB mode, the efficiency can be further increased due to the reduction of DC current I_{DC} as the output back-off of the power amplifier from its P1dB compression point is reduced [8]. Therefore, in terms of achieving high efficiency, power amplifiers are preferred to operating in the small back-off condition as long as the transmitted power spectral density (PSD) and error vector magnitude (EVM) meet the standard specifications with enough margins.

The expressions in (2.11) and (2.13) also demonstrate that the consideration of power amplifier efficiency is the same as the consideration of either amplifier output power or amplifier power dissipation. The amount of the total input power to the power amplifier in Fig. 2.1, either DC power or RF input power that is not converted into the RF output power, is dissipated as heat. Higher power output corresponds to higher energy efficiency. From (2.10), lower power dissipation leads to higher power output, which in turn results in higher power amplifier efficiency. From a design point of view, the design objective of maximum energy efficiency is identical to the design objective of either minimum power dissipation P_{D} or DC power P_{DC}. Therefore, the energy efficiency of the power amplifier has become a challenging requirement for most PA designers. Cell phone handset PAs have to operate efficiently to conserve battery power and base station PAs are also need to be efficient as possible due to cooling limitations [5].

The energy consumption of a system is highly related to the power consumption through a time interval T. There are many different definitions of the energy consumption in the literature. Considering that the energy consumption covered in this book only focus on the transmitter, especially for the PA, we define the energy consumption of the PA at the transmitter as shown in Fig. 2.1 during the time interval T as

$$E = T(P_{\mathrm{DC}} + P_{\mathrm{D}})\,[\mathrm{Joule}] \tag{2.15}$$

where P_{DC} is the DC consumption power of the PA and P_{D} represents the static power dissipated in the PA as heat. Substituting $P_{\mathrm{DC}} = P_{\mathrm{OUT}}/\eta_{\mathrm{B}}$ in (2.11) into (2.15), the energy consumption in (2.15) can be rewritten as:

$$E = T(\mu P_{\mathrm{OUT}} + P_{\mathrm{D}})\,[\mathrm{Joule}] \tag{2.16}$$

where $\mu = 1/\eta_{\mathrm{B}}$ is a factor, with η_{B} the efficiency of the transmit power amplifier. In the case that the PA has a relatively large power gain, we have shown the relationship between η_{B} and η_{PAE}, or $\eta = \eta_{\mathrm{B}} \approx \eta_{\mathrm{PAE}}$.

If P_{D} includes the power dissipated in all other circuit blocks of the transmitter and receiver, the energy consumption expression in (2.16) represents the energy consumption of a system, which is the same as one used in [9, 10]. From the PA standpoint of view, expressions of (2.15) and (2.16) precisely represent the energy consumption of the PA.

As the energy consumption of the PA has been defined, the energy efficiency of the PA needs to be defined next. The energy efficiency of the PA is the ratio of the benefit obtained after sustaining the energy cost and the amount of energy consumption in (2.15) and (2.16). The benefit is usually related to the amount of data reliably transmitted in the time interval T. Several performance functions have been used in the literature to evaluate this quantity, such as *system capacity*, and *system throughput* in [9, 10]. Throughput is the amount of data reliably transmitted over a communication channel in the certain time period T. The throughput metric is measured in bits/s and also depends on the signal-to-noise ratio (SNR) and the transmission channel condition. Compared to capacity, throughput is more practically used to evaluate the system performance.

Energy Efficiency: With a general function f(SNR) that represents throughput as the system benefit, the energy efficiency of the PA is defined as

$$\mathrm{EE} = \frac{T \times f(\mathrm{SNR})}{T \times (\mu P_{\mathrm{OUT}} + P_{\mathrm{D}})} = \frac{f(\mathrm{SNR})}{\mu P_{\mathrm{OUT}} + P_{\mathrm{D}}}\,[\mathrm{bits/Joule}] \tag{2.17}$$

EE can be improved by either increasing the numerator or decreasing the denominator in (2.17). In this book, however, we focus on increasing the energy efficiency by decreasing the denominator through the increase of the efficiency η of the PA. It is clear that EE can be improved by decreasing the factor $\mu = 1/\eta$ through the increase of the efficiency η of the PA, which in turn requests either to increase the output power P_{OUT} of the PA or to decrease the DC power P_{DC}. Even though EE can be improved by increasing the throughput in (2.17), the increase of the throughput takes the entire system involved, including a transmitter and receiver, multi-antennas, a channel condition and so on. Therefore, energy efficiency improvements through increasing the throughput are very complicated and are beyond the scope of this book.

2.3 Fundamentals of Modulation

The main objective of the modulation process is to shift the spectrum of the information-bearing baseband signal to a high-frequency band suitable for transmission. The band center of the modulated signal is located at the carrier frequency and its bandwidth is twice the bandwidth of the baseband signal.

2.3.1 *The Convolution Property*

One of the most important properties of the Fourier transform in linear time-invariant (LTI) systems is the convolution operation. With the convolution operation, the relation between the input $x(t)$ and output $y(t)$ of a continuous-time LTI system with impulse response $h(t)$ is given by

$$y(t) = x(t) * h(t) = \int_{-\infty}^{+\infty} x(\tau)h(t-\tau)d\tau \tag{2.18}$$

The relationship between the Fourier transform and inverse Fourier transform for the output signal $y(t)$ is expressed as

$$y(t) = F^{-1}\{Y(\omega)\} = \frac{1}{2\pi}\int_{-\infty}^{+\infty} Y(\omega)\mathrm{e}^{j\omega t}d\omega \tag{2.19}$$

$$Y(\omega) = F\{y(t)\} = \int_{-\infty}^{+\infty} y(t)\,\mathrm{e}^{-j\omega t}dt \tag{2.20}$$

Then, (2.20) can be written as

$$\begin{aligned} Y(\omega) &= \int_{-\infty}^{+\infty}\left[\int_{-\infty}^{+\infty} x(\tau)h(t-\tau)d\tau\right]\mathrm{e}^{-j\omega t}dt \\ &= \int_{-\infty}^{+\infty} x(\tau)\left[\int_{-\infty}^{+\infty} h(t-\tau)\mathrm{e}^{-j\omega t}dt\right]d\tau \\ &= \int_{-\infty}^{+\infty} x(\tau)\mathrm{e}^{-j\omega\tau}H(\omega)d\tau \\ &= H(\omega)\int_{-\infty}^{+\infty} x(\tau)\mathrm{e}^{-j\omega\tau}d\tau = H(\omega)X(\omega) \end{aligned} \tag{2.21}$$

In the derivation of (2.21), $H(\omega)$ and $X(\omega)$ are the Fourier transform of the system impulse response $h(t)$ and the Fourier transform of the input signal $x(t)$. Thus, the convolution of two signals in the time domain corresponds to the multiplication of their Fourier transforms in the frequency domain.

2.3.2 *Modulation Property*

By using duality, a property of the Fourier transform between the time and frequency domains, we can obtain the relationship in the frequency domain from the multiplication of two signals in the time domain. If the multiplication of one signal $x(t)$ by another $s(t)$ in the time domain is expressed as

$$y(t) = x(t) \times s(t) \tag{2.22}$$

Then, Fourier transform expression of (2.22) is given by

$$Y(\omega) = \frac{1}{2\pi}[X(\omega) * S(\omega)] \tag{2.23}$$

Such multiplication of one signal by another in the time domain is also called the *modulation process*, or *modulation*, where the signal with the lower frequency is referred to as the modulating signal while the other one with the higher frequency is called the carrier signal. The property connected by (2.22) and (2.23) is called the modulation property.

In a practical modulation process, a cosine waveform is usually used as a carrier signal. Let $x(t)$ be a modulation signal whose Fourier transform $X(\omega)$ is limited to the frequency of ω_b and $s(t) = \cos(\omega_c t)$ a carrier signal whose Fourier transform is expressed as $S(\omega) = \pi\delta(\omega - \omega_c) + \pi\delta(\omega + \omega_c)$. Then, from (2.23) we have Fourier transform of the signal $y(t)$ given by

$$Y(\omega) = \frac{1}{2\pi}X(\omega) * S(\omega) = \frac{1}{2}[X(\omega - \omega_c) + X(\omega + \omega_c)] \tag{2.24}$$

It is clear from (2.24) that the spectrum of the baseband signal $x(t)$ is completely shifted to higher frequencies centered at the carrier frequencies of $\pm\omega_c$ after the modulation process illustrated in Fig. 2.2. This frequency-shift process does not cause any distortion if the condition $\omega_c > \omega_b$ is met. In practice, it should be more than twice the frequency of ω_b (or $2\omega_b$). So the original signal $x(t)$ can completely recovered by multiplying the modulated signal $y(t)$ with a referenced carrier signal, which will be discussed later.

The carrier signal generally used in practice is a cosine signal, even though it is not a necessary restriction, such as a square wave signal. In the case of a cosine carrier with the frequency of ω_c, the modulated signal can be expressed as

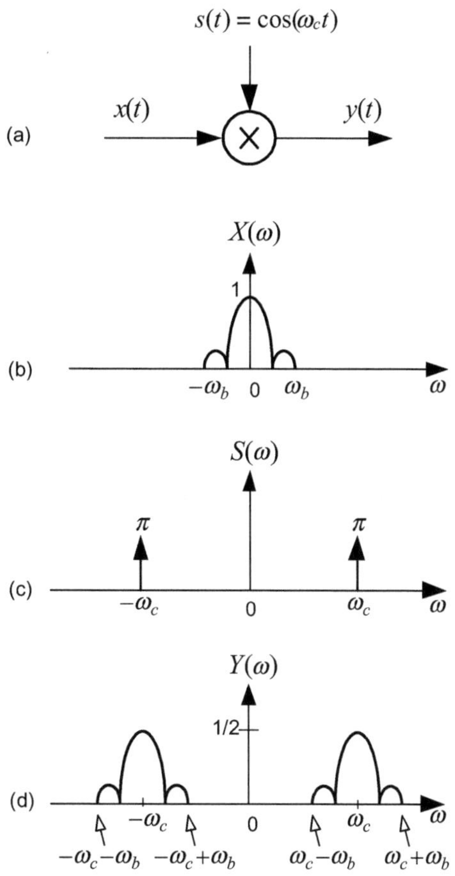

Fig. 2.2 Spectrum shift after amplitude modulation with a sinusoidal signal: (**a**) amplitude modulation with a sinusoidal signal, (**b**) spectrum of modulating signal $x(t)$, (**c**) spectrum of sinusoidal signal $s(t) = \cos(\omega_c t)$, and (**d**) spectrum of modulated signal $y(t)$

$$y(t) = A(t)\cos[\omega_c t + \phi(t)] = A(t)\cos[\theta(t)] \tag{2.25}$$

where $\theta(t)$ is the instantaneous phase and $A(t)$ is the instantaneous amplitude. The instantaneous frequency $\omega(t)$ is defined as

$$\omega(t) = \frac{d\theta(t)}{dt} = \omega_c + \frac{d\phi(t)}{dt} \tag{2.26}$$

The functions $\phi(t)$ and $d\phi(t)/dt$ are called the *phase deviation* and *frequency deviation*, respectively.

If the amplitude $A(t)$ is only proportional to the modulating signal $x(t)$, the expression of $y(t)$ in (2.25) is referred to as *amplitude modulation*, which is expressed as

$$y(t) = k_a x(t)\cos[\omega_c t + \phi_c] \tag{2.27}$$

where k_a is the amplitude deviation constant in volt and ϕ_c is the constant phase.

If the phase deviation $\phi(t)$ is proportional to the modulating signal $x(t)$ only, or $\phi(t) = k_{\mathrm{p}}x(t)$, the expression $y(t)$ in (2.25) is referred to as *phase modulation*, which is defined as

$$y(t) = A_c \cos\left[\omega_{\mathrm{c}}t + k_{\mathrm{p}}x(t)\right] \tag{2.28}$$

where k_{p} is the phase deviation constant in radian and A_{c} is the amplitude constant. Similarly the frequency deviation $d\phi(t)/dt$ is only proportional to the modulating signal $x(t)$, or $d\phi(t)/dt = k_{\mathrm{f}}x(t)$, the expression of (2.25) is referred to as *frequency modulation*, which is defined as

$$y(t) = A_{\mathrm{c}} \cos\left[\omega_{\mathrm{c}}t + k_{\mathrm{f}}\int_{-\infty}^{t} x(\tau)d\tau\right] \tag{2.29}$$

where k_{f} is the frequency deviation constant in hertz.

For the amplitude, phase and frequency modulations expressed in (2.27) to (2.29), the instantaneous frequencies are ω_{c}, $\omega_{\mathrm{c}} + k_{\mathrm{p}}dx(t)/dt$, and $\omega_{\mathrm{c}} + k_{\mathrm{f}}x(t)$, respectively. Since phase modulation and frequency modulation differ only in an integration operation, they are also called *angle modulation*. For detailed descriptions regarding fundamental modulations, the interested reader can reference the Ziemer and Tranter [11].

If not only the phase deviation $\phi(t)$ but also the amplitude $A(t)$ are proportional to the modulating signal $x(t)$, the modulated signal $y(t)$ is referred to as combination of both phase and amplitude modulations, such as a Quadrature Amplitude Modulation (QAM) modulation format. Due to its robust bit error rate (BER) performance and relatively simple architecture, the combination of phase and amplitude modulations shall be mostly used through this book.

2.4 Digital Baseband Modulation

In modern communications, information is usually transmitted in the form of a bit stream in order to achieve high-quality transmission performance. The bit stream, however, must be transferred into continuous-time waveforms that are suitable for transmission over a communications channel. This transformation is called mapping, in which one of finite energy waveforms $\{s_m(t), m = 1, 2, \ldots, M\}$ is selected to present one of $M = 2^k$ possible normalized symbols or bits $\{A_m = (2m - 1 - M), m = 1, 2, \ldots, M\}$ at a time for transmission over the channel.

2.4.1 2-Level Pulse Amplitude Modulation (2-PAM) and Binary Phase Shift Keying (BPSK)

In the case of $M = 2$ symbols, a two-level pulse amplitude modulation (PAM) is called the binary *Pulse Amplitude Modulation* (PAM) waveforms, whose waveforms are chosen as

$$s_m(t) = A_m g(t), \quad m = 1, 2 \tag{2.30}$$

where the symbol amplitudes A_m have the values of $A_m = \pm 1$, $m = 1, 2$ and the waveform $g(t)$ is a shaping pulse. Usually the special waveforms (or $A_1 = 1$ for $m = 1$, $A_2 = -1$ for $m = 2$) are chosen as

$$s_1(t) = -s_2(t), \quad 0 \le t \le T_b \tag{2.31}$$

where the bit duration $T_b = 1/f_b$.

In digital PAM, also called *Amplitude Shift Keying* (ASK), with the case of $M = 2$, the modulated signal is represented as

$$\begin{aligned} y(t) &= \sum_{n=-\infty}^{\infty} A_m g(t - nT_b) \cos(2\pi f_c t) \\ &= s_m(t) \cos(2\pi f_c t), \quad m = 1,\ 2\ \ 0 \le t \le T_b \end{aligned} \tag{2.32}$$

where f_c is the carrier frequency.

In digital phase modulation, called *Phase Shift Keying* (PSK), with the case of $M = 2$, the modulated signal is expressed as

$$y(t) = |s_m(t)| \cos[2\pi f_c t + \pi(m - 1)], \quad m = 1, 2\ \ 0 \le t \le T_b \tag{2.33}$$

We see from (2.32) and (2.33) that in the case of $M = 2$ digital PAM signals are identical to digital PSK. Furthermore, we note that the PAM and PSK modulated signals are dependent on the combination of the baseband amplitude and carrier phase. In practice, amplitude shapes of the phase-modulated signals play a very important role in determining the bandwidth of the RF-transmitted signals. An illustration of these two modulation types is shown in Fig. 2.3, where two kinds of pulse shapes or a squared waveform and one-half cycle of a sinusoid are used. In the former, the modulation process only changes the phase of the carrier signal. In the latter, it changes not only the phase of the carrier signal, but also its amplitude. Later, we shall see that the bandwidth efficiency of the modulated signal can significantly benefit from the property of such a smooth pulse shape. Figure 2.4 shows a block diagram of a 2-level (or $M = 2$) PAM or bi-phase PSK (BPSK) transmitter in the radio frequency (RF) or the intermediate frequency (IF) domain.

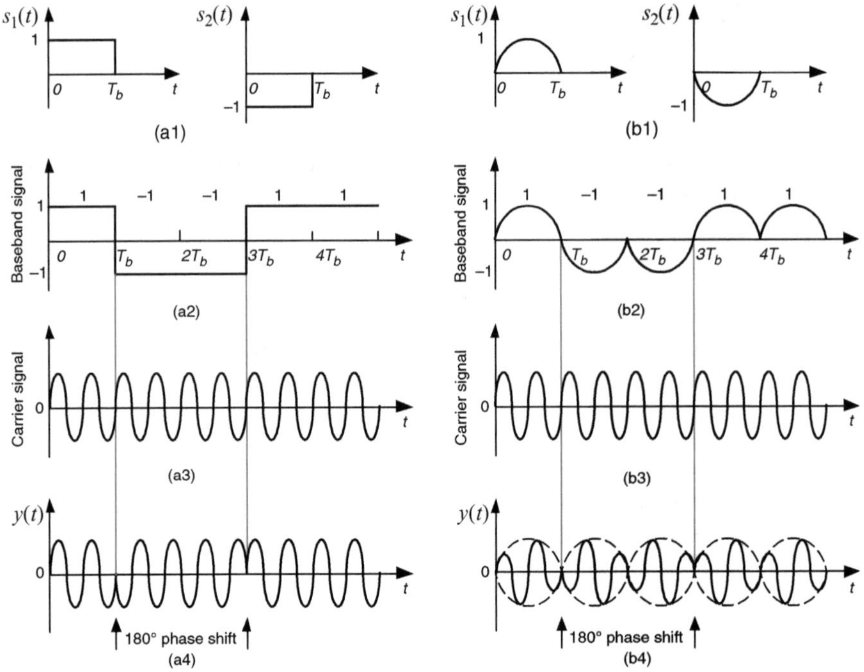

Fig. 2.3 Signal waveforms of PAM and PSK for $M = 2$: (**a1**, **b1**) pulse signals, (**a2**, **b2**) baseband signals, (**a3**, **b3**) carrier signals, and (**a4**, **b4**) modulated signals

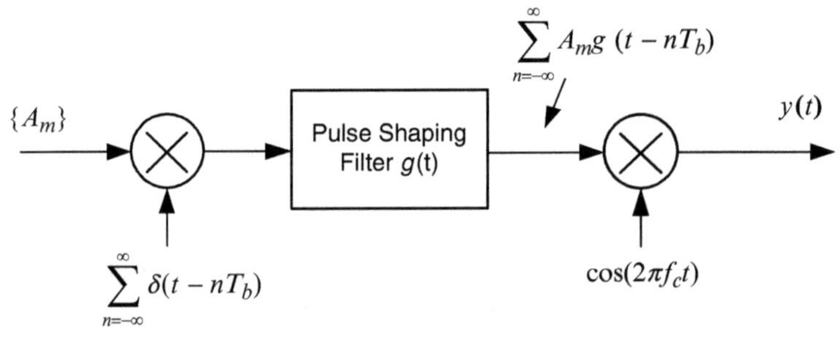

Fig. 2.4 Block diagram of a 2-level PAM or BPSK modulator

2.4.2 *Quadrature Amplitude Modulation (QAM) and Quadrature Phase Shift Keying (QPSK)*

In order to increase the bandwidth efficiency of 2-level PAM and BPSK signals, we can use two independent baseband signal streams to modulate a pair orthogonal carrier signals. The simplest quadrature modulation format is 4-ary quadrature

PAM, which is more often called 4-ary quadrature amplitude modulation (4QAM), or quadrature PSK (QPSK). The bandwidth efficiency of QPSK is twice the bandwidth efficiency of BPSK because the information transmitted over the same bandwidth is doubled. In practice, QPSK type modulation is widely used in many standards due to its robust BER performance compared with high-level QAM formats.

In general, QPSK signal can be expressed as

$$\begin{aligned} y(t) &= \sum_{n=-\infty}^{\infty} A_{mi} g(t - nT_{\mathrm{s}}) \cos(2\pi f_{\mathrm{c}} t) - \sum_{n=-\infty}^{\infty} A_{mq} g(t - nT_{\mathrm{s}}) \sin(2\pi f_{\mathrm{c}} t) \\ &= s_{mi}(t) \cos(2\pi f_{\mathrm{c}} t) - s_{mq}(t) \sin(2\pi f_{\mathrm{c}} t) \\ &= M_m(t) \cos\left[2\pi f_{\mathrm{c}} t + \theta_m(t)\right], \quad m = 1, 2 \;\; 0 \le t \le T_{\mathrm{s}} \end{aligned} \tag{2.34}$$

where the modulus and the phase are calculated by

$$\begin{aligned} M_m(t) &= \sqrt{s_{mi}^2(t) + s_{mq}^2(t)} \\ \theta_m(t) &= \tan^{-1}\left[s_{mq}(t)/s_{mi}(t)\right] \end{aligned} \tag{2.35}$$

In (2.34), A_{mi} and A_{mq}, which are independent from each other, are shown the symbol amplitudes of the in-phase (I) and quadrature (Q) branches, and $s_{mi}(t)$ and $s_{mq}(t)$ are the baseband waveforms of the I and Q branches. Here T_{s} is the symbol duration and is equal to twice the bit duration, or $T_{\mathrm{s}} = 2T_{\mathrm{b}}$.

A QPSK modulator consists of two BPSK modulators plus a serial-to-parallel converter and combiner as shown in Fig. 2.5. In a QPSK modulator, the input bit stream must be converted into the symbol stream on the I and Q branches. Such a

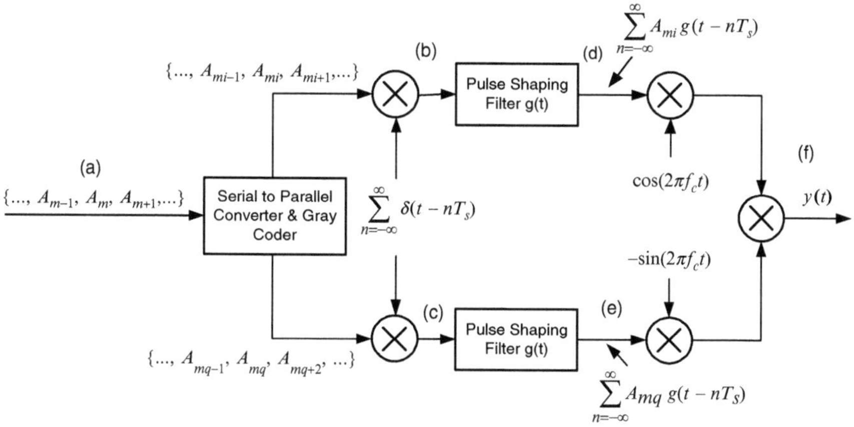

Fig. 2.5 Block diagram of QPSK modulator

conversion can be realized through either a serial-to-parallel converter or a mapping circuit, in which two consecutive bits are converted into a pair of symbols on the I and Q branches. In the conversion, the bit interval T_b before the conversion becomes the symbol interval T_s after the conversion, which is $T_s = 2T_b$. This means that the bit rate is twice the symbol rate for QPSK. Thus, the theoretical spectral efficiency of 2 bit/s/Hz can be achieved with QPSK modulation in some applications, where the 1-bit/s/Hz theoretical spectral efficiency of BPSK modulation is insufficient to provide the available bandwidth efficiency. In QPSK modulation, every pair of symbols on the I and Q branches determines the carrier phase state. The four possible symbols result in the four different carrier phase states. Usually, the mapping format from the four possible symbols to the four phase states is performed in accordance with the *Gary code* as listed in Table 2.3, in which the signs of each pair of adjacent symbols after coding differ by only one to the corresponding symbols as illustrated in Fig. 2.6. The advantage of the Gary code is to ensure that a single symbol error corresponds to a single bit error after the parallel-to-serial converter at the receiver [12].

All the waveforms corresponding to points from (a) to (f) in Fig. 2.5 are plotted in Fig. 2.7, where the rectangular pulse for $g(t)$ is assumed. The phase transition of the modulated carrier signal $y(t)$ in Fig. 2.7f is calculated using (2.35) based on the waveforms in Fig. 2.7d, e. The phase transition of the modulated carrier signal $y(t)$ can be also obtained from its constellation diagram in Fig. 2.8.

Table 2.3 Two-bit binary to Gary code

Decimal	Binary	Gary code	Gary code as decimal
0	00	00	0
1	01	01	1
2	10	11	3
3	11	10	2

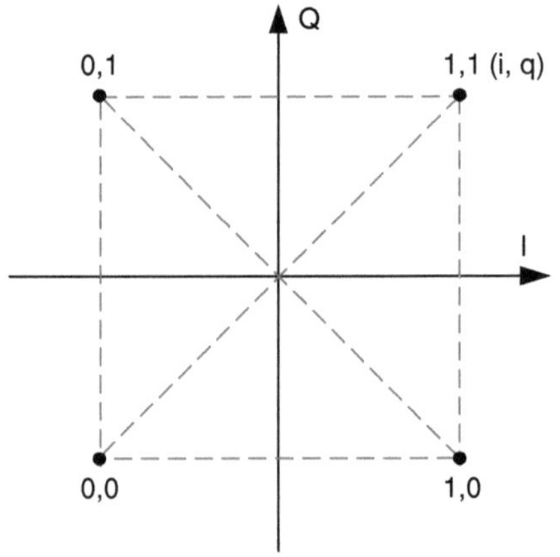

Fig. 2.6 Constellation of QPSK baseband signals

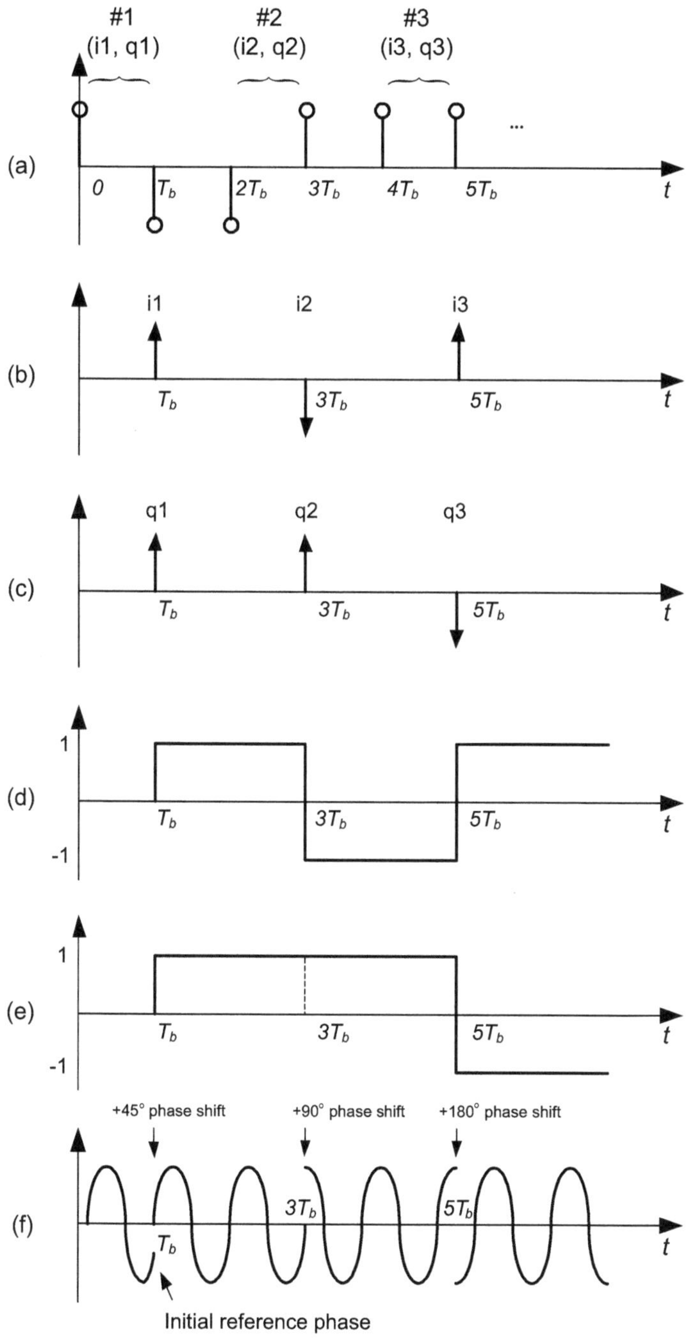

Fig. 2.7 Signal waveforms for QPSK in Fig. 2.4: (**a**) information bits, (**b**) symbol impulses of the I branch, (**c**) symbol impulses of the Q branch, (**d**) baseband waveform of the I branch, (**e**) baseband waveform of the Q branch, (**f**) phase-modulated waveform

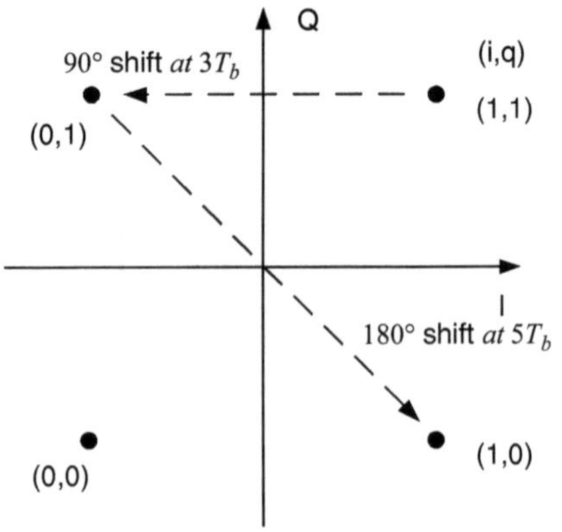

Fig. 2.8 Phase transitions on constellation of QPSK for Fig. 2.7f

2.4.3 Power Spectral Density of Baseband Signals

Most digital communication systems are highly band-limited because the usable spectrum resources are severely congested. As a result, system architects must consider the bandwidth-efficient modulation technique as the highest priority in the assigned channel. The bandwidth efficiency of a modulated signal is usually characterized by its power spectral density (PSD), which is the distribution of power as a function of frequency. Therefore the designers can choose the modulation format based on its PSD characteristic to meet the requirements of the channel bandwidth.

For a random process, $X(t)$, its PSD $\Psi(f)$ and autocorrelation function $R(\tau)$ are related through the Fourier transform, or the Fourier transform

$$\Psi(f) = F[R(\tau)] = \int_{-\infty}^{\infty} R(\tau)\,e^{-j2\pi f\tau} d\tau \tag{2.36}$$

and the inverse Fourier transform

$$R(\tau) = F^{-1}[\Psi(f)] = \int_{-\infty}^{\infty} \Psi(f)\,e^{j2\pi f\tau} df \tag{2.37}$$

If τ is equal to zero, (2.37) becomes

$$R(0) = \int_{-\infty}^{\infty} \Psi(f) df \tag{2.38}$$

In above expression, $R(0)$ represents the average power of the random process, which is equal to the area under $\Psi(f)$.

For a stationary process, the autocorrelation function of $X(t)$ is independent of time and defined as

$$R(\tau) = R(t_1 - t_2) = E[X(t_1)X(t_2)] = E[X(t)X(t+\tau)] \tag{2.39}$$

where t_1 and t_2 are two different time instants and $E[\bullet]$ presents the ensemble average.

Most bandwidth efficient modulation techniques use a pair of orthogonal carrier signals $\cos(2\pi f_c t)$ and $\sin(2\pi f_c t)$ to carry two independent baseband signals $x_I(t)$ and $x_Q(t)$ on the I and Q branches, which are equivalent to $s_{mi}(t)$ and $s_{mq}(t)$ in (2.34), respectively. This modulation scheme is called *quadrature modulation*. The expression for such a quadrature-modulated signal is given as

$$s(t) = x_I(t)\cos(2\pi f_c t) - x_Q(t)\sin(2\pi f_c t) \tag{2.40}$$

The baseband signals on the I and Q branches are represented in general form:

$$x_I(t) = \sum_{n=-\infty}^{\infty} a_n g(t - nT_s), \quad x_Q(t) = \sum_{n=-\infty}^{\infty} b_n g(t - nT_s) \tag{2.41}$$

where the information symbols a_n and b_n are independent with equiprobable values, T_s is the symbol interval, and $g(t)$ is the spectrum-shaping pulse.

Another representation of the signal in (2.40) is

$$\begin{aligned} s(t) &= \mathrm{Re}\{[x_I(t) + jx_Q(t)]e^{j2\pi f_c t}\} \\ &= \mathrm{Re}\{x_L(t)e^{j2\pi f_c t}\} \end{aligned} \tag{2.42}$$

where Re{ } stands for the real part of the complex signal in the bracket and $x_L(t)$ is the *equivalent lowpass signal* of the modulated signal $s(t)$.

To derive the expression for PSD, we assume that $s(t)$ is a stationary processing, and the baseband signals of $x_I(t)$ and $x_Q(t)$ have zero mean values. The autocorrelation function of $s(t)$ in (2.42) is

$$R_s(\tau) = \mathrm{Re}\{R_{x_L}(\tau)e^{j2\pi f_c t}\} \tag{2.43}$$

The PSD of the modulated signal in (2.40) is the Fourier transform of $R_s(\tau)$, or

$$\begin{aligned} \Psi_s(f) = F[R_s(\tau)] &= \int_{-\infty}^{\infty} \{\mathrm{Re}[R_{x_L}(\tau)e^{j2\pi f_c t}]\}e^{-j2\pi f\tau}d\tau \\ &= \int_{-\infty}^{\infty} \left\{\frac{1}{2}\left[R_{x_L}(\tau)e^{j2\pi f_c t} + R^*_{x_L}(\tau)e^{-j2\pi f_c t}\right]\right\}e^{-j2\pi f\tau}d\tau \\ &= \tfrac{1}{2}[\Psi_{x_L}(f - f_c) + \Psi_{x_L}(-f - f_c)] \end{aligned} \tag{2.44}$$

where $\Psi_{x_L}(f)$ is the PSD of the equivalent lowpass process $x_L(t)$. In (2.44), we used the property $R^*_{x_L}(\tau) = R_{x_L}(-\tau)$ and the quality property of the Fourier transform $e^{j2\pi f_c t}x(t) \leftrightarrow X(f - f_c)$.

Similar to (2.37), the PSD of the modulated signal expressed in (2.44) is the shifted PSD of the equivalent lowpass signal $x_L(t)$ to the center frequency $\pm f_c$ except the constant factor 1/2. It becomes useful to determine the PSD of the modulated signal through the PSD of the equivalent lowpass signal.

The derivations of the PSD of the equivalent lowpass signal $x_L(t)$ are complicated and fairly long and beyond the scope of this book. Here we just present its expression

$$\Psi_{x_L}(f) = \sigma^2 f_s |G(f)|^2 + \mu^2 f_s^2 |G(0)|^2 \delta(f) + 2\mu^2 f_s^2 \sum |G(mf_s)|^2 \delta(f - mf_s) \quad (2.45)$$

where σ^2 and μ are the variance and mean of the sequence of information symbols $\{a_n\}$ or $\{b_n\}$, respectively. $G(f)$ is the Fourier transform of the shaping pulse $g(t)$ and $f_s = 1/T_s$ is the symbol rate.

The expression (2.45) consists of three terms to emphasize the three different types of spectral components. The first term is the continuous spectral component, and its shape is completely dependent on the squared module of the Fourier transfer of the shaping pulse $g(t)$. The second term is the DC component. The third term consists of discrete harmonics, each spaced f_s apart in frequency.

If the information binary symbols $\{a_n\}$ or $\{b_n\}$ are assumed to be independent random variables with equal probability for the values of ± 1, the information symbols have zero mean and unit variance. In this case, all discrete spectral harmonics and DC components vanish and only the continuous component is left in (2.45). This condition is valid for most digital modulation techniques. Thus, the designer can simply choose a proper shaping pulse $g(t)$ to achieve a bandwidth-efficient transmission system. In this condition, (2.45) can be simplified to

$$\Psi(f) = \Psi_{x_L}(f) = f_s |G(f)|^2 \quad (2.46)$$

Here, we drop off the subscript x_L for the purpose of simplicity.

It is clearly seen from (2.46) that the PSD of the equivalent baseband signal $s_L(t)$ is proportional to the squared module of $|G(f)|^2$ of the Fourier transform of the equivalent baseband signal. Meanwhile, the PSD of the modulated signal, which is expressed in (2.44), is the shifted PSD of the equivalent lowpass signal $x_L(t)$ to the center frequency +/- f_c except the constant factor 1/2. Therefore finding a spectrum-shaping pulse with the property of the narrow main lobe and small side-lobes in the frequency domain plays an important role in achieving high bandwidth efficiency for the spectrum-shaping types of modulation techniques.

2.4.4 *Non-Overlapped Pulse Waveform Modulation*

From (2.46), we know that the PSD shape of the modulated signal depends only on the shape of the Fourier transform of the shaping pulse $g(t)$. The fundamental type of shaping pulse that is widely used in digital modulation techniques is a non-overlapped shaping pulse. This non-overlapped shaping pulse only lasts one

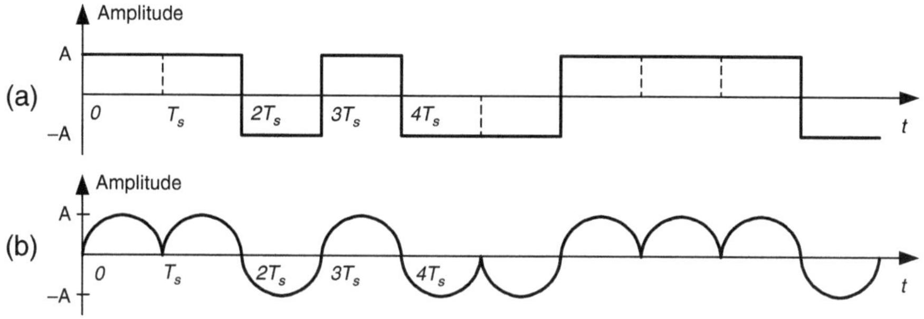

Fig. 2.9 Two different types of baseband waveforms: (**a**) signals consisting of non-return to zero (NRZ) and (**b**) signals consisting of one-half cycle of a sinusoid

symbol interval T_s in the time domain. For a BPSK signal, the symbol interval T_s equals the bit interval T_b. Two widely used non-overlapped pulses are the rectangular and the one-half cycle of a sinusoid. The former is used for the unfiltered BPSK/QPSK/OQPSK modulations, while the latter is used for MSK modulation and is also adopted in the ZIGBEE standard [13].

To see the PSD related with Fourier transform of the pulse $g(t)$, first we consider the rectangular pulse whose expression is given by

$$g(t) = \begin{cases} A, & 0 \le t \le T_s \\ 0, & \text{elsewhere} \end{cases} \tag{2.47}$$

This pulse is used to weight the sequence of random variables, each having zero mean and unit variance, as illustrated in Fig. 2.9a. The Fourier transform of the $g(t)$ is

$$\begin{aligned} G(f) &= \int_{-\infty}^{\infty} g(t)\mathrm{e}^{-j2\pi ft}dt = \int_0^{T_s} A\mathrm{e}^{-j2\pi ft}dt \\ &= AT_s \frac{\sin(\pi f T_s)}{\pi f T_s} \mathrm{e}^{-j\pi f T_s} \end{aligned} \tag{2.48}$$

Hence

$$|G(f)|^2 = A^2 T_s^2 \left(\frac{\sin(\pi f T_s)}{\pi f T_s} \right)^2 \tag{2.49}$$

Since the mean and variance of the random information sequences are zero and unit, respectively, we can use (2.46) to calculate the PSD of the QPSK signal:

$$\Psi_{\mathrm{QPSK}}(f) = A^2 T_s \left(\frac{\sin(\pi f T_s)}{\pi f T_s} \right)^2 \tag{2.50}$$

Next, we consider the one-half cycle sinusoid pulse expressed as

$$g(t) = \begin{cases} A\sin(\pi t/T_s), & 0 \le t \le T_s \\ 0, & \text{elsewhere} \end{cases} \tag{2.51}$$

The random waveform weighted with such a pulse is illustrated in Fig. 2.9b. Similar to the derivation above, the Fourier transform of the $g(t)$ is

$$\begin{aligned} G(f) &= \int_0^{T_s} A\sin(\pi t/T_s)\mathrm{e}^{-j2\pi f t}dt \\ &= \frac{2AT_s}{\pi}\frac{\cos(\pi f T_s)}{1-4f^2T_s^2}\mathrm{e}^{-j\pi f T_s} \end{aligned} \tag{2.52}$$

From (2.46) the power spectral density is

$$\Psi_{\text{MSK}}(f) = \frac{4A^2T_s}{\pi^2}\left(\frac{\cos(\pi f T_s)}{1-4f^2T_s^2}\right)^2 \tag{2.53}$$

It is seen in (2.50) and (2.53) that the main lobe of the power spectral density (PSD) of QPSK/OQPSK is located at $f = 1/T_s$, while the main lobe of MSK is located at$f = 1.5/T_s$. This means that the main lobe of MSK is 50% wider than that for QPSK/OQPSK. On the other hand, the side lobes in QPSK/OQPSK fall off at a rate of f^{-2}, while the side lobes in MSK drop at a rate of f^{-4}, which is faster than QPSK/OQPSK.

It is also noted that the wider the pulse width T_s in the time domain is, the narrower the main lobe in the frequency domain is. Hence, a pulse period lasting more than T_s would increase the bandwidth efficiency. The baseband signals can be generated by overlapping pulse sequences, which will be introduced in the next section.

2.5 Overlapped Pulse-Shaping Modulation

To increase the bandwidth efficiency by means of extending the period of the pulse lasting, it is also preferable to be either free of intersymbol interference or have minimal ISI. Besides achieving bandwidth efficiency, it is also important for the modulated signals to achieve energy efficiency when passing through nonlinear power amplifiers. For most modulation techniques with bandwidth and energy efficiency, data in the Q channel is delayed by a half-symbol interval $T_s/2$ relative to data in the I channel in order to reduce the envelope fluctuation. Equation (2.41) can be rewritten as

$$
\begin{aligned}
x_{\mathrm{I}}(t) &= \sum_{n=-\infty}^{\infty} a_n g(t - nT_{\mathrm{s}}) \\
x_{\mathrm{Q}}(t) &= \sum_{n=-\infty}^{\infty} b_n g(t - nT_{\mathrm{s}} - T_{\mathrm{s}}/2)
\end{aligned}
\tag{2.54}
$$

2.5.1 Overlapped Raised-Cosine Pulse-Shaping Modulation

A basic ideal for the overlapped raised-cosine modulation is to achieve ISI free at the decision instants in the time domain by overlapping two consecutive shaping pulses, in which each pulse lasts exactly twice symbol interval of $2T_{\mathrm{s}}$ in the time domain and has a narrow main lobe and fast roll-off of sidelobes of the Fourier transform in the frequency domain. The Raised-cosine pulse with the interval of $2T_{\mathrm{s}}$ is obtained by convolving a square pulse with a half-cycle of the sine waveform, each with interval of T_{s}.

The convolution of two pulse signals is to simply let one pulse signal pass through a filter $H(s)$ with an impulse response of $h(t)$, which is equal to another pulse signal, as shown in Fig. 2.10.

One early-published paper regarding the convolution pulse-shaping method is Quadrature Overlapped Raised-Cosine (QORC) modulation [14]. The input $x(t)$ to the filter is a square waveform and impulse response of the filter $h(t)$ is a half-cycle sine signal

$$
x(t) = \begin{cases} 1, & 0 \le t \le T_{\mathrm{s}} \\ 0, & \text{elsewhere} \end{cases} \tag{2.55}
$$

$$
h(t) = \begin{cases} \dfrac{\pi}{2T_{\mathrm{s}}} \sin\left(\dfrac{\pi t}{T_{\mathrm{s}}}\right), & 0 \le t \le T_{\mathrm{s}} \\ 0, & \text{elsewhere} \end{cases} \tag{2.56}
$$

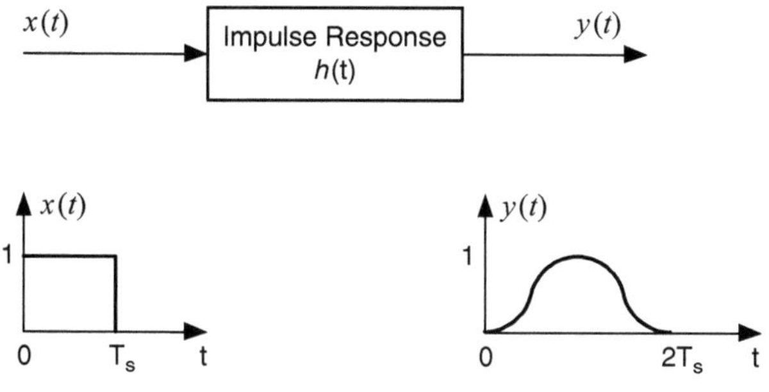

Fig. 2.10 Pulse-shaping filter

where amplitude $\pi/2T_s$ is used to normalize the filter output or raised-cosine pulse waveform. In fact, (2.55) is the pulse shape of QPSK signal while (2.56) is the pulse shape of MSK signal.

The output of the filter is the convolution of the input and impulse response

$$g(t) = x(t) * h(t) = \begin{cases} \frac{1}{2}\left(1 - \cos\frac{\pi t}{T_s}\right), & 0 \le t \le 2T_s \\ 0, & \text{elsewhere} \end{cases} \tag{2.57}$$

Thus, the duration of the raised-cosine pulse through the convolution operation becomes twice the duration T_s or $2T_s$.

Similar to the relationship between QPSK and Staggered QPSK (SQPSK), the symbol sequences on the Q branch for QORC can be delayed by a half-symbol interval offset $T_s/2$ relative to that on the I branch in order to reduce the envelope fluctuation of the modulated signal. This offset QORC is called the staggered QORC (SQORC) [14].

Figure 2.11 shows a block diagram of SQORC modulator, where a conceptual block diagram of raised-cosine (RC) pulse shaping can be implemented using either a filter as shown in Fig. 2.10 or a circuit illustrated in Fig. 2.12. In order to overlap raised-cosine pulses, each RC pulse-shaping needs one serial-to-parallel converter and two delay T_s devices.

Figure 2.13 shows the baseband waveforms of the SQORC signal. Figure 2.13a illustrates the SQORC baseband signal performed by the overlapping method implemented in Fig. 2.12. Figure 2.13b shows the eye diagram. It can be seen that there are no ISI at the sampling decision instants and no jitter at the jitter instants. The modulation technique with such a property is also called *Intersymbol Interference and Jitter-Free (IJF) OQPSK (IJF-OQPSK)*, which will be introduced in the next section. Figure 2.13c shows the constellation of the SQORC signal, where the maximum envelope fluctuation is 3 dB. The envelope is always 1 when consecutive symbols have alternative polarity on both the I channel and Q channel,

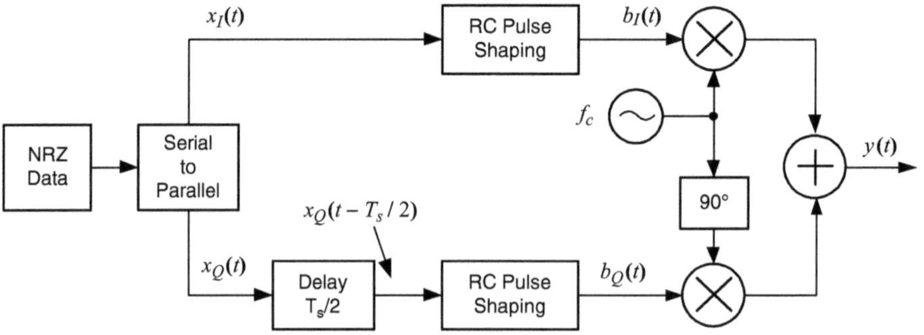

Fig. 2.11 Block diagram of QORC/SQORC modulator

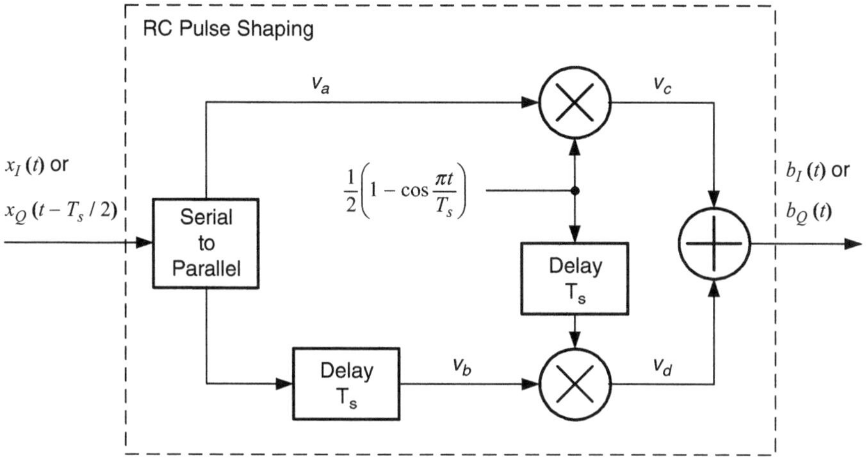

Fig. 2.12 Block diagram of RC pulse shaping

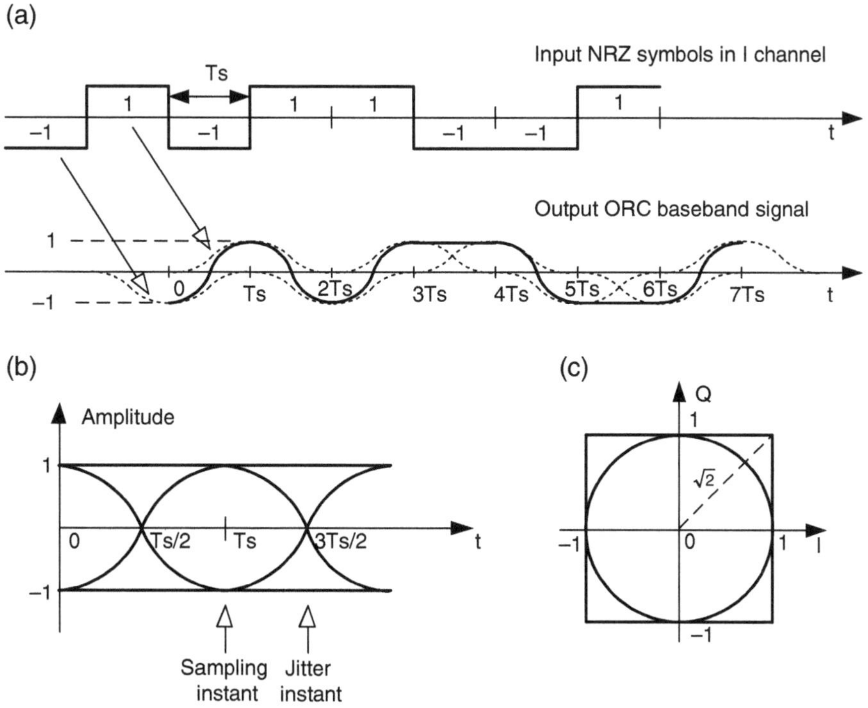

Fig. 2.13 Baseband waveform: (**a**) overlapped baseband signal, (**b**) eye diagram, and (**c**) constellation

while the maximum envelope amplitude occurs when consecutive symbols have the same polarity on both the I channel and Q channel. This results in the maximum envelope amplitude of $\sqrt{2}$, as shown in Fig. 2.13c. This amount of envelope fluctuation causes the regrowth of the PSD at the output of the power amplifier when it operates at the saturation or close to the saturation region.

Equation (2.57) in the time domain has the following relationship in the Fourier transform domain:

$$G(\omega) = X(\omega) \times H(\omega) \tag{2.58}$$

The square waveform $x(t)$ in (2.55) and impulse response $h(t)$ in (2.56) have the Fourier transform

$$X(\omega) = T_s \frac{\sin\left(\frac{\omega T_s}{2}\right)}{\frac{\omega T_s}{2}} e^{-j\omega T_s/2} \tag{2.59}$$

$$H(\omega) = \frac{\cos\left(\frac{\omega T_s}{2}\right)}{\left[1 - \left(\frac{\omega T_s}{\pi}\right)^2\right]} e^{-j\omega T_s/2} \tag{2.60}$$

Substituting (2.59) and (2.60) into (2.58), the Fourier transform of the filter output is

$$G(\omega) = \frac{\sin(\omega T_s)}{\omega\left[1 - \left(\frac{\omega T_s}{\pi}\right)^2\right]} e^{-j\omega T_s} \tag{2.61}$$

Hence

$$|G(f)|^2 = \frac{\sin^2(\omega T_s)}{\omega^2\left[1 - \left(\frac{\omega T_s}{\pi}\right)^2\right]^2} \tag{2.62}$$

Since the mean and variance of the random information sequences are zero and unit, respectively, we can use (2.46) to calculate the PSD for SQORC signal as

$$\Psi_{SQORC}(f) = f_s|G(f)|^2 = T_s \frac{\sin^2(2\pi f T_s)}{(2\pi f T_s)^2\left[1 - \left(\frac{2\pi f T_s}{\pi}\right)^2\right]^2} \tag{2.63}$$

The normalized PSD is

$$\frac{\Psi_{SQORC}(f)}{\Psi_{SQORC}(0)} = \frac{\sin^2(2\pi f T_s)}{(2\pi f T_s)^2\left[1 - \left(\frac{2\pi f T_s}{\pi}\right)^2\right]^2} \tag{2.64}$$

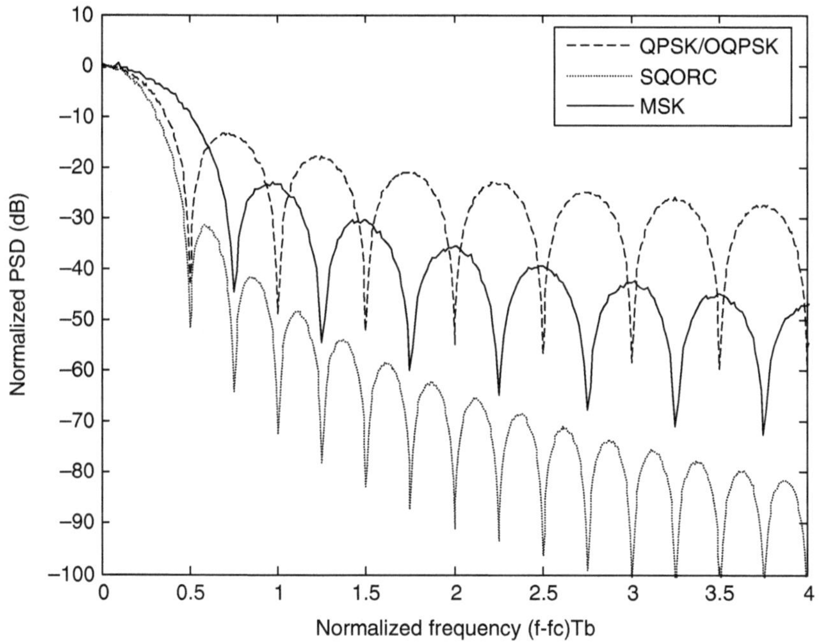

Fig. 2.14 Power spectral densities of OQPSK, MSK and SQORC in a linear channel

The power spectral density curves of the QPSK/OQPSK, MSK, and SQORC signals in a linear channel are illustrated in Fig. 2.14. The main lobe of the PSD of SQORC signal is located at $f = 1/T_s = 0.5/T_b$, which is the same as QPSK/OQPSK signals and the main lobe of MSK signal is located at $f = 1.5/T_s = 0.75/T_b$. The side lobes in SQORC fall off at a rate of f^{-6}, which is much faster compared with QPSK/OQPSK at a rate of f^{-2} and MSK at a rate of f^{-4}. In fact, the PSD of SQORC takes on the form of the product of the PSDs of QPSK/OQPSK and MSK because the Fourier transform of QORC/SQORC is the product of the Fourier transforms of QPSK/OQPSK and MSK. So the main lobe of SQORC retains the same width as the main lobe of QPSK/OQPSK and the side lobes of SQORC become narrower and fall off at a rate of f^{-6}, which is equal to the addition of their rates of f^{-2} and f^{-4}.

2.5.2 IJF-OQPSK Modulation

Intersymbol interference and jitter-free (IJF-OQPSK) [15] baseband signals are identical to SQORC baseband signals except for their implementation methods. The former uses a simple and precise logic switch circuit to generate its baseband waveforms instead of using overlapped raised-cosine pulse. This method resulted in fast applications of IJF-OQPSK/SQORC in satellite earth stations in the late 1980s. After that, a look-up table (LUT) based IJF-OQPSK and SQORC was widely used in digital communication systems to achieve energy and spectrally efficient transmissions.

As we have seen, overlapping a current raised-cosine pulse with $2T_s$ and a previous raised-cosine pulse forms a current SQORC baseband signal during the symbol interval T_s. In turn, it is determined by a current and previous symbol data. Thus a total of four types of baseband signals in one symbol interval T_s depending on combinations of the input symbol data are plotted in Fig. 2.15.

In Fig. 2.15, the shape of the current baseband signal during T_s is determined by the combination of the current and previous Non-Return-to-Zero (NRZ) data (symbols) $d_n d_{n-1}$ in either I channel or Q channel. An actual hardware implementation of IJF-OQPSK baseband signal is illustrated in Fig. 2.16. Each pair $d_n d_{n-1}$ of the current and previous NRZ symbols in the I and Q channels are used as addresses to turn on one of four switches in each symbol duration. The output baseband signal

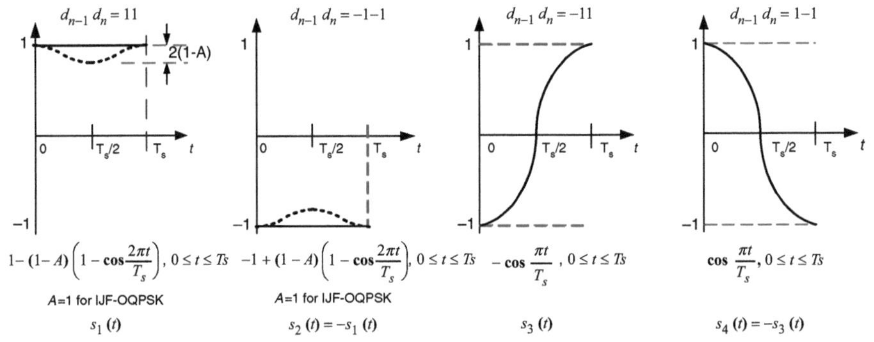

Fig. 2.15 Total four overlapped baseband segments in one symbol interval T_s, where the *solid-line waveforms* are for IJF-OQPSK and *dot-line* are for SQAM with $0.5 \le A \le 1.0$

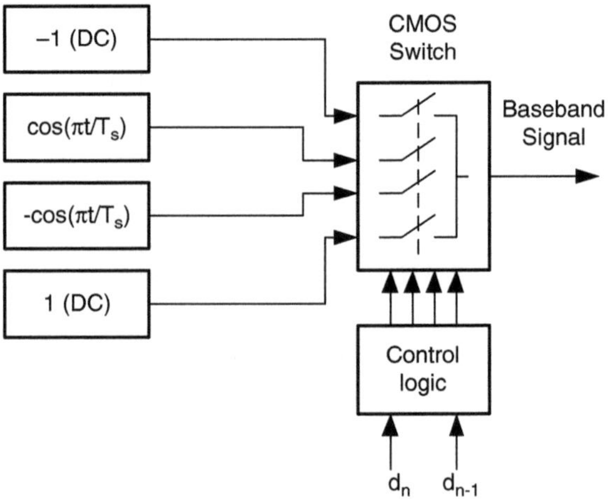

Fig. 2.16 IJF-OQPSK switch based baseband signal generator. Redrawn from [12]

is sent to the I–Q modulator to modulate the carrier signal after passing through a smooth lowpass filter to remove high frequency harmonics due to switching operation. The baseband waveform, eye diagram and constellation of the IJF-OQPSK are identical to those for SQORC as shown in Fig. 2.13. The mathematic expression PSD for IJF-OQPSK is also the same as that for SQORC as expressed in (2.64) and plotted in Fig. 2.14.

2.5.3 Other Overlapped Pulse-Shaping Modulations

SQORC and IJF-OQPSK signals have a 3-dB envelope fluctuation, as shown in Fig. 2.13c. As a result, the spectral side lobes slightly spread up due to AM–AM and AM–PM conversions of the power amplifies after they pass through nonlinear amplification channels. In order to reduce such a 3 dB envelope fluctuation and also keep the interval T_s of the pulse waveform unchanged, two modified overlapped pulse-shaping waveforms, called superposed quadrature amplitude modulation (SQAM) [16] and self-convolving minimum shift key (SCMSK) modulation [17], were proposed. In the former, the pulse waveform with the interval of $2T_s$ is generated by superposing two raised-cosine pulses, each having the interval of T_s and opposite polarity, to the raised-cosine pulse with the interval of $2T_s$ that is the same as a pulse-shaping waveform used in SQORC/IJF-OQPSK. In the latter, the pulse waveform with the interval of $2T_s$ is created by convolving a half-cycle of a sinusoidal pulse with the interval of T_s with itself. In the following, major concepts of these two modulation techniques are described simply. The interested reader can reference the Appendix C.3 for the detailed derivations.

SQAM: The SQAM was developed based on IJF-OQPSK modulation for the purpose of further reducing the maximum envelope fluctuation of the IJF-OQPSK by 3 dB. To form a SQAM pulse waveform, another two raised-cosine pulses, each having single symbol duration of T_s and adjustable amplitude parameter A, with negative polarity are superposed to the original raised-cosine pulse with double symbol interval of $2T_s$. The superposed pulse waveform is expressed by

$$s(t) = g(t) + d(t) \tag{2.65}$$

where

$$g(t) = \frac{1}{2}\left(1 + \cos\frac{\pi}{T_s}(t - T_s)\right) \tag{2.66}$$

$$d(t) = -\frac{1-A}{2}\left(1 - \cos\frac{2\pi t}{T_s}\right), \quad 0.5 \le A \le 1.0, \ 0 \le t \le 2T_s \tag{2.67}$$

In (2.67), A is an adjustable amplitude parameter. Figure 2.17 illustrates the SQAM pulse-shaping process by adding two raised-cosine pulses $d(t)$ to one raised-cosine pulse $g(t)$. Figure 2.18 illustrates the comparison among SQAM pulse waveforms with different parameters of A, where $A = 1$ corresponds to IJF-OQPSK signal.

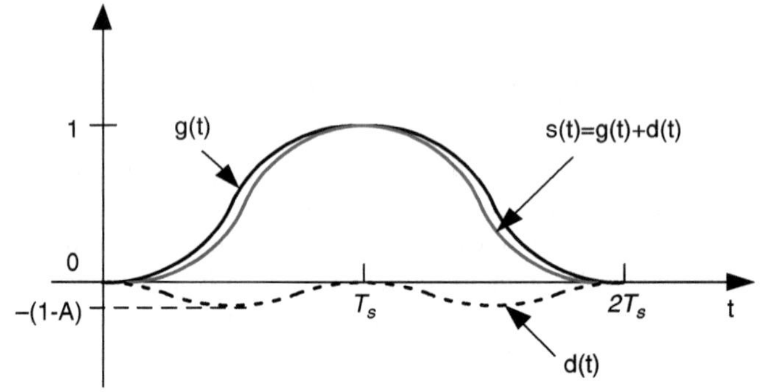

Fig. 2.17 SQAM pulse shaping by superposing two raised-cosine pulses with symbol interval of T_s to one with $2T_s$

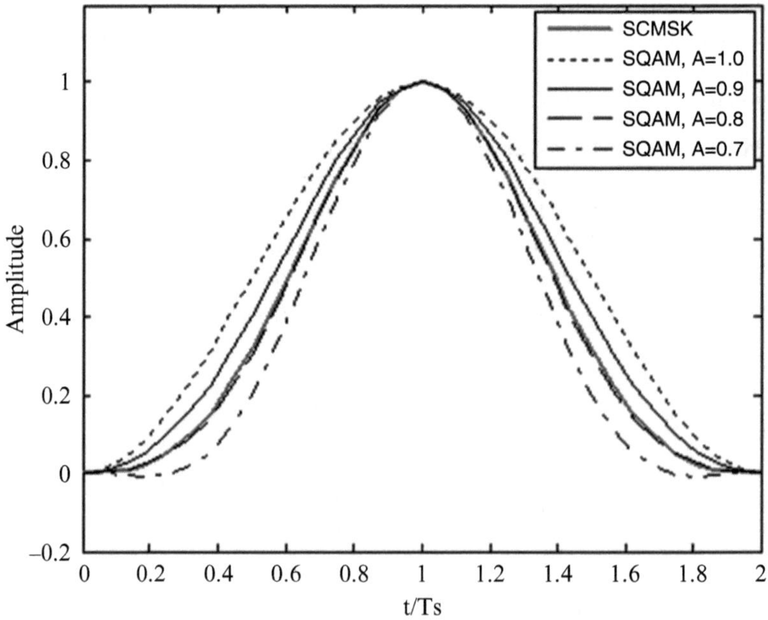

Fig. 2.18 Shaping pulses of SCMSK and SQAM in a twice symbol interval $2T_s$. Note that SQAM with $A = 1$ is equal to IJF-OQPSK pulse

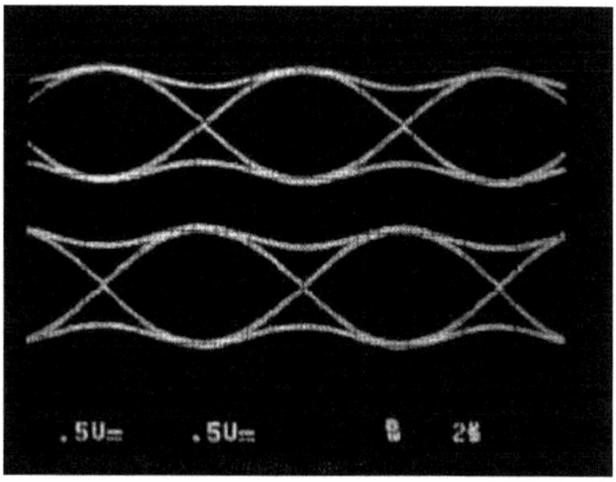

Fig. 2.19 Eye diagrams of SQAM baseband I–Q signals with $A = 0.8$ at the symbol rate of 135.416 kilosymbols per second (or the bit rate of 270.833 kilobits per second is divided by 2). Referenced from [18]

The SQAM modulator is the same as the IJF-OQPSK modulator except for the replacement of IJF-OQPSK waveform segments of $s_1(t)$ and $s_2(t)$ by SQAM waveform segments of $s_1(t)$ and $s_2(t)$, as shown in Fig. 2.15. To avoid the maximum envelope fluctuation of 3 dB that appears when two consecutive symbols have the same polarity in both the I and Q channels for IJF-OQPSK, the waveforms around the centers of the segments $s_1(t)$ and $s_2(t)$ for SQAM are reduced from 1 to $1-2 \times (1 - A)$ with $0.5 \le A \le 1.0$.

For example, the maximum envelope is significantly reduced from 3 to 0.7 dB when A changes from 1 to 0.7. Therefore, IJF-OQPSK is a special case of SQAM signal at $A = 1$. Figure 2.19 illustrates eye diagrams of SQAM-baseband signals with $A = 0.8$ at the symbol rate of 135.417 kilosymbols/s, which corresponds to the bit rate of 270.833 kbits/s for a GMSK signal in the 2G GSM system. A detailed description of the SQAM signal is given in Appendix C.

SCMSK: The idea of SCMSK pulse waveform generation was triggered by a SQORC pulse waveform generation introduced in (2.57), in which its pulse waveform is generated by convolving a rectangular pulse of QPSK in (2.55) with one half-cycle of the sinusoidal pulse of MSK in (2.56); and hence the PSD of the SQORC modulated signal takes the form of the production of the power spectral densities of QPSK and MSK. This fact reveals a way to find a new overlapped pulse whose PSD has fast roll-off sidelobes by convolving the pulse shaping waveforms of another two modulation signals, each having the symbol interval of T_s.

Similar to SQORC, a SCMSK pulse waveform is generated by convolving a half-cycle of sinusoidal pulse of MSK signal with itself, or by letting one half-cycle of the sinusoidal pulse pass through a filter with the impulse response of another half-cycle of the sinusoidal pulse. As a result, the sidelobes of SCMSK signal roll off as twice fast as the sidelobes of MSK signal, while the main lobe of SCMSK signal remains the same as the main lobe of MSK signal. The pulse waveform of SCMSK is generated as

$$g(t) = x(t) * h(t) \tag{2.68}$$

The input pulse to the filter and the impulse response of the filter are given by

$$x(t) = \begin{cases} \sin(\pi t/T_s), & 0 \le t \le T_s \\ 0, & \text{otherwhere} \end{cases} \tag{2.69}$$

$$h(t) = \begin{cases} \dfrac{2}{T_s}\sin(\pi t/T_s), & 0 \le t \le T_s \\ 0, & \text{otherwhere} \end{cases} \tag{2.70}$$

The convolution of $x(t)$ and $h(t)$ is easily derived by substituting (2.69) and (2.70) into (2.68) and it may be expressed in the form

$$g(t) = \begin{cases} (1/\pi)\sin(\pi t/T_s) - (t/T_s)\cos(\pi t/T_s), & 0 \le t \le T_s \\ -(1/\pi)\sin(\pi t/T_s) + (t/T_s - 2)\cos(\pi t/T_s), & T_s \le t \le 2T_s \end{cases} \tag{2.71}$$

The pulse waveform of SCMSK in the time domain within the twice symbol interval of $2T_s$ is shown in Fig. 2.18, where it is very close to the pulse waveform of SQAM with $A = 0.8$, which creates the best PSD and BER performances in both linear and nonlinear channels compared with other A parameters. Therefore, it is expected that their PSDs should be close to each other as well. Figures 2.20 and 2.21 show PSDs in a linear channel and a nonlinear channel, respectively. It can be seen that PSD of the SQAM signal with $A = 0.8$ rolls off the fastest, followed by the

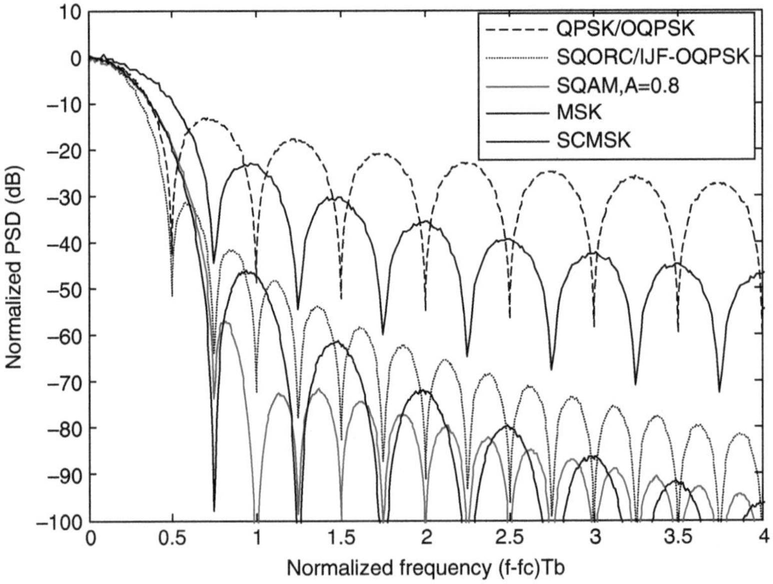

Fig. 2.20 Power spectral densities for QPSK/OQPSK, SQORC/IJF-OQPSK, SQAM, MSK, and SCMSK in a linear channel, where $T_b = T_s/2$ is the bit duration

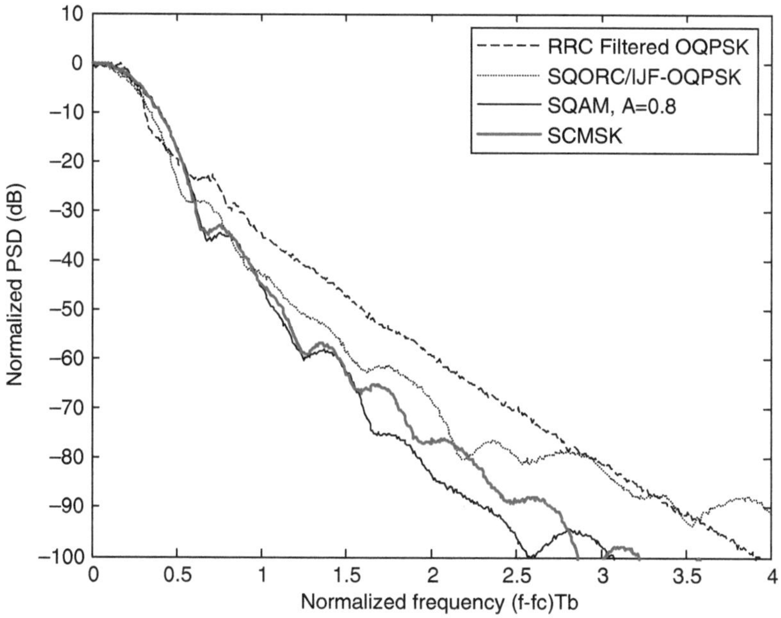

Fig. 2.21 Power spectral densities for OQPSK, SQORC/IJF-OQPSK, SQAM, and SCMSK in a nonlinear channel, where OQPSK is passed through a root-raised cosine filter with $\alpha = 0.35$, where $T_b = T_s/2$ is the bit duration

SCMSK signal in the linear channel. In a nonlinear channel, the SQAM signal with $A = 0.8$ rolls off as fast as the SCMSK signal up to $(f - f_b)T_b = 1.5$. The main lobe of SQAM signal $A = 0.8$ is the same as the SCMSK signal and both of them are equal to the normalized frequency $(f - f_b)T_b = 0.75$, which is wider than the main lobe of QPSK signal, which is equal to $(f - f_b)T_b = 0.5$. Figure 2.22 illustrates E_b/N_o degradation of SQAM signal versus the parameter of A compared with an ideal theoretical QPSK $E_b/N_o = 8.4\,\text{dB}$ at BER $= 10^{-4}$ in a linear channel. The best performance in linear and nonlinear channels is obtained for $A = 0.8$. With this value, E_b/N_o is degraded by 0.3 dB in a linear channel and 0.5 dB in a nonlinear channel [16]. It has been demonstrated that the SCMSK signal has the same E_b/N_o performance as SQAM signal in both linear and nonlinear channels because of their identical similarities in their pulse waveforms in Fig. 2.18.

It is seen from Fig. 2.18 in the time domain and Fig. 2.20 in the frequency domain that for the cases of SQAM signal with $A = 0.8$ and $A = 1$ (or IJF-OQPSK), the more smoothly the pulse reaches zero in the time domain, the faster the side lobes of the PSD roll off in the frequency domain. On the other hand, the narrower the pulse around the center is in the time domain, the wider the main lobe of the PSD is in the frequency domain. This phenomenon is easily understood from the property of the Fourier transform that the more smoothly the pulse reaches zero, the faster the high-frequency components decay, and the narrower the pulse around the center is, the wider the bandwidth occupied by the low-frequency components is.

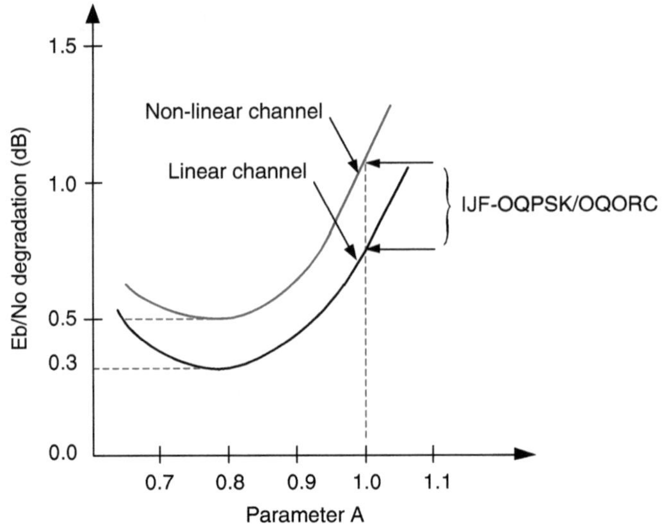

Fig. 2.22 E_b/N_o degradation of SQAM signal from an ideal theoretical QPSK in a linear channel requires $E_b/N_o = 8.4$ dB at BER $= 10^{-4}$. Redrawn from [16]

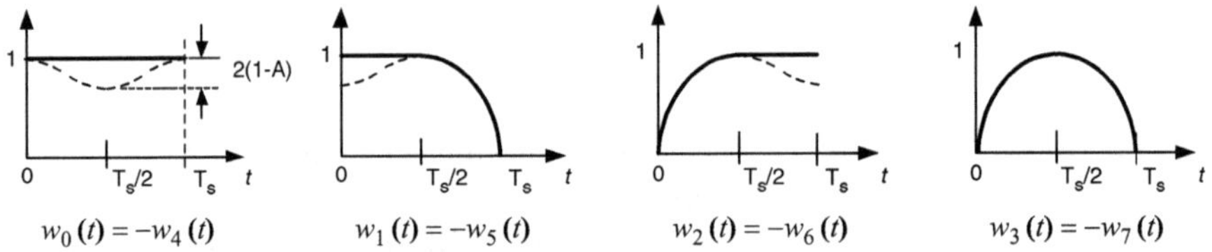

Fig. 2.23 Composite waveforms with one symbol interval T_s, where the solid-curves are for QORC and dashed-curves are for SQAM

2.5.4 Bit Error Rate in Coherent Demodulation

Like the demodulation for QPSK/OQPSK and MSK signal, an optimum cross-correlation-type receiver can be used to demodulate the overlapped pulse-shaping signals of QORC/SQORC and SQAM. The detailed receiver structures are described in [14, 19, 20], where an integrate-and-dump (I&D) filter is used on each correlation branch. Basically, for the cases of QORC/SQORC and SQAM signals there are all eight possible waveforms in any typical one symbol transmission interval $0 \leq t \leq T_s$, as shown in Fig. 2.23 and each of them occurs with equal probability. These eight waveforms are defined as

$$
\begin{aligned}
w_0(t) &= 1 - (1 - A)\left(1 - \cos\frac{2\pi t}{T_s}\right), \quad 0 \le t \le T_s \\
w_1(t) &= \begin{cases} 1 - (1 - A)\left(1 + \cos\frac{2\pi t}{T_s}\right), & 0 \le t \le \frac{T_s}{2} \\ \sin\frac{\pi t}{T_s}, & \frac{T_s}{2} \le t \le T_s \end{cases} \\
w_2(t) &= \begin{cases} \sin\frac{\pi t}{T_s}, & 0 \le t \le \frac{T_s}{2} \\ 1 - (1 - A)\left(1 + \cos\frac{2\pi t}{T_s}\right), & \frac{T_s}{2} \le t \le T_s \end{cases} \\
w_3(t) &= \sin\frac{\pi t}{T_s}, \quad 0 \le t \le T_s
\end{aligned}
\tag{2.72}
$$

where $A = 1$ corresponds to QORC/SQORC or IJF-OQPSK signals and values in the range $0 < A \le 1.0$ are used for a SQAM signal. Receivers for QORC/SQORC and SQAM signals traditionally perform symbol-by-symbol detection that employs simple integrate-and-dump (I&D) filters as detectors. This type of the detection is referred to as suboptimum detection in [14, 20]. In such a symbol-by-symbol detection, only four positive waveforms of the total eight waveforms are used to perform correlation detection at the suboptimum receiver, where each waveform individually multiplies the received signal on its branch before an integrate-and-dump filter as shown in Fig. 13 in [14]. Note that these four waveforms with the symbol period T_s have different energy within the symbol interval; and thus, the outputs of the I&D filters must be biased by the related waveform segment energy divided by 2 before the detector [20].

In general, the optimum correlation receiver maximizes the output signal energy during the symbol period T_s in a linear channel corrupted by the additive Gaussian noise by minimizing the equivalent noise bandwidth. This is equivalent to maximizing the signal-to-noise ratio (SNR). The objective of the correlation device is not to identify any one of the eight possible waveform segments. Instead it is to decide whether +1 or −1 was sent from the transmitter [14]. The optimum correlation receiver will be discussed in Chap. 3 in more detail.

It is shown in [14] that the simplest demodulator with the third-order Butterworth lowpass filter for QORC/SQORC yields a good BER performance in a linear channel corrupted with Gaussian noise, approximately 0.7 dB E_b/N_o degradation from an ideal system performance at $\text{BER} = 10^{-5}$ when compared with the I&D detection based demodulator. In fact, the Butterworth-type filter is used in many practical applications due to its simple implementation and slight BER degradation. Figure 2.24 illustrates the block diagram of a coherent demodulator with Butterworth filters for QPSK/OQPSK type of signals, including the overlapped pulse waveforms with pulse-shaping of QORC/SQORC and SQAM.

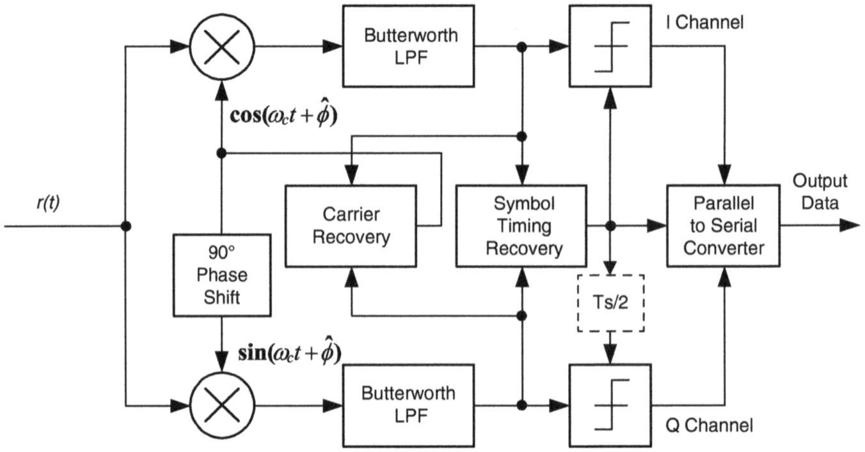

Fig. 2.24 Block diagram of a coherent demodulator with Butterworth filters for QPSK/OQPSK type of signals, where the delay $T_s/2$ is used for OQPSK signal only

The carrier phase estimate $\hat{\phi}$ for the received carrier phase ϕ is used in generating the local reference carrier signal for the coherent demodulator. If the phase error $\phi - \hat{\phi}$ is very small, the transmitted baseband signals on the in-phase and quadrature branches at the outputs of the lowpass filters are recovered. Then the symbol timing recovery circuit synchronizes the local symbol clock with the recovered baseband signals, and then samples the recovered baseband signals on the I–Q branches at the maximum eye-opening points to make decisions through the decision blocks. The sampling clock on the Q branch needs a half-symbol delay relative to the I branch if a type of offset QPSK modulation is used. Finally, the recovered symbol sequences on the I–Q branches are combined into serial output data through the parallel to serial converter.

A Butterworth lowpass filter is used here due to its mild properties of amplitude attenuation and group delay variation. A 3 dB bandwidth of the Butterworth lowpass affects the BER performance of the system, and therefore should be optimized. If the bandwidth is too wide, the SNR is reduced due to more additional Gaussian noise passing through the filter. On the other hand, if it is too narrow, ISI is produced due to severe group delay variation within the bandwidth. It has been found in [16] that the optimum 3-dB bandwidth is about $0.55f_s$ or $B_{3\,\mathrm{dB}}T_s = 0.55$, where $T_s = 1/f_s$ is the symbol duration and $B_{3\,\mathrm{dB}}$ is a 3 dB bandwidth.

One effective way to reduce the ISI in a narrow-bandwidth case is to add a group delay equalizer after the receive lowpass filter. Thus, the narrow receive filter cascaded with the group delay equalizer can greatly attenuate the noise and meanwhile significantly reduce ISI. Figure 2.25 shows the recovered eye diagrams of SQAM signal with $A = 0.8$ at the output of the receive lowpass filter through a nonlinear channel. It is noted that ISI is reduced after the group delay equalizer and eye diagrams open widely, where the second-order group delay equalizer is used.

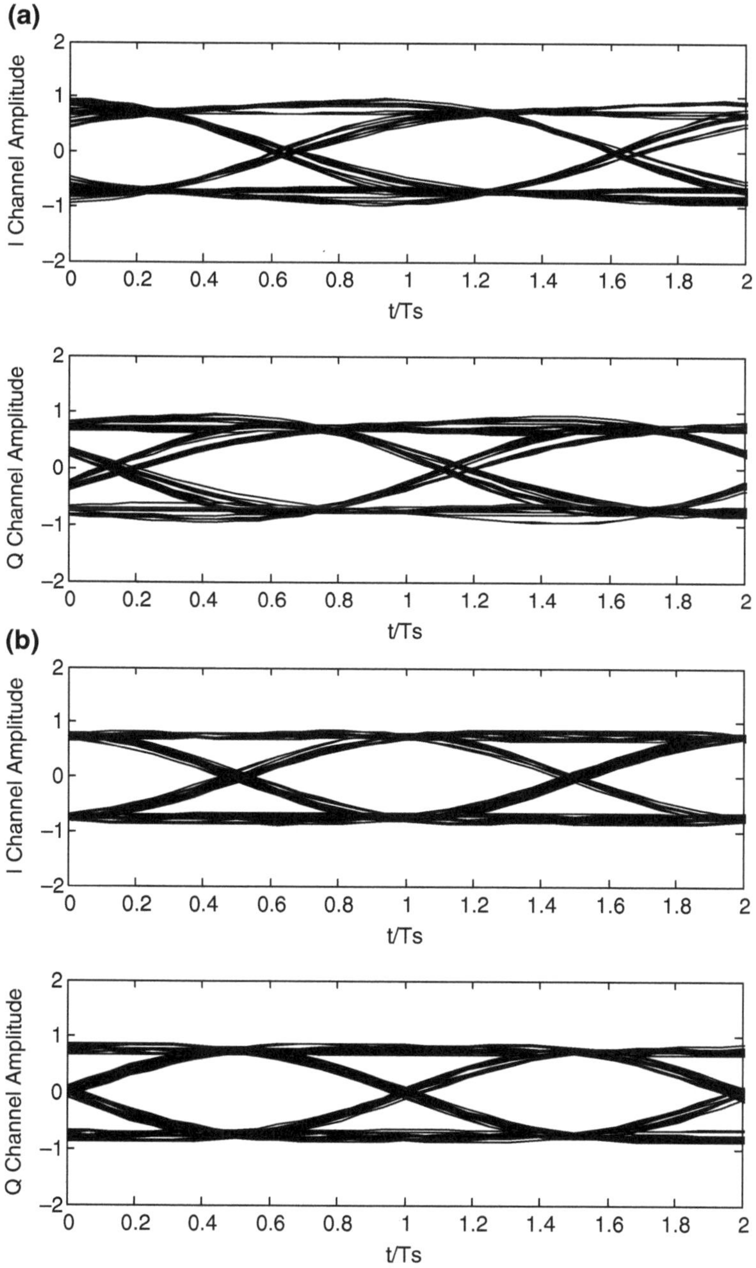

Fig. 2.25 Simulated eye diagrams at the output of the fourth-order Butterworth filter with $BT_s = 0.55$ through a nonlinear (hard-limited) channel. (**a**) SQAM with $A = 0.8$ and without group delay equalizer, (**b**) SQAM with $A = 0.8$ and with group delay equalizer

To simulate the BER performance of the SQAM system in a nonlinear channel, we use an ideal hard-limiter to simulate the effect of the nonlinear amplification on the SQAM signal. The ideal hard-limiter represents a good first-order approximation of the saturated high-power amplifier [15]. Figure 2.26 shows the simulated BER performance results of the SQAM signal through a nonlinear channel, where the parameter $A = 0.8$ is used for the SQAM signal due to its best BER performance, as shown in Fig. 2.22. It is also shown in Fig. 2.26 that both SQAM and SCMSK signals have the same BER performance through a nonlinear channel, where the BER performance is degraded by 0.5 dB in the case where no group delay equalizer is cascaded with the Butterworth lowpass filter, and by 0.3 dB in the case where the second-order group delay equalizer is cascaded with the Butterworth lowpass filter. A 0.2 dB improvement is achieved with the group delay equalizer.

2.6 Minimum Bandwidth and ISI-free Nyquist Pulse Shaping

In practical applications, most communication channels are band-limited to some specified bandwidth $2B_w$ Hz for each user in order to efficiently utilize the total available channel bandwidth W. For this case, the channel bandwidth can be equivalent to its baseband bandwidth B_w, or its equivalent lowpass frequency response is non-zero for $|f| \leq B_w$ and zero for $|f| > B_w$. When such a channel is ideal (its amplitude and group delay responses are both constant) for $|f| \leq B_w$, what is the maximum data rate that can be transmitted through the channel without causing intersymbol interference (ISI) in the desired channel and adjacent channel interference (ACI) in the adjacent channels? If such a channel is not ideal for $|f| \leq B_w$, how do we design a compensator or equalizer to compensate for the channel such that the channel impairments can be minimized at the output of the compensator?

The answer to the first question is to design a spectrum shaping pulse that satisfies the *Nyquist criterion* or zero ISI. The solution for the second problem is to design an equalizer to minimize ISI, including the channel impairments caused by both amplitude and phase distortions.

2.6.1 *Nyquist Minimum Transmission Bandwidth with ISI-Free*

The higher the transmission data rate is, the wider channel bandwidth the transmission system occupies. Nyquist [21] investigated the relationship between the theoretical minimum transmission bandwidth and ISI. He demonstrated that if synchronous *impulse streams* at a transmission rate of f_s symbols per second are

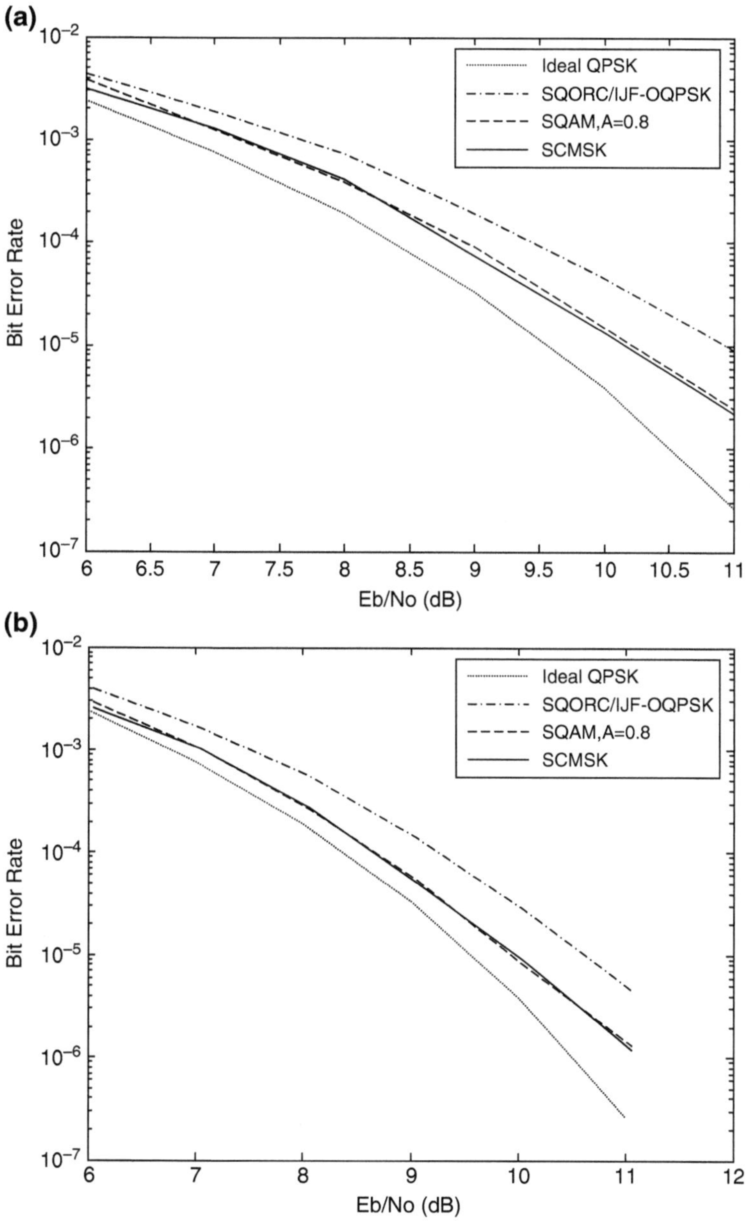

Fig. 2.26 Bit error rate performance of SQORC/IJF-OQPSK, SQAM, and SCMSK signals in a nonlinear channel: (**a**) without group delay equalizer, and (**b**) with group delay equalizer. (*Note*: 0.3 dB degradation for SQAM/SCMSK and 0.9 dB for IJF-OQPSK at 10e−4)

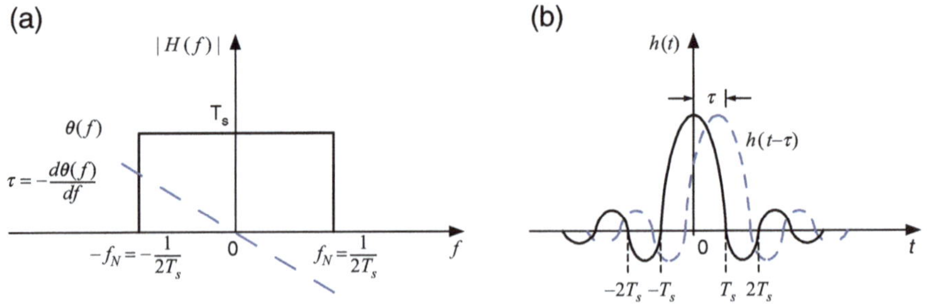

Fig. 2.27 Nyquist channel and its impulse response: (**a**) ideal brick-wall (Nyquist channel) transfer function, and (**b**) impulse response

transmitted through a lowpass channel filter with the minimum bandwidth $f_N = f_s/2$ Hz, whose transfer function is characterized by an ideal brick-wall amplitude response and linear phase response, then the output signals of such a channel filter can be detected independently without ISI at the detection instants. The minimum bandwidth $f_N = f_s/2$ is called *Nyquist minimum transmission bandwidth* or *Nyquist frequency*.

Such an ideal lowpass channel transfer function $H(f)$ is in the shape of a brick-wall, as shown in Fig. 2.27a. For baseband transmission systems when the amplitude and phase responses of the overall transfer function $H(f)$ are the same as the one shown in Fig. 2.27a, the transfer function $H(f)$ and its impulse response $h(t)$ are derived as follows:

$$H(f) = \begin{cases} T_s, & |f| \leq f_N = \dfrac{1}{2T_s} \\ 0, & |f| > f_N = \dfrac{1}{2T_s} \end{cases} \tag{2.73}$$

$$\begin{aligned} h(t) &= F^{-1}[H(f)] = \int_{-T_s/2}^{T_s/2} T_s e^{j2\pi f t} dt \\ &= \frac{\sin\left(\dfrac{\pi t}{T_s}\right)}{\dfrac{\pi t}{T_s}} = \mathrm{sinc}\left(\frac{\pi t}{T_s}\right) \end{aligned} \tag{2.74}$$

where $f_N = 1/(2T_s) = f_s/2$ is known as *Nyquist frequency* and is also equal to the cut-off frequency of the channel filter, where T_s is the symbol duration and f_s is the symbol rate. In binary transmission systems, the symbol duration T_s is equal to the bit duration T_b or $T_s = T_b$. In multilevel M or multi-state M transmission systems, the symbol duration is calculated by $T_s = T_b \log_2 M$. For example, in the QPSK or 4-QAM modulation, we have $M = 4$, and $T_s = 2T_b$.

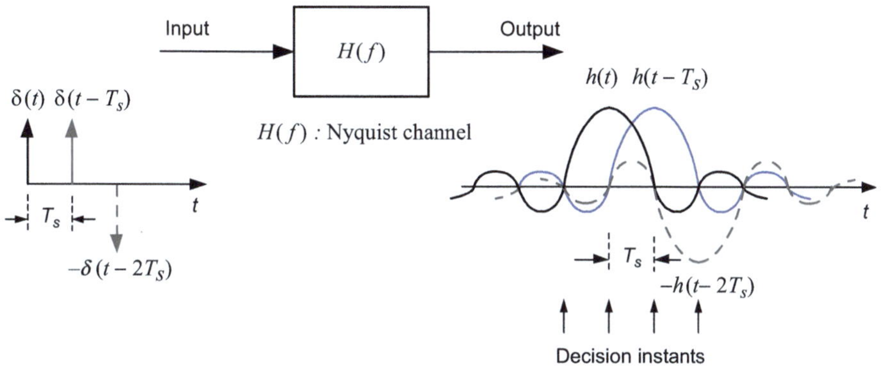

Fig. 2.28 Concept of Nyquist channel achieving zero ISI transmission for input impulse streams

In the derivation above, a zero linear phase for the filter $H(f)$ is assumed. For each input impulse, the impulse response $h(t)$ can be calculated from (2.74) and has its values at the integer multiples of the symbol interval of T_s below

$$h(nT_s) = \begin{cases} 1, & \text{for } n = 0 \\ 0, & \text{for } n = \pm 1,\ \pm 2,\ \pm 3, \ldots \end{cases} \tag{2.75}$$

Thus, if each impulse response at the output of the received channel filter is of the form as given in (2.75), the received sequences can be detected without ISI. If the ideal Nyquist filter has non-zero phase but linear phase (*dashed line* in Fig. 2.27), the impulse response is shifted by an amount of the filter delay, which is equal to $\tau = -d\phi(f)/df$ and is constant over all frequencies. In this case, the output impulse responses still meet the Nyquist criterion because all impulse streams are delayed by the same delay τ.

When a transmission channel meets the criterion of a Nyquist channel, or the transmission symbol rate of f_s is less than and equal to twice the Nyquist frequency of f_N ($f_s \le 2f_N$), the outputs of the transmission channel are ISI-free at all decision-making instants for the inputs of impulse streams with the rate of $f_s = 1/T_s$, as shown in Fig. 2.28. On the other hand, if the transmission symbol rate is greater than twice the Nyquist frequency, or $f_s > 2f_N$, the outputs of the transmission channel have ISI at these decision instants.

The names "Nyquist filter" and "brick-wall filter" are often used as alternatives to describe the Nyquist channel for satisfying ISI-free transmission at the decision instants. The impulse response of the Nyquist filter is also called "Nyquist pulse".

It should be noted that the inputs to the Nyquist filter are impulse streams in order to satisfy ISI-free transmission. However, in most communication systems the inputs are rectangular (or NRZ) pulses instead of impulse streams. For these rectangular-pulse inputs to satisfy ISI-free transmission, a $x/\sin(x)$-shaped amplitude compensator has to be cascaded with the Nyquist filter. The frequency response of the cascaded amplitude compensation with the Nyquist filter is shown in Fig. 2.29. The reason for that is explained as follows:

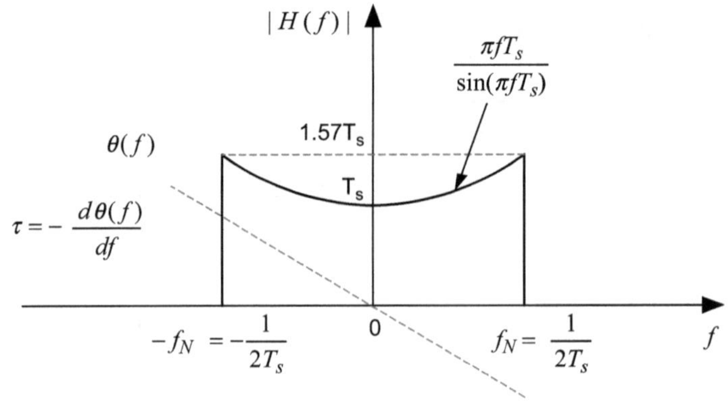

Fig. 2.29 Frequency response of Nyquist channel for rectangular pulse transmission with amplitude compensation

For an impulse input, its Fourier transform is constant over all frequencies. Thus, the output of the Nyquist filter is simply the inverse Fourier transform of the Nyquist filter. For a rectangular-pulse input, its Fourier transform is the form of sin (x)/x. *To satisfy the Nyquist ISI-free transmission, the rectangular pulse should be transferred into the impulse before entering the Nyquist filter. If such a rectangular pulse is passed with an amplitude compensator with the transfer function* x/sin(x), *then the output of the amplitude compensator becomes an impulse and now satisfies ISI-free transmission.*

A practical approach to the amplitude compensation is to use a second-order lowpass with the damping factor less than 0.707 so that its amplitude response increases with the frequency and approximates the shape of x/sin(x) within the Nyquist bandwidth f_N. This lowpass filter can be added either before or after the Nyquist filter.

Unfortunately, the ideal Nyquist channel filter with the minimum bandwidth is not realizable, and therefore it cannot be implemented with any hardware circuits or components. In the case of approximating it, the approximated Nyquist channel filter may need a larger-order number of filter. In addition to approximating the amplitude response, it is very difficult to achieve a linear phase filter as requested for the Nyquist channel filter.

Fortunately, a practically and widely used function that satisfies the free-ISI is the raised-cosine filter, whose amplitude response is in the form of the raised-cosine shape. The amplitude response of the raised-cosine (RC) filter is expressed as

$$|H_{\mathrm{rc}}(f)| = \begin{cases} T_{\mathrm{s}} & 0 \le |f| \le \dfrac{1-\alpha}{2T_{\mathrm{s}}} \\ T_{\mathrm{s}} \cos^2\left[\dfrac{\pi T_{\mathrm{s}}}{2\alpha}\left(|f| - \dfrac{1-\alpha}{2T_{\mathrm{s}}}\right)\right] & \dfrac{1-\alpha}{2T_{\mathrm{s}}} \le |f| \le \dfrac{1+\alpha}{2T_{\mathrm{s}}} \\ 0 & |f| > \dfrac{1+\alpha}{2T_{\mathrm{s}}} \end{cases} \tag{2.76}$$

where α is the roll-off factor and determines the raised-cosine filter shape and bandwidth. For $\alpha = 0$, an ideal brick-wall filter having a minimum bandwidth equal to the Nyquist frequency of $f_N = 1/2T_s$ is achieved. However, this brick-wall filter cannot be realized in practice. In practical applications, values between $0.2 \leq \alpha \leq 1.0$ are usually chosen. The Nyquist frequency f_N is also called the *excess bandwidth* and is expressed as a percentage of f_N. For example, $\alpha = 0.35$ corresponds to excess bandwidth of 35%, while $\alpha = 1$ corresponds to excess bandwidth of 100%.

The impulse response $h_{rc}(t)$ or the inverse Fourier transform (2.76) is

$$h_{rc}(t) = \frac{\sin(\pi t/T_s)}{\pi t/T_s}\frac{\cos(\pi\alpha t/T_s)}{1 - 4\alpha^2 t^2/T_s^2} \tag{2.77}$$

Figure 2.30 illustrates the amplitude response of the raised-cosine filter and the corresponding impulses for $\alpha = 0$, 0.3, and 1. The single-side bandwidth is equal to $f = (1 + \alpha)f_N$. It is clearly seen that the tails of $h_{rc}(t)$ decay faster with the increase of α value. The lower the tail of $h_{rc}(t)$ decays, the bigger the ISI is at the decision instants with severe phase jitter.

The expression of the raised-cosine filter (2.76) can also be rewritten

$$|H_{rc}(f)| = \begin{cases} T_s & 0 \leq |f| \leq \dfrac{1-\alpha}{2T_s} \\ \dfrac{T_s}{2}\left\{1 + \cos\left[\dfrac{\pi T_s}{\alpha}\left(|f| - \dfrac{1-\alpha}{2T_s}\right)\right]\right\} & \dfrac{1-\alpha}{2T_s} \leq |f| \leq \dfrac{1+\alpha}{2T_s} \\ 0 & |f| > \dfrac{1+\alpha}{2T_s} \end{cases} \tag{2.78}$$

If the input to the raised-cosine filter is the rectangular (NRZ) pulse streams, a $x/\sin(x)$-shaped amplitude compensator must be cascaded with the raised-cosine filter. For a rectangular pulse $g(t)$ with the interval T_s and amplitude $1/T_s$, its Fourier transform is

$$\begin{aligned} G(f) &= \int_{-T_s/2}^{T_s/2} g(t)e^{-j2\pi f t}dt = \frac{1}{T_s}\int_{-T_s/2}^{T_s/2} e^{-j2\pi f t}dt \\ &= \frac{\sin(\pi f T_s)}{\pi f T_s} \end{aligned} \tag{2.79}$$

Thus, the amplitude response of the amplitude compensator expressed as the inverse of $G(f)$ in (2.79) is added to the raised-cosine filter (2.78) to form the cascaded amplitude response for ISI-free pulse transmission:

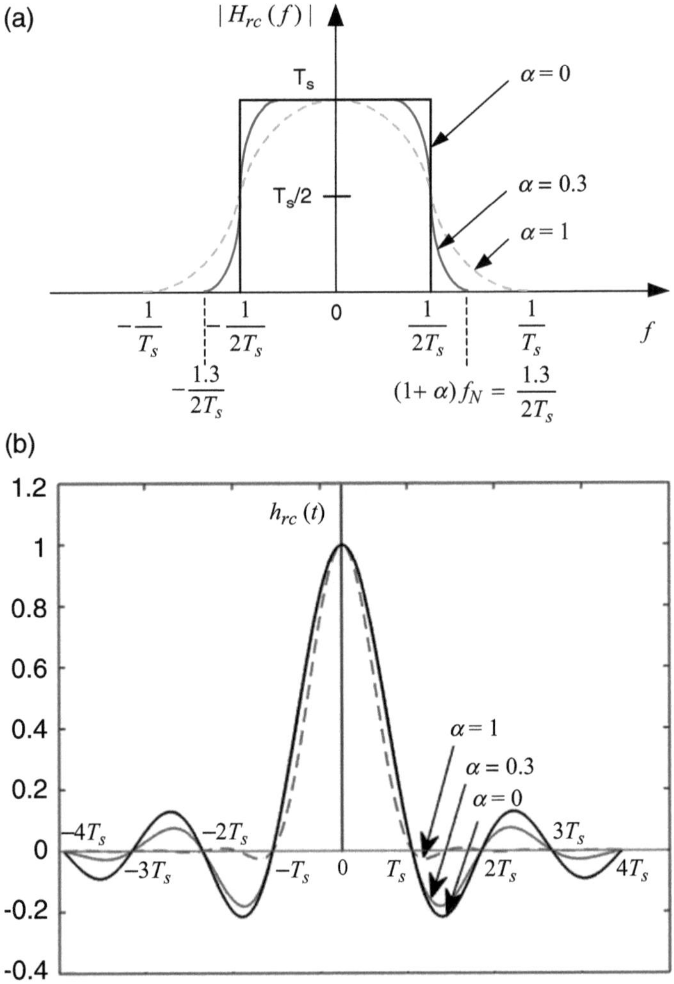

Fig. 2.30 Raised-cosine filter characteristics: (**a**) amplitude response, and (**b**) impulse response

$$
|H(f)| = \begin{cases} \dfrac{\pi f T_s}{\sin(\pi f T_s)} T_s & 0 \le |f| \le \dfrac{1-\alpha}{2T_s} \\ \dfrac{\pi f T_s}{\sin(\pi f T_s)} \dfrac{T_s}{2} \left\{ 1 + \cos\left[\dfrac{\pi T_s}{\alpha} \left(|f| - \dfrac{1-\alpha}{2T_s} \right) \right] \right\} & \dfrac{1-\alpha}{2T_s} \le |f| \le \dfrac{1+\alpha}{2T_s} \\ 0 & |f| > \dfrac{1+\alpha}{2T_s} \end{cases} \tag{2.80}
$$

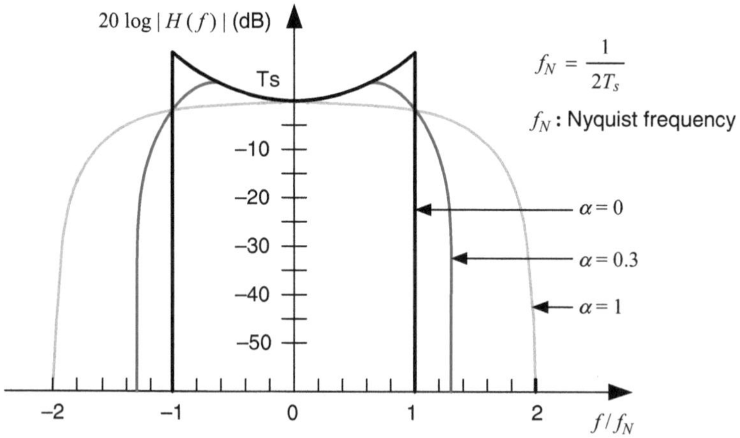

Fig. 2.31 Amplitude response of raised-cosine filter cascaded with $x/\sin(x)$ shaped amplitude compensator for rectangular pulse ISI-free transmission. (*Note*: Attenuation is $4 - 6 = -2$ dB at $f/f_N = 1$ for $\alpha \neq 0$)

Figure 2.31 illustrates the amplitude response of the cascaded filter with different α values. For $\alpha = 0$, an unrealizable minimum-bandwidth filter or Nyquist filter is obtained, as shown by the dark black line. Theoretically, the attenuation has an infinite value beyond the bandwidth. Practically, attenuation might be in the range from 20 to 50 dB, depending on implementation approach. In an analog filter design approximation, the attenuation level outside the channel should not be designed too large due to severe group delay variation of the analog filter with high order. In a digital FIR filter design approximation, however, the attenuation level can be large by increasing the number of FIR filter coefficients because of its linear property of the group delay.

The simulated eye diagrams for NRZ data filtered by a raised-cosine filter either with $x/\sin(x)$-shaped amplitude aperture compensation or without it are illustrated in Fig. 2.32. In the case with $x/\sin(x)$ compensation (Fig. 2.32a,b) eye diagrams are ISI-free at the maximum opening instants. In the case without $x/\sin(x)$ compensation (Fig. 2.32c), the eye diagram has ISI at the maximum opening instants. Therefore, the amplitude aperture compensator is necessary for rectangular (NRZ) pulse streams to achieve ISI-free transmission. It is noted that for $\alpha = 1$ in Fig. 2.32a, the eye diagram has zero data polarity transmission jitter at $t/T_s = 0.5$ or 1.5. With $\alpha = 0.3$ there is a significant data polarity transmission jitter, which results in severe timing jitter in the recovered timing-clock signal. Furthermore, large tails corresponding to small α values exhibit more sensitivity to timing errors and thus result in severe BER degradation due to ISI. On the other hand, a large α value will result in a large excess bandwidth.

Considering that the overall channel filter consists of a transmitter filter and a receiver filter, the raised-cosine filter in (2.76) can be split into two parts:

$$H_{\mathrm{rc}}(f) = H_{\mathrm{t}}(f) H_{\mathrm{r}}(f) \tag{2.81}$$

If the receiver filter is matched to the transmitter filter or $H_{\mathrm{r}}(f) = H_{\mathrm{t}}(f)$, the desired magnitude response of the filter, called the *square root of raised-cosine (SRRC)* filter, is obtained:

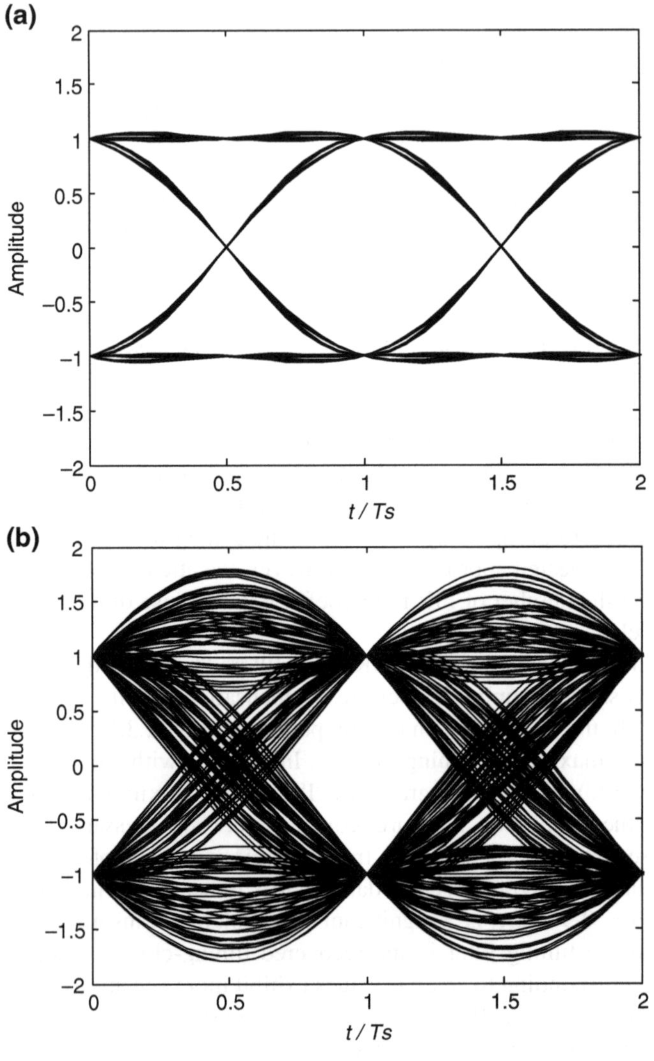

Fig. 2.32 Eye diagrams of NRZ data filtered by raised-cosine filter: (**a**) roll-off factor $\alpha = 1$ with $x/\sin(x)$, (**b**) roll-off factor $\alpha = 0.3$ with $x/\sin(x)$ and (**c**) roll-off factor $\alpha = 1$ without $x/\sin(x)$

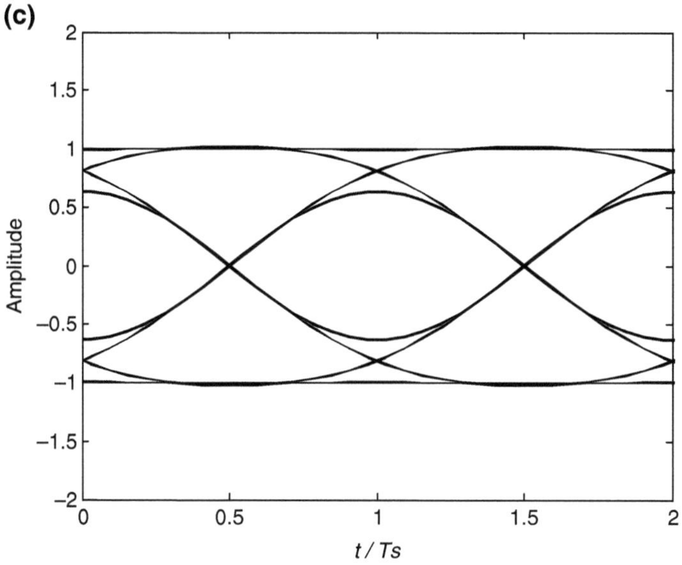

Fig. 2.32 (continued)

$$|H_{\mathrm{t}}(f)| = |H_{\mathrm{r}}(f)| = H_{\mathrm{srrc}}(f) = \sqrt{|H_{\mathrm{rc}}(f)|}$$
$$= \begin{cases} \sqrt{T_{\mathrm{s}}} & 0 \le |f| \le \dfrac{1-\alpha}{2T_{\mathrm{s}}} \\ \sqrt{T_{\mathrm{s}}}\cos\left[\dfrac{\pi T_{\mathrm{s}}}{2\alpha}\left(|f| - \dfrac{1-\alpha}{2T_{\mathrm{s}}}\right)\right] & \dfrac{1-\alpha}{2T_{\mathrm{s}}} \le |f| \le \dfrac{1+\alpha}{2T_{\mathrm{s}}} \\ 0 & |f| > \dfrac{1+\alpha}{2T_{\mathrm{s}}} \end{cases} \tag{2.82}$$

The abbreviation $\sqrt{\alpha}$ is used to stand for SRRC in some literature. In this case, the receiver filter $H_{\mathrm{r}}(f)$ is called the *match filter* to the transmitter filter $H_{\mathrm{t}}(f)$. If the input data to the transmitter filter $H_{\mathrm{t}}(f)$ is NRZ data, the amplitude aperture compensator with the form $x/\sin(x)$ should be cascaded with $H_{\mathrm{t}}(f)$ together for ISI-free transmission. It should be noted that the amplitude aperture compensator $x/\sin(x)$ is unnecessarily needed in the digital FIR filter designs for either a RC filter or a SRRC filter due to the input of the approximate impulse streams after performing up-sampling rate of N by inserting $N-1$ zero between two adjacent data sequences.

In practical designs, considering the physical realization of the filter, (2.82) is usually written as

$$H_{\mathrm{t}}(f) = \sqrt{|H_{\mathrm{rc}}(f)|}\mathrm{e}^{-j2\pi f\tau} \tag{2.83}$$

where τ is some certain constant delay that ensures the peaks of the impulse response $h_{\mathrm{t}}(t) = F^{-1}[H_{\mathrm{t}}(f)]$ occurring after $t \ge 0$.

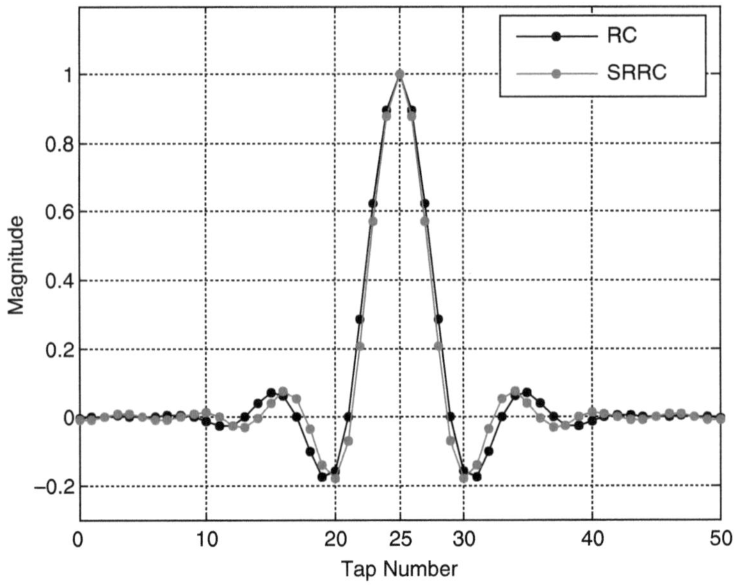

Fig. 2.33 Normalized impulse responses of raised-cosine filter and square root raised-cosine filter with $\alpha = 0.3$

The impulse response of the SRRC filter is obtained by taking the inverse Fourier transform of (2.82), which is expressed as

$$h_{\text{srrc}}(t) = \frac{1}{\sqrt{T_s}} \frac{1}{1 - (4\alpha t/T_s)^2} \left\{ \frac{\sin[(1-\alpha)\pi t/T_s]}{\pi t/T_s} + \frac{4\alpha \cos[(1+\alpha)\pi t/T_s]}{\pi} \right\} \tag{2.84}$$

The normalized impulse responses of the RC and SRRC filters are depicted in Fig. 2.33, where four samples per symbol duration T_s are marked by dots. It is clear that the impulse response of the SRRC filter has non-zero crossing at every four sample instants starting from the middle peak instant, while that of the RC filter has zero crossing at these instants. At these sampling instants, which are also called decision-making instants, other adjacent signals reach their peak values. Thus, the SRRC-filtered signal at the transmitter has ISI, while the signal at the output of the matched filter or another SRRC filter at the receiver is ISI-free due to the composite full response of the overall system.

2.6.2 Analog Filter Approximation to SRRC Filter

Analog filters play an important role in spectrally efficient transmission systems. To achieve spectrally efficient transmission, some communication systems use the analog filter to approximate a raised-cosine filter as the pulse-shaping filter at the transmitter or the matched filter at the receiver due to its simple structure and low cost. This was especially true in the before 1990, when digital signal-processing chips were not affordable, available, or practical due to their high cost and speed limitations. For example, in the late 1980s most digital microwave transmission systems and digital satellite/earth station transmission systems were equipped with analog raised-cosine filters to achieve spectrally efficient transmission. With the development of digital signal-processing technologies over the past two decades, the raised-cosine filter in modern digital communications can be easily realized by a digital FIR filter with linear phase or constant group delay within a 3-dB corner frequency. However, analog filters are still needed to filter out out-band noise, interference signals, and image components, such as a reconstruction filter after a digital-to-analog converter at the transmitter, and a channel selection filter or an anti-aliasing filter before the analog-to-digital converter at the receiver. In today's wireless communication systems, such as 2G GSM and 3G wideband code-division multiple access (WCDMA) systems, the analog filter still has an advantage over the digital filter in wireless handset devices due to its low cost and low current consumption. For example, in WCDMA systems, a fifth-order analog filter is used at the receiver to achieve functionality in both channel selection filtering and approximately matched filtering [22].

In this section, we introduce one example for practical analog filter design approximation to the amplitude response of SRRC filter. Due to the non-constant group delay characteristics of the analog filter, an allpass filter as a group delay equalizer needs to be cascaded with the analog filter to reduce group delay variation within the required bandwidth. This analog approach can meet the need for low-cost, low-power consumption, and low-form-factor user equipment (UE). For detailed design procedures of group delay equalization, the interested reader can refer to Appendix D.

2.6.2.1 Amplitude Approximation

In a white Gaussian noise (WGN) channel, either a RC or a SRRC filter is used to achieve spectrum shaping and ISI-free transmission in band-limited channels. Previously (from the 1990s to the 2000s), the RC and SRRC filters were mostly approximated by using Butterworth and Chebyshev analog filters, or other types of analog filters whose parameters are optimized by computer to minimize the amplitude error between the amplitude response of the analog filter and that of the target RC or SRRC filter. Even today, in the 3GPP WCDMA system, the analog filter is still used to approximate the SRRC filter as part of the channel selection filtering at the receiver, matched to the transmitter-side SRRC filter to achieve the minimal ISI and the maximum adjacent channel interference (ACI) rejection [22–24].

When the analog filter is used to approximate the SRRC filter, the attenuation requirement of the analog filter at the transmitter is determined by the adjacent channel power ratio (ACPR), which is the ratio of the average power in the main channel and adjacent channels, while the attenuation requirements of the analog filter at the receiver are decided by both in-channel SNR and adjacent channel interference (ACI). The attenuation of the analog filter, on the other hand, is primarily determined by filter prototype, 3-dB corner frequency, and filter order. Four common filter prototypes are Butterworth, Chebyshev, Inverse Chebyshev, and elliptic. In terms of a given order, the larger the attenuation of the filter, the worse the group delay variation of the filter. The elliptic filter provides the largest attenuation and a sharper transition region among the four types of filters but the worst group delay variation within a 3-dB corner frequency. The Butterworth filter shows the smallest group delay variation, but the smallest attenuation as well.

The 3-dB corner frequency of the analog filter is generally determined by the Nyquist frequency f_N or the half-symbol rate $f_s/2$ as described in Sect. 2.6.1. An actual 3-dB corner frequency can be obtained through an approximation to a SRRC filter with a specified roll-off factor, corresponding to a certain symbol rate. A design example at the transmitter will be introduced in the following section while the other one at the receiver will be illustrated in Sect. 7.5.2.

2.6.2.2 Group Delay Compensation

In general, due to natural characteristics of the large group delay fluctuation of the analog filter within the 3-dB corner frequency, an allpass filter needs to be cascaded to reduce group delay fluctuation as much as possible; meanwhile it does not the change magnitude response of the analog filter. Because of the spectral-efficiency requirement, most communication channels are usually characterized as band-limited linear filters if all active circuits operate at linear regions. Then, the transfer functions of such channels may be expressed by their frequency response

$$H_c(j\omega) = \left|H_c(j\omega)\right| e^{j\theta_c(\omega)} \tag{2.85}$$

where $\left|H_c(j\omega)\right|$ is the amplitude response and $\theta_c(\omega)$ is the phase response. Another characteristic, which is more useful to describe the channel's behavior, is the group delay, which is obtained from the negative derivative of the phase, or

$$\mathrm{GD}_c(\omega) = -\frac{d\theta_c(\omega)}{d\omega} \tag{2.86}$$

A channel is said to be ideal if $\left|H_c(j\omega)\right|$ and $\mathrm{GD}_c(\omega)$ both are the constant within the transmitted signal bandwidth. Otherwise, the transmitted signal may be distorted through such a channel, depending on how much worse $\left|H_c(j\omega)\right|$ and $\mathrm{GD}_c(\omega)$ are. The signal distortion caused by non-ideal $\left|H_c(j\omega)\right|$ is called *amplitude distortion*, and that caused by non-ideal $\mathrm{GD}_c(\omega)$ is called *delay distortion.*

The group delay equalizer or compensator can only compensate for the delay distortion by introducing extra delay in such a way that the overall group delay variation is minimized as much as possible. Suppose the transfer function of the equalizer is expressed by its frequency response

$$H_e(j\omega) = |H_e(j\omega)|e^{j\theta_e(\omega)} \tag{2.87}$$

To compensate for the group delay of $H_c(j\omega)$, the equalizer needs to be cascaded with $H_c(j\omega)$. Thus, the overall transfer function $H(j\omega)$ is

$$\begin{aligned} H(j\omega) &= H_c(j\omega) \times H_e(j\omega) \\ &= |H(j\omega)|e^{j\theta(\omega)} \end{aligned} \tag{2.88}$$

where the amplitude and phase responses are expressed by

$$\begin{aligned} |H(j\omega)| &= |H_c(j\omega)| \times |H_e(j\omega)| \\ &= |H_c(j\omega)| \end{aligned} \tag{2.89}$$

and

$$\theta(\omega) = \theta_c(\omega) + \theta_e(\omega) \tag{2.90}$$

In (2.89), the amplitude response of the group delay equalizer is assumed to be a constant and is normalized to 1. The group delay of $H(j\omega)$ is given by

$$\begin{aligned} \mathrm{GD}(\omega) &= -\frac{d\theta(\omega)}{d\omega} = -\frac{d\theta_c(\omega)}{d\omega} - \frac{d\theta_e(\omega)}{d\omega} \\ &= \mathrm{GD}_c(\omega) + \mathrm{GD}_e(\omega) \end{aligned} \tag{2.91}$$

where $\mathrm{GD}_e(\omega)$ is the group delay of the equalizer. For an ideal case, $\mathrm{GD}(\omega)$ is the constant or nearly constant $\mathrm{GD}(\omega) \approx C$ within the interested frequency band. To achieve such a nearly constant group delay, computer optimization programs can be used to minimize the peak-to-peak variation of the overall group delay within the interested bandwidth B_w or

$$\Delta\mathrm{GD}_{\mathrm{PP}}(\omega) = \mathrm{GD}_{\mathrm{MAX}}(\omega) - \mathrm{GD}_{\mathrm{MIN}}(\omega), \quad \omega \le B_w \tag{2.92}$$

The question is how small $\Delta\mathrm{GD}_{\mathrm{PP}}(\omega)$ should be. Usually, $\Delta\mathrm{GD}_{\mathrm{PP}}(\omega)$ is determined by the requirement of EVM tolerance at the receiver. In general, a first-order allpass filter or a second-order allpass filter is used as a fundamental unit of the group delay equalizer. A higher-order group delay equalizer can be constructed by cascading such multiple fundamental sections together, in which the first-order section is needed to achieve the odd order of the equalizer.

A characteristic analysis of the first-order and second-order allpass filter sections as a group delay equalizer is introduced in Appendix D.

We start with an analog filter design approximation to a SRRC filter in the transmitter. To achieve ISI-free transmission, the analog filter needs not only to approximate the amplitude response of the SRRC filter, but also to have small group delay fluctuation. Depending on practical applications, a filter with small group delay fluctuation can be created by cascading a one-stage or multiple-stage group delay equalizer with the approximated SRRC filter. Designing a group delay equalizer may be complicated compared with the design of the filter because several parameters in the equalizer need to be carefully adjusted in order to achieve smaller group delay variation.

Design Example 2.1 In an SCPC satellite earth station system, data are transmitted through a QPSK modulation format at the rate of 64 kbps. A SRRC filter with $\alpha = 0.5$ is used either at the transmitter to perform spectrum-shaping transmission in a limited bandwidth of 45 kHz or at the receiver to match the SRRC filter of the transmitter to attenuate outside channel Gaussian noise and adjacent channel interferers as well as to minimize the in-channel ISI before the decision. Design an analog lowpass at the transmitter to approximate such an ideal SRRC filter without a shape of $x/\sin(x)$ as amplitude compensation. If it is needed, a second-order lowpass filter with a *damping factor* $\zeta < 0.707$ as an amplitude compensation shape of $x/\sin(x)$ can be cascaded with the SRRC filter, which is described in Sect. 2.6.4. A digital SRRC filter is assumed in the receiver to match the SRRC filter of the transmitter.

Solution To achieve such an approximation to a SRRC filter, we choose a fourth-order Butterworth lowpass filter due to its mild group delay characteristic to approximate the amplitude response of the SRRC filter with $\alpha = 0.5$ and a second-order allpass filter as a group delay equalizer to compensate for the group delay fluctuation of the Butterworth lowpass filter in order to achieve small ISI as much as possible.

For the QPSK signal transmission at the bit rate $f_b = 64\,\text{kbps}$, the Nyquist frequency f_N is equal to 16 kHz due to $f_N = f_s/2 = f_b/4$. After approximation to the amplitude response of the SRRC with $\alpha = 0.5$ and $f_N = 16\,\text{kHz}$, the fourth-order Butterworth lowpass with the cut-off frequency of 17.2 kHz can achieve the best amplitude approximation, as shown in Fig. 2.34a.

The analog filter closely approximates the ideal SRRC filter down to the 10-dB attenuation point. Beyond –10 dB, distortion caused by approximation errors may be ignored in practice, depending on the requirement of the desired signal leakage to the adjacent channels. To achieve smaller group delay variation, a second-order allpass filter is chosen as the group delay equalizer; its group delay response is shown in Fig. 2.34b. The group delay fluctuation of the fourth-order Butterworth lowpass filter is significantly reduced from 12 μs to less than 1 μs within the cutoff frequency of 17.2 kHz after the equalizer. For the detailed design procedure and circuit implementation, the interested reader can again refer to Appendix D.

Figure 2.35 illustrates the simulated eye diagram at the output of the RX digital SRRC filter, which is matched to the TX analog approximation SRRC filter designed above. It can be seen that the received signal has very small ISI at the decision-making instants. Therefore, such an analog approximation to the SRRC filter is satisfied in a low-cost and low-power consumption application.

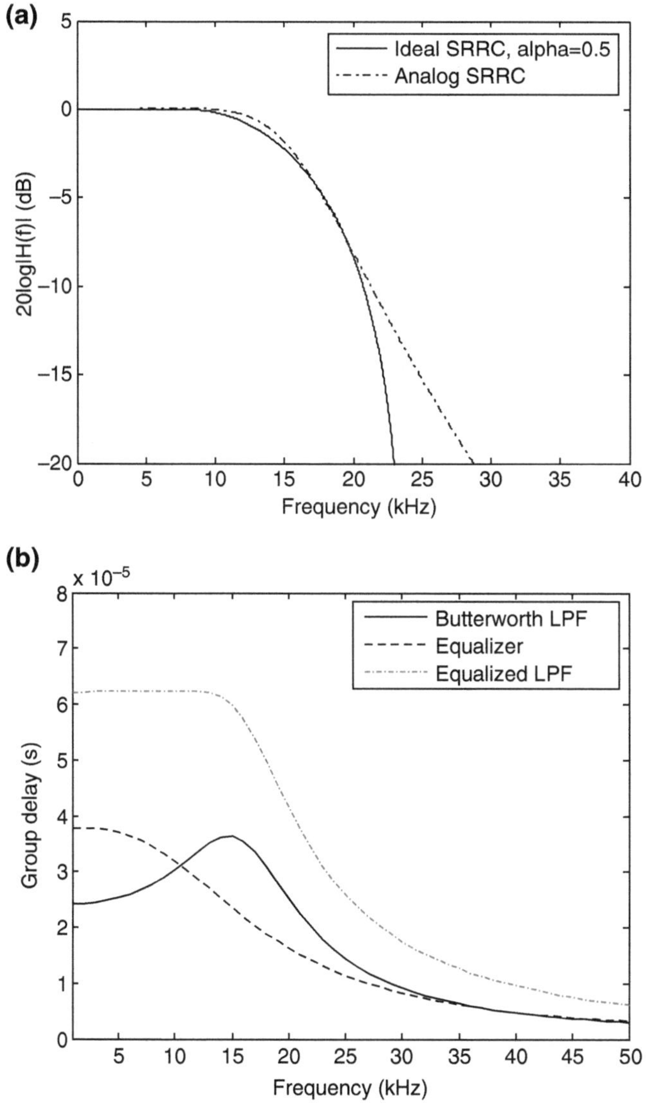

Fig. 2.34 Frequency response of the fourth-order analog filter approximation to a SRRC filter with $\alpha = 0.5$: (**a**) amplitude response and (**b**) group delay response

2.6.3 Digital Filter Approximation to Raised-Cosine Filter

In digital communication systems, the strictly band-limited and ISI-free requirements of a wireless communication channel demand the use of a pulse-shaping filter. These requirements are very difficult to meet by using an analog filter approximation to a RC filter plus a group delay equalizer. Therefore, a digital filter

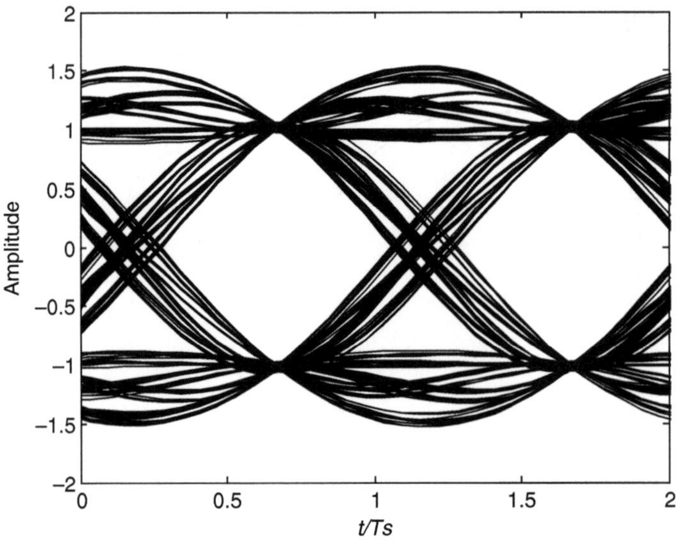

Fig. 2.35 Received eye diagram at the output of the RX digital SRRC filter with $\alpha = 0.5$, which is matched to the TX SRRC filter approximated by a fourth-order Butterworth lowpass filter cascaded with a second-order equalizer

approximation to an RC filter is still a dominant design approach due to its accurate approximation to both amplitude and phase characteristics, especially in the transmitter.

From the impulse response of the raised-cosine filter in time domain, it can be seen that the impulse response has an *infinite duration,* even though it decays very little as time increases. Hence, it is impossible to implement either a RC or SRRC filter with an infinite duration of the impulse response. Also it is unnecessary due to the power assumption and cost. The simplest way to approximate an ideal RC or SRRC filter is to design a FIR filter, due to its symmetric impulse responses and linear phase. The coefficients of the FIR filter approximation to a RC or SRRC filter can be obtained from either its impulse response in the time domain or its frequency response in the frequency domain. Some basic design procedures using these methods will be introduced in the following sections.

2.6.3.1 Filter Design by Window

The simplest method of FIR filter design in the time domain is called the *window method.* A basic method to approximate an ideal filter is to truncate the ideal impulse response $h(n)$. After truncation, the designed filter with impulse response $h_d(n)$ is given by

$$h_{\mathrm{d}}(n) = \begin{cases} h(n), & n = 0, \pm 1, \ldots, \pm (N-1)/2 \\ 0, & \text{otherwise} \end{cases} \tag{2.93}$$

If a finite duration rectangular window $w(n)$ is used, the impulse response $h_{\mathrm{d}}(n)$ in (2.93) can be expressed as

$$h_{\mathrm{d}}(n) = h(n)w(n) \tag{2.94}$$

where

$$w(n) = \begin{cases} 1, & n = 0, \pm 1, \ldots, \pm (N-1)/2 \\ 0, & \text{otherwise} \end{cases} \tag{2.95}$$

Besides the rectangular window, other windows with a less abrupt truncation property may also be used to achieve small side-lobes in the frequency domain. However, this is achieved at the price of a wider main lobe. It is also well known that the *Gibbs phenomenon* can be moderated through the use of such a window with the property of a smooth transition to zero [25]. A useful sinusoid window can be used to truncate the impulse response of the RC or SRRC, and expressed as

$$w(n) = \begin{cases} \sin\left(\dfrac{\pi[n + (N-1)/2]}{N-1}\right), & n = 0, \pm 1, \ldots, \pm (N-1)/2 \\ 0, & \text{otherwise} \end{cases} \tag{2.96}$$

For the rectangular window, we can calculate the impulse response of the digital RC filter from (2.77), or

$$\begin{aligned} h_{\mathrm{rc}}(n) &= h_{\mathrm{rc}}(t)\big|_{t=nT_{\mathrm{sam}}} \\ &= \frac{\sin(\pi n T_{\mathrm{sam}}/T_s)}{\pi n T_{\mathrm{sam}}/T_s} \frac{\cos\left(\dfrac{\pi \alpha n T_{\mathrm{sam}}}{T_{\mathrm{s}}}\right)}{1 - \left(\frac{2\alpha n T_{\mathrm{sam}}}{T_{\mathrm{s}}}\right)^2}, \quad n = 0, \pm 1, \ldots, \pm (N-1)/2 \end{aligned} \tag{2.97}$$

where N is the length of the filter. The minimum number of the samples per symbol is 2, corresponding to the sampling duration $T_{\mathrm{sam}} = T_{\mathrm{s}}/2$. Usually four samples per symbol are used in most transmitter filter designs, or $T_{\mathrm{sam}} = T_{\mathrm{s}}/4$.

Similar to the design of $h_{\mathrm{rc}}(n)$, the calculation of the impulse response of the digital SRRC filter can be obtained from (2.84):

$$\begin{aligned} h_{\text{srrc}}(n) &= h_{\text{srrc}}(t)\big|_{t=nT_{\text{sam}}} \\ &= \frac{1}{\sqrt{T_{\text{s}}}} \frac{1}{1-(4\alpha nT_{\text{sam}}/T_{\text{s}})^2} \left\{ \frac{\sin\left[(1-\alpha)\pi nT_{\text{sam}}/T_{\text{s}}\right]}{\pi nT_{\text{sam}}/T_{\text{s}}} \right. \\ &\quad \left. + \frac{4\alpha\cos\left[(1+\alpha)\pi nT_{\text{sam}}/T_{\text{s}}\right]}{\pi} \right\}, \quad n = 0, \pm 1, \ldots, \pm (N-1)/2 \end{aligned} \tag{2.98}$$

For practical implementations, the impulse response of the RC filter or SRRC filter should be delayed by $(N-1)/2$. Thus, we need to transfer (2.97) and (2.98) to $h_{\text{rc}}(n) = h_{\text{rc}}[n-(N-1)/2]$ and $h_{\text{srrc}}(n) = h_{\text{srrc}}[n-(N-1)/2]$, $n = 0, 1, \ldots, N-1$, respectively, after obtaining them.

2.6.3.2 Filter Design by Impulse Invariance

In the filter design based on the concept of *impulse invariance* [25], we know that a discrete-time system can be defined by sampling the impulse response of its corresponding continuous-time system. In the impulse invariance design procedure for a bandlimited system, the impulse response of the discrete-time filter is chosen to be proportional to equally spaced samples of the impulse response of the corresponding continuous-time filter: i.e.,

$$h(n) = T_{\text{sam}} h_{\text{c}}(nT_{\text{sam}}) \tag{2.99}$$

where T_{sam} is a sampling interval. Note that we use T_{sam} as the scaling factor to multiply $h_{\text{c}}(nT_{\text{sam}})$ in order to normalize its frequency response.

In this design we are interested in the relationship between the frequency responses of the discrete-time and continuous-time filters. The frequency response of the discrete-time filter is related to that of its continuous-time filter by

$$H\left(e^{j\omega}\right) = \sum_{k=-\infty}^{\infty} H_{\text{c}}[j(\omega - k\omega_{\text{sam}})] \tag{2.100}$$

where $\omega_{\text{sam}} = 2\pi f_{\text{sam}} = 2\pi/T_{\text{sam}}$ is the sampling frequency in radians/s. If the continuous-time filter is bandlimited to ω_{B}, or

$$H_{\text{c}}(j\omega) = 0, \quad \omega_{\text{B}} \le |\omega| \le \pi \tag{2.101}$$

Then, (2.100) becomes

$$H\left(e^{j\omega}\right) = H_{\text{c}}(j\omega), \quad |\omega| \le \omega_{\text{B}} \tag{2.102}$$

or

$$H\left(e^{j\omega}\right) = H_c(f) \quad |f| \leq f_B \tag{2.103}$$

If $H_c(f)$ is sampled at equally spaced points in the frequency domain with a frequency step $\Delta f = f_{\text{sam}}/N$, where N is odd number, then (2.100) can be rewritten as

$$H\left(e^{j\omega}\right) = H_c(m\Delta f) = H_c(mf_{\text{sam}}/N) \tag{2.104}$$

The relationship between the frequency response and impulse response of the digital raised-cosine (RC) filter with the length of N is given by

$$H_{rc}\left(e^{j\omega}\right) = \sum_{n=-(N-1)/2}^{(N-1)/2} h_{rc}(n)e^{-j\omega nT_{\text{sam}}} \tag{2.105}$$

Substituting (2.104) into (2.105), we can express (2.105) as

$$H_{rc}(mf_{\text{sam}}/N) = \sum_{n=-(N-1)/2}^{(N-1)/2} h_{rc}(n)e^{-j2\pi mn/N} \tag{2.106}$$

Then, the inverse transform of (2.106) is

$$h_{rc}(n) = \sum_{m=-(N-1)/2}^{(N-1)/2} H_{rc}\left(\frac{mf_{\text{sam}}}{N}\right)e^{j2\pi mn/N}, \quad n = 0, \pm 1, \ldots, \pm (N-1)/2 \tag{2.107}$$

If the RC filter is realized by cascading the transmitter filter with the receiver filter, which is matched to the transmitter filter, each of them is expressed as

$$H_t(f) = H_r(f) = \sqrt{H_{rc}(f)} \tag{2.108}$$

Similar to the derivation of (2.107), the impulse response of the SRRC filter is obtained by solving

$$\begin{aligned} h_t(n) &= h_r(n) \\ &= \sum_{m=-(N-1)/2}^{(N-1)/2} \sqrt{H_{rc}\left(\frac{mf_{\text{sam}}}{N}\right)}e^{j2\pi mn/N}, \quad n = 0, \pm 1, \ldots, \pm (N-1)/2 \end{aligned} \tag{2.109}$$

2.6.3.3 Digital Design Implementation

Now we give an example by using these two different design methods to design a SRRC filter at the transmitter to achieve spectrally efficient transmission through a linear channel.

Design Example 2.2 Create a digital FIR implementation of the transmitter *SRRC* filter with $\alpha = 0.3$ and $N = 63$, and implement it in a *Xilinx* chip with an 11-bit fixed point.

Solution In order to simplify hardware design, we choose the sampling rate $f_{\text{sam}} = 4/T_s$, or four samples per symbol interval, making the total length of the FIR filter with $N = 63$ spans (63 + 1)/4 = 16 symbols. We both window and impulse invariance methods to design the FIR filter; their impulse responses and frequency responses are illustrated in Fig. 2.36. It can be seen that the impulse responses obtained by using two different design methods are identical, while their frequency responses are very close to each other up to the normalized frequency $f/f_N = (1 + \alpha) = 1.3$. To estimate the design accuracy relative to an ideal filter, we also plot the frequency response of the SRRC FIR filter, with $N = 263$ in Fig. 2.36b, which is used to closely approach such an ideal SRRC filter. Frequency responses of the FIR filters with both window and impulse invariance methods are very close to that of the SRRC FIR filter, with $N = 263$ until –30 dB attenuation. Therefore, using either of these two methods to design the digital SRRC filter can achieve a satisfactory approximation to an ideal SRRC filter.

For a comparison between different window truncation functions, Fig. 2.37 illustrates the frequency responses of the designed SRRC filter with different windows. It is clear that significant small side-lobes are obtained at the price of a slightly wider main lobe.

The coefficients of the impulse response of the SRRC filter calculated from (2.98) and (2.109), respectively, are listed in Table 2.4: each coefficient has two different values; where the first one is obtained by using the window method and the second one is obtained by using the impulse invariance method. Because the impulse response is symmetric, the impulse response values of the transmitter $h_t(n)$ at the positive index are the same as those at the negative index. Therefore, they are not listed except for the coefficients of h_1 and h_{31}.

Assume that 11-bit fixed point, denoted by $C_{10}C_9...C_0$, is used to represent the coefficients in the Xilinx implementation, where C_{10} is a sign bit. After the center coefficient h_0 is normalized to 1023 in decimal, the rest values are listed in Table 2.5. Xilinx Virtex2 Chip has a Coregen function of the FIR filter. By using these coefficients, the FIR Coregen can realize such a SRRC filter.

2.6.4 Amplitude Compensation for a SINC Function

As we discussed in earlier, the *x*/sin(*x*) amplitude compensation is required to cascade with the raised-cosine filter at the transmitter as expressed in (2.80) if the

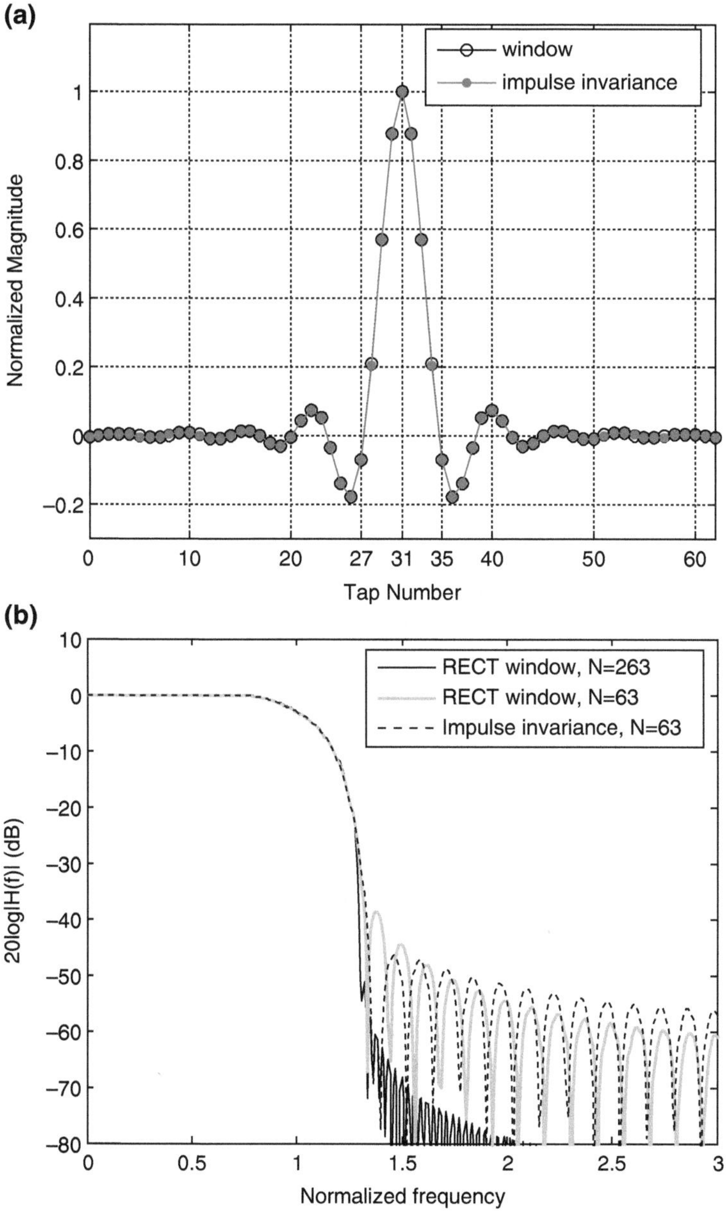

Fig. 2.36 Impulse response and frequency response of FIR filter with a tap length of $N = 63$ to approximate to an idea SRRC filter with $\alpha = 0.3$: (**a**) impulse response, and (**b**) frequency response

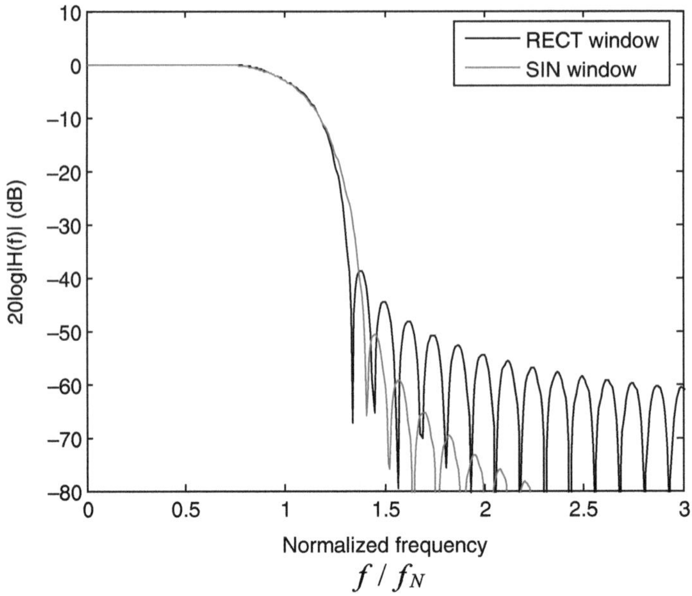

Fig. 2.37 Frequency responses of FIR SRRC filter with $\alpha = 0.3$ and tap length $N = 63$ truncated by rectangular and sinusoid windows

Table 2.4 Coefficients of impulse response of a FIR filter with 63-tap

h_{-31}	h_{-30}	h_{-29}	h_{-28}	h_{-27}	h_{-26}	h_{-25}
0	6e−4	6e−4	−7e−4	−2.1e−3	−1.7e−3	1.5e−3
−3.6e−3	−2.8e−3	1.2e−3	4.6e−3	3.6e−3	−1.7e−3	−6.7e−3
h_{-24}	h_{-23}	h_{-22}	h_{-21}	h_{-20}	h_{-19}	h_{-18}
5e−3	5e−3	−9e−4	−9.4e−3	−1.3e−2	−5.6e−3	1.1e−2
−6.2e−3	5e−4	8.0e−3	9.0e−3	1.7e−3	−7.9e−3	−1.01e−2
h_{-17}	h_{-16}	h_{-15}	h_{-14}	h_{-13}	h_{-12}	h_{-11}
2.54e−2	2.34e−2	−1e−4	−3.36e−2	−5.23e−2	−3.47e−2	1.85e−2
−1.3e−3	1.1e−2	1.32e−2	−1.7e−3	−2.37e−2	−3.06e−2	−5.8e−3
h_{-10}	h_{-9}	h_{-8}	h_{-7}	h_{-6}	h_{-5}	h_{-4}
7.74e−2	9.64e−2	4.46e−2	−6.57e−2	−1.724e−1	−1.889e−1	−5.15e−2
4.16e−2	7.48e−2	5.23e−2	−3.52e−2	−1.42e−1	−1.801e−1	−6.93e−2
h_{-3}	h_{-2}	h_{-1}	h_0	h_1	...	h_{31}
2.364e−1	5.924e−1	8.863e−1	1.0	8.863e−1	...	0.0
2.058e−1	5.687e−1	8.784e−1	1.0	8.784e−1	...	−3.6e−3

input to the raised-cosine filter is a NRZ signal. In the digital FIR implementation, however, the amplitude aperture compensator $x/\sin(x)$ may be unnecessarily needed because the input impulse streams are performed with an up-sampling rate of N by inserting $N - 1$ zero between two adjacent data sequences before passing through a digital RC or SRRC filter, especially in the case of $N > 4$.

Table 2.5 Coefficients represented by 11-bit fixed points

h_{-31}	h_{-30}	h_{-29}	h_{-28}	h_{-27}	h_{-26}	h_{-25}
0	1	1	−1	−2	−2	2
−4	−3	1	5	4	−2	−6
h_{-24}	h_{-23}	h_{-22}	h_{-21}	h_{-20}	h_{-19}	h_{-18}
5	5	−1	−10	−13	−6	11
5	1	8	9	2	−8	−11
h_{-17}	h_{-16}	h_{-15}	h_{-14}	h_{-13}	h_{-12}	h_{-11}
26	24	0	−34	−54	−36	19
1	11	14	−2	−26	−31	5
h_{-10}	h_{-9}	h_{-8}	h_{-7}	h_{-6}	h_{-5}	h_{-4}
79	98	46	−67	−176	−193	−53
43	77	54	−36	−145	−184	−71
h_{-3}	h_{-2}	h_{-1}	h_0	h_1	...	h_{31}
242	606	907	1023	907	...	0
211	582	900	1023	900	...	−4

The discrete signal at the output of the RC or SRRC filter is transferred to the continuous signal through a digital-to-analog converter (DAC). If the output of the RC or SRRC filters or the input of the DAC is a sequence of samples, $y_d(n)$, an impulse train $y_i(t)$ to the input of the zero-order hold block can be formed as [25] follows:

$$y_i(t) = \sum_{n=-\infty}^{\infty} y_d(n)\delta(t - nT_{\text{sam}}) \tag{2.110}$$

where T_{sam} is the sampling interval associated with the sequence $y_d(n)$. Thus, the output signal $y_z(t)$ of the zero-order hold block is the convolution of the input $y_i(t)$ of the zero-order hold block and the impulse response $h_z(t)$ of the zero-order hold in the time domain, or

$$\begin{aligned} y_z(t) &= y_i(t) * h_z(t) \\ &= \left(\sum_{n=-\infty}^{\infty} y_d(n)\delta(t - nT_{\text{sam}})\right) * h_z(t) \\ &= \sum_{n=-\infty}^{\infty} y_d(n)h_z(t - nT_{\text{sam}}) \end{aligned} \tag{2.111}$$

The impulse response $h_z(t)$ of the zero-order hold is expressed as

$$h_z(t) = \begin{cases} 1, & 0 < t < T_{\text{sam}} \\ 0, & \text{otherwise} \end{cases} \tag{2.112}$$

Its Fourier transform is

$$H_z(\omega) = \frac{\sin(\omega T_{\text{sam}}/2)}{\omega/2} \mathrm{e}^{-j\omega T_{\text{sam}}/2} \tag{2.113}$$

Thus, the Fourier transform of (2.111) becomes

$$\begin{aligned} Y_z(\omega) &= Y_i(\omega) H_z(\omega) \\ &= Y_i(\omega) \frac{\sin(\omega T_{\text{sam}}/2)}{\omega/2} \mathrm{e}^{-j\omega T_{\text{sam}}/2} \end{aligned} \tag{2.114}$$

To recover the input signal $Y_i(\omega)$, the output signal $Y_z(\omega)$ of the zero-order hold needs to be passed through a reconstruction filter with a transfer function of $H_r(\omega)$. The Fourier transform of the reconstruction filter output is

$$\begin{aligned} Y(\omega) &= Y_z(\omega) H_r(\omega) \\ &= Y_i(\omega) H_z(\omega) H_r(\omega) \end{aligned} \tag{2.115}$$

It can be seen from (2.115) that the production of the last two items should be equal to the ideal lowpass filter $H_l(\omega)$ in order to recover $Y_i(\omega)$, or

$$H_z(\omega) H_r(\omega) = H_l(\omega) \tag{2.116}$$

$$H_l(\omega) = \begin{cases} T_{\text{sam}}, & |\omega| < \omega_c \\ 0, & \text{otherwise} \end{cases} \tag{2.117}$$

Thus, the reconstruction filter is

$$\begin{aligned} H_r(\omega) &= \frac{H_l(\omega)}{H_z(\omega)} \\ &= H_c(\omega) H_l(\omega) \\ &= \begin{cases} \dfrac{\omega T_{\text{sam}}/2}{\sin(\omega T_{\text{sam}}/2)} \mathrm{e}^{j\omega T_{\text{sam}}/2}, & |\omega| < \omega_c \\ 0, & \text{otherwise} \end{cases} \end{aligned} \tag{2.118}$$

Note from (2.118) that the reconstruction filter $H_r(\omega)$ is obtained by cascading a compensation filter $H_c(\omega)$ and an ideal lowpass filter $H_l(\omega)$. Its advance time shift of $T_{\text{sam}}/2$ seconds can be compensated if the ideal filter $H_l(\omega)$ has a delay time shift of τ, in which τ meets the condition $\tau \geq T_{\text{sam}}/2$. The compensation filter $H_c(\omega)$ is

$$H_c(\omega) = \frac{1}{T_{\text{sam}}} \frac{\omega T_{\text{sam}}/2}{\sin(\omega T_{\text{sam}}/2)} \mathrm{e}^{j\omega T_{\text{sam}}/2} \tag{2.119}$$

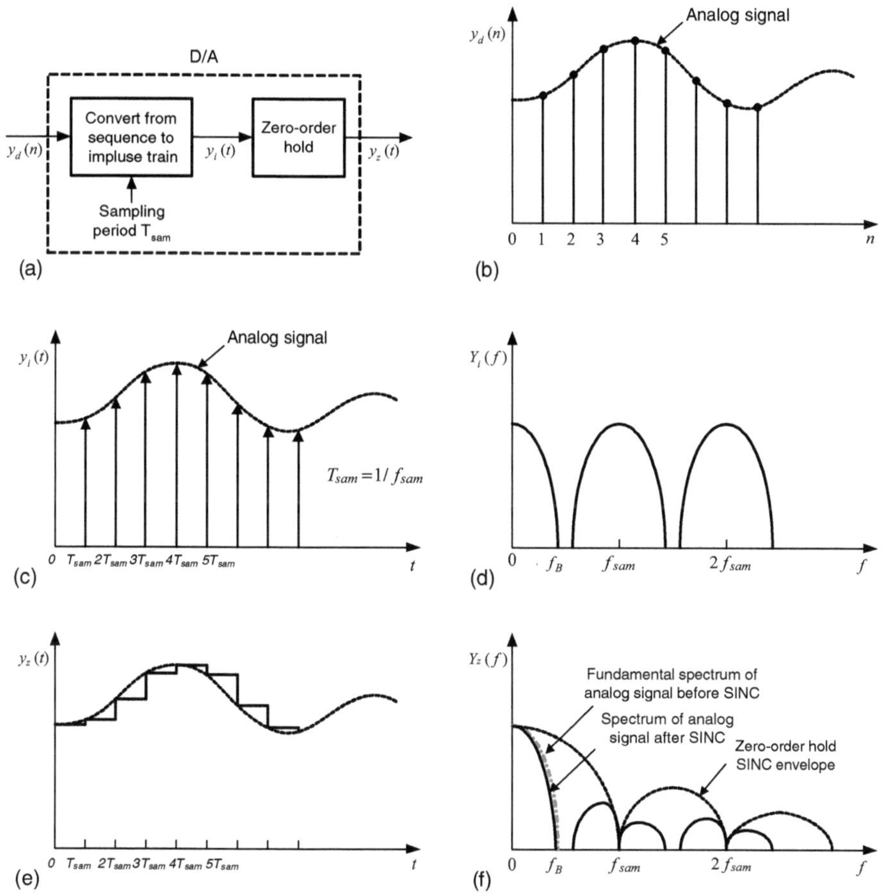

Fig. 2.38 Input and output of DAC in time and frequency domains: (**a**) block diagram of DAC, (**b**) discrete-time sequence, (**c**) impulse train, (**d**) spectrum of the impulse train, (**e**) output of the zero-order hold, and (**f**) Spectrum of the zero-order hold output. f_B is the maximum frequency for a bandlimited analog signal, and f_{sam} is the sampling frequency. Referenced from [25, 26]

The $x/\sin(x)$ shape amplitude compensation given in (2.119) is scaled by $1/T_{sam}$, which is used to cancel a scaling factor of T_{sam} in the ideal filter $H_1(\omega)$ as expressed in (2.117). It is also noted that such an advance time shift of $T_{sam}/2$ in (2.118) can be ignored without any effect on the performance.

At this point, it can be seen from (2.113) that the zero-order hold block introduces $\sin(x)/x$ shape amplitude distortion in the frequency domain. Thus, the amplitude of the signal spectrum at the output of the DAC is multiplied by a function $\sin(x)/x$ or Sinc(x). As a result, Sinc(x) function acts as amplitude distortion for the DAC output signal.

Figure 2.38 illustrates the amplitude distortion caused by the zero-order hold characteristic of the DAC, showing the input and the output of the DAC in the time

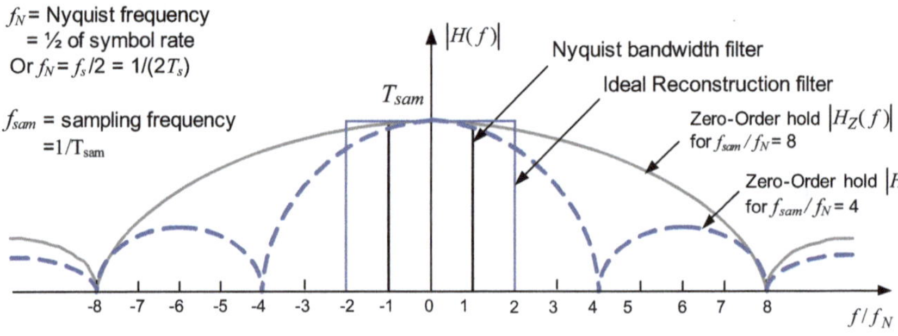

Fig. 2.39 Amplitude response of zero-order hold with sampling frequency $f_{\mathrm{sam}} = 4f_{\mathrm{s}} = 8f_{\mathrm{N}}$. The maximum bandwidth of raised-cosine filter is $2f_{\mathrm{N}}$, or $f/f_{\mathrm{N}} = 2$

domain and their corresponding spectrum patterns in the frequency domain. Figure 2.38f shows that the frequency response of the zero-order hold acts as a lowpass filter that attenuates not only image components but also the fundamental component of the desired signal. Attenuation on the desired signal, however, is frequency dependent, where the attenuation becomes severe when the sampling frequency f_{sam} decreases. For example, the fundamental component before the zero-order hold circuit is shown in the dash-dot line, while after the zero-order hold circuit, the fundamental component is represented by the solid line.

Figure 2.39 shows the magnitude of the frequency response of the zero-order hold circuit when the sampling frequency is either four times or eight times the Nyquist frequency. It is obvious that an amplitude compensator is needed when the sampling frequency is twice as large as the symbol rate or four times as large as the Nyquist frequency, as shown by the dashed line. To compensate for a sin(x)/x shape distortion, it is natural for the amplitude compensator to have the opposite frequency response of the zero-order hold function, or a x/sin(x)-shaped frequency response. Thus, the combination of their cascaded frequency response is constant. Usually it is enough for the overall amplitude response to be constant within the bandwidth of $(1+\alpha)f_{\mathrm{N}}$.

From Fig. 2.39 we can see that this magnitude compensator may be neglected since the amplitude drops slightly as the sampling frequency increases. For example, the amplitude drops by only $2\sqrt{2}/\pi$ (or –0.91 dB) at the symbol rate or twice the Nyquist frequency ($f/f_{\mathrm{N}} = 2$, which is equivalent to $\omega = \pi/(2T_{\mathrm{s}})$ in (2.113) when the sampling frequency of f_{sam} is four times the symbol rate of f_{s} ($f_{\mathrm{sam}} = 4f_{\mathrm{s}} = 8f_{\mathrm{N}}$), as indicated by the light-dark solid line. It drops about $2/\pi$ (or –3.92 dB) when the sampling frequency of f_{sam} is twice the symbol rate f_{s} ($f_{\mathrm{sam}} = 2f_{\mathrm{s}} = 4f_{\mathrm{N}}$), as indicated by the light-dark dashed line. In the former case, this amplitude compensation may be neglected due to –0.91 dB attenuation. In the latter case, the amplitude compensation is needed because of -3.92 dB attenuation. The spectrum bandwidth of the raised-cosine filtered signal is within the range of $f_{\mathrm{N}} < f \leq 2f_{\mathrm{N}}$, corresponding to the alpha value range $0 < \alpha \leq 1$.

Increasing the sampling clock rate of the DAC not only reduces the attenuation effect of the zero-order hold on the desired signal, but also lowers the quantization noise floor and relaxes attenuation requirements for the reconstruction filter [26]. A DAC with a higher clock rate also increases the design cost and power consumption. In some cases, it is very complicated to design a DAC with a high clock rate due to a wider bandwidth of the transmission data. For example, in the ultra-wideband (UWB) system [27] it costs much more to design a DAC with an over-sampling rate of 1056 MHz because the bandwidth of an OFDM signal is about 264 MHz.

In the case where the sampling rate is very difficult to increase, the amplitude compensation technique is an effective method to deal with the amplitude distortion caused by the SINC-function. Amplitude compensation can be achieved with either the digital or the analog filter. In the former case, the pre-distortion is created before the DAC, while in the latter the post-compensation is generated after the DAC. In both cases, the amplitude response of the compensator is the inverse of the SINC-function or 1/Sinc(x). In practice, the overall amplitude response is required to be flat only within the bandwidth of the desired signal, which is equal to $(1+\alpha)f_N$.

For detailed design information regarding the pre-distortion–based equalizer or compensator, the interested reader can refer to [26]. Here, we introduce the post-compensation method in more detail. In the post-compensation method, an analog filter whose frequency response is approximately equal to the inverse of the SINC-function is inserted either before or after the reconstruction filter. To reach such an inverse shape of the SINC-function, the analog filter needs to have a peak around the Nyquist frequency. The second-order analog filter can realize such a peak with a proper dumpling factor. The transfer function of the second-order lowpass filter is

$$H_c(s) = \frac{\omega_n^2}{s^2 + 2\zeta\omega_n s + \omega_n^2} \tag{2.120}$$

where ω_n is the *natural frequency* of the filter and ζ is the *damping factor*. As we know, the amplitude of the frequency response of the second-order lowpass filter has a peak around the natural frequency when $\zeta < 0.707$ is met. Therefore, we can use this peak property of the frequency response to approximate the inverse of the SINC-function.

Considering that a baseband signal is passed through either a RC filter or SRRC filter approximated by a FIR filter, we only need to compensate the amplitude distortion caused by the SINC-function up to the bandwidth $B_w = (1+\alpha)f_N$. Hence, the amplitude *gain* of the compensation transfer function relative to the DC amplitude should be approximately equal to the amplitude *attenuation* of the SINC-function up to B_w. Usually the natural frequency $f_n = \omega_n/2\pi$ is set greater than B_w, depending on the ratio of the sampling frequency to the signal bandwidth.

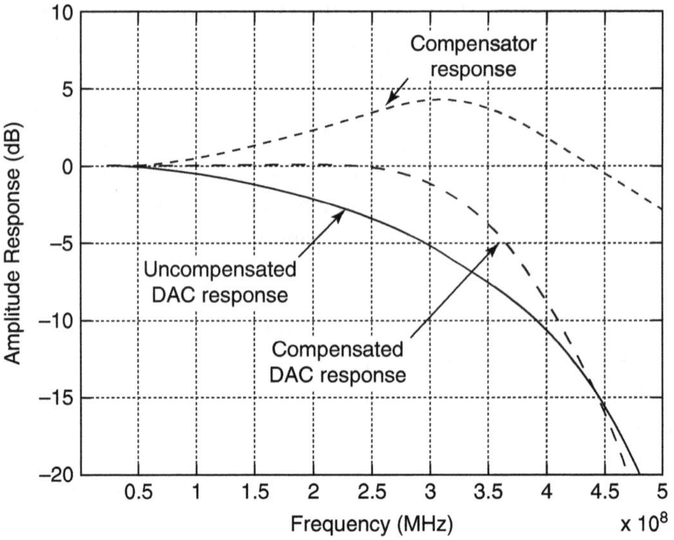

Fig. 2.40 Amplitude response curves of the SINC function and amplitude compensator, where sampling frequency is 528 MHz and Nyquist frequency is 264 MHz

For example, in the UWB OFDM system the single-side bandwidth of the OFDM baseband signal is about 260 MHz [27]. Due to the channel spacing of 528 MHz and the signal bandwidth of 260 MHz, we can have the minimum clock frequency (or sampling frequency) of 528 MHz operate for the DAC. The attenuation of the SINC-function at the half-sampling frequency of $f_{sam}/2 = 264$ MHz is –3.92 dB from (2.113). Thus, the spectrum attenuation of the OFDM baseband signal approximates to –3.92 dB at the bandwidth edge frequency of 260 MHz. To compensate for amplitude distortion, we set the natural frequency $f_n = 350\,\text{MHz}$ and the damping factor $\zeta = 0.33$ in (2.120) so that the compensator has a gain of about 3.7 dB at the half-sampling frequency of 264 MHz, as shown in Fig. 2.40. After the amplitude compensation, the amplitude response of the compensated DAC at a half the sampling frequency is about $-3.92 + 3.7 = -0.22$ dB.

Figure 2.41 illustrates the PSD of the OFDM signal with a bandwidth of 260 MHz at the output of the reconstruction filter. It can be seen that the PSD of the OFDM signal around the bandwidth edge frequency of 260 MHz in Fig. 2.41b increases after the amplitude compensation so that the overall spectrum becomes flat. Compared with digital-filter–based compensation, analog-filter–based compensation is simple in structure, inexpensive, and has low power consumption.

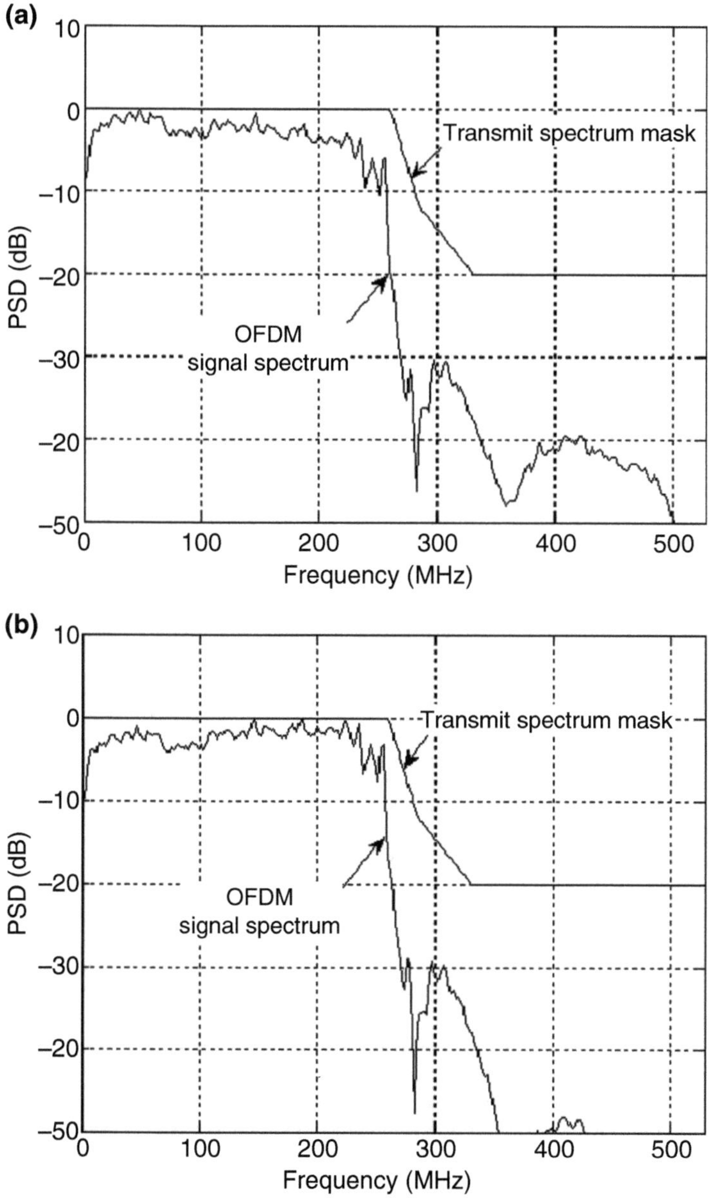

Fig. 2.41 Power spectral density of Multiband OFDM signal at the transmitter: (**a**) without compensator and (**b**) with compensator, where sampling frequency is 1056 MHz

References

1. Sklar, B. (2002). *Digital communications: Fundamentals and applications* (p. 168). India: Pearson Education Asia.
2. Wang, A. Y., & Sodini, C. G. (2006). On the energy efficiency of wireless transceivers. In *ICC 2006 Proceedings* (pp. 3783–3788).
3. McCune, E. (2015). A technical foundation for RF CMOS power amplifiers. *IEEE Solid-State Circuits Magazine, 8*(2), 75–82.
4. Joung, J., Ho, C. K., Adachi, K., & Sun, S. (2015). A survey on power amplifier centric techniques for spectrum- and energy-efficient wireless communication. *IEEE Communication Surveys and Tutorials, 17*(1), 315–333.
5. Cripps, S. C. (1999). *RF power amplifiers for wireless communications* (p. 49). Norwood, MA: Artech House.
6. Wimpenny, G. (2012, February 6). Envelope tracking power amplifier characterization (White Paper). Nujira limited. Retrieved from www.nujira.com
7. Application Report, "GC5325 Envelope Tracking," *Texas Instruments*, SLWA058B, April 2010.
8. Cripps, S. C. (2002). *Advanced techniques in RF power amplifier design* (pp. 4–5). Norwood, MA: Artech House.
9. Zappone, A., & Jorswieck, E. (2015). Energy efficiency in wireless networks via fractional programming theory. *Foundations & Trends in Communications and Information Theory, 11* (3–4), 185–396.
10. Buzzi, S., Chih-Lin, I., Klein, T. E., Vincent Poor, H., Yang, C., & Zappone, A. (2016). A survey of energy-efficient techniques for 5G networks and challenges ahead. *IEEE Journal on Selected Areas in Communications, 34*(4), 697–709.
11. Ziemer, R. E., & Tranter, W. H. (2002). *Principles of communications: Systems modulation and noise* (5th ed.). New York: Wiley.
12. Feher, K. (1995). *Wireless and digital communications: Modulation & spread spectrum applications*. Upper Saddle River, NJ: Prentice-Hall PTR.
13. ZigBee Alliance. (2006, December). ZigBee Specifications (Version 1.0). Retrieved from http://www.zigbee.org/
14. Austin, M. C., & Chang, M. U. (1981). Quadrature overlapped raised-cosine modulation. *IEEE Transactions on Communications, COM-29*(3), 237–249.
15. Le-Ngoc, T., & Feher, K. (1983). Performance of IJF-OQPSK modulation schemes in a complex interference environment. *IEEE Transactions on Communications, COM-31*(1), 137–144.
16. Seo, J. S., & Feker, K. (1985). SQAM: a new superposed QAM modem technique. *IEEE Transactions on Communications, COM-33*(3), 296–300.
17. Gao, W., Ju, D., & Wu, Y. Self-convolving minimum shift keying (SCMSK) modem for satellite system. In *Proceedings of IEEE Singapore ICCS'88* (pp. 341–345).
18. Gao, W., & Feher, K. (1996). All digital reverse modulation architecture based carrier recovery implementation for GMSK and compatible FQPSK. *IEEE Transactions on Broadcasting, 42*(1), 55–62.
19. Proakis, J. G. (1995). *Digital communications* (3rd ed.). New York: McGraw-Hill.
20. Simon, M. K. (2001, June). *Bandwidth-efficient digital modulation with application to deep-space communications*. Deep-space communications and navigation series.
21. Nyquist, H. (1928). Certain topic in telegraph transmission. *Transactions of the AIEE, 47*(2), 617–644.

22. Tenbroek, B., Strange, J., Nalbantis, D., Jones, C., Flowers, P., Brett, S., et al. (2008). Single-chip tri-band WCDMA/HSDPA transceiver without external SAW filters and with integrated TX power control. In *ISSCC* 2008 (pp. 202–204).
23. Jussila, J. (2003, June). Analog baseband circuits for WCDMA direct conversion receivers. PhD Dissertation, Helsinki University of Technology, Finland.
24. Li, Z., Li, M., Zhao, D., Ma, D., Ni, W., & Ouyang, Z. (2010). TD-SCDMA/HSDPA transceiver and analog baseband chipset in 0.18-μm CMOS process. *IEEE Transactions on Circuits and Systems-II, 57*(2), 90–94.
25. Oppenheim, A. V., Schafer, R. W., & Buck, J. R. (1999). *Discrete-time signal processing*. Upper Saddle River, NJ: Prentice Hall.
26. Yang, K. (2006, April 13). Flatten DAC frequency response. *END Magazine*, 65–74.
27. Multiband OFDM physical layer proposal for IEEE 802.15 Task Group 3a (2004, September 14). MBOA-SIG multiband OFDM alliance SIG.

Chapter 3
Bandwidth-Efficient Modulation With OFDM

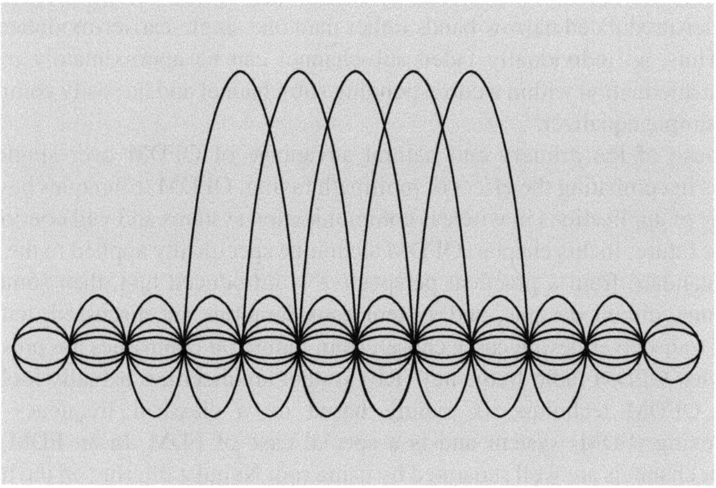

3.1 Introduction

An orthogonal frequency division multiplexing (OFDM) technique has been developed for wideband data transmission through multipath fading channels without the need for complex equalizers. The concept of OFDM dates back to the 1960s, when Chang [1] first proposed the synthesis of orthogonal signals for multichannel data transmission in 1968. Wideband transmission systems are more vulnerable to multipath fading because the fading notches have a higher chance of dropping into the transmission bandwidth. As its name implies, OFDM is a scheme of splitting a single data sequence at a high bit rate into many parallel

© Springer International Publishing Switzerland 2017

W. Gao, *Energy and Bandwidth-Efficient Wireless Transmission*, Signals and Communication Technology, DOI 10.1007/978-3-319-44222-8_3

sub-data streams at a low symbol rate to conventionally modulate orthogonal subcarriers in order to space these subcarriers close together in a certain bandwidth. OFDM has continuously developed into a very popular scheme for wideband digital communication systems, such as 802.11a/g/n/ac-based wireless local area networks (WLANs), digital television, audio broadcasting, and 4G mobile LTE communication standards.

In modern digital communications, conventional single-carrier modulation schemes for high-data-rate transmissions like high-order QAM formats have faced difficulties in coping with severe channel transmission conditions—such as frequency-selective fading due to changeable multipaths—even when using complex adaptive equalization techniques. Sometimes the equalizer is unable to effectively compensate for multipath distortion when there are deeper fade notches dropping within the desired signal bandwidth. To overcome the effect of multipath fading, channel equalization in an OFDM system is relatively simple because the bandwidth of OFDM can be viewed as the orthogonal arrangement of many subcarrier-modulated narrow bands rather than one single-carrier-modulated wideband. Thus, an individually faded sub-channel can be approximately treated as constant attenuation within a corresponding sub-channel and be easily compensated with a simple equalizer.

Because of the primary and natural advantage of OFDM over single-carrier schemes in combating the effect of multipath fading, OFDM techniques have found a variety of applications in wireless communication systems and will continue to do so in the future. In this chapter, OFDM technique specifically applied to the 802.11a Wi-Fi standard from a practical perspective is introduced first, then some digital algorithms, circuit designs, and system considerations are discussed; and finally two RF transceiver design cases chosen from industrial companies are presented to show what OFDM radio frequency (RF) system architectures actually look like.

The OFDM technique is mainly based on a classical frequency-division multiplexing (FDM) system and is a special case of FDM. In an FDM system, adjacent channels are well separated by using root Nyquist filtering on the baseband signals and a guard interval in the transmission band. In the 1960s, Saltzberg [2] and Chang [1] studied and evaluated the performance of OFDM systems to achieve bandwidth efficiency by overlapping the spectra of the sub-channels compared with an FDM-based single channel per carrier (SCPC) system. SCPC refers to transmitting a single signal at a given frequency channel with a relatively narrow bandwidth. In FDM access (FDMA) technology, an SCPC system using low-cost equipment is commonly used for data and voice applications. For example, in the early 1980s the SCPC systems developed by SPAR Aerospace Ltd. in Canada were mainly used for voice and data communications in remote areas where the existing communication networks supporting the large-capacity communications could not reach because of the small capacity requirements of remote areas. SCPC communication networks, however, are suitable for providing such small-capacity communications for these remote areas. In voice transmission, each user's voice signal is first converted to the digital data signal with a rate of 32 kbits/s, and then the digital data sequences are used to modulate a 70-MHz intermediate frequency (IF) carrier using a BPSK scheme. The modulated IF signal is further up-converted to the RF at 6 GHz. In data transmission, each user's data with a rate 64 kbits/s modulates a 70-MHz IF carrier using a QPSK scheme. The modulated IF signal is further up-converted to the RF at 6 GHz for transmission over a satellite channel. For voice communications, a pair of channel frequencies is utilized, one for each direction of transmission. The transmission can be turned on by voice activation to save power on the satellite. A typical voice channel conversation is active only about 40% of the time in any one direction because a pair of conversations occurs between two directions. Typically, the channel spacing is 45 kHz among all sub-channels. Thus, a 36-MHz transponder in a satellite supports the capacity of 800 SCPC channels (36 MHz/45 kHz). Each earth station can support a small numbers of users, typically fewer than ten.

In this chapter, we use the 802.11a Wi-Fi standard as an example to introduce some basic concepts and methods of generating an OFDM baseband signal at a transmitter, down-converting an RF signal into the OFDM baseband signal, and then demodulating the OFDM baseband signal at the receiver. The Wi-Fi is defined by the Wi-Fi Alliance as any "Wireless Local Area Network" (WLAN) product based on the IEEE 802.11 standards.

3.2 Generation of the 802.11a OFDM Signal

An OFDM signal is generated by first splitting a single data sequence with a high data transmission rate into many sub-data sequences with a low data transmission rate, each called a sub-channel data sequence, through a serial-to-parallel converter; and then each sub-channel data sequence modulates a very low IF orthogonal carrier signal, called a subcarrier signal, to construct a modulated OFDM signal in the baseband time domain by summing all subcarrier-modulated signals together. The OFDM signal is further up-converted with a local oscillator to the RF signal for transmission. The primary difference between OFDM generation and single-carrier modulation generation is that the OFDM signal consists of *a sum of orthogonal subcarriers* that are modulated by using a phase shift keying (PSK) or quadrature amplitude modulation (QAM) scheme with parallel sub-channel data sequences or symbols. The OFDM technology application started with third-generation WLANs, or the 802.11g and 802.11a standards, in 2002 for use at high data transmission rates of 54 Mbps at frequency bands of 2.4 and 5 GHz, respectively. The fourth-generation WLAN, or the 802.11n standard, was launched in 2007 to enable data transmission rates up to 600 Mbps for use with medium-resolution video streaming by means of multi-input and multi-output (MIMO) technology. In 2012, the fifth-generation of the 802.11ac standard was released to support Ethernet data transmission rates up to 3.6 Gbps, with the first version less than 1.8 Gbps. From the early 802.11g/a standard, to the 802.11n standard, to the latest 802.11ac standard, the newer standards have been backward-compatible with the previous standards. In the following sections, a 16-QAM OFDM signal adopted in the 802.11a-based WLAN system is used as an example to better illustrate the concept of the OFDM signaling format. The generation of an OFDM signal in the continuous time domain

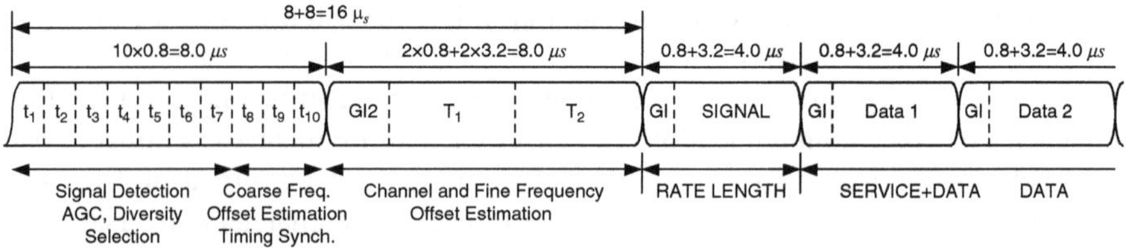

Fig. 3.1 OFDM baseband signal frame in 802.11a

is first introduced as a fundamental concept, and then its generation in the discrete time domain is presented.

In general, a transmitted OFDM baseband signal comprises contributions from several different subframes in the time domain (as shown in Fig. 3.1) and is expressed as [3]

$$s_{\text{FRAME}}(t) = s_{\text{PREAMBLE}}(t) + s_{\text{SIGNAL}}(t - T_{\text{SIGNAL}}) + \bar{s}_{\text{DATA}}(t - T_{\text{DATA}}) \quad (3.1)$$

The three subframes PREAMBLE, SIGNAL, and DATA fields constitute one frame, and the time offsets determine the starting time of the corresponding subframe: $T_{\text{SIGNAL}} = 16\,\mu\text{s}$, and $T_{\text{DATA}} = 20\,\mu\text{s}$.

3.2.1 *Preamble Field*

The preamble subframe that consists of ten short training symbols denoted by t_1 to t_{10} and two long training symbols denoted by T_1 and T_2 is used for synchronization, as shown in Fig. 3.1. The preamble is followed by the SIGNAL field and then DATA field. A short OFDM training *segment* that corresponds to 4 short symbols and consists of 12 subcarriers in the frequency domain is modulated by the elements of the *complex* sequence S and generated by the following equation:

$$s_{\text{SHORT}}(t) = w_{\text{SHORT}}(t) \sum_{k=-N_{\text{ST}}/2}^{N_{\text{ST}}/2} S_k \exp(j2k\Delta ft) \quad (3.2)$$

where S_k is given by

$$\begin{aligned} S_{-26:26} = \sqrt{13/6} \times \{&0,0,1+j,0,0,0,-1-j,0,0,0,1+j,0,0,0,-1-j,0,0,0,1\\ &+j,0,0,0,00,0,0,-1-j,0,0,0,-1-j,0,0,0,1+j,0,0,0,1+j,0,0,0,1\\ &+j,0,0,0,1+j,0,0\} \end{aligned} \quad (3.3)$$

A factor of $\sqrt{13/6}$ is used to normalize the average power of the resulting OFDM symbol, which utilizes 12 of 52 subcarriers. Here, the modulation process expressed by (3.2) is equivalent to an inverse Fourier transform of a set of coefficients S_k given in (3.3). The most common way to perform the inverse Fourier transform is by using an inverse fast Fourier transform (IFFT) algorithm. Adding zero to both sides of (3.3) or from subcarriers -27 to -32 and from 27 to 31 achieves 64-point FFT values in the frequency domain, as listed in Table G.2 of Annex G [3]. These 64-point FFT values are used as the frequency domain inputs to an IFFT algorithm processor that is discussed below. After the IFFT operation, 64-point short training sequences are generated in the time domain listed in Table G.3 of Annex G [3]. These 64-point sequences comprise *four short training symbols*. The 10 complex short training symbols are extended periodically for 160 samples by concatenating with an additional 4 short symbols and then another 2 short symbols.

A long OFDM training symbol that comprises 53 subcarriers, including a zero at direct current (DC) in the frequency domain, is binary-modulated by the elements of the *real* sequence L and generated by the following equation:

$$s_{\text{LONG}}(t) = w_{\text{LONG}}(t) \sum_{k=-N_{\text{ST}}/2}^{N_{\text{ST}}/2} L_k \exp(j2k\Delta f(t - T_{\text{GI2}})) \tag{3.4}$$

where L_k is given by

$$\begin{aligned} L_{-26:26} = \{&1, 1, -1, -1, 1, 1, -1, 1, -1, 1, 1, 1, 1, 1, 1, -1, -1, 1, 1, -1, 1, \\ &-1, 1, 1, 1, 1, 01, -1, -1, 1, 1, -1, 1, -1, 1, -1, -1, -1, \\ &-1, -1, 1, 1, -1, -1, 1, -1, 1, -1, 1, 1, 1, 1\} \end{aligned} \tag{3.5}$$

where T_{GI2} is equal to 1.6 μs. Repeating 64 samples of one long training sequence and then cyclically extending the last 32 samples results in a 161-sample vector, as shown in Table G.6 [3]. After being multiplied by the window function, the samples of two long training symbols with the cyclic prefix are appended to the short sequence section.

The preamble sequence is formed by concatenating ten short training symbols and two long training symbols

$$s_{\text{PREAMBLE}}(t) = \bar{s}_{\text{SHORT}}(t) + \bar{s}_{\text{LONG}}(t - T_{\text{SHORT}}) \tag{3.6}$$

where $\bar{s}_{\text{SHORT}}(t)$ is generated by concatenating two *segments* $(s_{\text{SHORT}}(t) + s_{\text{SHORT}}(t))$ of four short symbols in (3.2), and one half-segment of four short symbols, $\bar{s}_{\text{LONG}}(t)$ is constructed by concatenating two long symbols $(s_{\text{LONG}}(t) + s_{\text{LONG}}(t))$ in (3.4) and one cyclic prefix, and $T_{\text{SHORT}} = 8.0$ μs is the duration of ten short training symbols. Figure 3.2 shows the preamble sequence in the time domain, where the length of ten short symbols at the beginning is equal to $160(10 \times 16)$ samples, and the two long symbols plus the cyclic prefix following the ten short symbols is equal to $160(2 \times 64 + 32)$ samples. It can be seen that the

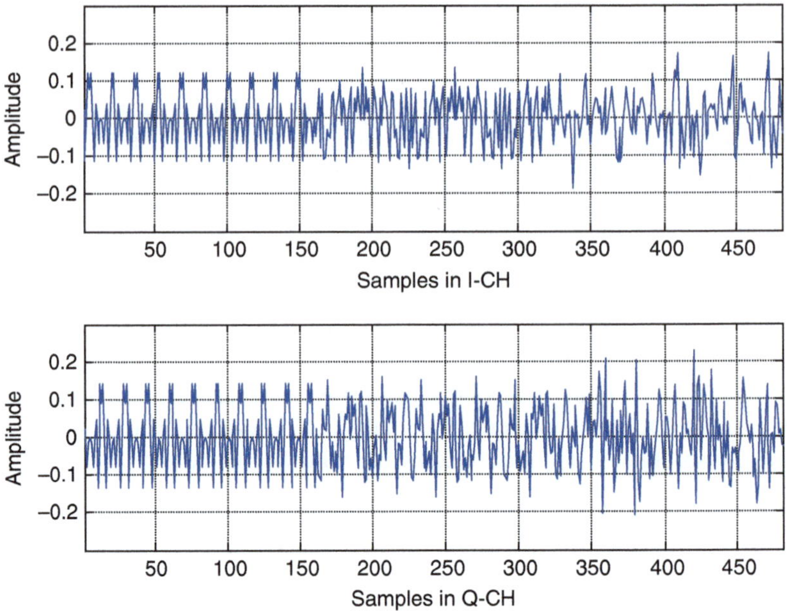

Fig. 3.2 Baseband signals in the I and Q channels, consisting of 10 short training symbols from sample 1 to sample 160, 2 long training symbols from 161 to 320, 1 SIGNAL symbol from 321 to 400 and 1 data symbol from 401 to 480

short training symbols repeat ten times and long training symbols repeat two times plus the cyclic prefix in both I and Q channels. The timing synchronization in the receiver is carried out based on the cross-correlation between the received training sequence and a locally regenerated training sequence, which is introduced in the following sections.

3.2.2 *Signal Field*

The SIGNAL field following the OFDM training symbols comprises 24 bits that contain the RATE, including the modulation type and coding rate, and LENGTH information of the transmitted frame for correct detection in the receiver. These 24 bits are encoded at the rate 1/2 convolutional encoder to yield the 48 bits and then are interleaved. The 48 interleaved bits are BPSK-mapped to yield real coefficients in the frequency domain given in Table G.11 [3], where bits 0–23 in Table G.9 are mapped to the coefficients -26 to -1 in Table G.11, and bits 24–47 are mapped to the coefficients 1–26. Four pilot values are inserted at locations -21, -7, 7, and 21. The coefficients from -32 to -27 and from 27 to 31 are set to zero for a 64-point IFFT operation.

Similar to performing an IFFT operation for the preamble sequences described above, the 80 samples of the SIGNAL field in the time domain are derived by taking the IFFT calculation in Table G.11, extending it cyclically, and multiplying the window function. The SIGNAL filed samples in Table G.12 are then appended to the long training preamble, as shown in Fig. 3.2.

3.2.3 Data Field

In an 802.11a WLAN system, data are scrambled first and then coded with a convolutional encoder of coding rates $R = 1/2, 2/3$, or $3/4$, depending on the desired data transmission rate. All encoded data bits shall be interleaved by a block interleaver with a block size corresponding to the number of bits in a single OFDM symbol. After being interleaved, the data are mapped to a certain modulation format of BPSK, QPSK, 16-QAM, or 64-QAM, depending on the RATE specified in the SIGNAL field. Through a serial-to-parallel converter, the stream of the mapped complex numbers is converted into groups of 48 complex data. These 48 parallel complex data plus 4 pilot real data form a total of 52 parallel modulation data to individually modulate each subcarrier in one OFDM symbol interval, as shown in Fig. 3.3.

In an OFDM modulator (shown in Fig. 3.3), the input serial data stream $\{b_k\}$ is mapped into a complex sequence $\{d_n\}$ of 16-QAM symbols at baseband and then divided into groups of $N_{\mathrm{SD}} = 48$ complex numbers in parallel. In the mapping stage, 4-bit consecutive serial data stream $\{b_k\}$ are mapped into one complex sequence $\{d_n\}$ due to 16-QAM mapping. Each complex symbol $d(n) = d_i(n) + jd_q(n)$ is represented by the in-phase (I) and quadrature (Q) components, which modulates a pair of orthogonal subcarriers $\cos(2\pi k\Delta ft)$ and $\sin(2\pi k\Delta ft)$, respectively.

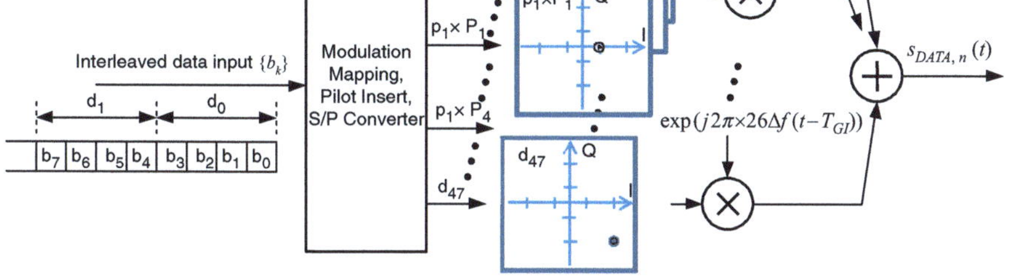

Fig. 3.3 Block diagram of OFDM modulator in continuous time domain

Table 3.1 Timing-related main parameters

Parameter	Value
N_{SD}: number of data subcarriers	48
N_{SP}: number of pilot subcarriers	4
N_{ST}: number of total subcarriers	52($N_{SD}+N_{SP}$)
N_{SYM}: number of OFDM symbols in a frame	Depending on MAC requesting
N_{FFT}: point of FFT/IFFT operation	64
Δf: subcarrier frequency spacing	0.3125 MHz (20 MHz/64)
f_{sam}: sampling frequency	20 MHz
T_{FFT}: IFFT/FFT period	3.2 μs ($1/\Delta f$)
T_{GI}: guard interval	0.8 μs ($T_{FFT}/4$)
T_{GI2}: training symbol GI duration	1.6 μs (TFFT/4)
T_{SYM}: symbol interval	4 μs ($T_{GI}+T_{FFT}$)
T_{SHORT}: short training sequence duration	8 μs ($10 \times T_{FFT}/4$)
T_{LONG}: long training sequence duration	8 μs ($T_{GI2}+2 \times T_{FFT}$)
$T_{PREABLE}$: PLCP preamble duration	16 μs ($T_{SHORT}+T_{LONG}$)
T_{SIGNAL}: signal duration	4.0 μs ($T_{GI}+T_{FFT}$)

The subcarrier frequency of Δf is equal to 312.5 kHz. Generally, the complex symbol corresponding to subcarrier k of OFDM symbol n can be expressed with $d_k(n)$. Table 3.1 shows some main parameters used in 802.11a.

As shown in Table 3.1, each OFDM symbol contains 4 pilot subcarriers and 48 data subcarriers. The locations of the remaining 64 subcarriers are inserted with zero. Hence, the nth OFDM symbol is expressed as [3]

$$\begin{aligned} s_{\mathrm{DATA},n}(t) &= \sum_{k=0}^{N_{SD}-1} d_k(n)\exp(j2\pi M_k \Delta f(t - T_{GI})) \\ &\quad + p_{n+1} \sum_{k=-N_{ST}/2}^{N_{ST}/2} P_k \exp(j2\pi k \Delta f(t - T_{GI})) \end{aligned} \tag{3.7}$$

where the function, M_k, defines a mapping from the logical subcarrier number 0–47 into frequency offset index −26 to 26 as shown in (3.7), while skipping the pilot subcarrier locations and the 0th (DC) subcarrier. The subcarriers beyond the index −26 to 26 up to −32 to 32 are set to zero.

$$M_k = \begin{cases} k-26 & 0 \le k \le 4 \\ k-25 & 5 \le k \le 17 \\ k-24 & 18 \le k \le 23 \\ k-23 & 24 \le k \le 29 \\ k-22 & 30 \le k \le 42 \\ k-21 & 43 \le k \le 47 \end{cases} \tag{3.8}$$

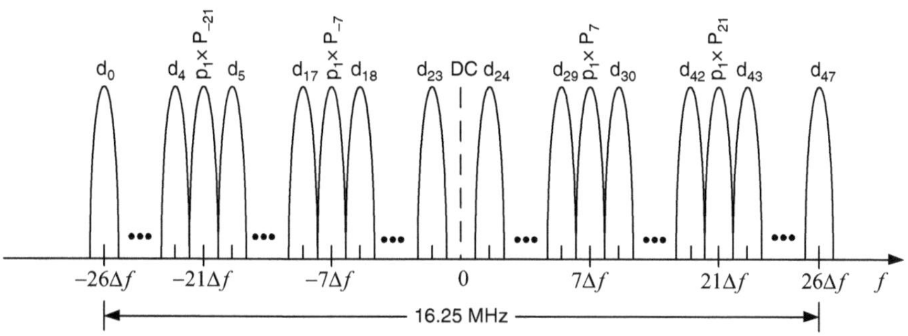

Fig. 3.4 Subcarrier frequency location for first data symbol, where pilot signals are located at the frequency indexes of −21, −7, 7, and 21

The four pilot subcarriers for the *n*th OFDM symbol are produced via Fourier transform of real sequence P, located at the index −21, −7, 7, and 21 and given by

$$\begin{aligned} P_{-26:26} = \{&0,0,0,0,1,0,0,0,0,0,0,0,0,0,0,0,0,0,1,0,0,0,0,0,0,0,0,0,0,0,\\ &0,0,1,0,0,0,0,0,0,0,0,0,0,0,0,0,-1,0,0,0,0,0\} \end{aligned} \tag{3.9}$$

To avoid the discrete lines at these four pilot locations, the polarity of the pilot subcarriers is BPSK modulated by a pseudo-binary sequence, p_{n+1}, which is a cyclic extension of the 127 elements given by

$$p_{0:126} = \{1,1,1,1,-1,-1,-1,1,\ldots,-1,-1,-1\} \tag{3.10}$$

Compared with other *M-ary* QAM formats, these pilot signals with the BPSK scheme require the lowest signal-to-noise ratio (SNR) to achieve a reliable demodulation and are therefore easily coherently demodulated at the receiver. For the detailed contents of (3.10), see (25) in [3]. Each sequence element is used to multiply four pilot subcarriers for the *n*th OFDM symbol for achieving random polarity. For example, the first element of $p_0 = 1$ multiplies four pilot subcarriers of the SIGNAL symbol, while the elements from p_1 on are used to multiply 4 pilot subcarriers for the DATA symbols.

Figure 3.4 shows that these four pilots, or P_{-21}, P_{-7}, P_7, and P_{21}, are individually multiplied with the first element p_1 in the first data symbol. In order to prevent the carrier signal from feeding through at the transmitter and to keep the received signal from being saturated at the receiver, the 0th subcarrier corresponding at DC component is not used.

After the orthogonal subcarriers are individually modulated by the data symbols, the spectrums of the baseband signals are distributed in the frequency range from the most negative subcarrier of −8.125 MHz (−26 × 0.3125) to the most positive subcarrier of 8.125 MHz (26 × 0.3125). The effectively utilized bandwidth is the multiplication of the subcarrier frequency spacing of 0.3125 MHz and the total number of subcarriers of 52, or $0.3125 \times 52 = 16.25$ MHz. Note that the locations of 12 subcarriers from -27 to -32 and from 27 to 32 are not used to avoid overlaps between the adjacent 20-MHz bands; otherwise the bandwidth for the total 64 subcarriers is 20 MHz. The spectrum of the subcarriers in the baseband frequency domain is then frequency-transferred into the spectrum of the RF frequency domain after the frequency up-conversion with the RF local oscillator.

Compared to FDM systems, OFDM system subcarriers are orthogonal to the other subcarriers in the frequency domain. Orthogonal means the peak of one subcarrier occurs at the nulls of the others. Therefore, OFDM signals are inter-carrier–interference (ICI) free and are densely compacted in the frequency domain. Orthogonal arrangement in the frequency domain makes OFDM systems' spectral efficiency greater compared to FDM systems.

As with intersymbol interference (ISI) in the time domain, being ICI-free in the frequency domain avoids performance degradation in the demodulation of the receiver. Figure 3.5 shows four subcarriers in both time and frequency domains, where all subcarriers are supposed to be modulated by data symbols with the same amplitude level, but in practice each subcarrier can be modulated by the data symbols with different amplitudes and phases. Note that each subcarrier has an exact integer number of cycles in the interval of one symbol, and the number of cycles between adjacent subcarriers differs by exactly one. In the frequency domain, at the peak of each subcarrier the ICI contributed from the other subcarriers is zero. This property in either the time domain or frequency domain is due to orthogonality among the subcarriers.

As shown in Fig. 3.5a, each subcarrier waveform is obtained by multiplying a sinusoidal signal with a squared-pulse window with the duration of one symbol in the time domain. The multiplication in the time domain becomes convolution in the frequency domain. The spectrum or Fourier transform of the windowed sinusoidal signal is a *Sinc* function $\tau \sin(\omega\tau/2)/(\omega\tau/2)$, where τ is the duration of the window or the duration of one OFDM symbol. Thus, four subcarriers result in four *Sinc*-shaped waveforms in the frequency domain as shown in Fig. 3.5b.

Mathematically, the expression of the complex baseband OFDM signal in (3.7) is in fact nothing more than the inverse Fourier transform of N_{SD} QAM input symbols [4]. In the discrete time domain, the inverse Fourier transform is the inverse discrete Fourier transform (IDFT). In practical hardware implementation, IDFT can be efficiently implemented by the inverse fast Fourier transform (IFFT) in order to reduce the mathematical operations used in the calculation of IDFT. Because the expression in (3.7) is completely identical to the IFFT operation, the transmitter needs the IFFT operation to transfer OFDM symbols from the frequency domain to the time domain. Similar to the transmitter, the receiver uses the fast Fourier transform (FFT) to transfer the OFDM symbol from the time domain back to the frequency domain in order to recover the original data.

By using the IDFT expression, the nth OFDM symbol in (3.7) can be expressed in the discrete time domain:

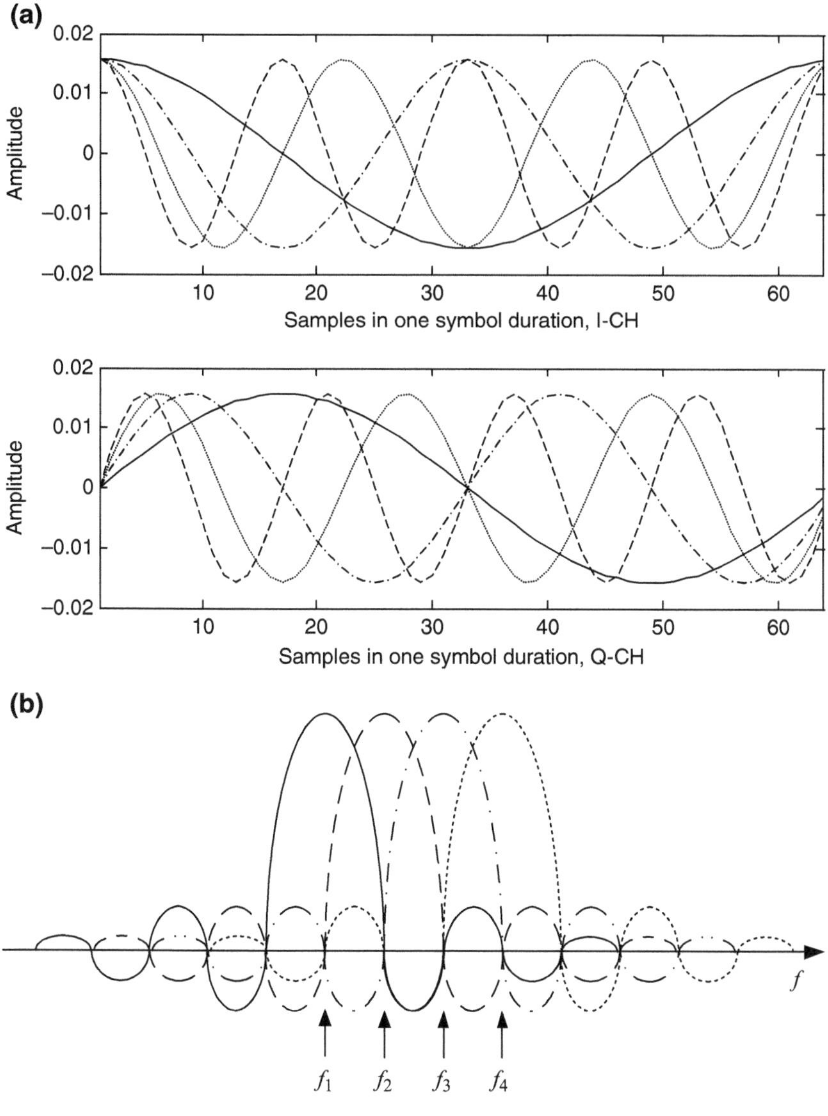

Fig. 3.5 Waveforms of four subcarriers, (**a**) in time domain within one symbol, and (**b**) in frequency domain, where the *solid-line waveform* represents the lowest subcarrier at f_1, while the *dot-line waveform* stands for the highest subcarrier at f_4

$$s_{\text{DATA}}(n) = \sum_{k=0}^{N_{\text{SD}}-1} d_k(n)\exp\left(j2\pi\frac{kn}{N_{\text{FFT}}}\right) + p(n+1)\sum_{k=0}^{N_{\text{ST}}-1} P_k\exp\left(j2\pi\frac{kn}{N_{\text{FFT}}}\right) \tag{3.11}$$

where the time t in the continuous time domain is replaced by a sample number n in the discrete time domain, and the subcarrier frequency spacing Δf is substituted by an N_{FFT}-point of IFFT operation.

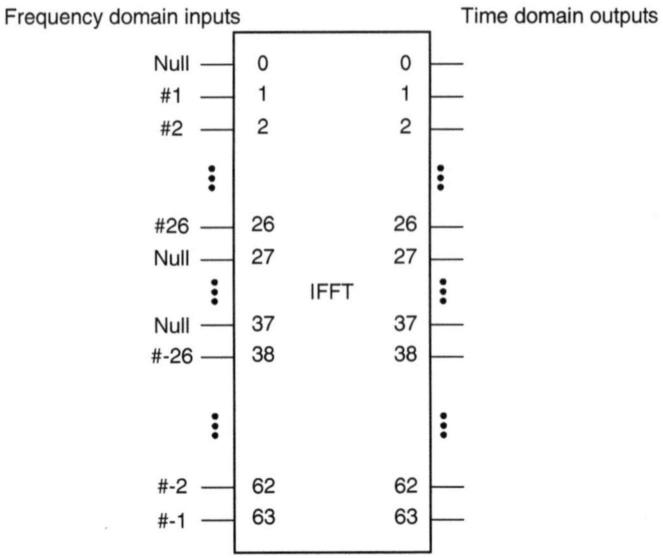

Fig. 3.6 Inputs and outputs of IDFT

In 802.11a WLAN specifications, a 64-point ($N_{\mathrm{FFT}} = 64$) is used for IFFT/FFT operations. Here, a 16-QAM OFDM symbol for the 36-Mbps data rate is used as an example. In 16-QAM mapping, every 4 bits are mapped into 1 complex symbol. The interleaved 192 bits shown in Table G.21 of [3] are mapped into 48 symbols, given in Table G.22 of [3]. Like the expression in (3.7), the first mapped symbol is tagged with the index −26 or subcarrier $-26\Delta f$, the second with the index −25 (with an increasing step of 1) and the last one (48th symbol) is labeled with the index 26, note that four pilot symbols are inserted at the index of −21, −7, 7, and 21, and zero is inserted at the index 0 (or DC) as shown in Table G.22 of [3]. To perform an IFFT calculation, the symbols with the index 1–26 are sent to the same numbered IFFT inputs, while the symbols with index −26 to −1 are copied into IFFT inputs 38–63. The rest of the inputs, 27–37 and the 0 (DC), are set to zero, as illustrated in Fig. 3.6. The indexes shown in Fig. 3.6 represent the center frequencies of the subcarriers, such as #−26, corresponding to the subcarrier frequency of $-26\Delta f$ shown in Fig. 3.4. After the IFFT, data at the index 0 represent the first output in one symbol duration of time domain, while data at the index 63 represent the last output in one symbol duration of time domain. The real and imagined parts of the first OFDM complex symbol in one symbol interval with guard interval (GI) are illustrated in Fig. 3.7.

The I and Q baseband signals shown in Fig. 3.7 are obtained by summing all 52 subcarriers modulated with 52 symbols, each having different amplitudes and phases. The composition of OFDM subcarriers with different amplitude and phases makes OFDM signals behave like Gaussian noise in the time domain. It will be shown in the later section of this chapter that OFDM signals have larger peak-to-average power ratios (PAPR), which are similar to the PAPR of Gaussian noise.

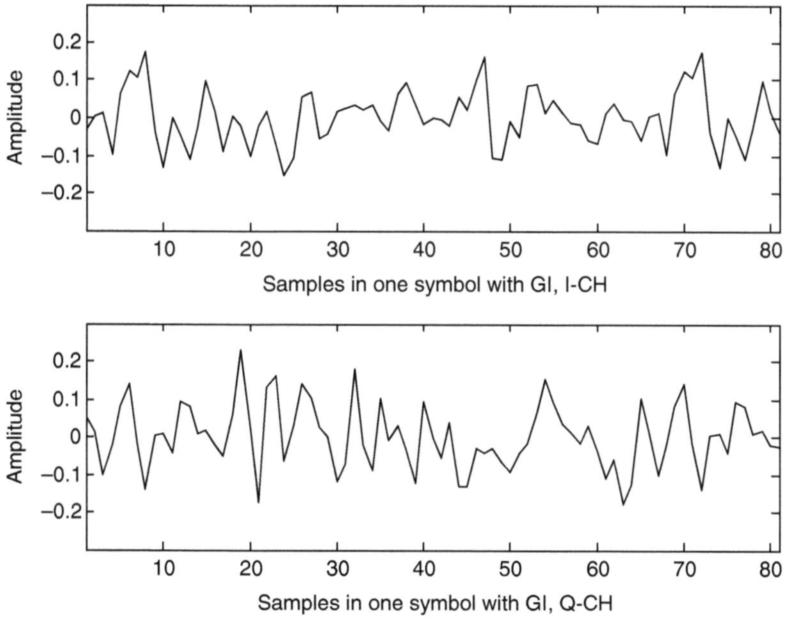

Fig. 3.7 Baseband signals in I and Q channels, where there are 64 samples in one IFFT symbol and 16 samples in one guard interval GI

To keep the orthogonal property and eliminate ICI between subcarriers in multipath fading, a guard time is introduced for each OFDM symbol. In the 802.11a system, the last 16 samples of an OFDM IFFT symbol with 64 samples in the time domain are added to the beginning of the OFDM IFFT symbol to form one extended OFDM IFFT symbol with the cyclic prefix having a total of 80 samples as shown in Fig. 3.7. The interval of the cyclic prefix is called the guard interval T_{GI}, which is equal to $T_{\mathrm{FFT}}/4$ or 16 samples.

Figure 3.8 illustrates three subcarriers in the I channel arriving at the input of the receiver through a two-ray channel, where the dashed curve is a delayed replica of each solid curve. Each delayed curve is attenuated by α due to a longer traveling path relative to the main path. In the figure, it is assumed that the data for all three subcarriers have the same magnitudes, but may have different polarities during the current and previous symbols. For example, the phase of subcarrier n has a continuous transition before the cyclic prefix is added (as shown Fig. 3.8) at points (a) and (b) from the previous symbol to the current symbol, assuming that they have the same symbol polarities. A $-90°$ phase transition, however, occurs at point (a) after the cyclic prefix is added to the beginning of the current symbol as a thick segment. The phase of subcarrier $n+1$ changes $180°$ at points (a) and (b) before the cyclic prefix is added, assuming that they have different polarities. The phase, however, is continuous at point (a) after the cyclic prefix is inserted into the beginning of the current symbol.

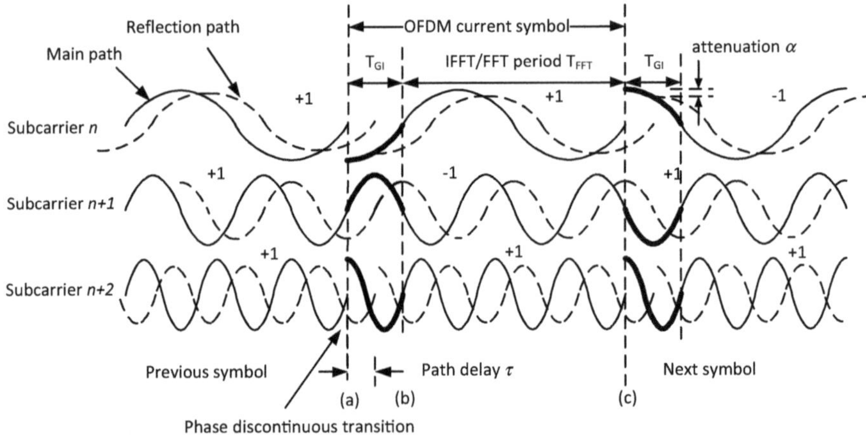

Fig. 3.8 OFDM signals in I-channel with three subcarriers that are modulated by data series through a two-ray multipath channel, where data series are listed on each subcarrier at different symbol durations

As long as the delay τ is smaller than the guard time T_{GI}, multipath signals cannot cause ICI in the frequency domain after an FFT operation in the receiver. In this case, there are no phase transitions during the FFT period T_{FFT}. The sum of sinusoidal signals with the same frequency but with different magnitudes and phases is still a pure sinusoidal signal with the same frequencies and resultant magnitudes and phases. Therefore, the summation of each subcarrier does not destroy the orthogonality between the subcarriers. However, the summation does damage the orthogonality between the subcarriers if the multipath delay is larger than the guard time. In this case, the phase transitions of the delayed subcarriers fall within the FFT interval at the receiver and the summation of sinusoidal signals is no long a pure sinusoidal signal. The disadvantage of the cyclic prefix is that it takes up system capacity and therefore reduces the overall data rate because it occupies the positions of information data in the time domain.

After cyclically extending and then windowing each IFFT symbol, the complex OFDM symbols are split into the real and image parts through a P/S converter, and converted to the analog signals through DACs. Figure 3.9 illustrates a detailed block diagram for a 802.11a WLAN system with 16-QAM and 64-point IFFT/FFT operation.

The total N_{sym} data symbols can be expressed by concatenating each data symbol in the time domain as shown in (3.7) after the cyclic prefix is added:

$$\bar{s}_{\mathrm{DATA}}(t) = \sum_{n=0}^{N_{\mathrm{sym}}-1} \bar{s}_{\mathrm{DATA},n}\left(t - nT_{\mathrm{sym}}\right) \tag{3.12}$$

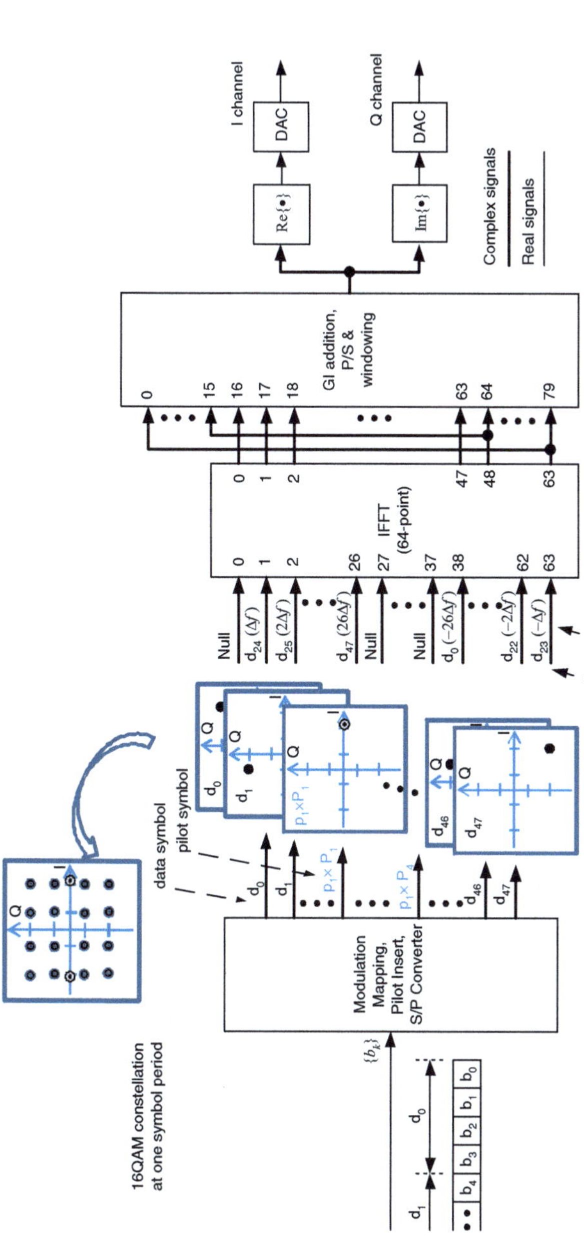

Fig. 3.9 Block diagram of a detailed OFDM modulator in the transmitter, where data symbols use 16-QAM format and pilots use BPSK format

where $\bar{s}_{\mathrm{DATA},n}(t)$ represents the nth extended OFDM symbol $s_{\mathrm{DATA},n}(t)$ by the cyclic prefix. The OFDM data will be concatenated with the preamble symbols and SIGNAL filed symbol to form a complete frame as expressed in (3.1).

3.2.4 Spectral Side-Lobe Reduction With Windowing

It can be seen from Fig. 3.8 that the subcarrier n on the main path has sharp phase transitions at the symbol boundaries of points (a) and (c) due to the cyclic prefixes inserted at the beginning of the symbols. Such discontinuous phase transitions can be also seen on the subcarrier $n+2$. Discontinuous phase transition of the carrier or subcarrier signal causes the spectral side-lobes to drop slowly. As a result, the spectral side-lobes with slowly roll-off could cause adjacent channel interference (ACI) due to power leakage into adjacent channels.

To make the spectral side-lobes roll off more rapidly, the phase transitions of the subcarriers at the boundaries between the OFDM symbols including the prefix should be as smooth as possible. The smoother the phase transition, the faster the side-lobes roll off. An effective method to make the phase transition smoother is to apply a windowing function to the individual OFDM symbols. Windowing an OFDM symbol forces the subcarrier amplitude to go smoothly to zero at the symbol boundaries, which equivalently makes the subcarrier phase transition go smoothly to zero. A widely used window function in the WLAN OFDM standards is the Tukey window [5], also known as the tapered cosine window, which is defined as

$$w(t) = \begin{cases} 0.5\{1 - \cos(\pi t/T_{\mathrm{TR}})\} & 0 \le t \le T_{\mathrm{TR}} \\ 1 & T_{\mathrm{TR}} < t \le T_{\mathrm{SYM}} \\ 0.5\{1 + \cos[\pi(t - T_{\mathrm{SYM}})/T_{\mathrm{TR}}]\} & T_{\mathrm{SYM}} < t \le T_{\mathrm{SYM}} + T_{\mathrm{TR}} \\ 0 & \text{otherwise} \end{cases} \tag{3.13}$$

where T_{TR} is the transition interval and T_{SYM} is the OFDM symbol interval, including the cyclic prefix interval as shown in Fig. 3.10. When the transition interval T_{TR} vanishes, the windowing function becomes a rectangular pulse with a duration of T_{SYM}. When T_{TR} does not vanish, the windowing function is a Tukey windowing waveform with a transition duration of T_{TR} at both sides.

In the 802.11a system, an OFDM symbol duration of T_{SYM} consists of an IFFT period of T_{IFFT} and a cyclic prefix interval of T_{prefix} or a guard interval of T_{GI} or $T_{\mathrm{SYM}} = T_{\mathrm{IFFT}} + T_{\mathrm{GI}}$. To properly apply the windowing function to OFDM symbols, a cyclic suffix must be appended to the end of the OFDM symbol by replicating the first N_{suffix} samples of a FFT/IFFT symbol. Here N_{suffix} samples occupy the interval of T_{suffix}. The last N_{prefix} samples of the OFDM symbol are inserted at the beginning of the FFT/IFFT symbol. Then, the OFDM symbol is multiplied by a Tukey window $w(t)$ to smooth the phase transitions of the OFDM subcarriers. However, in order to comply with the 802.11a standard, a symbol

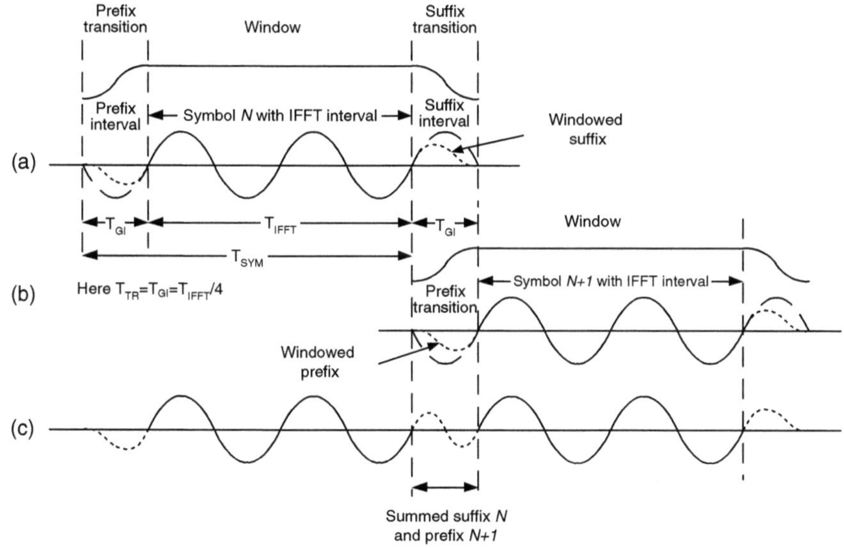

Fig. 3.10 Windowed OFDM symbols in time domain: (**a**) the Nth OFDM symbol with prefix and suffix intervals, (**b**) the $(N+1)$th OFDM symbol with prefix and suffix intervals, and (**c**) cascaded symbol N and symbol $N+1$

cannot be arbitrarily lengthened. Instead, the cyclic suffix shown in Fig. 3.10a must overlap in time and be summed with the cyclic prefix of the following symbol shown in Fig. 3.10b. It can be seen from Fig. 3.10a that the window function that decreases from an amplitude of 1 to 0 over the suffix transition duration forces the phase transition of the windowed OFDM symbol N continuously, and smoothly goes from its original value to zero; and from Fig. 3.10b, that the window function that increases from amplitude 0 to 1 over the prefix transition duration forces the phase transition of the windowed OFDM symbol $N+1$ continuously, and smoothly goes from zero to its final value. After these two OFDM symbols are overlapped in the transition duration, a continuous and smooth phase transition from one symbol to the next is generated as shown in Fig. 3.10c.

Figure 3.11 shows the curves of power spectral density versus different transition durations for the 802.11a OFDM signal sampled at a rate of 40 MHz, where the OFDM symbols are generated by taking a 128-point IFFT operation, and each OFDM symbol has both a 32-sample prefix and 32-sample suffix. It can be seen from Fig. 3.11 that power spectral density (PSD) with a rectangular window corresponding to $T_{TR}=0$ has a little margin at the frequency offsets of ± 20 MHz due to discontinuous phase transitions between OFDM symbols, and would violate the PSD mask of -40 dBc at the frequency offset of ± 30 MHz (beyond view range). When the transition duration T_{TR} is less than T_{GI} and is not equal to zero (such as a ten-sample-long duration), the spectral side-lobes drop faster because of continuous phase transitions between OFDM symbols. When the transition duration T_{TR} is equal to either the prefix or suffix interval, the spectral

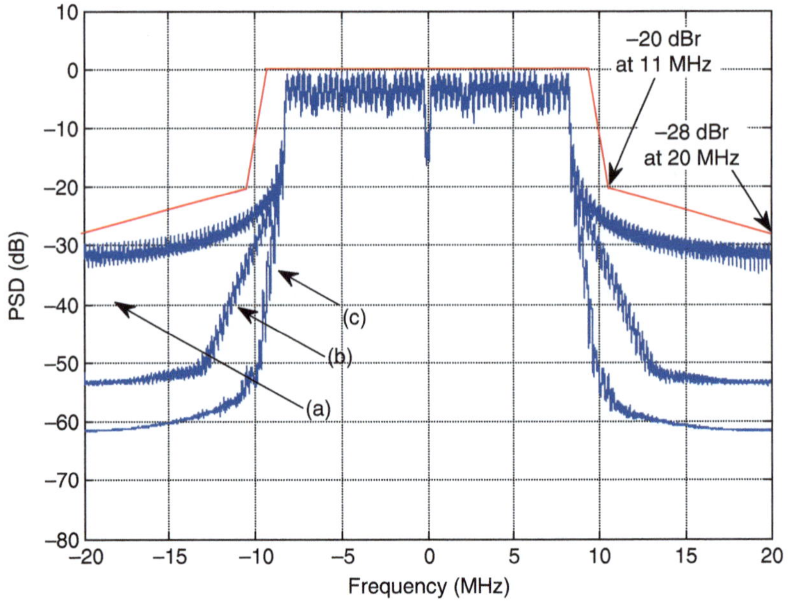

Fig. 3.11 Power spectral density of the 802.11a OFDM signal with a windowing function having different transition durations: (**a**) rectangular window, (**b**) Tukey window with roll off length of ten samples at each side, and (**c**) Tukey window with roll off length of 32 samples at each side

side-lobes roll off even faster due to smoother continuous phase transition between OFDM symbols. In practice, the transition duration is determined by the actual application's purpose and implementation requirements.

In addition to windowing, using digitally filtering techniques can also attenuate the spectral side-lobes. However, windowing is much simpler and has a low-cost implementation without distorting the amplitudes and phases of the OFDM subcarriers compared to filtering methods. Therefore, the widowing method is widely used in hardware implementation designs.

Transmit Modulation Accuracy: The basic principle of the digital modulation is a frequency transfer process in which digital bits are carried by an RF carrier by varying the carrier's magnitude and phase through the modulation. The modulated carrier signal has the same power spectral density (PSD) shape as that of the baseband modulation signal, but occupies twice baseband signal bandwidth. For a quadrature modulation, the digital bits are usually mapped or converted into complex numbers representing different modulation constellation points on the I and Q plane before modulating a pair of quadrature carriers.

In OFDM signal transmission, each mapped complex symbol modulates a corresponding complex subcarrier as described previously to form the modulated OFDM signal in the subcarrier frequency domain. The subcarrier frequency is located from negative frequency to positive frequency and is symmetric around zero frequency. After passing through a pair of DACs, the subcarrier OFDM I–Q signals are transferred onto the RF signal by multiplying a pair of RF orthogonal carriers. In non-OFDM signal transmission, such as WCDMA signal transmission, the spread code baseband I–Q signals are directly converted into the RF signal without modulating subcarriers by multiplying a pair of RF orthogonal carriers after being passed through a pair of construction filters.

In either OFDM or non-OFDM signal transmission, the baseband modulation I–Q signals carried by the RF signal may be distorted due to the I–Q imbalance, DC offsets, LO phase noise, and nonlinear amplification along a transmission chain either before or after the RF modulation. The quality of the RF modulated signal can be measured by the error vector magnitude (EVM). EVM, however, is calculated at the baseband domain by comparing the vector difference between the actual signal vector and the reference signal vector. Hence, the RF modulated signal should be down-converted to the baseband signal before EVM calculation. Most vector signal analyzers and spectrum analyzers can perform EVM measurement in either time or frequency domains. The concepts of the EVM measurement and calculation are introduced in Appendix B. The interested reader can refer to Appendix B in detail.

3.2.5 RF Transmitter Description

After IFFT operation, the baseband I–Q signals are up-converted to the RF signal through a RF quadrature modulator for transmission. In each frame, the preamble, including short and long symbols, is generated using an OFDM BPSK modulation with the specified waveform. The baseband signal of the SIGNAL field in the time domain is created through IFFT operation with BPSK mapping. The baseband signal of the DATA field is produced via IFFT operation with an *M-ary* QAM modulation mapping format, depending on the data rate. In this example, the 16-QAM modulation mapping format is used for the rate of 36 Mbits/s. These subframes are concatenated in the time domain to yield a frame expressed in (3.1). Finally, the real and imaginary parts of the complete frame modulate a pair of LO quadrature carrier signals for the frequency transfer from a BB domain to a RF domain after passing through DACs and then lowpass filters. The power amplifier (PA) driver has a variable gain range to achieve different applications and its output signal can be further amplified via a PA to meet the needs of long-distance transmission.

A general block diagram of the transmitter for the 802.11a OFDM system is illustrated in Fig. 3.12. Some major specifications for the transmitter and receiver

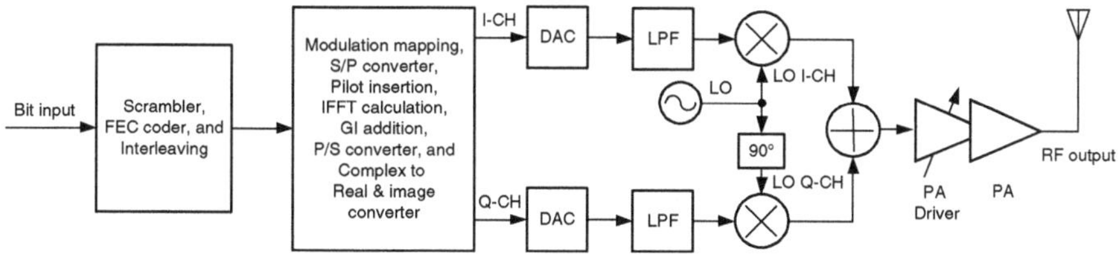

Fig. 3.12 Block diagram of the transmitter

Table 3.2 Major parameters of 802.11a transmitter and receiver

Data rate (Mbits/s)	Modulation mapping per carrier	Error correction code rate	TX EVM (dB)	RX minimum sensitivity (dBm)
6	BPSK	1/2	−5	−82
9	BPSK	3/4	−8	−81
12	QPSK	1/2	−10	−29
18	QPSK	3/4	−13	−77
24	16-QAM	1/2	−16	−74
36	16-QAM	3/4	−19	−70
48	64-QAM	2/3	−22	−66
54	64-QAM	3/4	−25	−65

Table 3.3 Operating channel numbers, bands, and channel center frequencies in US

County	Band (GHz)	Operating channel numbers	Channel center frequencies (MHz)	Max output power (mW/dBm)
United States	Low band (5.15–5.25)	36	5180	40/16
		40	5200	
		44	5220	
		48	5240	
United States	Middle band (5.25–5.35)	52	5260	200/23
		56	5280	
		60	5300	
		64	5320	
United States	Upper band (5.725–5.825)	149	5745	800/29
		153	5765	
		157	5785	
		161	5805	

are listed in Table 3.2. An 802.11a OFDM system shall operate in the 5-GHz band with a bandwidth of 20 MHz within three bands as listed in Table 3.3. The transmitter and receiver share the same frequency band but operate at different times with a slot time of less than 9 μs.

Two major specifications requested by the transmitter are the transmitting spectrum mask and the transmitting modulation accuracy or error vector magnitude (EVM). The PSD of the transmitted signal shall fall within the spectral mask to avoid the interference with the adjacent channels, while the EVM shall be less than the required error with a certain margin for a certain modulation format to ensure that the transmitted signal is high quality.

Both PSD and EVM are affected by some impairments in the transmit chain and some trade-offs as well. In order to achieve the required PSD and EVM characteristics with enough margins, some special attention needs to be paid to the signal and circuit designs. The impairments and tradeoffs are described in the next section.

- *Effect of Peak Factor Reduction on PSD and EVM.*
  Due to the very high peak-to-average power ratio (PAPR) of the OFDM signal, which reduces the efficiency of the RF PA and increases the complexity of the ADCs and DACs, peak-to-average power ratio reduction is usually taken in the digital domain. However, either PAPR reduction or peak clipping also causes PSD regrowth and leads to EVM degradation. Therefore, the trade-off between peak clipping and PSD spreading should be made.

- *Nonlinearity Effect of Transmit Chain on PSD and EVM.*
  Nonlinearity of the circuits in the transmit path, such as a RF quadrature modulator and a PA, may cause PSD regrowth and EVM degradation. Both the RF modulator and PA driver should be designed with high linearity to prevent either PSD regrowth or EVM degradation, while the PA should be designed to operate at a class AB mode to achieve relatively high efficiency. The pre-distortion (PD) linearization techniques may be needed in either digital or analog domain to achieve both spectral and energy efficiency in the class AB mode.

- *Memory Effect of PA on PSD symmetry.*
  In a memory PA, the third-order intermodulation (IM3) tones may be asymmetric and their response is dependent of the frequency offset from the carrier frequency. It is undesired to have the memory effects on PSD when one side has lower side-lobes of PSD with high margins and the other side has higher side-lobes with low margins. The memory effects of PA can arise from multiple sources including bias circuit effects self-heating, and trapping effects. Therefore, it is necessary to minimize the memory effects of the PA.

- *Effect of I–Q Gain and Phase Imbalance of RF Modulator on EVM.*
  The gain and phase imbalances on LO I–Q paths and I–Q mixers of RF modulator in Fig. 3.12 are two dominant sources [6] that cause EVM degradation. The simplest method to correct them is to use the I–Q calibration. During calibration, imbalance errors can be detected through a loop back from the RF transmitter path to the receiver baseband path and then be digitally compensated at the baseband [6, 7].

- *Effect of VCO Phase Noise on EVM.*
  The phase noise of VCO can degrade the EVM performance after the baseband I–Q signals modulate a pair of LO quadrature carriers that are obtained from VCO in the transmitter. Therefore, the low phase noise of VCO is required to achieve the low EVM value, especially for high-order QAM modulation formats, such as 64-QAM for the data rate of 54 Mbits/s.

Figure 3.13 illustrates the measured PSD of an OFDM 64QAM signal with the data rate of 54 Mbits/s and 20 MHz bandwidth at the output of the 802.11n transceiver chip and Fig. 3.14 shows its constellation. The measured EVM is about -39 dB at the transmitted power of -5 dBm, where dots at the center of each quadrature cross bars (total 64) present the measured data symbols. The two dots on the X axis are pilot symbols used for transmission channel

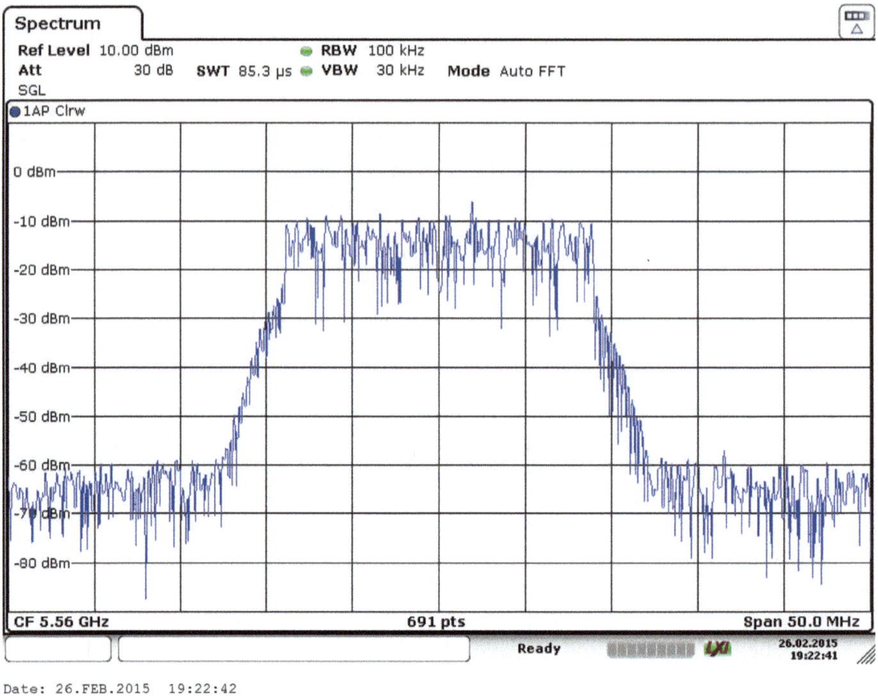

Fig. 3.13 PSD of 64QAM in the legacy 802.11a mode of 802.11n, where a test channel is i112 at 5.56 GHz with $B_{\mathrm{w}} = 20$ MHz, and output PWR $= -5$ dBm

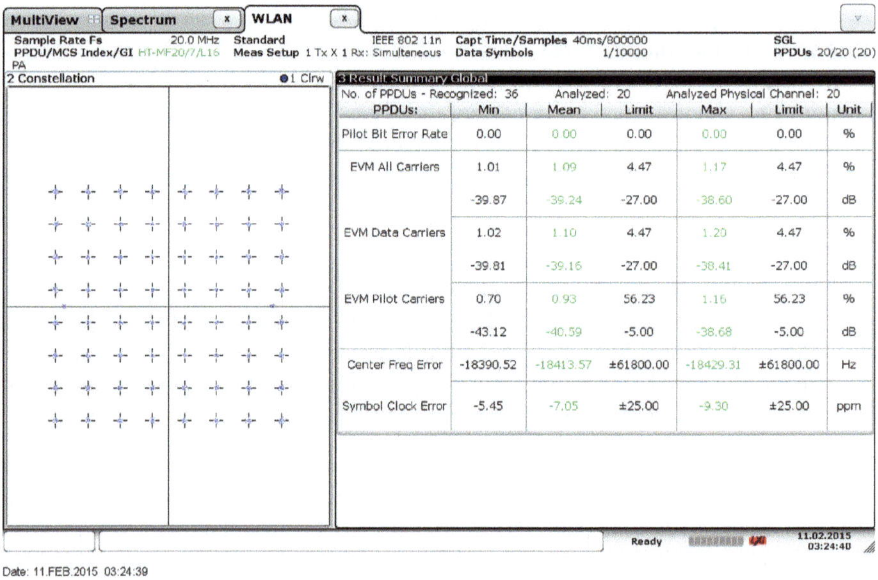

Fig. 3.14 EVM of 64QAM at the rate m7 in the legacy 802.11a mode of 802.11n

estimation. Such a low RMS EVM value of -39 dB represents that the transmitter behaves very low I–Q imbalance, very low VCO phase noise, and very low nonlinear distortion.

3.2.6 Peak-to-Average Power Ratio (PAPR)

As discussed previously, an OFDM signal in one OFDM symbol period is generated by summing N (here 52) parallel subcarriers that are individually modulated by N data symbols, each having different phases and amplitudes. This process results in a large peak-to-average power ratio. The PAPR of a given signal $s(t)$ is defined as the ratio of the peak power of $s(t)$ to its average power

$$\mathrm{PAPR} = 10\log_{10}\left(\frac{P_{\mathrm{peak}}}{P_{\mathrm{avg}}}\right) \tag{3.14}$$

For a sinusoidal signal $s(t) = A\cos(2\pi F_{\mathrm{c}}t)$, the PAPR can be estimated as:

$$\begin{aligned}\mathrm{PAPR} &= 10\log_{10}\left(\frac{P_{\mathrm{peak}}}{P_{\mathrm{avg}}}\right) = 10\log_{10}\left(\frac{\max\{s(t)\times s^*(t)\}}{E\{s(t)\times s^*(t)\}}\right)\\ &= 10\log_{10}\left(\frac{\max\{|s(t)|^2\}}{E\{|s(t)|^2\}}\right) = 10\log_{10}\left(\frac{\max\{|s(t)|^2\}}{\frac{1}{T_{\mathrm{c}}}\int_0^{T_{\mathrm{c}}}|s(t)|^2 dt}\right)\end{aligned} \tag{3.15}$$

where $F_{\mathrm{c}} = 1/T_{\mathrm{c}}$ is the carrier frequency, T_{c} is the period, and the * operator represents complex conjugate. P_{peak} can be simply calculated as:

$$P_{\mathrm{peak}} = \max\{|A^2\cos^2(2\pi F_{\mathrm{c}}t)|\} = A^2 \tag{3.16}$$

The average power of $s(t)$ is obtained as:

$$\begin{aligned}P_{\mathrm{avg}} &= \frac{1}{T_{\mathrm{c}}}\int_0^{T_{\mathrm{c}}} A^2\cos^2(2\pi F_{\mathrm{c}}t)dt\\ &= \frac{1}{T_{\mathrm{c}}}\int_0^{T_{\mathrm{c}}} A^2\left[\frac{1}{2}+\frac{1}{2}\cos(4\pi F_{\mathrm{c}}t)\right]dt = \frac{A^2}{2}\end{aligned} \tag{3.17}$$

The PAPR of $s(t)$ is then estimated as

$$\mathrm{PAPR} = 10\log_{10}\left(\frac{A^2}{A^2/2}\right) = 3\,\mathrm{dB} \tag{3.18}$$

For a complex sinusoidal signal $s(t) = A\,e^{j2\pi F_c t}$, the peak value of the signal is

$$P_{\text{peak}} = \max\left\{Ae^{j2\pi F_c t}Ae^{-j2\pi F_c t}\right\} = A^2 \quad (3.19)$$

The mean squared value of the signal is

$$P_{\text{avg}} = E\left\{\left|Ae^{j2\pi F_c t}\right|^2\right\} = \frac{A^2}{T_c}\int_0^{T_c}\left|e^{j2\pi F_c t}\right|^2 dt = A^2 \quad (3.20)$$

Then, the PAPR of a complex sinusoidal signal is

$$\text{PAPR} = 10\log_{10}\left(\frac{P_{\text{peak}}}{P_{\text{avg}}}\right) = 0\,\text{dB} \quad (3.21)$$

Equation (3.15) can be also used to calculate the PAPR of the RF OFDM signal, where $|s(t)|$ represents the envelope of the RF OFDM signal.

In practice, power complementary cumulative distribution function (CCDF) curves provide important and critical information regarding to the PAPR and the probability for that particular PAPR value. A CCDF curve shows how many percentage of time the signal spends at or above a given PAPR value. In most standard specifications, the percentage of 0.01% is used to determine the PAPR value of the signal. Figure 3.15 shows CCDF curves of three different signals,

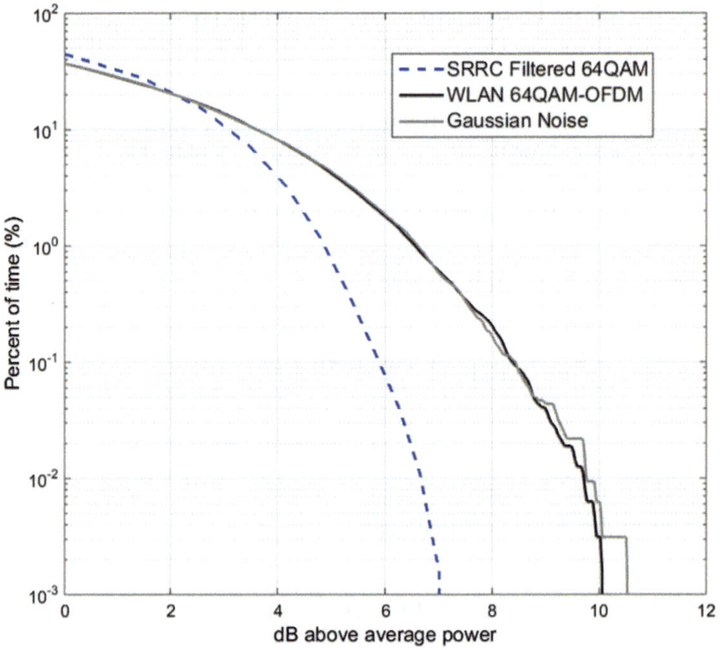

Fig. 3.15 CCDF curves of a 64-QAM signal, a WLAN 64-QAM OFDM signal and Gaussian noise

which are single carrier modulated by the SRRC filtered 64QAM baseband signal, a 64QAM OFDM signal with 52 subcarriers and Gaussian noise. At the percentage of time 0.01% on *y*-axis, the PAPR values of the single carrier 64QAM, WLAN 64QAM OFDM signal, and Gaussian noise are about 6.5, 9.7 and 9.8 dB, respectively. For the WLAN 64QAM OFDM signal, this means the signal power exceeds the average power by at least 9.7 dB for 0.01% of the time.

It can be seen that a 64QAM OFDM signal used in the 802.11a standard system appears to have higher PAPR value than a single carrier based 64QAM signal and similar PAPR value to Gaussian noise. OFDM signals, therefore, can also be called Gaussian noise-like signals because both signals have similar characteristics in the time and frequency domains.

As a rule of thumb, in order to avoid degradations of the adjacent channel power ratio (ACPR) and EVM, the power amplifier should operate at back-off by a PAPR value from its P1dB compression point. Backing-off the amplifier output power, however, significantly decreases energy efficiency. In order to achieve high efficiency, it is preferred that the input modulated signal of the PA has constant envelope or small PAPR value.

The modulation format of a signal affects its power characteristics. Using CCDF curves, we can fully characterize the power statistics of different modulation formats and decide how many dB back-off from the P1dB point of a power amplifier is needed. From Fig. 3.15, the OFDM signal needs a 9.7-dB back-off from the PA's P1dB point while the filtered 64QAM signal requires 6.5-dB back-off. Therefore, it is more challenge for the OFDM signal to achieve a high efficiency compared with the filtered 64QAM signal without causing the PA compression. The PA compression results in the degradations of ACPR and EVM due to the output signal compression.

In order to achieve high efficiency of power amplifiers, it is necessary to reduce the PAPR value, allowing smaller back-off from the P1dB point and therefore transmitting higher average power. PAPR can be reduced by using Crest Factor Reduction (CFR) technique, which is often used to limit the peak values of the transmitted signals in wireless communications and other applications.

Crest factor (CF) is the ratio of the peak amplitude of the waveform to its root-mean-square (RMS) value, while PAPR value is the ratio of the peak amplitude *squared* (giving the peak *power*) of the waveform to its RMS value *squared* (giving the average *power*). Thus, the PAPR is the square of the CFR. When expressed in decibels, CF and PAPR are equivalent due to the way decibels are calculated for power ratio versus amplitude ratio.

Considering the fact that a larger PAPR occurs infrequently, it is possible to attenuate these peaks only at the cost of a slight amount of self-distortion. There are a few peak reduction techniques to reduce the peaks. Two relatively simple methods are clipping and peak window, and peak cancellation.

Clipping and Peak Window (CPW): The procedure of the clipping and peak window is to clip the peaks of the signal first whenever the peaks are above a threshold level. As a result, a kind of self-distortion is introduced, which significantly increases the out-of-band PSD spreading and degrades the ACPR and EVM.

To remedy the out-of-band spreading due to clipping, a different approach with a certain non-rectangular window is used to multiply the peaks of the signal [4]. Some windows, like a Gaussian shaped, Cosine and Kaiser, can be used provided that they have good spectral shapes. To minimize the out-of-band spreading, the window should be as narrowband as possible.

The problem of clipping and peak window is to result in a certain amount of out-of-band spectral spreading due to its nonlinear peak clipping. To avoid the out-of-band spreading, a linear peak cancelation technique called peak cancellation is preferred. This technique was first published in [8], and then independently developed in [9], and described in [4].

Peak Cancellation (PC): Instead of multiplying the peaks of the signal with a peak window function having amplitude less than 1 in the CPW method, peak cancellation is performed by subtracting a time-shifted and scaled reference function from the peaks of the signal. Such subtractions in the time domain may be repeatedly carried out many times to reach the desired PAPR value. The subtraction between the signal and reference function in the time domain results in the addition of their spectral functions in the frequency domain. As long as the reference function has approximately the same bandwidth as the transmitted signal, the peak cancellation does not cause any out-of-band spectral spreading.

One of suitable reference functions is a *Sinc function,* which is obtained from a rectangular function shape having approximately the same bandwidth as the transmitted OFDM signal in the frequency domain. To limit the infinite length of the *Sinc* function to being the same as the interval of one OFDM symbol plus the cyclic prefix, the *Sinc* function is multiplied with a Tukey window function, which is the same as one used for windowing OFDM symbols. Thus, the windowed reference function has the same bandwidth as the OFDM signals [4, 9]. Therefore, the PC method will not degrade the out-of-band spectrum properties.

3.3 Synchronization of 802.11a OFDM Signal

In an OFDM receiver, synchronization needs to be performed before the OFDM demodulation. Synchronization includes two major synchronization tasks: symbol timing and carrier frequency synchronizations. First, since the propagation delay from the transmitter to the receiver and time difference between them are generally unknown in the receiver, symbol boundary and symbol timing must be derived from the received OFDM signal to minimize the effects of intersymbol interference (ISI) and other interferences. Second, since the propagation delay in the transmitted signal and random carrier phase generated in the transmitter may result in a carrier frequency offset and phase shift, the carrier phase shift and frequency offset should be estimated in the receiver with coherent detection.

Fortunately, both symbol timing and carrier frequency synchronizations can be performed by means of the properties of the training sequences and pilot signals carried by the transmitted OFDM signal. In this section, these two kinds of synchronization techniques are introduced.

3.3.1 Symbol Timing Synchronization

In the 802.11a WLAN, each frame starts with the preamble filed, which consists of ten short training symbols and two long training symbols used for synchronization. Symbol timing synchronization includes frame detection and symbol timing synchronization. They all rely on the training sequences. In the short training symbols as shown in Fig. 3.1, the first six symbols may be used for the signal detection and AGC setting, and the last four symbols can be utilized for coarse frequency offset estimation and timing synchronization. In addition, the symbol timing synchronization should be done within these ten short training symbols.

Since the training symbols are known in the receiver, the correlation property can be exploited for both symbol and carrier frequency synchronization by performing the correlation between the received signal and the known training symbols in the receiver. The known training symbols can be either locally generated ones or delayed the received training symbols. However, it would be better to use the locally generated training symbols due to being free of noise. The correlation is called *a cross-correlation* when the local training symbols are used to correlate with the received signal. It is called *an auto-correlation* when the delayed training symbols of the received signal are used to perform the correlation.

At the receiver, the received signal $r(n)$ is correlated with one local short symbol $s(n)$ and the correlator output $R_{\mathrm{cr}}(n)$ in the discrete time domain is written as

$$R_{\mathrm{cr}}(n) = \sum_{k=0}^{L-1} r(n-k)s^{*}(k) \tag{3.22}$$

where the symbol * represents the complex conjugate and L is the number of the samples in one short symbol.

Figure 3.16 shows a block diagram of the cross-correlation between the received signal $r(n)$ and one local short symbol $s(n)$, where T is the sample interval. This structure is a matched filter that correlates the input signal with the known short symbol. Since the preamble filed in the 802.11a standard has 10 short symbols with each having 16 samples, the cross-correlation output $R_{\mathrm{cr}}(n)$ contains 10 peaks as shown in Fig. 3.17. Each peak indicates that one of the ten short symbols in the received signal is completely aligned with the local short symbol at the corresponding sampling point or the last sample of each short symbol in the received signal. Thus, the symbol timing information can be obtained. Because of this unique property of the training symbol pattern, the symbol timing synchronization with the cross-correlation method is reliable even in noise environments.

The frame can be detected by comparing the peak of the cross-correlation with a threshold within the range of several peaks, such as the first three peaks. To indicate the tenth peak, or last peak, an auto-correlation can be performed by correlating the received signal with itself having a delay L of one short symbol, or

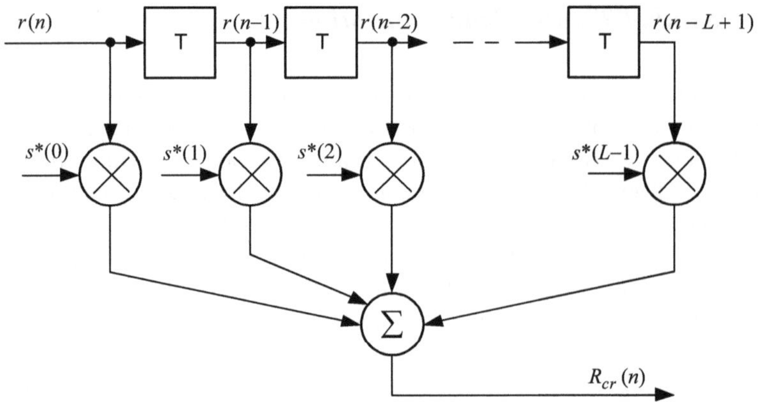

Fig. 3.16 Block diagram of a cross-correlator

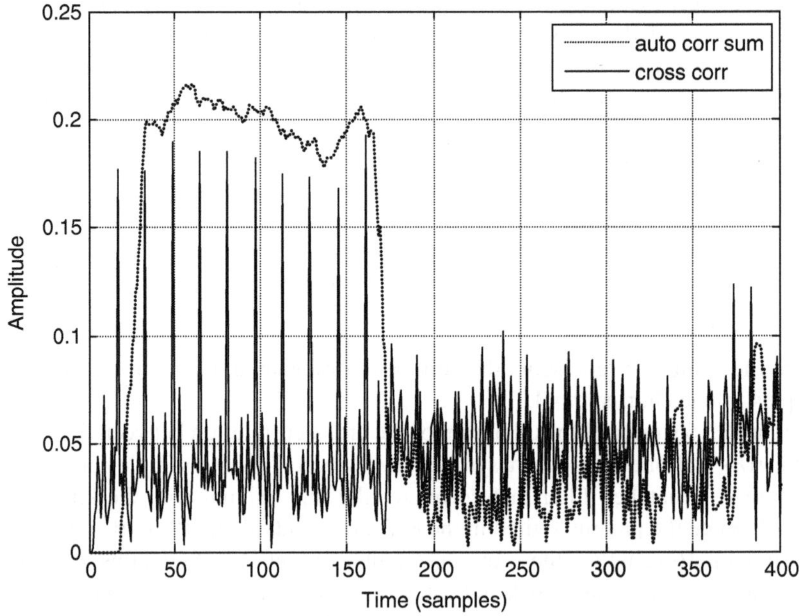

Fig. 3.17 Correlation with one local short symbol at SNR = 15 dB, where the received signal consists of ten short symbols, two long symbols, and one SIGNAL field

$$R_{\mathrm{au}}(n) = \sum_{k=0}^{L-1} r(n-k)r^*(n-k-L) \tag{3.23}$$

Figure 3.18 shows a block diagram of auto-correlation, where a moving sum block has the length of L, or the number of samples in one short symbol. The

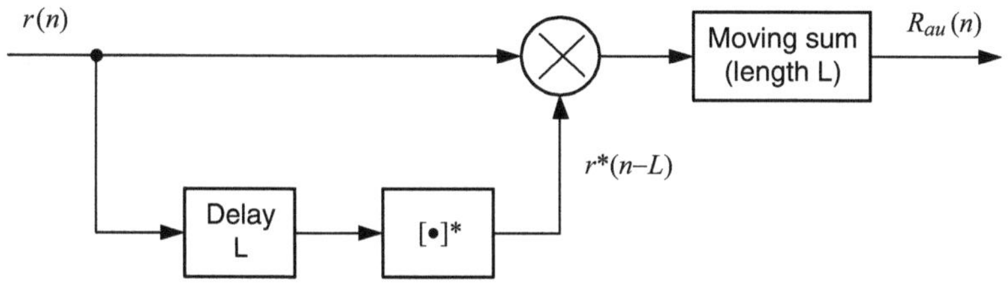

Fig. 3.18 Block diagram of an auto-correlator

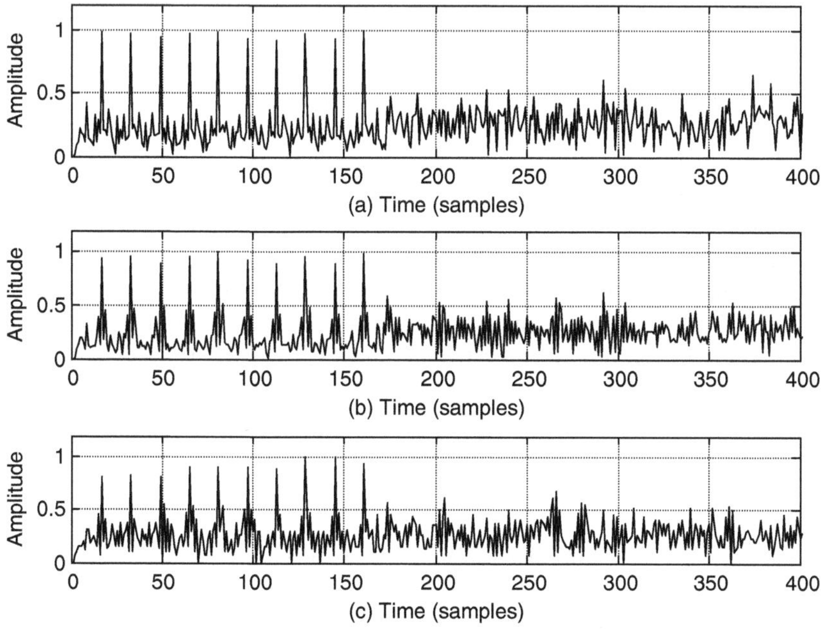

Fig. 3.19 Cross-correlation normalized amplitude output at SNR = 15 dB: (**a**) actual values for both input signal and one local short symbol, (**b**) actual value for input signal, sign value for one local short symbol, and (**c**) sign values for both input signal and one local short symbol

magnitude of auto-correlation with the length L of moving sum is illustrated as a dish line in Fig. 3.17. If the cross-correlation peaks are within the plateau, the last peak is used as the beacon position to indicate the starting point of the next symbol. Therefore, the frame detection, symbol timing, and AGC setting can be determined based on the combination of the magnitudes of the peak and plateau.

In practice, the auto-correlation and cross-correlation can be performed by taking signs of the received signal and local short symbol rather than their actual values, which leads to reducing the complexity of the hardware implementation. Figure 3.19 shows the amplitude outputs of the cross-correlation, where Fig. 3.19a illustrates the correlation output using both input signal and the local short symbol with the actual values, Fig. 3.19b displays the correlation output utilizing the input signal with actual values and local short symbol with the sign values, and Fig. 3.19c demonstrates the correlation output employing both input signal and local short symbol with the sign values. It can be seen that the difference between peaks is small as shown in Fig. 3.19b, and slightly becomes large as shown in Fig. 3.19c. There are differences between the two side-lobes around each peak. However, these side-lobes don't affect the peak detection, but the design complexity is reduced. Such a low-complexity design can be also applied to the auto-correlator shown in Fig. 3.18. Advantages of using sign values in the correlation design are that the amplitude of the output is independent of its input signal level such that a threshold level is easily determined and the multipliers can be replaced with the logic gates to reduce the design complexity.

Symbol timing synchronization can be also obtained by using the cyclic prefix in each OFDM symbol, in which the first 16 samples are identical to the last 16 samples. Actually, this partial repeat property in each OFDM symbol can be exploited for both symbol timing and carrier frequency synchronization [4, 10, 11]. This kind of utilization is useful in the long frame size based transmission, where timing and frequency variation varies fast with the time. It is also particularly suited to some applications where no special training sequences are available. For packet transmission, like 802.11 a/n/ac standards, the cyclic prefix can be used to fine track any variations on symbol timing and carrier frequency in a frame if needed. As we will see later in this section, the carrier frequency offset can be also estimated with such a matched filter structure.

3.3.2 Carrier Frequency Synchronization

In order to achieve the best performance in the receiver, similar to the single carrier signal detection, a coherent detection technique is needed for the OFDM signal detection. This includes channel estimation, and carrier frequency offset and phase estimation.

Carrier frequency offsets (CFO) are mainly caused by the frequency differences between the transmitter and receiver oscillators, and Doppler shifts through the transmission channel. CFO affects all subcarrier equally, and is usually classified into two categories: *integer subcarrier spacing CFO*, and *fractional subcarrier spacing CFO* [12, 13]. Generally, a frequency offset consists of both integer and fractional numbers, in which either one can be zero. Fractional CFO results in the loss of orthogonality between the subchannels, and thus causes ICI and degrades the BER performance in the receiver. Integer CFO does not introduce ICI, but does introduce a frequency rotation of data subcarriers and a phase shift proportional to OFDM symbol number [12], which can cause wrong decisions. Therefore, an integer CFO should be corrected before the decision. Actually, an OFDM system is much more sensitive to the frequency offset than a single carrier system.

There are different techniques to estimate and compensate for the frequency offset using time domain or frequency domain approaches called *pre-FFT* and *post-FFT* synchronization, respectively.

Pre-FFT Synchronization: this kind of synchronization can be further classified into two categories: *non-data-aided* (NDA) and *data-aided* (DA).

- NDA method exploits similarities between the cyclic prefix (CP) part and the corresponding data part of a received OFDM symbol to estimate CFO [14, 15]. This can be done by correlating the CP and the corresponding OFDM symbol to estimate both symbol or frame timing and frequency offsets. This method requires no additional training symbols, thus improving transmission efficiency. Since each OFDM symbol contains the CP, the frequency offset estimation can be continuously estimated on each OFDM symbol to handle the impact of multipath fading if the channel environment changes fast.
- DA method utilizes the training symbols inserted at the beginning of every OFDM frame to estimate CFO and perform symbol or frame timing synchronization, such in the 802.11 WLAN standards [3]. This method provides a wider CFO estimation range than the NDA method does in the range -1.0 to 1.0 subcarrier spacing [16], depending on the length of the training symbol, even though it reduces transmission efficiency due to the insertion of the training symbols.

Post-FFT Synchronization: This type of synchronization usually performs the estimation of the remaining CFO left by pre-FFT synchronization in the frequency domain because the primary CFO synchronization should be carried out before the FFT operation. The remaining CFO can be estimated in the frequency domain by either correlating the received pilot subcarriers with a shifted version of the known pilot subcarriers [17] or correlating the first OFDM symbol with the second OFDM symbol that is repeated the first OFDM symbol [18]. Depending on spacing between pilot subcarriers, this approach can estimate CFO range up to several integers of subcarrier spacing and is only effectively performed after coarse timing synchronization and coarse frequency offset estimation have been established during pre-FFT synchronization.

Focusing on the 802.11a WLAN system, this book only introduces CFO estimation based on the training symbols in the time domain. The coarse frequency estimation is performed during the interval of the short training symbols while the fine frequency estimation is carried out during the interval of the long training symbols.

As described in the previous section, the last four short training symbols in the 802.11a standard specification can be used to estimate the coarse frequency offset while the next two long training symbols can be utilized to estimate the fine frequency offset by means of their identical characteristics between the two long symbols. In a continuous time domain, the received RF signal is down converted with a pair of quadrature local oscillation signals to the complex baseband signal. After passing through the lowpass filters, the baseband signal ignoring noise can be written as

$$y(t) = x(t)\mathrm{e}^{j2\pi\Delta ft} \tag{3.24}$$

where $x(t)$ is the transmitted signals and Δf is the frequency offset. Given that a short training symbol is periodic with $\Delta t = T_{\mathrm{ST}} = 0.8\,\mu\mathrm{s}$, in which T_{ST} is the interval of one short training symbol, the delayed signal is

$$\begin{aligned} y(t - T_{\mathrm{ST}}) &= x(t - T_{\mathrm{ST}})\mathrm{e}^{j2\pi\Delta f(t - T_{\mathrm{ST}})} \\ &= x(t - T_{\mathrm{ST}})\mathrm{e}^{j2\pi\Delta ft}\mathrm{e}^{-j2\pi\Delta fT_{\mathrm{ST}}} \end{aligned} \tag{3.25}$$

Thus,

$$y(t) \times y^*(t - T_{\mathrm{ST}}) = |x(t)|^2\mathrm{e}^{j2\pi\Delta fT_{\mathrm{ST}}} \tag{3.26}$$

Taking angles of both sides of (3.26) gives

$$2\pi\Delta fT_{\mathrm{ST}} = \arg[y(t)y^*(t - T_{\mathrm{ST}})] \tag{3.27}$$

Or

$$\Delta f = \frac{1}{2\pi T_{\mathrm{ST}}}\arg[y(t)y^*(t - T_{\mathrm{ST}})] \tag{3.28}$$

Thus, the frequency offset can be solved by dividing the angle of two consecutive training symbols delayed by T_{ST} with $2\pi T_{\mathrm{ST}}$. In the discrete time domain, (3.28) is given by

$$\Delta f = \frac{1}{2\pi N_{\mathrm{ST}}T_{\mathrm{sam}}}\arg[y(n)y^*(n - N_{\mathrm{ST}})] \tag{3.29}$$

or

$$\Delta f = \frac{1}{2\pi N_{\mathrm{ST}}T_{\mathrm{sam}}}\tan^{-1}\left(\frac{\mathrm{Im}[y(n)y^*(n - N_{\mathrm{ST}})]}{\mathrm{Re}[y(n)y^*(n - N_{\mathrm{ST}})]}\right) \tag{3.30}$$

where $N_{\mathrm{ST}} = 16$ is the number of samples in the duration T_{ST} of one short training symbol, T_{sam} is the interval of the sampling frequency, and $T_{\mathrm{ST}} = N_{\mathrm{ST}}T_{\mathrm{sam}}$. In order to improve the accuracy in the presence of noise, the output of the auto-correlation in (3.30) is typically performed using a moving sum with the length of N_{ST} prior to the computation of angle, which is expressed as

$$\Delta f = \frac{1}{2\pi N_{\mathrm{ST}}T_{\mathrm{sam}}}\tan^{-1}\left(\frac{\sum_{k=0}^{N_{\mathrm{ST}}}\mathrm{Im}[y(k)y^*(k - N_{\mathrm{ST}})]}{\sum_{k=0}^{N_{\mathrm{ST}}}\mathrm{Re}[y(k)y^*(k - N_{\mathrm{ST}})]}\right) \tag{3.31}$$

This estimate is the same as the maximum likelihood estimate of the frequency offset proposed in [15]. It is noted that the maximum-likelihood estimate for the frequency offset is independent of the noise power.

The estimated frequency offset is limited within a range of $[-\pi, \pi)$ by the angle function above. Substituting the range in the above equation, the minimum (or negative) value of frequency offset can be estimated at the sampling frequency $f_{\mathrm{sam}} = 1/T_{\mathrm{sam}} = 20\ \mathrm{MHz}$

$$\Delta f_{\min} = \frac{-\pi}{2\pi \times 16 \times 1/\left(20 \times 10^6\right)} = -625\ \mathrm{kHz} \tag{3.32}$$

and the maximum value is

$$\Delta f_{\max} < \frac{\pi}{2\pi \times 16 \times 1/\left(20 \times 10^6\right)} = 625\ \mathrm{kHz} \tag{3.33}$$

Figure 3.20 shows the frequency offset estimation procedure for the frequency offset of 300 kHz in the range of the short and long preamble sequences for the 802.11a system. The estimated average frequency offset value over the last four short symbols in MATLAB simulation is 300.5 kHz. The actual frequency offset is

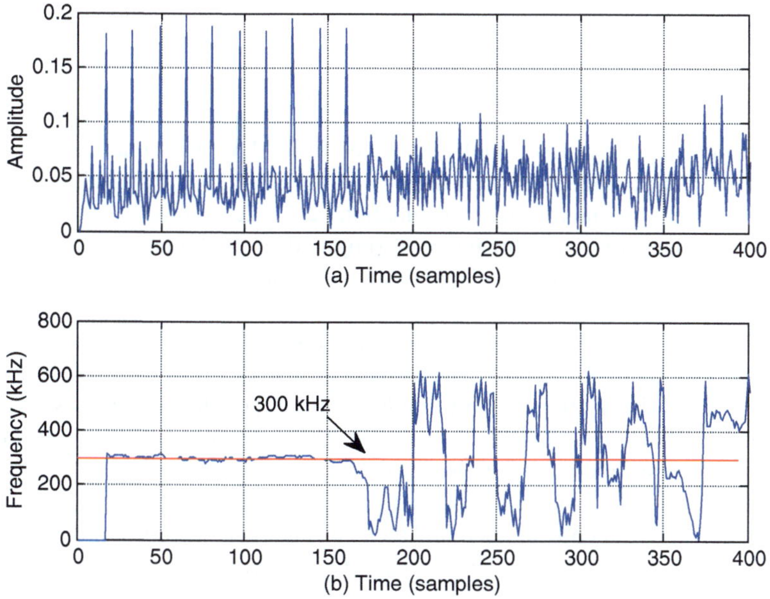

Fig. 3.20 Auto-correlation for frequency offset estimation over the preamble symbols (ten short and two long symbols) at SNR = 15 dB: (**a**) cross-correlation output, and (**b**) frequency offset estimation with auto-correlation, where an average frequency offset $\Delta f = 300.5\,\mathrm{kHz}$ in (**b**) is estimated over last four peaks in (**a**)

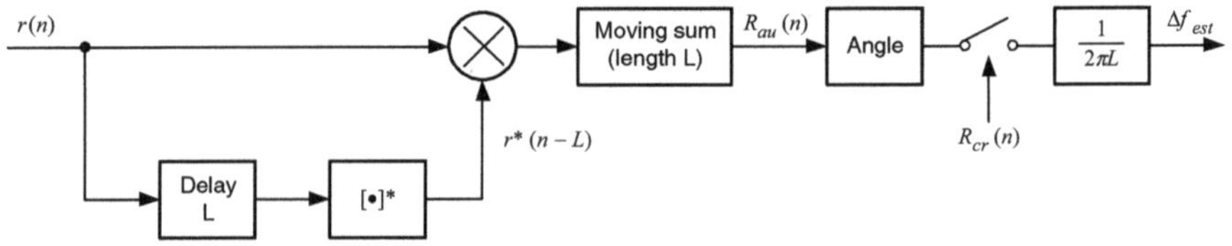

Fig. 3.21 Block diagram of the frequency offset estimator

shown with a wider line in Fig. 3.20b. The peaks in Fig. 3.20a indicate the values of the frequency offset estimation should be sampled at these moments.

Figure 3.21 shows the block diagram of the frequency offset estimator based on the preamble of 802.11a WLAN standard, where the cross-correlation signal $R_{\mathrm{cr}}(n)$ is generated in Fig. 3.16, and its output is shown in Fig. 3.20a.

Two long training symbols with 64 samples each can be used to estimate the fine frequency offset after the coarse frequency offset estimation. Substituting the number of N_{ST} samples in one short symbol with $N_{\mathrm{LG}} = 64$ in one long symbol into (3.32) and (3.33), the minimum and maximum frequency offsets respectively are

$$\Delta f_{\min} = \frac{-\pi}{2\pi \times 64 \times 1/\left(20 \times 10^6\right)} = -156.25\ \text{kHz} \tag{3.34}$$

and

$$\Delta f_{\max} < \frac{\pi}{2\pi \times 64 \times 1/\left(20 \times 10^6\right)} = 156.25\ \text{kHz} \tag{3.35}$$

The fine frequency offset can be estimated from (3.31) by replacing the number N_{ST} of one short training symbol with the number N_{LT} of one long training symbol. Thus, the final frequency offset estimation $\Delta f = \Delta f_{\mathrm{ST}} + \Delta f_{\mathrm{LT}}$ is utilized for frequency offset correction for the remaining part of the frame after long training symbols. Here Δf_{ST} and Δf_{LT} represent the coarse and fine frequency offset estimations, respectively.

Figure 3.22 illustrates the procedure of the residual frequency offset estimation over two long training symbols. The residual frequency offset of 100 kHz is added to the received signal after 10 short training symbols and the estimation starts at the first long symbol with an initial value of zero. The cross-correlation is performed by correlating the received signal with a local long preamble symbol having 64 samples while the auto-correlation sum is carried out by correlating the received signal with a 64-sample delayed copy of itself and then calculating a moving sum with a length of 64 samples. The cross-correlation creates two peaks and the auto-correlation and sum produces a plateau, which is unique to the preamble period and covers two peaks. When the correlation peaks are within the plateau, these two peaks are used as the beacon positions to form a boundary range, where the frequency offset is sampled and then is averaged. The simulation shows the residual frequency offset of 101.1 kHz is estimated under SNR = 15 dB.

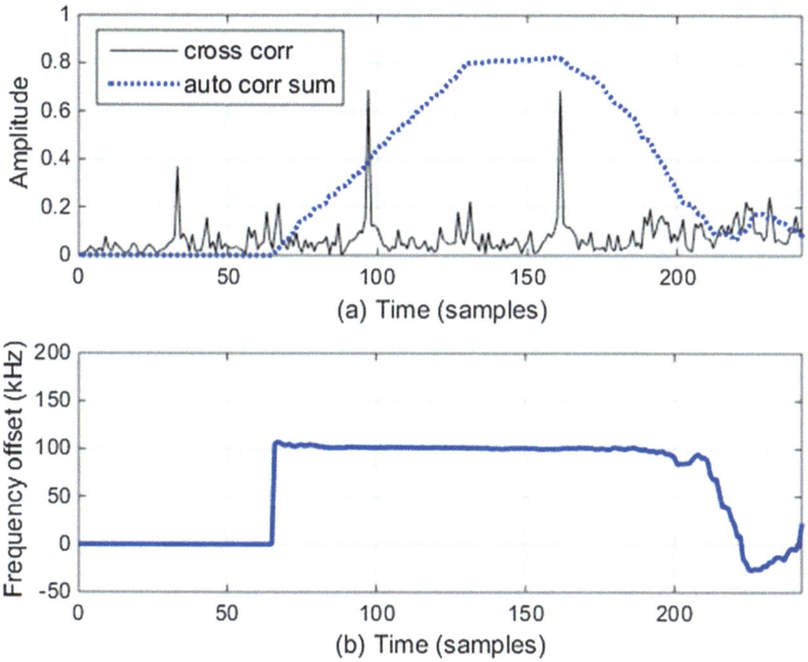

Fig. 3.22 Auto-correlation for residual frequency offset estimation over two long training symbols at SNR = 15 dB, where a residual frequency offset is set to 100 kHz after ten short training symbols and estimated frequency offset is 101.1 kHz: (**a**) auto and cross correlation, and (**b**) frequency offset estimate

The maximum-likelihood estimate for the normalized frequency offset and frame synchronization proposed by [15] for non-data-aided (NDA) is given by

$$\hat{\varepsilon} = \frac{1}{2\pi}\tan^{-1}\frac{\sum_{k=\theta-L+1}^{\theta}\mathrm{Im}\{y(k)\times y^*(k-N)\}}{\sum_{k=\theta-L+1}^{\theta}\mathrm{Re}\{y(k)\times y^*(k-N)\}} \tag{3.36}$$

where L is the guard interval, N is samples of FFT operation in one data symbol, and $\hat{\varepsilon}$ is equal to ΔfT. It is noted that $\hat{\varepsilon}$ is independent of noise. It is noted that (3.36) is identical to (3.31) except that the former uses the normalized frequency offset while the latter utilizes the actual frequency offset. In (3.36), the starting position θ of the OFDM symbol should be used to estimate the frequency offset. A simpler way to estimate θ is to find the maximum sum of the auto-correlation and is given by [15]

$$\hat{\theta} = \underset{\theta}{\mathrm{argmax}} \sum_{k=\theta-L+1}^{\theta}\left[\left|\mathrm{Re}\{y(k)y^*(k-N)\}\right| + \left|\mathrm{Im}\{y(k)y^*(k-N)\}\right|\right] \tag{3.37}$$

The frequency offset $\hat{\varepsilon}$ and the time instant $\hat{\theta}$ are estimated simultaneously, and the maximum value $\hat{\theta}$ in each OFDM symbol indicates the frequency offset.

In the cyclic prefix (CP) or guard interval based frequency detection, from (3.36) the detectable range of the frequency offset is limited by $|\varepsilon| \leq 0.5$, or $\pm 1/2$ of the subcarrier spacing. The estimation suffers from subcarrier ambiguity when the frequency offset $|\varepsilon|$ is greater than 0.5. For example, if the interval of one OFDM symbol is 3.2 μs, such as 802.11a WALN, the maximum frequency offset can be detected up to

$$\Delta f < \frac{\varepsilon}{T_{\text{FFT}}} = \frac{0.5}{3.2 \times 10^{-6}} = 156.25 \text{ kHz} \tag{3.38}$$

And the minimum frequency offset can be detected down to

$$\Delta f \geq -\frac{\varepsilon}{T_{\text{FFT}}} = -\frac{0.5}{3.2 \times 10^{-6}} = -156.25 \text{ kHz} \tag{3.39}$$

Compared with the detectable range for data-aided (DA) based frequency offset estimation as given in (3.32) and (3.33), the detectable range for non-data-aided (NDA) based frequency offset estimation as shown in (3.38) and (3.39) is four times smaller.

A low complexity ML estimator for frequency offset and symbol timing can be realized by replacing $y(k)$ and $y(k - N)$ with their sign version $c(k)$ and $c(k - N)$ in (3.36) and (3.37), respectively. Thus, they are expressed as [15]

$$\hat{\varepsilon}_c = \frac{1}{2\pi} \tan^{-1} \frac{\sum_{k=\theta-L+1}^{\theta} \text{Im}\{c(k) \times c^*(k-N)\}}{\sum_{k=\theta-L+1}^{\theta} \text{Re}\{c(k) \times c^*(k-N)\}} \tag{3.40}$$

$$\hat{\theta}_c = \underset{\theta}{\text{argmax}} \sum_{k=\theta-L+1}^{\theta} \left[\left|\text{Re}\{c(k)c^*(k-N)\}\right| + \left|\text{Im}\{c(k)c^*(k-N)\}\right| \right] \tag{3.41}$$

Figure 3.23 illustrates a block diagram of both frequency offset and symbol timing estimators based on the ML algorithm, where the maximum (or peak) value

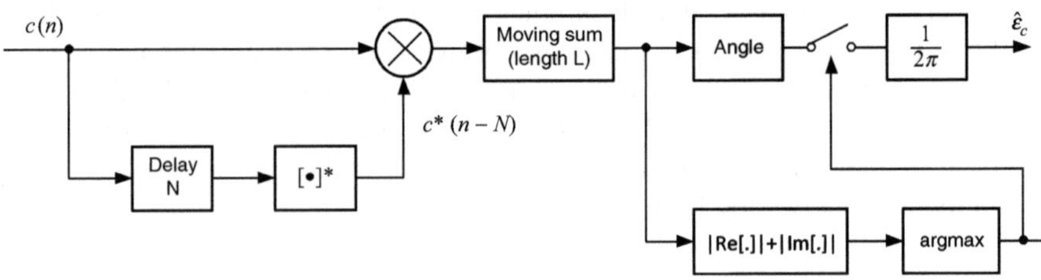

Fig. 3.23 Block diagram of a low complexity ML based estimator for frequency offset and symbol timing position. Redrawn from [15]

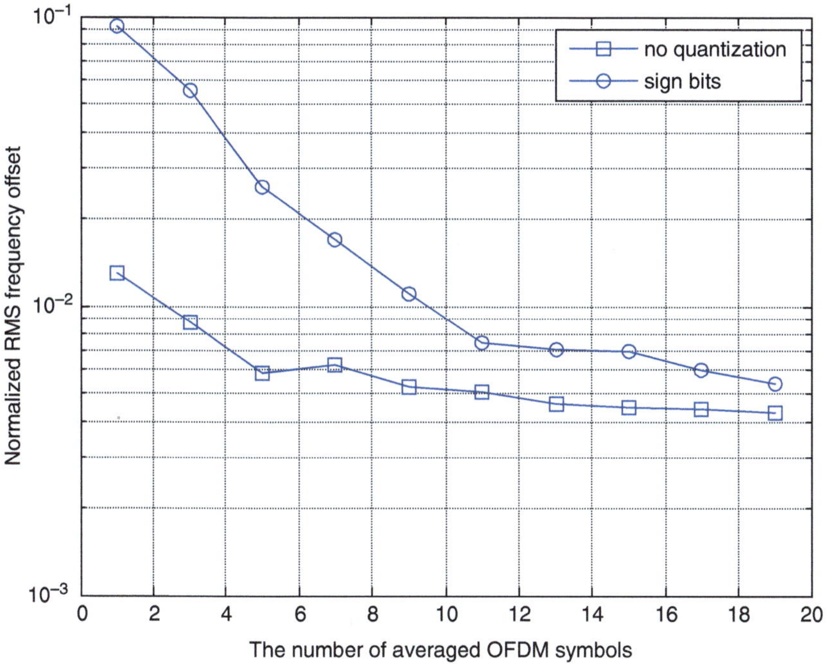

Fig. 3.24 RMS frequency offset versus the number of averaged OFDM symbols. Redrawn from [15]

$\hat{\theta}_c$ gives the OFDM symbol start position and also detects the frequency offset at that time moment. The estimation can be averaged over several OFDM symbols and then be used for the rest of OFDM symbols.

To demonstrate the capability of the ML estimation algorithm for the frequency offset in a NDA system, the root mean square (RMS) value of the residual frequency offset error was simulated in multipath fading channels and the results in a two-ray Rician fading channel are illustrated in Fig. 3.24 [15], where it contains one direct path and one Rayleigh fading path with 10 μs time delay and equal power relative to the direct path. The parameters used in the simulation are listed in Table 3.4. It can be seen that the performance of the sign-bit based frequency offset detector can be significantly improved by averaging frequency offset over several OFDM symbols.

If the CP is heavily disturbed by severe multipath fading, the estimation accuracy will be significantly degraded, causing degradation of the BER performance. In order to increase the frequency offset estimation accuracy, especially in such severe multipath fading channels, a pilot subcarrier aided approach can be used to further perform the estimation of the remaining CFO by correlating the received pilot subcarriers with a shifted version of the known pilot subcarriers [17]. This method is also called as Post-FFT synchronization and is only effectively performed to track the residual CFO after coarse timing and coarse frequency

Table 3.4 Parameters used in the simulation [15]

Parameter	Value
Number of subcarrier	1024
Modulation	16-QAM
Symbol rate	8×10^6 symbols/s
Symbol period	128 μs
Guard interval	16 μs
Sampling frequency	9 MHz
Frequency offset	0.48 of subcarrier spacing
Carrier frequency	1.5 GHz
Vehicle speed	100 km/h
SNR	10 dB

synchronizations have been established. This is because the FFT operation can only be correctly performed after most timing errors and frequency offset errors are compensated.

3.3.3 *Channel Estimation Technique*

An OFDM system divides a wideband spectrum of a single carrier FDM system into a number of overlapping but orthogonal narrowband subchannels and hence coverts a frequency selective channel into almost non-frequency selective subchannels. The frequency response of each subchannel can be approximately treated as constant within the subchannel when frequency selective fading occurs in the desired bandwidth. Therefore, the multipath fading channel can be compensated by using a frequency domain equalizer as simple as a one-tap equalizer. Furthermore, the use of CP, which is achieved by duplicating the last portion of an OFDM symbol as its head, avoids ISI caused by multipath fading. These special design features ensure OFDM systems behave robustly against multipath fading compared with single carrier systems.

In this section, only commonly used methods in data-aided channel estimation are described. In data-aided channel estimation, known information to the receiver is inserted in OFDM symbols so that the current transmission channel characteristics can be estimated by comparing the difference between the received known information and locally duplicate information. Two arrangements of sending known information with data together are commonly used:

- Sending known training symbols together with OFDM symbols in the time domain.
- Sending known pilots together with data in the frequency domain.

The former arrangement is usually called channel estimation with training symbols. The latter one is known as pilot aided channel estimation.

The channel estimation that employs training symbols periodically sends training symbols so that the channel estimation can be periodically updated. In the IEEE 802.11a standard, for example, the two long training symbols are inserted to the beginning of the OFDM frame, just following the ten short training symbols, and used to estimate the channel in either frequency domain or time domain. When the channel estimation is performed in the frequency domain, the channel frequency response is estimated. The channel impulse response, on the other hand, is estimated when it is carried out in the time domain.

In the pilot aided channel estimation, the pilots are multiplexed with the data in the frequency domain. For such frequency domain estimation, the channel at each pilot subchannel is estimated and then interpolated via different methods, such as linear interpolation (LI). The interpolated channel estimation is used to compensate for data subcarrier channels between two continuous pilot subcarriers.

In the 802.11a WLAN system, two long training symbols can be used for both channel estimation and fine frequency offset estimation. Considering the time limitation within two long training symbols, we will only introduce pilot aided channel estimation in this section and leave two long training symbols for the use of fine frequency estimation. Reader who is interested in the channel estimation based on training symbols can reference literature [14, 19].

3.3.3.1 Pilot Aided Channel Estimation

The transmitted OFDM signal in the discrete-time domain, excluding CP, can be expressed with N-point inverse discrete Fourier transform (IDFT) as:

$$x[n] = \mathrm{IDFT}_N\{X(k)\} = \sum_{k=0}^{K-1} X(k)e^{j2\pi kn/K} \quad n = 0, 1, \ldots, N-1 \tag{3.42}$$

where $\{X(k)\}$ is the modulation data in frequency domain, k is the subcarrier index (0, 1, …, $K-1$) with K being the total number of subcarriers, and n is the nth OFDM symbol. The CP is inserted to the beginning of each OFDM symbol $x(n)$ to form a new OFDM symbol $s(n)$ with the CP.

The transmitted signal is sent to the receiver through a multipath fading channel and the received signal $r(n)$ can be denoted by

$$r(n) = s(n) \otimes h(n) + w(n) \tag{3.43}$$

where $h(n)$ is the discrete channel impulse response (CIR) of the channel and $w(n)$ is the *i.i.d.* complex zero-mean additive white Gaussian noise (AWGN) whose real and imaginary parts both have variance σ_n^2.

At the receiver side, after fulfilling symbol timing synchronization, frequency offset correction, and removing the CP from $r(n)$, the received signal $r(n)$ is represented by $y(n)$ as follows:

$$y(n) = x(n) \otimes h(n) + w(n) \tag{3.44}$$

Here we suppose that the CP interval is longer than the length of the channel impulse response (CIR), which does not cause ISI between OFDM symbols. A DFT is further performed on the received samples $y(n)$ in (3.44) to de-multiplex the multicarrier signals

$$\begin{aligned} Y(k) &= \mathrm{DFT}\{y(n)\} = \frac{1}{N}\sum_{n-0}^{N-1} y(n)\mathrm{e}^{-j2\pi kn/N}, \quad k = 0, 1, \ldots, N-1 \\ &= X(k)H(k) + W(k) \end{aligned} \tag{3.45}$$

In the frequency domain, the received pilot signals $\{Y_{\mathrm{p}}(k)\}$ are extracted from $\{Y(k)\}$ and then the channel estimation can be obtained by

$$\hat{H}_{\mathrm{p}}(k) = \frac{Y_{\mathrm{p}}(k)}{X_{\mathrm{p}}(k)} + \frac{W(k)}{X_{\mathrm{p}}(k)} \tag{3.46}$$

where $\{X_{\mathrm{p}}(k)\}$ are pilot symbols transmitted at the transmitter. Equation (3.46) indicates the estimation of the channel frequency response is simply the division of the received symbol by the transmitted pilot symbol. The channel transfer function at data symbols $\{\hat{H}_{\mathrm{d}}(k)\}$ can then be obtained by using linear interpolation algorithm from two adjacent pilot symbols $\{\hat{H}_{\mathrm{p}}(k)\}$ as shown in (3.46). With the calculation of the channel response $\{\hat{H}_{\mathrm{d}}(k)\}$ at data symbols, the estimated transmitted data samples $\left\{\hat{X}_{\mathrm{d}}(k)\right\}$ can be recovered by simply dividing the received signal by the estimated channel response

$$\hat{X}_{\mathrm{d}}(k) = \frac{Y(k)}{\hat{H}_{\mathrm{d}}(k)} + \frac{W(k)}{\hat{H}_{\mathrm{d}}(k)}, \quad k = 0, 1, \ldots, N-1 \tag{3.47}$$

Even though there are many methods used for estimating the channel responses, pilot symbol assisted channel estimation for OFDM signals in the frequency domain is presented here due to its simple, fast and accurate features.

Equation (3.46) is actually identical to the *Least Squares* (LS) estimation [20, 21]. In general, LS method is utilized to get the initial channel estimates at the pilot subcarriers and the initial channel estimates are then further used at the OFDM data subcarriers via interpolation methods, depending on the variation of the channel with time. The goal of choosing an interpolation method is to have lower computation complexity, but at the same time this method can achieve relatively higher accuracy for a given system.

3.3.3.2 Linear Interpolation

The simplest interpolation is a linear interpolation [20, 21], in which the channel frequency response (CFR) between pilot subcarriers is simply estimated from a straight line between two adjacent $\{\hat{H}_p(k)\}$ that are estimated by using two successive pilot symbols. Using LI, the channel estimation at the data subcarrier k, where $mL \le k < (m+1)L$, is given by

$$\hat{H}_{\mathrm{d}}(k) = \frac{l}{L}\left[\hat{H}_{\mathrm{p}}(m+1) - \hat{H}_{\mathrm{p}}(m)\right] + \hat{H}_{\mathrm{p}}(m), \quad 0 \le l < L, \; m = 0, 1, \ldots, N_{\mathrm{p}} - 1 \tag{3.48}$$

where N_p is the number of the total pilot subcarriers, $L - 1$ is the number of data subcarriers between adjacent pilots, and l is the distance between the mth pilot subcarrier and the kth data subcarrier.

For example, two extreme cases are given here. If l is equal to 0, (3.48) becomes

$$\hat{H}_{\mathrm{d}}(k) = \hat{H}_{\mathrm{p}}(m) \tag{3.49}$$

If l is equal to L, (3.48) is

$$\hat{H}_{\mathrm{d}}(k) = \hat{H}_{\mathrm{p}}(m+1) \tag{3.50}$$

In this example, the first equation shows the CFR at the kth data subcarrier is equal to the CFR at the mth pilot subcarrier, while the second one indicates the CFR at the kth data subcarrier is equal to the CFR at the $(m+1)$th (or next) pilot subcarrier.

To further simplify the design implementation, piecewise constant [22] as shown in Fig. 3.25 can be used for the CFR estimation at data subcarriers. In this method constant CFR is estimated at the mth pilot subcarrier and is used for the left side half data subcarriers between $(m-1)$th and mth pilot subcarriers and the right side half data subcarriers between the mth and $(m+1)$th subcarriers.

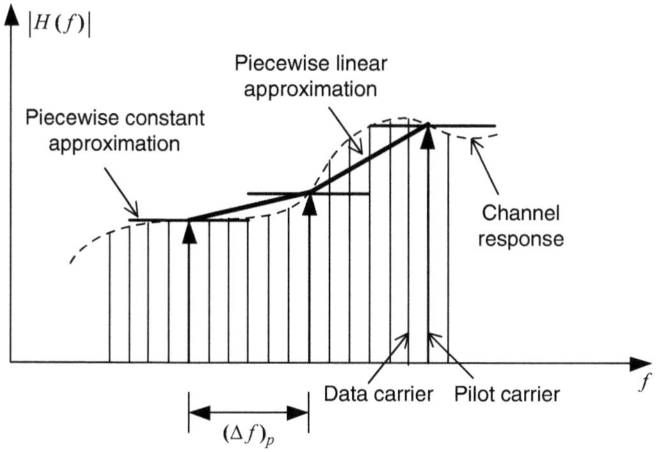

Fig. 3.25 Spectrum of received OFDM signal with amplitude fading distortion. Redrawn from [22]

If the estimation difference between two adjacent pilot subcarriers is large, linear interpolation can be used. Otherwise the piecewise constant can be utilized.

3.3.3.3 Adaptive Equalization Techniques

Since LI estimation does not accurately perform channel estimation, especially in severe frequency selective fading across the entire band; an adaptive equalizer with one complex tap can be used to achieve further accurate channel compensation after LI estimation. After LI estimation, the constellation diagram of the recovered signal can be usually recognized in the type of a modulation format, but still not completely recovered yet, depending on the transmission channel condition. To completely compensate for any distortions caused by multipath fading, a one-tap complex adaptive linear equalizer can be employed after the LI estimation.

There are two different criteria used to determine the values of the linear equalizer coefficients. The basic idea is to minimize ISI or distortion at the output of the equalizer when the equalizer coefficients are set to some certain values determined by these two different criteria:

1. The minimization of the peak distortion.
2. The minimization of the mean-square error.

Assume there are L data subcarriers in an OFDM symbol. Then, an adaptive equalizer output at the nth OFDM symbol in the frequency domain can be expressed as

$$y_n(l) = c_n(l)v_n(l), \quad l = 1, 2, \ldots, L \tag{3.51}$$

where $c_n(l)$ is the coefficient of the equalizer on the lth data subcarrier branch at the nth OFDM symbol and $v_n(l)$ is the input signal to the equalizer on the lth data subcarrier at the nth OFDM symbol. For the 802.11a case, L is equal to 48 due to a total of 48 data subcarriers in one OFDM symbol. Figure 3.26 illustrates a block diagram of the equalizer with one complex tap on the lth data subcarrier branch.

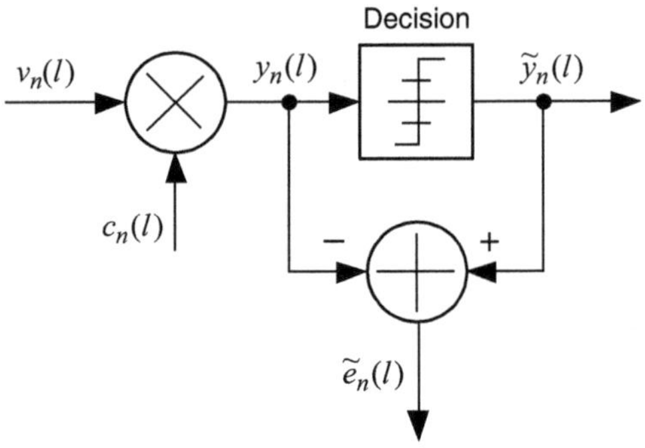

Fig. 3.26 A block diagram of one complex-tap equalizer on the lth data subcarrier branch in a frequency domain

Zero-Forcing (ZF) Equalizer: The error signal on the lth data subcarrier branch at the nth OFDM symbol is

$$e_n(l) = d_n(l) - y_n(l) \tag{3.52}$$

where $d_n(l)$ is the desired data signal and is usually unknown in the receiver except for the pilots.

For the first criterion above, the zero-forcing algorithm completely eliminates the ISI at the output of the equalizer by using an inverse filter to the transfer function of the distorted channel, regardless of noise in the channel. The zero-forcing algorithm is obtained by adjusting the coefficient of the equalizer to force the cross-correlation between the error signal $e_n(l) = d_n(l) - y_n(l)$ and the desired data signal $d_n(l)$ to be zero when ignoring the ratio of the signal-to-noise at the output of the equalizer. This requires

$$E\left[e_n(l)d_n^*(l)\right] = E\left[(d_n(l) - y_n(l))d_n^*(l)\right] = 0 \tag{3.53}$$

To meet the condition of (3.53), the coefficient of the equalizer is adaptively updated as follows [23]

$$\mathrm{c}_{n+1}(l) = \mathrm{c}_n(l) + \lambda e_n(l)d_n^*(l) \tag{3.54}$$

where $c_n(l)$ is the coefficient of the equalizer on the lth data subcarrier branch at the nth OFDM symbol, $c_{n+1}(l)$ is one of the equalizer on the lth data subcarrier branch at the $(n + 1)$th OFDM symbol, and λ is a step size that controls the rate of the coefficient adjustment. The decision output signal of $\widetilde{y}_n(l)$, however, can be used to replace the training signal $d_n(l)$ due to its reliability after using the pilot aided channel estimation and *LI* approach. Thus, (3.54) can be written as

$$\mathrm{c}_{n+1}(l) = \mathrm{c}_n(l) + \lambda\widetilde{e}_n(l)\widetilde{y}_n^*(l), \quad l = 1, 2, \ldots, L \tag{3.55}$$

where $\widetilde{e}_n(l) = \widetilde{y}_n(l) - y_n(l)$ is used for replacing $e_n(l) = d_n(l) - y_n(l)$ as the error signal at the nth OFDM symbol. The expression in (3.55) is a practical *zero-forcing algorithm*. At the pilots, the equation (3.46) is preferred because of its one-time accurate estimation compared to (3.55).

Least-Mean Square (LMS) Equalizer: The LMS algorithm is determined with the mean square error (MSE) criterion, which minimizes the mean square value of the error signal at the output of the equalizer by adjusting the coefficient of the equalizer. It is obvious that the optimal coefficient value of the equalizer is dependent of the SNR at the input of the equalizer. In the MSE criterion, the mean square error function ζ is minimized by adjusting the coefficient $c_n(l)$ of the equalizer and the function ζ is given by:

$$\begin{aligned}\xi &= E\left[|e_n(l)|^2\right] = E\left[e_n(l)e_n^*(l)\right] \\ &= E\{e_n(l)[d_n(l) - c_n(l)v_n(l)]*\}\end{aligned} \tag{3.56}$$

One widely used minimization method is called *steepest descent*, in which the coefficient is adjusted as follows:

$$c_{n+1}(l) = c_n(l) + \lambda\left(-\frac{\partial \xi}{\partial c^*}\right)\bigg|_{c=c_n(l)} \tag{3.57}$$

where λ is the step size that determines the rate of convergence as well as the iterative stability. To minimize ξ, the derivative of ξ with respect to c^* in the bracket above should be set to zero. A simple and well known *LMS* algorithm can be obtained as:

$$c_{n+1}(l) = c_n(l) + \lambda\widetilde{e}_n(l)v_n^*(l), \quad l = 1, 2, \ldots, L \tag{3.58}$$

where the error signal $\widetilde{e}_n(l) = \widetilde{y}_n(l) - y_n(l)$ is used for replacing $e_n(l) = d_n(l) - y_n(l)$ in the case where the desired sequence $d_n(l)$ is not available and the decision sequence $\widetilde{y}_n(l)$ is more reliable. Comparing (3.58) with (3.55), we can see the difference between the ZF algorithm and the LMS algorithm is that $v_n(l)$ in (3.58) is the input sequence to the equalizer while $\widetilde{y}_n(l)$ in (3.55) is the decision sequence at the output of the equalizer. The ZF algorithm gives a faster convergence than the LMS algorithm, while the LMS algorithm gives a higher SNR than ZF algorithm. In a high input SNR condition, their SNRs at the output of the equalizer are almost same.

Besides the LI method, there are also other channel estimation algorithms introduced in [24], such as Gaussian, cubic-spline, and Wiener filter scheme. The 802.11a WLAN system estimates the channel frequency response by using four pilot subcarriers in each OFDM symbol as a first step. The channel frequency estimations are then interpolated to get the CFR at data subcarriers. These pilot signals are located at subcarriers −21, −7, 7 and 21 in the 52 subcarriers and are BPSK modulation by a pseudo binary sequence to prevent the generation of spectral lines. A precise estimation can be further achieved by using the adaptive equalizer with one complex-tap as described above after the LI approach.

3.4 Design Challenges for RF Transceivers

To meet the need for higher data rates and greater system advantages in multipath fading channels, the 802.11a standard ratified in 1999 first time adopts OFDM signals in 802.11 family and is capable of providing data rates of 54 Mbps and producing high throughput and a high level of performance in the 5-GHz band because of less sharing with other users, such as Bluetooth radios, cordless phones, and microwave oven compared with the 802.1b operation band of 2.4 GHz.

The 802.11a standard consists of two main layers: Media Access Control (MAC) layer and Physical (PHY) layer. These two layers allow a functional separation of the standard, but more importantly allow a single data protocol to be used with several different RF transmission techniques.

Physical Layer: The PHY layer of the 802.11a standard defines OFDM transmission technique because of splitting a single channel data with a high rate into multiple subchannel data with a low rate to modulate corresponding subcarriers in the baseband domain in order to mitigate the effects of multipath fading. As a result, the compensation for distortions due to multipath fading becomes as simple as using a frequency domain equalizer with one complex tap.

As indicated in Table 3.3, the 802.11a standard operates in three RF bands named as U-NII lower band, U-NII middle band, and U-NII upper band and supports multiple 20 MHz channels among them. There are eight channels with a bandwidth of 20 MHz in U-NII lower and middle bands and four channels with a bandwidth of 20 MHz in U-NII upper band. These three bands require different output powers at the transmitter, which are 40 mW (16 dBm), 200 mW (23 dBm), and 800 mW (29 dBm), respectively. To support multiple data rates in a 20 MHz band for different applications, the 802.11a system adopts M-ary QAM formats with different error-correcting code rates as listed in Table 3.2.

MAC Layer: The MAC is a set of rules to determine how to access the medium and data link components. The MAC drives every user data transmission into the air and provides the core framing operation and interaction with a wired network backbone. It also executes a TDMA access method, where the transmitter and receiver operate at different time slots or at the TDMA mode. In general, the MAC Layer manages and maintains communications between 802.11 stations (radio network cards and access points (AP)) by coordinating access to a shared radio channel and utilizing protocols that enhance communications over a wireless medium. In other words, as the "brain" of the network, the MAC layer controls the 802.11a PHY Layer to perform all tasks of carrier sensing, transmission and reception of 802.11a frames.

System Partition: To minimize noise interference between digital circuits and analog circuits, a design decision with two-chip solution known as RF transceiver and digital baseband chips, rather than integrating them into a single chip, was made at the beginning phase of the design process [25]. MAC and digital baseband functions are implemented in the baseband ship. Therefore, an appropriate interface should be considered to connect the analog signals and digital signals between the two chips. To minimize the pin count and substrate noise that might be introduced by relatively large switching power in the ADCs/DACs, an analog interface is preferred [25], in which ADCs/DACs are in the baseband chip.

3.4.1 RF Transceiver

Like other RF transceivers, the 802.11a RF transceiver mainly consists of a transmitter, a receiver, and a frequency synthesizer. The RF transceiver performs to up-convert the I and Q analog baseband signals to a RF signal for the transmission or down-convert the RF signal to the I and Q analog baseband signals for the reception by means of the TDMA mode.

3.4.1.1 Architecture and Frequency Planning

The most common architecture in the transmitter is direct up-conversion because it avoids the need for additional mixers and LO circuits used for additional up conversions, which are required in the traditional superheterodyne. Therefore, the direct up-conversion transmitter achieves a low-cost solution with a high level of integration and is very suitable for wideband modulation transmission applications such as the 802.11 family. One of the drawbacks that most direct conversion transmitters face, or LO leakage, can be avoided in the 802.11a because no subcarrier at DC (0th subcarrier) is used. Another possible drawback in the direct conversion transmitter is the frequency pulling effect of the internal (or external) PA output signal on VCO phase noise due to the fact that the VCO operates at twice the frequency of the RF frequency if a pair of quadrature LO signals are obtained by dividing the VCO frequency by 2. The second order (or even-order) harmonic distortion (HD2) of the RF signal, whose frequency is equal to a VCO operation frequency, is mainly generated by the up-conversion mixers due to imbalance loads on the positive and negative outputs of the differential mixer. When the loads are balanced, even-order cancellation of the harmonics is achieved. Otherwise, even-order harmonics are further amplified either by the power amplifier or the power amplifier driver. Even-order harmonics can be also generated by the power amplifier due to the nonlinearity of the PA.

When the HD2 content presented in the RF amplification signals leaks into the PLL due to finite isolation, and then is divided by the PLL's divider to occur at the input of the phase and frequency detector (PFD) together with the divided VCO signal, it may result in interference and noise within the loop bandwidth at the output of the PFD. As a result, the pulling or disturbance on the VCO signal causes the VCO phase degradation and drastically degrades the transmitter EVM.

The leakage from the PA output to the VCO can happen either through the power supply and substrate, or other paths making it difficult to diagnose. To alleviate the pulling effect on the VCO in the direct-conversion transmitter, the optimization of the floor plan, package, and PCB layout is required to maximize the isolation between the PA output and PLL such as attenuation on the HD2 by adding an appropriate bypass capacitor at the power supply of the PA. One effective approach to minimizing the effect of HD2 on VCO phase noise disturbance or VCO frequency pulling is to add a passive lowpass filter at the output of the PA during the PA design. It has been verified that a third-order passive LPF with a π-type achieves

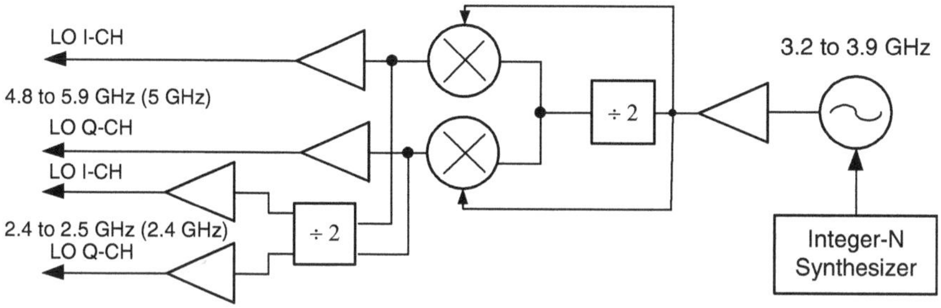

Fig. 3.27 Block diagram of LO generation for 2.4 and 5 GHz band. Referenced from [6]

30 dB attenuation at least at the HD2 frequency for an 802.11n WLAN system and has only about 0.2 dB insertion loss in-band without causing EVM degradation.

Another effective approach to eliminating the frequency pulling or phase noise disturbance on the VCO is to ensure the frequencies of the RF signal and its harmonics far away from the VCO frequency. As illustrated in Fig. 3.27, VCO operates at the frequency range of 3.2 to 3.9 GHz, which is two-third of the RF channel frequency for the 802.11a band. The 3.2–3.9 GHz synthesizer output is divided by 2 and then mixed with the 3.2–3.9-GHz synthesizer output again to generate a pair of quadrature LO signals at 4.8–5.9 GHz [6] at the cost of additional mixers. Another method to avoid the pulling effect is to utilize dual up-conversion [25–27] for the transmitter as shown in Fig. 8.8, where the RF channel frequency (4.8–5.85 GHz) is generated with two stages of the frequency conversion by mixing an LO frequency at 2/3 the channel frequency (3.2–3.9 GH) with 1/2 LO quadrature frequencies (1.6–1.95 GHz) for the 802.11a WLAN systems. Since the transmitter output frequency and its harmonic frequencies are away from the VCO frequency, no pulling effect on the VCO signal could happen. Furthermore, any LO leakage to the antenna will be far away from the in-band signal and appear as out-of-band tone, which will not interfere with other receivers operating in the 5-GHz band.

The most common choice in the receiver architecture is direct down-conversion due to low cost, low power consumption, and small die size. The quadrature LO frequencies at the receiver are obtained as the same as those at the transmitter. The I and Q baseband signals are generated by mixing the received RF signal with the quadrature LO signals after high order harmonic components are attenuated by lowpass filters on the I and Q channels. In either a direct up conversion or dual-up conversion architecture, only one frequency synthesizer is required, even for dual-band 2.4-GHz 802.11n and 5-GHz 802.11a WLAN systems.

It should be noted that at the Mobile World Congress 2015 in Barcelona, Broadcom introduced Industry's First 5G Wi-Fi 2 × 2 MIMO combo chip, or BCM4359 of the 802.11ac standard, with real simultaneous dual-band (RSDB) to support dual-band (both 5 GHz and 2.4 GHz bands) simultaneously. When operating at a RSDB mode, two frequency synthesizers are required in the BCM4359 chipset.

3.4.1.2 Transmitter Chain Design Challenges

Since the 802.11a OFDM symbol consists of 52 subcarriers; its peak-to-average power ratio (PAPR) is very high up to more than 11 dB for a 54 Mbps data rate. Even PAPR is reduced by a few dB with PAR reduction techniques, it is still high such that a PA should have a large back-off from its P1dB compression point, and thus reduces its efficiency. For this reason, most power amplifiers designed for 802.11a WLAN systems are biased in the class-AB operation range in order to achieve relatively high efficiency.

In the direct up-conversion transmitter, the up-conversion mixers are designed for high linearity and low LO feed-through (LOFT) overs a wide gain control range [6, 28]. The up-converted RF signal is then amplified through multi stages of programmable gain amplifiers to achieve the desired RF output power. Up to today, most RF PAs with high power output and switches are still not integrated into cellular and WLAN CMOS/Bi-CMOS transceiver ICs. Thus, the PA driver that usually designed as the final stage and biased in the class-AB mode is capable of driving a 50 Ω load through a balun and is internally matched to a differential load of 100 Ω [6]. The class-AB nature of the PA driver can achieve relatively high efficiency compared with the class-A mode PA meanwhile can keep good linear characteristic. It was reported in [6] that the transmitter can achieve the output P1dB powers of +14 and +16 dBm, respectively in the A-band and G-band. To achieve the required output power as listed in Table 3.3 in the WALN applications, an external PA is needed.

Two important specifications in the transmitter are error vector magnitude (EVM) and transmitted spectral mask or adjacent channel power ratio (ACPR). EVM is used to evaluate the performance of the transmitted signal in-channel, while ACPR is utilized to assess interference performance to adjacent channels. Both of them are highly determined by PA nonlinearity. Distortion caused by PA nonlinearity can be partially compensated by either pre-distortion or post-distortion technique.

EVM: Different impairment sources in the transmit chain can contribute to overall EVM, which can be approximately expressed with each major impairment source

$$\mathrm{EVM_{TOT}}\ (\%) \approx \sqrt{\mathrm{EVM^2_{PA\_NL}} + \mathrm{EVM^2_{LO\_PN}} + \mathrm{EVM^2_{IQ\_IMB}} + \mathrm{EVM^2_{CFR}}} \quad (3.59)$$

where each EVM item in the square root is caused by PA nonlinearity, LO phase noise, I–Q imbalance, and crest factor reduction, respectively. It is assumed that matching between an EVM test point and transceiver chipset output is ideal. Otherwise an additional EVM item related to matching needs to be added inside the square root.

In (3.59), $\mathrm{EVM_{PA\_NL}}$ can be compensated by PA pre-distortion methods, $\mathrm{EVM_{IQ\_IMB}}$ can be reduced by the I–Q calibration in the digital domain. However, the terms of $\mathrm{EVM_{LO\_PN}}$ and $\mathrm{EVM_{CFR}}$ cannot be compensated. In the 802.11a WLAN system, $\mathrm{EVM_{LO\_PN}}$ can be calculated by integrating LO phase noise from a pilot effective tracking frequency to the 20 MHz channel bandwidth based on the measured LO phase noise versus frequency offset. The pilot tracking range is about

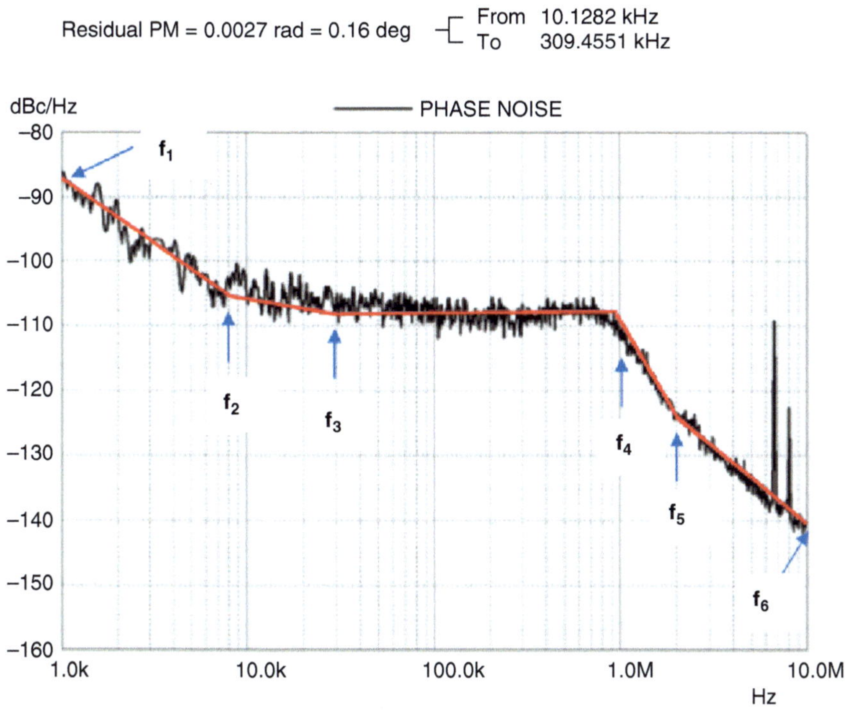

Fig. 3.28 Measured phase noise of PLL at transmitter output with the frequency 5.24 GHz. Referenced from [6]

Table 3.5 Approximate phase noise straight line segments of PLL at 5.24 GHz

i	1	2	3	4	5	6
f_i (Hz)	1×10^3	7×10^3	31×10^3	1×10^6	2×10^6	10×10^6
$L(f)$ (dBc/Hz)	−87	−105	−108	−108	−125	−140
a_i(dBc/decade)	−21.30	−4.64	0	−56.47	−21.46	N/A
$L(f_i)$ (dBc)	−87	−105	−108	−108	−125	−140
MS PN_i (dBc) (from i to $i+1$)	N/A	−55.0	−60.0	−45.1	−51.8	−60.3

10% of the subcarrier spacing 312.5 kHz, or 31 kHz. Within this range, the common frequency and phase errors are removed before demodulation in the receiver. Thus, EVM is a function of integrated phase noise beyond the tracking bandwidth and inside the channel bandwidth (add 3 dB to correct SSB to DSB).

For example, the measured phase noise (PN) of the PLL for the 802.11a WALN at 5.24-GHz frequency is shown in Fig. 3.28 [6]. The integrated PN within a defined frequency range can be read on the screen of the spectrum analyzer when the integrated PN measurement setting is turned on. Here we can use the straight line segments to approximate the measured PN curve and then integrate the phase noise in a desired frequency range. PN line segments from 31 kHz to 10 MHz are used to calculate the EVM and are listed in Table 3.5 with the integrated range given below:

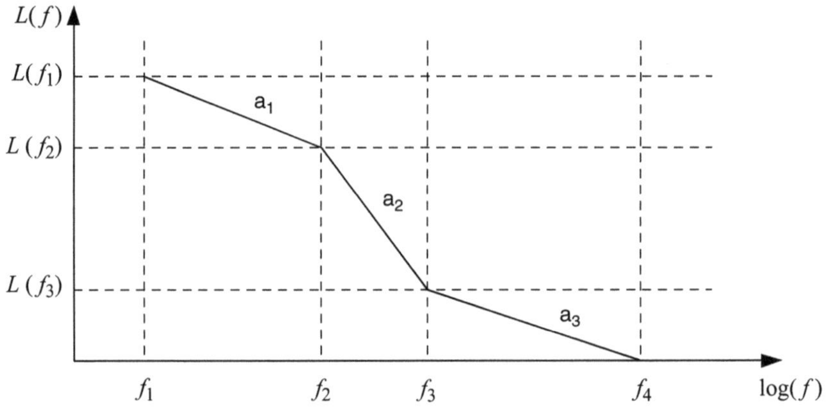

Fig. 3.29 A typical phase noise spectrum function $L(f)$

- Starting frequency of the integration: 31 kHz (or 10% of subchannel spacing 312.5 kHz)
- Stopping frequency of the integration: 10 MHz, a single side band (SSB)

To translate the phase noise to EVM in an OFDM system, we assume the following conditions are met:

- Linear impairment errors are corrected with equalization.
- Pilot tracking completes up to 10% of subcarrier spacing (or 31 kHz for 802.11a WLAN), and frequency offset and phase error are compensated during the tracking period.

Then EVM contributed by only the PN can be approximated within a desired frequency range by using the following equation

$$\mathrm{EVM}\,(\mathrm{dB}) = \mathrm{PN}\,(\mathrm{dBc}) \tag{3.60}$$

where the PN is obtained by integrating the phase noise from the pilot tracking ending frequency of 31 kHz to the channel bandwidth of 10 MHz for the 802.11a WLAN signal.

The PN is calculated based on the phase noise spectrum $L(f)$, which is defined as the attenuation in dB from the peak value $S_c(f)$ of the power spectral density of a clock signal at the clock frequency f_c to a value of $S_c(f)$ at the frequency offset f as shown in Fig. 3.29 [29]. Thus, the phase noise spectrum $L(f)$ can be expressed as:

$$L(f)(\mathrm{dB}) = 10\log\left[\frac{S_c(f)}{S_c(f_c)}\right] \tag{3.61}$$

$L(f)$ in (3.61) represents the ratio of two spectral amplitudes at the frequencies f and f_c. In practice, the RF carrier tone at the frequency of f_c is first down-converted to the baseband in a spectrum analyzer and then the phase noise spectrum $L(f)$ is measured at the baseband as illustrated in Fig. 3.28, where the measured curve presents the phase noise spectrum $L(f)$ in dBc/Hz versus the frequency offset from zero frequency.

The phase noise spectrum $L(f)$ can usually be approximated by a linear piecewise function based on the measured $L(f)$ curve when the frequency axis is in log scale. In such a case, $L(f)$ can be written as [29]:

$$L(f) = \sum_{i=1}^{K-1} [a_i(\log(f) - \log(f_i)) + L(f_i)] \tag{3.62}$$

where a_i is the slope of the line segment from f_{i-1} to f_i and is given by

$$a_i = \frac{L(f_{i+1}) - L(f_i)}{\log(f_{i+1}) - \log(f_i)} \tag{3.63}$$

The mean square (MS) of the phase noise PN can be calculated by:

$$\mathrm{PN_{MS}} = 2\int_0^\infty 10^{\frac{L(f)}{10}} df \tag{3.64}$$

where a constant of 2 is used due to single side band (SSB) to double side band (DSB) conversion. The root mean square (RMS) of the phase noise can be obtained through (3.64) as

$$\mathrm{PN_{RMS}}\,(\mathrm{rad}) = \sqrt{\mathrm{PN_{MS}}} = \sqrt{2\int_0^\infty 10^{\frac{L(f)}{10}} df} \tag{3.65}$$

Substituting a_i in (3.63) into $L(f)$ in (3.62), and then substituting $L(f)$ into (3.64), we have the MS phase noise expression as follows:

$$\begin{aligned} \mathrm{PN_{MS}} &= \sum_{i=1}^{K-1} 2 \times 10^{\frac{L(f_i)}{10}} f_i^{\frac{-a_i}{10}} \left(\frac{a_i}{10} + 1\right)^{-1} \left(f_{i+1}^{\frac{a_i}{10}+1} - f_i^{\frac{a_i}{10}+1}\right) \\ &= \sum_{i=1}^{K-1} \mathrm{PN}_i \end{aligned} \tag{3.66}$$

where PN_i is the MS phase noise in the range from f_{i-1} to f_i. The RMS phase noise is given by

$$\mathrm{PN_{RMS}}\,(\mathrm{rad}) = \sqrt{\mathrm{PN_{MS}}} \tag{3.67}$$

Thus, the integrated phase noise values in the different frequency ranges are listed in Table 3.5. EVM contributed from the phase noise in the frequency range from f_3 to f_6, or 31 kHz to 10 MHz, can be calculated with (3.67) by using the MS phase noise values in these ranges

Table 3.6 Approximate phase-noise straight line segment of PLL at 5.24 GHz

i	**1**	**2**
f_i (Hz)	10×10^3	310×10^3
$L(f)$ (dBc/Hz)	−105	−110
PN_i (dBc) (from i to $i+1$)	N/A	−50.77

$$\begin{aligned}\mathrm{EVM} &= 20\log(\mathrm{PN_{RMS}}) = 20\log\left(\sqrt{\mathrm{PN_4} + \mathrm{PN_5} + \mathrm{PN_6}}\right) \\ &= 20\log\left(\sqrt{10^{-45.1/10} + 10^{-51.8/10} + 10^{-60.3/10}}\right) \\ &= -44.15\,(\mathrm{dB})\end{aligned} \tag{3.68}$$

Even with the integrated frequency range from 1 kHz to 10 MHz, EVM contributed by the phase noise is equal to −43.70 dB, which is still relatively small. The measured EVM for the legacy 802.11a mode based on such RMS phase noise is actually −40 dB at −5 dBm TX output power [6]. The difference between them is due to other impairment contributions affecting the EVM, such as I–Q imbalance, quantization error of DACs, and any nonlinearity from the mixers and PA driver.

To further verify the accuracy of this approximation method, we approximate the phase noise in a certain frequency range from about 10 to 310 kHz as shown in Fig. 3.28, where the measured RMS phase noise PM in radian shown on the screen is 0.0027 rad. One linear piecewise can be used to approximate the phase noise in this range, which is listed in Table 3.6, where $\mathrm{PN}_i = -50.77$ is calculated by using (3.66).

By substituting $\mathrm{PN_2} = -50.77$ into (3.67), we have the RMS phase noise in radian as follows:

$$\mathrm{PN_{RMS}} = \sqrt{10^{-50.77/10}} = 0.0029\,(\mathrm{rad}) \tag{3.69}$$

The approximated phase noise is very close to the measured value of 0.0027 rad as shown in Fig. 3.28. Therefore, (3.66) and (3.67) can approximate either the phase noise of the clock signal or the PLL output signal as well as EVM contributed by the phase noise.

In addition, the period jitter J_{PER} of the clock at the frequency of f_c can be calculated from the RMS phase noise $\mathrm{PN_{RMS}}$ as follows:

$$J_{\mathrm{PER\_RMS}}(s) = \frac{\mathrm{PN_{RMS}}}{2\pi f_c} \tag{3.70}$$

where the unit of jitter is second.

Another impairment that affects EVM is the I–Q gain and phase imbalance errors. Unlike the PN, the I–Q imbalance errors can be minimized by calibration, which is usually performed during the calibration procedure. The cancellation of the I–Q imbalance will be introduced in Chap. 6. The calibrated EVM due to the I–

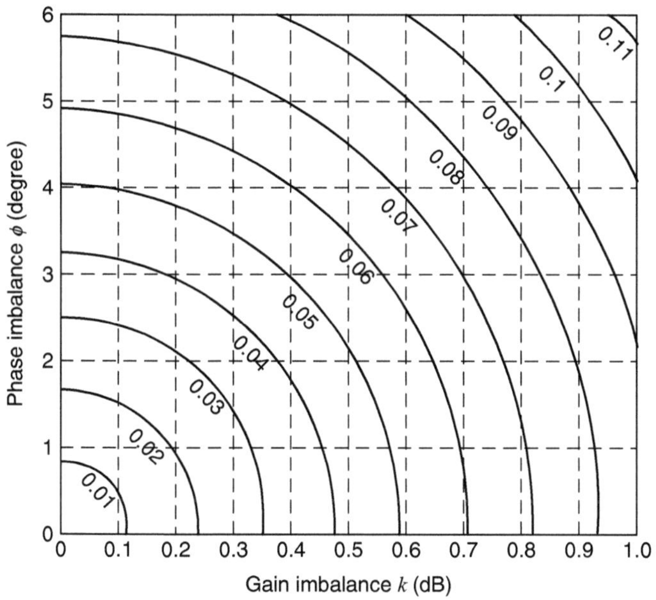

Fig. 3.30 EVM variation versus transmitter I–Q gain and phase imbalance for *M*-QAM modulation format. Redrawn from [30]

Q imbalance errors should be lower than the overall required EVM in order to leave enough margins. Figure 3.30 illustrates EVM versus I–Q imbalance error for *M*-QAM modulation format in a single carrier system. In this figure, other impairments or imperfections, such as LO phase noise and PA nonlinearity, are neglected or minimized. The minimum EVM curve shown in Fig. 3.30 is 0.01 (or 1%) at a gain imbalance of about 0.12 dB and a phase imbalance of 0.8°, which leads to $\mathrm{EVM} = 20\log(0.01) = -40$ dB due to the I–Q imbalance errors. The EVM curves shown in Fig. 3.30 can be also applied to *M*-QAM modulation format in an OFDM system even though it is derived from a single-carrier system.

The last term in (3.59) is caused by intentionally reducing PAPR to improve the efficiency of power amplifiers, and cannot be minimized due to the nonlinear distortion. The CFR scheme can be used for low-rate data frames (6–24 Mbit/s) without violating the EVM requirement of the 802.11a specification. For high-rate data frames (36–54 Mbit/s), however, the CFR scheme is usually not adopted because of the EVM degradation, which equivalently results in the degradation of the required higher SNR for reliable detection [25].

In addition, one dominate impairment that degrades EVM in the transmit chain is nonlinearity of the PA. In order to achieve high efficiency, the PA is preferred to operate close to its P1dB point. However, the EVM performance in the transmitter degrades dramatically when the PA operates close to its P1dB point. Therefore, the transmitter's EVM expressed in (3.59) is dominated by the item $\mathrm{EVM}_{\mathrm{PA\_NL}}$ when the PA operates near its P1dB point. To determine an optimal back-off (BO) from the P1dB point, we measure EVM versus the BO from P1dB by using the legacy 802.11a WLAN signal with a rate of MCS7 and HT-20 at the carrier frequency of 5500 MHz on a Wi-Fi transceiver chipset of the 802.11ac standard as shown in Fig. 3.31. In the measurements, an external PA is added at the output of the RF

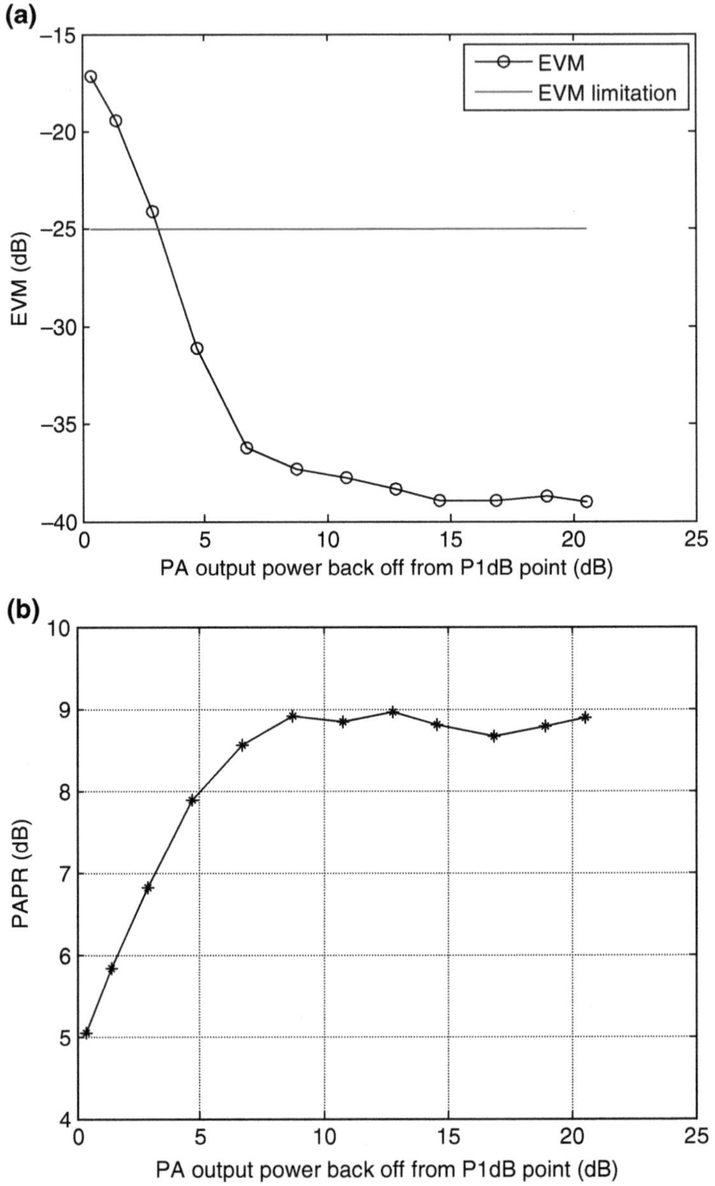

Fig. 3.31 Measured EVM and PAPR versus PA output power back-off from the P1dB point: (**a**) Measured EVM versus PA output power back-off from the P1dB point, and (**b**) PAPR versus PA output power back-off from the P1dB point

transceiver to achieve a larger coverage range. The P1dB measured at the external PA output is about 2 dB less than the saturation power.

It can be seen from Fig. 3.31b that the maximum PAPR in the linear region is about 9 dB and PAPR is reduced when PA BO decreases due to PA's clipping. As a rule of thumb, a PA is preferable to operating at a BO from P1dB by a PAPR value in order to avoid EVM degradation. Thus, the preferred output power of the PA is calculated by

$$\begin{aligned} P_{\mathrm{OP}}(\mathrm{dBm}) &\approx \mathrm{P1dB} - \mathrm{PAPR} \\ &= \mathrm{P1dB} - 9 \end{aligned} \tag{3.71}$$

At this 9-dB BO, EVM degrades about 2 dB compared with a minimum EVM as shown in Fig. 3.31a.

DC Offset and I–Q Imbalance Calibration: In a quadrature modulation structure, the in-phase and quadrature baseband signals modulate a pair of orthogonal carrier signals and are then summed to form a modulated RF signal for transmission.

As described in the previous chapters, a direct up-conversion modulator is subject to several error sources of DC offsets, quadrature phase and gain imbalances. The DC offset errors determine the carrier suppression and the gain and phase imbalances affect sideband suppression. If these errors are not small enough, they will impair the accuracy of the RF modulated signal, or EVM of the RF modulated signal, especially for higher order modulation schemes such as 64-QAM. DC offset and I–Q imbalance errors can be minimized through the calibration procedure before transmission. A calibration loop back is usually built into the RF transceiver chip and calibration process is performed after the RF transceiver is powered on. The I–Q imbalance calibration will be discussed in Chap. 6.

Transmit Spectrum Mask: As mentioned above, besides EVM specification, another critical specification in the transmit chain is the transmit spectrum mask (TSM) that limits the transmitted power spectral density (PSD) within. The main purpose of defining TSM is to minimize the adjacent channel power ratio (ACPR), which is the ratio of the power in the adjacent channel to the power in the main channel. The 802.11a specification defines the transmitted spectrum mask to be 40 dB down in the alternate channels beyond the frequency offset of ± 30 MHz from the desired channel center. Meanwhile considering a maximum 10-dB PAPR value of the OFDM signal, we might leave as much as a 10-dB back-off from DAC's full scale. Thus, a total of 50 dB down is required beyond ± 30 MHz frequency offset. A 9-bit DAC has a dynamic range of 54 dB or achieves a SNR of 54 dB to cover a total of 50-dB down range. A 4-dB margin is obviously not enough. So a 10-bit DAC is needed to provide a 10-dB margin. When DAC employs 160 MHz oversampling frequency as introduced in [25], an additional 1.5-bit resolution due to oversampling ratio of 8 can be equivalently achieved, which leads to an additional $\mathrm{SNR} = 1.5 \times 6 = 9$ dB. Thus a 9-bit DAC combined with oversampling frequency of 160 MHz results in a dynamic range of 63 dB,

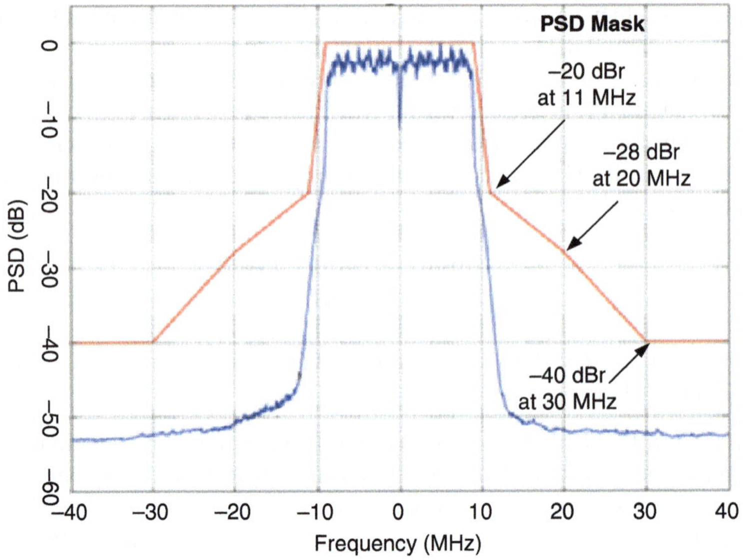

Fig. 3.32 Power spectral density of the 802.11a WLAN OFDM signal with a data rate of 54 Mbps and bandwidth of 20 MHz at the carrier frequency of 5500 GHz and PA operates at a 9 dB back-off from the P1dB point

which gives a 13-dB margin. As mentioned in [25], an 8-bit DAC with such an oversampling frequency would be sufficient enough to give a 7-dB margin, but a 9-bit DAC provides an adequate margin for the following circuit blocks to relax their design requirements.

Nonlinearity of the PA has a significant effect on the transmitted spectral density. Similar to the PA nonlinearity effect on EVM, the PA output power should be backed off by the PAPR value from the P1dB point to leave an enough margin from the transmitted spectral mask. Figure 3.32 shows the measured PSD at the PA output for the 802.11a WLAN OFDM signal with a data rate of 54 Mbps, bandwidth of 20 MHz, PAPR value of 9 dB, and a 64-QAM modulation format. The upper curve represents the required transmit spectrum mask and the lower curve is the measured PSD when the PA operates at a 9-dB back-off from the P1dB point. The spectrum mask is divided into four frequency ranges, from 9 to 11 MHz range, from 11 to 20 MHz range, from 20 to 30 MHz range and greater than 30 MHz range. It can be seen that the minimum margin is greater than 11 dB across overall mask ranges.

To achieve relatively high efficiency, however, the PA should operate close to the P1dB point. To determine an appropriate back-off from the P1dB point, we measure the PSD margin values versus the BO from the P1dB point in these spectrum mask ranges described above and illustrate these curves in Fig. 3.33. It can be seen that PSD does not degrade until BO is less than 5 dB from P1dB. At a 5-dB BO, the minimum margin among three different mask frequency ranges is about 12 dB from the spectral mask.

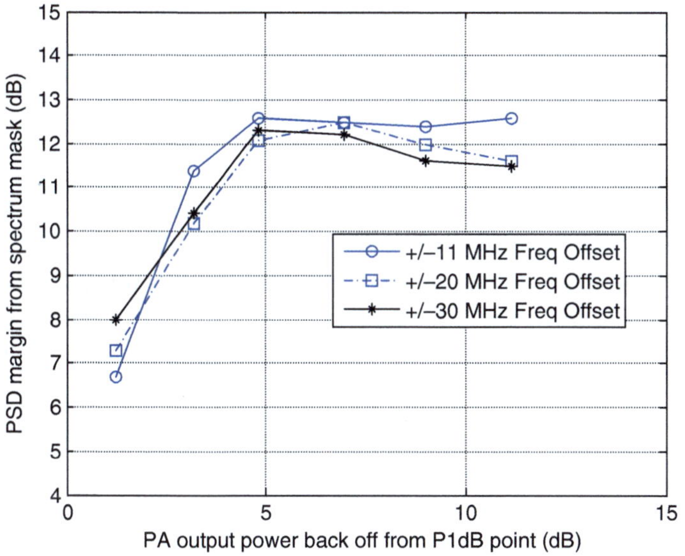

Fig. 3.33 PSD margin from transmit spectral mask versus PA output power back-off from the P1dB point

Compared with the curve of EVM versus BO shown in Fig. 3.31a, the requirement of PA BO for PSD is less strict. For example, a 6-dB BO from the P1dB gives a 10-dB margin for EVM while a 3-dB BO achieves a 10-dB margin for PSD. Therefore, the requirement of PA BO for EVM is more stringent than the requirement for PSD, and PA BO should be determined by the tolerance of EVM degradation.

3.4.1.3 Receiver Chain Design Challenges

Generally, a RF receiver consists of low noise amplifiers (LNA), quadrature mixers, variable gain amplifiers (VGA) and lowpass filters (LPF) and performs three major functions as follows:

- Properly amplify or attenuate the received RF signal to achieve the I-Q baseband signals with the desired level at the inputs of ADCs after frequency down-conversion.
- Attenuate outside channel interferes and noise through filters to achieve the maximum SNR before the decision.

To achieve the optimal receiver performance, the 802.11a WLAN defines two important specifications for different data rates, which are the minimum input level sensitivity and adjacent channel rejection (ACR). The sensitivity is defined as the minimum input level measured at the antenna connector when the packet error rate (PER) shall be less than 10% in the receiver. The sensitivity level is rate-dependent and is specified in a range from −82 dBm for the lowest data rate of 6 Mbits/s to −65 dBm for the highest date rate of 54 Mbits/s. The ACR is defined as the measured power difference between the adjacent channel interfering and the desired channel signal when 10% PER is reached by setting the desired signal level 3 dB above the rate-dependent sensitivity and raising the power of the adjacent channel interfering signal.

Sensitivity System Requirements: In the 802.11 WLAN time-division multiplexing (TDM) systems, the receiver sensitivity test is carried out when the transmitter is turned off. The sensitivity level or minimum level measured at the antenna connector can be calculated as:

$$P_{\text{sen}}\,(\text{dBm}) = -174\,(\text{dBm/Hz}) + \text{NF}\,(\text{dB}) + 10\log(B_{\text{enb}})(\text{Hz}) + \text{SNR}(\text{dB}) \tag{3.72}$$

where NF is the receiver noise figure (NF), B_{enb} [38] is the double-sideband noise bandwidth of the receiver, SNR is the signal-to-noise ratio to achieve the required PER of 10%, and kT at a room temperature of 17 °C (290 K) is −174 dBm/Hz, in which k is a Boltzmann constant and T is the absolute temperature in degrees Kelvin. The sensitivity equation above is derived in detail in Sect. 7.6.1 of Chap. 7. The noise bandwidth B_{enb} [38] is usually not equal to a 3-dB bandwidth $B_{3\,\text{dB}}$ of the receiver chain, but in practice the 3-dB bandwidth could be used to approximate the noise bandwidth in the system simulation. The relationship between B_{enb} and $B_{3\,\text{dB}}$ depends on the number of poles in the transfer function. Typically, B_{enb} is equal to about $1.1 \times B_{3\ \text{dB}}$ when the number of poles is greater than 3 and a single pole roll off is dominated. In practice, NF or P_{sen} is measured at either the LNA input or antenna connector, depending on actual applications. The difference between measurement at the LNA input and measurement at the antenna input lies in a loss of the front-end block between them.

In (3.72), the first three parameters are the total noise power in the receiver equivalent noise bandwidth of B_{enb} and the last parameter of SNR depends on the modulation and error correction coding formats. NF is the most important parameter to be carefully considered during the design of the 802.11a receiver system because it is the only parameter that can be minimized by optimizing the front-end circuit designs. The 802.11a WALN standard defines the sensitivity levels for different data rates in [3], in which NFs of 10- and 5-dB implementation losses are assumed. With today's processing technologies, however, most RF transceivers can achieve the sensitivity levels much lower than those specifications defined in [3].

Usually, BER or PER is originally related to the bit energy to noise density ratio of E_b/N_o and the relationship between E_b/N_o and S/N is given as follows [31]:

$$\frac{E_b}{N_o} = \frac{S}{N} \times \frac{B_{\text{enb}}}{f_b} \tag{3.73}$$

where B_{enb} is the receiver noise bandwidth and f_b is the bit rate. From the equation above, we also have

$$\frac{S}{N} = \frac{E_b}{N_o} \times \frac{f_b}{B_{enb}} \tag{3.74}$$

which can be expressed in decibels,

$$\left[\frac{S}{N}\right]_{dB} = \left[\frac{E_b}{N_o}\right]_{dB} + 10\log\frac{f_b}{B_{enb}} \tag{3.75}$$

From (3.75), we can see that E_b/N_o is independent of the bit rate and noise bandwidth while S/N is related to them. The ratio of f_b/B_{enb} is approximately equal to the spectral efficiency in bit/s/Hz. If the noise bandwidth B_{enb} is replaced with the channel bandwidth B_w, the ratio f_b/B_w is the spectral efficiency in bits/s/Hz. Hence, the spectral efficiency or bandwidth efficiency is defined as:

$$\eta_s = \frac{f_b}{B_w}\ (\text{bit/s/Hz}) \tag{3.76}$$

Here we give one example of calculating the sensitivity of the 802.11a WALN OFDM signal with a data rate of 54 Mbits/s to achieve a BER of 10^{-5}.

Design Example 3.1 For a data rate of 54 Mbits/s with a coding rate of 3/4 in the 802.11a, the minimum required E_b/N_o to achieve a system BER of 10^5 is about 12 dB in an additive white Gaussian noise (AWGN) channel [4]. A 12 dB E_b/N_o includes a 5-dB coding gain. Since there are a total of 52 subcarriers and the subchannel spacing is 312.5 kHz, the signal occupied bandwidth is 52×312.5 kHz = 16.25 MHz. Assuming the noise bandwidth B_{enb} is equal to 18.3 MHz, which corresponds to 1.1 times the 3-dB bandwidth of 16.6 MHz for the receiver channel selection filter. What is the sensitivity level needed by the receiver?

Solution The required S/N to achieve the BER of 10^{-5} is

$$\begin{aligned}\left[\frac{S}{N}\right]_{dB} &= \left[\frac{E_b}{N_o}\right]_{dB} + 10\log\frac{f_b}{B_{enb}} \\ &= 12 + 10\log\frac{54\ \text{Mbits/s}}{18.3\ \text{MHz}} = 16.7\,(\text{dB})\end{aligned} \tag{3.77}$$

Considering a 3-dB implementation loss due to the I–Q imbalance error, carrier frequency, and timing synchronization errors, a typical OFDM system requires approximately 20-dB (16.7 + 3) of SNR. Hence, we have the sensitivity level to achieve BER = 10^{-5} as follows:

$$P_{sen} = -174(\text{dBm/Hz}) + \text{NF} + 10\log\left(18.3 \times 10^6\right) + 20(\text{dB}) \tag{3.78}$$

To achieve the reference sensitivity of −65-dBm for the 802.11a 64QAM modulation format with a data rate of 54 Mbits/s, we have the required NF at the antenna connector from (3.78) as follows:

$$\mathrm{NF} = -65 + 81.7 = 16.7(\mathrm{dB}) \tag{3.79}$$

With antenna switch insertion loss (IL) ranging from 0.8 to 1.2-dB, and diplexer IL ranging from 0.7 to 1-dB in the 5-GHz band for separating 2- and 5-GHz WLAN bands, the NF referred to the LNA input is 14.5–15.2-dB, which gives plenty of headroom for RF IC designers to pass the reference sensitivity. On the current markets, most 802.11a WLAN transceiver chips can achieve a 5–6-dB NF. With a 6-dB NF, the sensitivity referred to the input of the receiver is

$$\begin{aligned} P_{\mathrm{sen}} &= -174(\mathrm{dBm}) + 6(\mathrm{dB}) + 10\log\left(18.3 \times 10^6\right) + 20(\mathrm{dB}) \\ &= -75.4(\mathrm{dBm}) \end{aligned} \tag{3.80}$$

As a consequence, most transceivers pass the reference sensitivity test with at least an 8-dB margin.

A realistic 802.11a WLAN system requires about a SNR of 20 dB to pass 10% PER [32]. Hence, a PER of 10% is equivalent to BER of 10^{-5} for the 802.11a WALN system to achieve the same SNR of 20 dB in an AWGN channel.

Dynamic Range requirement: The dynamic range of the receiver is capable of handling strong signals well as it is able to reliably pick up weak ones even in the presence of nearby strong interferers. In the 802.11a WLAN specifications, a maximum receive signal level is −30 dBm measured at the antenna connector and a minimum receive signal level for the data rate of 6 Mbits/s is −82 dBm. However, a range from −20 to −92 dBm is appropriate for most transceivers with enough dynamic ranges. The received signal level measured at the antenna input is in the range from its minimum level to its maximum level; the final signal level reaching to the input of the ADCs should be around a certain desired level after appropriate amplification. To reduce the dynamic range requirements of the ADC, the lowpass filter and the variable gain amplifier (VGA) along with the receive chain provide filtering function to the blocking signals and the adjacent channel signals and amplifying function to the desired signal, respectively.

To handle such a large signal variation, the maximum gain of the receiver can be determined by amplifying the weakest signal at the antenna connector to the desired level at the ADC input. Considering non 50 Ω impedances for the VGA circuits, we use the voltage gain to determine the maximum gain of the receiver. Assuming that 1 V peak-to-peak differential voltage with a sinusoidal waveform is required at the output of the last stage before the ADC in the receiver chain and the weakest signal of −92 dBm is measured at the antenna connector port, their corresponding RMS voltage with a 50 ohm load are $0.3536V_{\mathrm{rms}}$ and $5.623\mu V_{\mathrm{rms}}$, respectively. Thus, these two values expressed in $\mathrm{dB}V_{\mathrm{rms}}$ are equal to $20\log 10(0.3536) = -9\mathrm{dB}\,V_{\mathrm{rms}}$

and $20\log10(5.623 \times 10^{-6}) = -105\,\text{dB}V_{\text{rms}}$, respectively. The required voltage values expressed in dBV_{rms} are given as

$$\text{At ADC input}: -9\ \text{dB}\,V_{\text{rms}} - 10\,\text{dB} - 5\,\text{dB} = -24\,\text{dB}\,V_{\text{rms}} \tag{3.81}$$

$$\text{At antenna input}: -105\,\text{dB}\,V_{\text{rms}} - 2\,\text{dB} = -107\,\text{dB}\,V_{\text{rms}} \tag{3.82}$$

In (3.81), a 10-dB PAPR and a 5-dB fading variation are considered as headroom, while in (3.82) a 2-dB loss from the antenna connector to the input of the LNA is added. Thus, the maximum gain of 83 dB is needed to handle the weakest signal. The total gain of 83 dB consists of the LNA gain, the mixer gain, and the VGA gain. The gain distribution along the receiver chain depends on the received signal power at the antenna port and should be set to achieve optimal SNR and required minimum P1dB to signal power ratio (P1dB/S). P1dB/S minimizes the nonlinearity effect on the performance of the receiver and should be greater than the signal PAPR value.

In the 802.11a system, the minimum dynamic range of the ADCs is mainly determined by adjacent channel interferers. For an 802.11a WLAN OFDM signal with a data rate of 54 Mbits/s, the adjacent channel (AC) interferers at a frequency offset of 20 MHz are 1 dB smaller than the desired signal, the alternative adjacent channel (AAC) interferers at the frequency of 40 MHz are 15 dB larger than the desired signal. The following calculations assume a third-order lowpass filter with a 3-dB cutoff frequency of 10 MHz with 5% tuned accuracy. Thus these adjacent channel interferers are minimally attenuated by 5 and 10 dB, respectively at the channel edges that are close to the signal channel. When the adjacent channel-interfering signals are properly scaled to the input range of the ADCs, the ADCs need to feature a minimum dynamic range for different interfering signals, respectively

$$\text{At AC}: -1\,\text{dB} - 5\,\text{dB}\,(\text{filter att.}) + 20\,\text{dB}(\text{SNR}) = 14\,\text{dB} \tag{3.83}$$

$$\text{At AAC}: 15\,\text{dB} - 10\,\text{dB}\,(\text{filter att.}) + 20\,\text{dB}\,(\text{SNR}) = 25\,\text{dB} \tag{3.84}$$

Clearly, (3.84) is more stringent than (3.83) with respect to the required dynamic range. In the following calculation for the dynamic range requirement of the ADCs, (3.84) is used to determine the required dynamic range.

Theoretically the SNR of 20 dB in (3.83) and (3.84) is the minimum requirement to meet $\text{BER} = 10^{-5}$ or $\text{PER} = 10\%$, but an extra SNR of 10 dB is needed to achieve relatively high performance, including some margins for combating multipath fading. Furthermore, headroom of 10 dB should be reserved due to a 10-dB PAPR for the 64QAM OFDM signal. As a result, the final required dynamic range of the ADC is

$$25\,\text{dB} + 10\,\text{dB}\,(\text{ext. SNR}) + 10\,\text{dB}\,(\text{headroom}) = 45\,\text{dB} \tag{3.85}$$

The dynamic range of 45 dB requires at least an 8-bit ADC with a 20-MHz bandwidth, corresponding to a 48-dB dynamic range. The use of oversampling at the ADCs gains an additional effective bit after filtering. For example, an 80-MHz ADC can gain one additional effective bit. Thus, the 8-bit ADC combined with an oversampling frequency of 80 MHz gives a dynamic range of 54 dB.

Adjacent Channel Rejection: The adjacent channel rejection test in [3] defines the selectivity and rejection requirements of the channel filter. An interfering signal is applied to the antenna port in either AC with a frequency offset of ±20 MHz or AAC with a frequency offset of ±40 MHz from a desired signal, respectively. In the test, total interfering power in either AC or AAC is increased until 10% PER is reached, where the desired signal is set to 3 dB above its rate-dependent sensitivity level. The power difference between the interfering and the desired channel is defined as the adjacent channel rejection (ACR). The ACR is also rate-dependent.

Adjacent channel rejection is mainly determined by the channel filter rejection in the adjacent channel band. If the bandwidth of the channel filter is too narrow, more interfering signal in the adjacent channel is rejected. However, it may also distort the desired signal. On the other hand, if the bandwidth of the channel filter is too wide, less interfering signal is rejected and more white Gaussian noise is passed through. Therefore, the bandwidth of the channel selection filter would be appropriate to achieve a larger SNR in the presence of the adjacent channel interferers.

In practice, a slightly higher ACR figure is typically required to achieve a PER of less than 10%. For a receiver architecture employing direct conversion, the selectivity is achieved in the baseband filters that usually are partitioned into analog and digital filters. An analog filter with low order such as an anti-alias filter can be used to partially attenuate the interfering signals to reduce the dynamic range requirement of the ADCs, while a digital filter achieves channel selectivity after the ADCs. The optimum split between the analog and digital domain depends on the resolution of the ADC and the implementation of the digital filter. In the presence of a large interfering signal, it is preferable to implement the analog filters early in the baseband chain to prevent the VGAs from saturating [33].

The attenuation of the adjacent channel-interfering signals in the analog domain may also depend on the sampling frequency of the ADCs. A higher oversampling frequency can relax the attenuation requirement at the adjacent channels. In order to avoid the anti-alias interference, the analog filter should attenuate the adjacent channel-interfering signal by a certain amount of dB in the frequency range from the sampling frequency minus 10 MHz or $f_{sam} - 0.5 \times B_w$ upwards, where B_w is equal to 20 MHz.

3.4.2 *Digital Baseband and MAC Processor*

The digital baseband (DBB) and MAC processor chip is a back-end unit, where DBB mainly performs the implementations of encoding and decoding procedures, modulation and demodulation schemes, IFFT and FFT operations, signal detection, AGC setting, channel compensation and channel selection filtering, while the MAC processor manages and maintains communications between 802.11 stations, including radio cards and access points by coordinating access to a shared radio channel and utilizing protocols that enhance communications over a wireless medium. The 802.11 MAC, often viewed as the "brains" of the network, uses an 802.11 Physical (PHY) layer, such as an 802.11a RF and DBB transceiver, to perform the tasks of carrier detection, transmission, and reception of 802.11 frames.

3.4.2.1 Digital Baseband Designs

The digital baseband unit basically implements the fundamental functions of the 802.11a physical layer as mentioned above and also applies digital algorithms to compensate for analog impairments and to perform a variety of calibrations for delivering a high-performance and high-efficient transmission. Some challenge algorithms are as follows:

- Crest factor reduction
- Digital pre-distortion
- I–Q imbalance calibration
- Closed loop power control

Crest factor Reduction: To achieve highly efficient transmission, one of effective approaches is to reduce the PAPR value of the transmitted OFDM signal. The PAPR reduction extent may depend on the data rate because of different EVM requirements of the 802.11a specification. For low-rate data frames (6–18 Mbit/s), OFDM symbols can be clipped substantially without violating the EVM requirement. For high-rate data frames (24–54 Mbit/s), whose EVM degradations are heavily related to the PAPR reduction extent, the PAPR reduction should be made carefully to leave an acceptable margin if needed. For such high-rate data frames, a kind of pre-distortion technique rather than clipping is used to create an effectively linear range of the PA [25]. This pre-distortion technique uses a lookup table to dynamically scale up samples with large amplitude, which are expected to be compressed by the nonlinearity of the PA. If the compression of the PA occurs, the final PAPR at the PA output is almost not clipped. As a result, the PA can operate much closer to its P1dB compression point in order to achieve highly energy-efficient transmission.

Digital Pre-distortion: An efficient method to extend the linear range of the PA is digital pre-distortion (DPD), which is usually implemented in either a lookup table or a digital signal processor in the digital domain. Digital pre-distortion can accurately generate the pre-distorted baseband I and Q signals to compensate for the nonlinear distortion caused by the PA. In practice, it is a challenge for RF IC designers to implement the DPD in the digital domain to effectively compensate for the nonlinear distortion because the digitally predistorted I and Q baseband signals have 3-5 times bandwidth of the original I–Q baseband signals. The distorted I–Q signals with the extended bandwidth are strictly limited by the bandwidth of the transmitter, including DAC's bandwidth, especially for wide bandwidths of the 802.11 legacy. Such a limitation of the bandwidth may affect the predistortion accuracy. A digitally simplified predistortion algorithm, however, as mentioned in the paragraph of crest factor reduction, has been reported to be implemented into the 802.11a transceiver chipset [25].

I–Q Imbalance Calibration: The I–Q gain and phase imbalance errors at the transmitter can degrade not only the sideband suppression, but also the EVM of the transmitted signal. This sideband suppression is also called the image rejection ratio (IRR), which quantifies the factor that describes how much the mirror signal is suppressed or removed after up-conversion by combining the I–Q channel signals. During up-conversion the mirror signal is usually folded onto the bandwidth of the desired signal. As a result, the folded mirror signal distorts the desired signal and causes spectrum regrowth. In addition, the I–Q imbalance error at the receiver degrades the EVM of the received signal, which in turn results in BER degradation. Therefore, I–Q imbalance errors should be minimized by means of calibration at both the transmitter and receiver. The calibration procedure will be introduced in in detail Chap. 7.

Closed Loop Power Control: Closed loop power control provides a desired transmitted output power independent of process, temperature, and power supply variations, as well as load mismatches. A RF Wi-Fi transceiver has capability of detecting the transmitted signal strength indicator (TSSI) and the received signal strength indicator (RSSI). TSSI is an indication of the transmitted signal power level at the output of the power amplifier (PA) while RSSI is an indication of the received signal power level in the desired channel after the antenna. To determine TSSI, the PA provides an integrated power detector to the DBB and MAC chip to form closed loop power control within the transmitter system. The transmitter gain in 0.5 dB step is adjusted until the PA output power matches a target level. Thus, the actual PA output power can be precisely controlled within the desired output level.

In addition, when a client or station wants to access the wireless network, it needs to communicate an access point (AP) to which it is associated. An AP and the set of its associated clients are referred to as a "basic service set" (BSS). In WLANs that cover a large area, multiple APs are needed in order to provide contiguous coverage. For the transmission from an AP to a client, the receiver of the client can detect the received signal strength by means of RSSI to determine its appropriate AGC setting along the receive chain, as well as the signal-to-interference-plus-noise ratio (SINR) at the receiver. Meanwhile, the client needs to send the acknowledgment packet back to the AP after successfully receiving the packet from the AP. The SINR information at the receiver is sent with the acknowledge packet back to the AP such that the AP can decide whether to increase the transmitted power level or decrease that at the AP. From the AP standpoint of view, power control mechanism belongs to closed loop control due to feedback power information from the client.

In traditional cellular systems, the goal of power control is usually to maintain the signal-to-interference-plus-noise ratio (SINR) at the receiver as low as possible to achieve a reliable detection through closed loop mechanism within the BSS, and meanwhile to improve system capacity within the BSS since reducing the unwanted signal level to other clients.

3.4.2.2 MAC Basics

The MAC architecture mainly consists of the protocol control unit (PCU), the host interface unit (HIU), and the descriptor-based direct memory access (DMA) engine. The PCU manages all low-level timing-critical aspects of the 802.11 MAC layer function and controls the transfer of frame data between the digital BB and MAC unit, the HIU provides connectivity to the host processor over a PCI bus, and the DMA engine controls information and frame data transfer between the PCU and HIU units [25]. The PCU is the most important unit of the 802.11 MAC layer; its main function is described simply below.

Before transmitting frames, a station must first have access to the medium, which is a radio channel shared with other stations. The 802.11 standard defines two functions of medium access: the distributed coordination function (DCF), which is mandatory, and the point coordination function (PCF). With DCF, WLAN stations contend for access and attempt to transmit frames when there is no other station transmitting. If a station is sending a frame, the other stations must wait until the channel is released.

An important aspect of the DCF is a random back-off timer in a station if it detects a busy medium. If the channel is in use, the station must wait a random period of time before attempting to access the medium again. This prevents multiple stations from trying to send data to access the medium at the same time. When the number of active users increases, the back-off timer significantly reduces the number of collisions and corresponding retransmissions [34].

To ensure complete reception at the receiver, the receiving station needs to send an acknowledgement (ACK) back to the transmitting station if it detects no errors in the received frames. If the transmitting station does not receive an ACK after a specified period of time, it will assume that there was a collision caused by possible severe interference and retransmit the previous frame. If the transmitting station still does not receive the ACK after retransmitting the previous frame within the specified number for retransmitting the frame, it would either transmit the next frame or change the modulation format with a lower-order modulation due to a poor SNR channel.

The PCU implements multiple encryption methods, including the traditional wireless encryption protocol (WEP). The PCU encrypts the frame when the encryption function is enabled and generates the proper checksum value. The PCU also provides power management functions. When working in a multimode network, the WLAN chip set can be programmed to sleep automatically and awake just before the next beacon is scheduled to arrive. By analyzing the incoming beacon, the PCU can determine whether to remain awake for an additional frame or resume sleeping [25].

3.4.3 Radio Front-End Modules

Currently, most WLAN transceivers in portable electronic devices are capable of supporting 802.11 a/b/g/n/ac standards in a dual-band format of 2.4 and 5 GHz. In practical applications, they may need a front-end module (FEM) including an external PA and an external LNA to achieve a large coverage area. Figure 3.34 illustrates a dual-band (2.4 and 5 GHz) application of an 802.11ac Wi-Fi transceiver that is backwards compatible with 802.11b/a/g/n. Traditionally, existing dual-band Wi-Fi products, such as the iPhone 6, can only work on one band, either the 2.4 GHz or the 5 GHz band, at a time despite having dual-band hardware capability. Broadcom, a chip maker, announced at Mobile World Congress in 2015 the first 802.11ac (or 5G Wi-Fi) combo chip in the world, which sports real simultaneous dual band (RSDB) for mobile devices, potentially improving the speed as well as the quality of the connection.

In Fig. 3.34, the matching networks that consist of L and C components are used to create optimal impedance matching to achieve both minimum EVM values across bands in the transmitters or minimum sensitivity levels across bands in the receivers. A single-chip integrated FEM in either the 2.4- or 5-GHz range usually

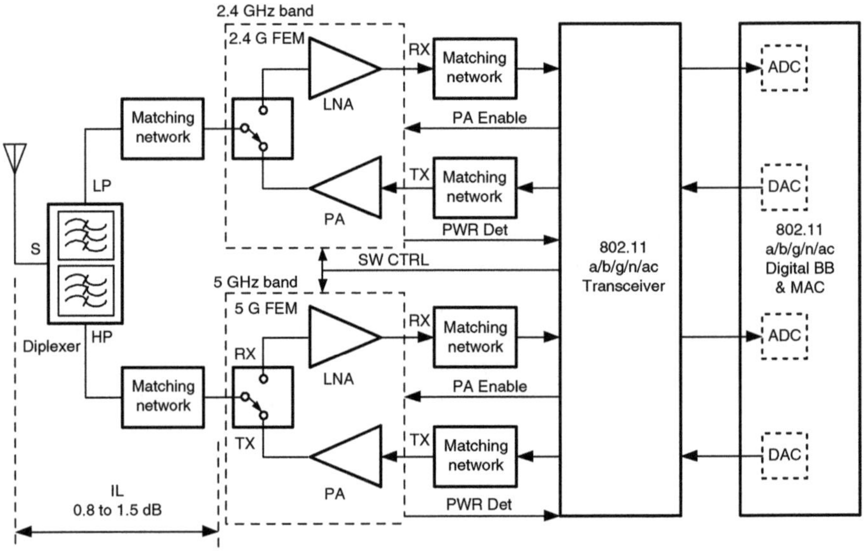

Fig. 3.34 Block diagram of a dual-band single input single output (SISO) WLAN transceiver with external front-end modules

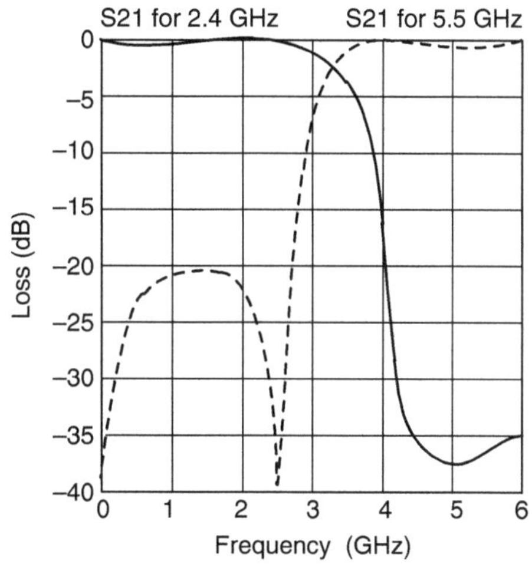

Fig. 3.35 A typical diplexer forward transmission for WLAN at 2.4 and 5.5 GHz

includes a highly linear PA with a power detector, a low NF LNA with bypass capability, and an antenna switch. The gain for the PA can be in the range 25–31 dB, depending on actual applications, and the gain for the LNA is usually around 12 dB with a noise figure in the range from 2–3 dB, depending on the operation frequency bands. A diplexer is a passive device and implements frequency domain multiplexing. Typically, the diplexer consists of a lowpass filter connecting a port LP to a third port S and a highpass filter connecting a port HP to a third port S. The signal power on the port LP is transferred to the port S and vice versa, while the signal power on the port HP is transferred to the port S and vice versa. In WLAN applications, the diplexer is capable of minimum attenuation of about 16–18 dB for the 2.4-GHz signal at the 5-GHz signal band or for the 5-GHz signal at the 2.4-GHz signal band, as shown in Fig. 3.35.

Design Example 3.2 If an RF receiver with a low sensitivity level is required to cover a long range, an external LNA with a low NF can be cascaded with the RF receiver to achieve a lower overall noise figure. In this cascaded case, the external LNA is considered the first stage, while the RF receiver is treated as the second stage. Assume that the external LNA has a noise figure of $NF_{EXT\_LNA} = 3.0\,dB$ and a power gain of $G_{EXT\_LNA} = 12$ dB and is added prior to the receiver with a noise figure of $NF_{RX} = 6$ dB. Assume that the receiver has an equivalent noise bandwidth of 18.3 MHz and needs SNR = 20 dB to achieve the BER of 10^{-5}, which are both obtained in the design example 3.1. (1) What is the noise figure of the two-stage cascaded receiver? (2) What is the sensitivity level referred to an antenna connector or a port S of the diplexer in Fig. 3.34 if the diplexer and matching network together cause an insertion loss (IL) of 1.5 dB?

Solution

1. Assume that there is perfect matching between the external LNA and the RF receiver. The cascaded NF_{CAS} in a two-stage circuit is calculated by

$$\begin{aligned}\mathrm{NF_{CAS}} &= \mathrm{NF_{EXT\_LNA}} + \frac{\mathrm{NF_{RX}} - 1}{\mathrm{G_{EXT\_LNA}}} \\ &= 10^{3/10} + \frac{10^{6/10} - 1}{10^{12/10}} = 2.18\end{aligned} \tag{3.86}$$

So the cascaded $\mathrm{NF_{CAS}}$ in dB is

$$\mathrm{NF_{CAS}}\big|_{\mathrm{dB}} = 10\log\mathrm{NF_{CAS}} = 3.4\,(\mathrm{dB}) \tag{3.87}$$

With such a two-stage cascaded receiver, the noise figure is reduced from 6 to 3.4-dB, or more than a 2.5 dB improvement, which is equivalent to improving the sensitivity level by 2.5 dB at the antenna port compared with the receiver without the external LNA.

This phenomenon of noise figure improvement indicates that the NF of the RF receiver can be reduced by cascading an external LNA with a low NF and a large gain value with it. The noise contributed by the RF receiver decreases as the gain preceding the stage increases, implying that the first stage in a cascade is the most critical. Hence, the LNA noise figure in the first stage active circuitry must be low with a reasonable current consumption. On the other hand, however, any attenuation (loss) in the front-end passive circuitry, such as filters or matching networks, causes the noise figure of the following stage to be increased by the same amount of loss when referred to the input of that stage.

2. With a maximum 1.5-dB IL due to a diplexer and matching network, the NF referred to the antenna connector is $3.4 + 1.5 = 4.9$ dB. The sensitivity referred to the antenna connector is calculated by

$$\begin{aligned}P_{\mathrm{sen}} &= -174 + 4.9 + 10\log\left(18.3 \times 10^{6}\right) + 20 \\ &= -76.5(\mathrm{dBm})\end{aligned} \tag{3.88}$$

In the transmitter, the average output power of the PA should be back-off from its P1dB compression point in order to avoid both EVM degradation and PSD regrowth. As we have known, the back-off required by a practical PA is dependent on the PAPR value of the transmitted OFDM signal. This means that the larger the PAPR, the larger the back-off is required. Since the rate-dependent EVM specification in the 802.11a WLAN system is more stringent than the rate-independent spectrum mask, a certain amount of PA back-off should be determined by the rate-dependent EVM. Generally, the PA back-off should be set in a linear amplification range without significant EVM degradation. In practice, as a rule of thumb, the back-off should be approximately equal to PAPR without significant EVM degradation. If the back-off is less than the PAPR, the transmitted signal will be clipped by the PA, which results in EVM degradation. It can be seen from Fig. 3.31a that for the signal with a PAPR of 9 dB the PA's back-off at the PAPR value slightly degrade EVM about 1–2 dB.

The back-off requirements for the PSD and EVM in the 801.11 WLAN systems are different. For example, to have a 10-dB margin from both EVM specification and spectrum mask in an 802.11a system with a rate of 54 Mbps, the back-off of the PA for the EVM is about 7 dB, as shown in Fig. 3.31a ,while the back-off for the PSD is about 3 dB, as shown in Fig. 3.33. Therefore, the back-off of the PA is more stringent for the EVM than for the PSD in order to achieve acceptable performance.

3.5 Design Applications

After some fundamental design ideas and issues are described for WLAN-based OFDM techniques, several actual products with WLAN based applications are introduced in this section. The demand for higher and higher bandwidths on 802.11 WLAN networks has fueled 802.11 WLAN standard developments—from the early 802.11b standard with 20-MHz bandwidth to the 802.11n standard with 40-MHz bandwidth up to the latest 802.11ac standard with 80-MHz (mandatory) and 160 MHz (optional) bandwidths. In early 2015, only a handful of chip suppliers such as Broadcom, Marvell, Qualcomm-Atheros, Redpine, and Quantenna were able to provide 802.11ac solutions. The 802.11ac standard that operates at the 5-GHz frequency band is the latest development tendency in the industry and is backward compatible with the 802.11a and 11n standards at the 5-GHz frequency band. The 802.11ac-based RF transceivers are discussed in the following section.

3.5.1 Marvell's WLAN 802.11ac Transceiver

Marvell presented a fully integrated three-stream MIMO 802.11ac WLAN SoC that integrates all functions of the 802.11a/b/g/n/ac WLAN standards to achieve a record over-the-air TCP/IP throughput of 1.1 Gb/s at the ISSCC conference in 2014, as shown in Fig. 3.36 [35]. Both transmitter and receiver in the 2.4- and 5-GHz bands utilize direct-conversion architecture. A single all-digital PLL (ADPLL) is used to generate the LO signals in both 2.4- and 5-Hz bands. A single lowpass filter is shared by both transmitter and receiver, and a single analog baseband block is utilized between the 2.4- and 5-GHz-chains to reduce silicon size.

In the transmitter, the pre-power amplifier (PPA) is biased in a Class-AB mode to achieve high power efficiency and good linearity compared with a PA biased in a Class-A mode. The mandatory transmission bandwidths in the 802.11ac are up to 80 MHz compared with the 802.11n standard's maximum bandwidth of 40 MHz. When such a wide-band signal is applied to the PPA, the PA memory effects can result in asymmetry in the PSD. Usually, the memory effects of the PA can arise

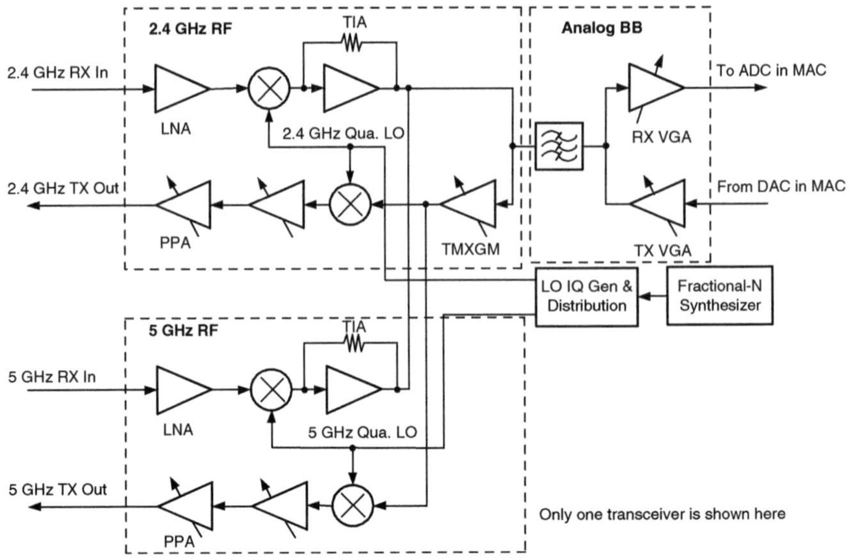

Fig. 3.36 Block diagram of the dual-band 3-stream MIMO 802.11ac WLAN

from multiple sources, including bias circuit effects, self-heating, and trapping effects. One significant source of memory effects in this transceiver is bias voltage draft with the frequencies of the baseband modulation signals [35]. To reduce the memory effects, a wideband bias scheme has been proposed to ensure that the PPA bias has low impedance over baseband frequencies and high impedance at the carrier frequency. For an 802.11ac signal centered at 5,775 MHz with an 80-MHz bandwidth, after the I–Q calibration, TX EVM of a 256-QAM signal achieves an EVM floor of −37 dB up to the PA output power of −5 dBm, as shown in Fig. 3.37.

The inputs of the LNAs in both the 2.4- and 5-GHz bands are single-end, which reduces the number of RF input pins, especially for MIMO architectures. A fifth-order Chebyshev lowpass filter is implemented with programmable gain and bandwidth, supporting all signal bandwidths up to 80 MHz for IEEE 802.11 a/b/g/n/ac standards. The measured receiver noise figures for the 2.4- and 5-GHz bands are 3 and 4.3 dB, respectively. The RF transceiver IC occupies a total die area of 46 mm^2 in a digital 40-nm CMOS process, of which 47% is occupied by the analog and RF circuits.

3.5.2 MediaTek's 802.11a/b/g/n/ac WLAN SoC

Mediate presented a 2 × 2 MIMO 802.11ac Wave 1 (stage 1) Wi-Fi + BT combo SoC chip with integrated dual-band power amplifiers, low noise amplifiers, and T/R switches in 2014 [36]. The proposed broadband TX transmitter can deliver a high output power of 17.5 dBm for the 802.11ac Wave 1 VHT80 (80 MHz bandwidth)

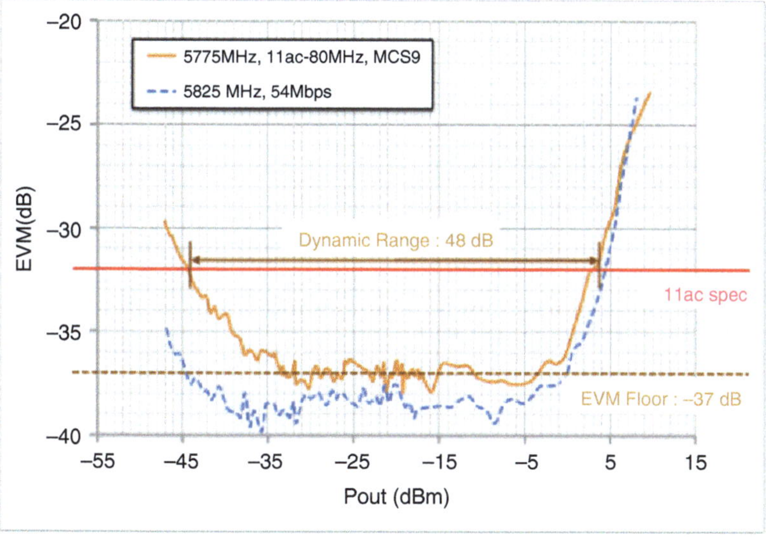

Fig. 3.37 Measured TX EVM at the 5 GHz transmitter output. *Top curve*: 5775 MHz, 802.11ac-80 MHz band, MCS9, *bottom curve*: 5825 MHz, 802.11a 54 Mbps rate. Coped from [35] at Copyright IEEE @ 1998

256-QAM MCS9 (modulation and coding scheme), and is extendable to the next stage 802.11ac Wave 2 VHT160 256-QAM by using a technology called Multi-User MIMO (MU-MIMO). The maximum throughput achieved by using VHT80 Wave 1 with two-spatial stream mode is 580 Mbps in an AWGN channel. The 802.11ac standard operates in the 5-GHz UNII band and is backward compatible with the 802.11a standard. Figure 3.38 shows a top-level block diagram of a 2×2 MIMO 802.11ac Wave 1 WiFi + BT combo SoC chip with integrated dual-band PAs, LNAs, and T/R switches. For simplicity, only one of the two dual-band transceiver block diagram is shown.

MediaTek's WLAN SoC has some particularly worthy qualities of: integrated dual-band PAs, LANs, and T/R switches; simple first-order LPFs after a pair of 10-bit TX DACs with a high sampling frequency of 960 MHz; PA linearization implementation with a digital pre-distortion (DPD) algorithm. These characteristics are described below.

The direct up-conversion (DUC) transmitter and direct down-conversion (DDC) receiver architectures are chosen for both the 2.4- and 5-GHz frequency bands due to their simple implementations. The dual-band PAs and T/R switches are integrated on-chip to further reduce the number of external components, which is the key to low cost and small board size. Hence, MediaTek has become one of several WLAN/BT vendors that have successfully integrated less demanding PAs onto the transceivers. However, it is more challenging for RF IC designers to minimize VCO pulling or VCO phase disturbance caused by a larger output power of the on-chip

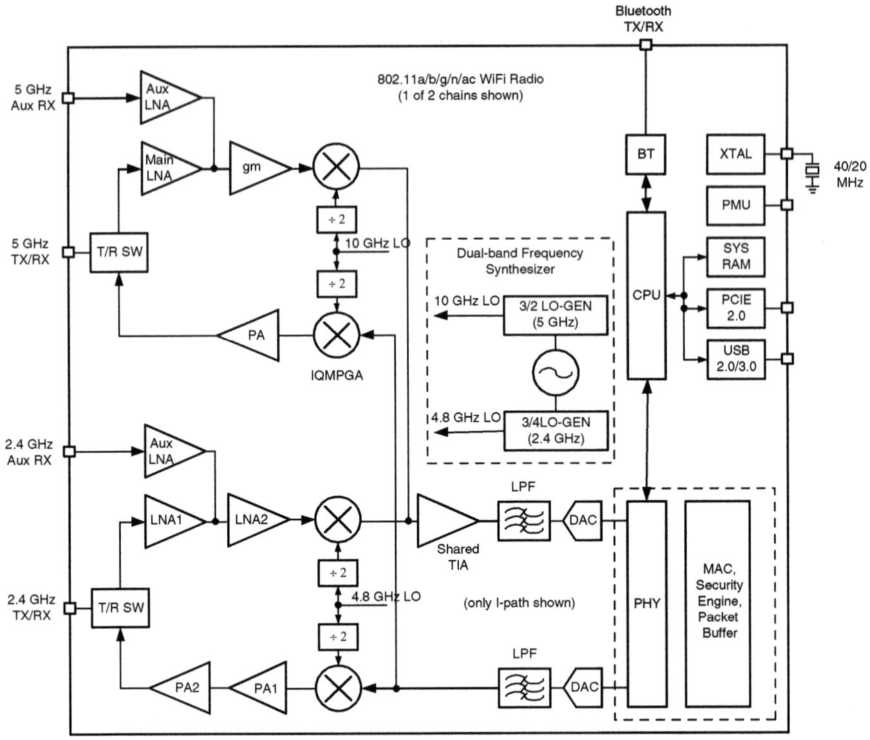

Fig. 3.38 Block diagram of a 2 × 2 MIMO 802.11a/b/g/n/ac WLAN SoC [36]

PA due to the second-order harmonic leakage back to the VCO loop at a DUC TX transmitter.

In order to meet the stringent EVM requirement of the 802.11ac Wave 1 VHT80 256-QAM, the proposed transmitter adopts what is called "broadband" TX architecture, in which a pair of 10-bit DACs are designed to operate at a high sampling rate of 960 MHz and are followed by a simple first-order passive RC lowpass filter for DAC image suppression on the I–Q channels, respectively. A 3-dB corner frequency of the lowpass filter can be larger than the channel bandwidth due to the high sampling rate of the DAC. Therefore, the broadband TX transmitter mitigates the I–Q mismatch errors and eliminates the need for more complex I–Q calibration. The baseband I–Q signals after the LPFs modulate a pair of orthogonal LO signals to form a RF-modulated signal as the input to the power amplifier. At the transmitter, the I–Q modulator and program gain amplifier (PGA) are designed as a cascaded structure called IQMPGA to reduce power consumption and space. The IQMPGA provides a 24-dB gain with a step of 6 dB within ±1-dB accuracy.

The PA is the most critical block in the TX path and dominates the TX EVM budget because of its nonlinearity distortion and TX current consumption due to high output power. With an 80-MHz bandwidth in VHT80, the memory effect of the PA becomes severe and becomes an additional contributor to EVM. Although the PA's nonlinearity distortion accompanied by the memory effect can be compensated by using a complex memory polynomial DPD (as described in Chap. 5), the approach taken in this work, however, uses a simple memoryless polynomial DPD to compensate for nonlinearity distortion only [37]. The PA bias is designed as a Class-AB mode to achieve high linearity and energy efficiency. Because of the larger PAPR value of the OFDM signal with the 256-QAM scheme, the PA bias circuit is designed to adaptively track the peak amplitude of the PA input signal to boost the PA bias voltage, hence improving the PA linearity whenever the input OFDM signal reaches its peak amplitude [36].

To obtain the AM-AM and AM-PM characteristics of the PA, a closed-loop is designed with a dedicated loopback path from the PA output inside the RF transceiver to the RX path [37]. The closed-loop calibration is executed during power-on. The pre-defined ramp signals that are generated from the digital baseband modulate a pair of orthogonal LO signals to form the RF signal. The RF signal is passed through the PA and is partially fed back to the receiver path through a coupler at the output of the PA. The feedback RF signal is then down-converted to the baseband I–Q signals, which are then digitally sampled through ADCs after being passed through the analog lowpass filters on the I–Q channels. The AM-AM and AM-PM characteristics of the PA are estimated by comparing the difference between the transmitted and received digital baseband signals. The AM-AM and AM-PM characteristics of the DPD can be obtained by reversing the estimated AM-AM and AM-PM characteristics of the PA and then stored in the AM-AM and AM-PM lookup tables (LUTs). During a normal transmission mode, the complex baseband signals are pre-distorted by multiplying the outputs of the DPD AM-AM and AM-PM LUTs, respectively. Finally, the pre-distorted RF signal is passed through the PA to minimize the nonlinearity effect of the PA on EVM degradation and PSD regrowth. The measured TX result shows that a single-stage Wi-Fi PA delivered a high output power of about 17.5 dBm for the 802.11ac Wave 1 VHT80/256QAM/MCS9 at the EVM specification of −32 dB [37]. The die size of the SoC is 27.8 mm^2 in 55-nm CMOS technology.

References

1. Chang, R. W., & Gibby, R. A. (1968). A theoretical study of performance of an orthogonal multiplexing data transmission scheme. *IEEE Transactions on Communications, COM_16*(4), 529–540
2. Saltzberg, B. R. (1967). Performance of an efficient parallel data transmission system. *IEEE Transactions on Communication Technology*, COM-15(6), 805–811.
3. *Part 11: Wireless LAN medium access control (MAC) and physical layer (PHY) specifications* (1999). IEEE Std. 802.11a.
4. Nee, R. V., & Prasad, R. (2000). *OFDM for wireless multimedia communications*. Boston: Artech House.
5. Harris, F. J. (1978). On the use of windows for harmonic analysis with the discrete Fourier transform. *Proceedings of the IEEE, 66*, 51–83.

6. Behzad, A., Carter, K. A., Chien, H.-M., Wu, S., Pan, M.-A., Lee, C. P., et al. (2007). A fully integrated MIMO multiband direct conversion CMOS transceiver for WALN applications (802.11n). *IEEE Journal of Solid-State Circuits, 42*(12), 2795–2805.
7. Gao, W., & Shih, D. (2011). Compensation for gain imbalance, phase imbalance and DC offsets in a transmitter. *US Patent Application* (Document Number: 20080063113), issued date: 10/2011.
8. Wild, A. D. (1997, September). *The peak-to-average power ratio of OFDM*. M.Sc. thesis, Delft University of Technology, Delft, Netherlands
9. May, T., & Rohling, H. (1998). Reducing the peak-to-average power ratio in OFDM radio. In *Proceedings of IEEE VTC'98,* Ottawa, Canada, May 18-21, 1998 (pp. 2474–2478).
10. Beek, J. V. D., Sandell, M., & Borjesson, P. O. (1997). ML estimation of timing and frequency offset in OFDM systems. *IEEE Transactions on Signal Processing, 45*(3), 1800–1805.
11. Schmidl, T. M., & Cox, D. C. (1997). Robust frequency and timing synchronization for OFDM. *IEEE Transactions on Communications, 45*(12), 1613–1621.
12. Manhas, P., Thakrai, S., & Arora, A. (2014). Synchronization issues in wireless OFDM systems: a review. *International Journal of Engineering Research & Technology (IJERT), 3* (3), 993–995.
13. Morelli, M., & Moretti, M. (Dec., 2008) Integer frequency offset recovery in OFDM transmissions over selective channels. *IEEE Transactions on Wireless Communications, 7*(12), 5220–5226.
14. Beek, J. V. D., Edfors, O., Sandell, M., Wilson, S. K., Borjesson, P. O. (1995). On channel estimation in OFDM systems. In *IEEE 45th VTC, 25–28 July, 1995* (Vol. 2, pp. 815–819).
15. Hsieh, M. H., & Wei, C. H. (1999). A low-complexity frame synchronization and frequency offset compensation scheme for OFDM systems over fading channels. *IEEE Transaction on Vehicular Technology, 49*(5), 1596–1609.
16. Wang, K., Singh, J., & Faulkner, M. (2004). FPGA implementation of an OFDM WLAN synchronizer. In *IEEE International Conference on Field-Programmable Technology* (pp. 89–94). Delta.
17. Zou, H., McNair, B., & Daneshrad, B. (2001). An integrated OFDM receiver for high-speed mobile data communications. *IEEE Global Telecommunications Conference, 5*, 3090–3094.
18. Moose, P. H. (1994). A technique for orthogonal frequency division multiplexing frequency offset correction. *IEEE Transactions on Communications, 42*(10), 2908–2914.
19. Heiskal, J., & Terry, J. (2002). *OFDM wireless LANs: A theoretical and practical guide*. Indianapolis, IN: Sams.
20. Jeon, W. G., Paik, K. H., & Cho, Y. S. (2000, September). An efficient channel estimation technique for OFDM systems with transmitter diversity. In *Proceedings of the IEEE International Symposium on Personal, Indoor and Mobile Radio Communications, London, UK* (Vol. 2, pp. 1246–1250).
21. Ozdemir, M. K., & Arslan, H. (2007). Channel estimation for wireless OFDM systems. *IEEE Communications Surveys & Tutorials, 9*(2), 18–48.
22. Rinne, J., & Renfors, M. (1996). Pilot spacing in OFDM systems on practical channels. *IEEE Transactions on Consumer Electronics, 42*(4), 959–962.
23. Proakis, J. G. (1995). *Digital communications*. New York: McGraw-Hill.
24. Kang, S. G. (2003). A comparative investigation on channel estimation algorithms for OFDM in mobile communications. *IEEE Transactions on Broadcasting, 49*(2), 142–149.
25. Meng, T. H., McFarland, B., Su, D., & Thomson, J. (2003). Design and implementation of an all-CMOS 802.11a wireless LAN chipset. *IEEE Communication Magazine, 41*(8), 160–168.
26. Moslehi, M., Foli, E., Hedayati, H., & Entesari, K. (2014). A 1.6 GHz/4.8 GHz dual-band CMOS fractional-N frequency synthesizer for S-band radio applications. In *IEEE Radio Frequency Integrated Circuit Symposium* (pp. 429–432).
27. Abdollahi, S., Weber, D., Dogan, H. & Su D. (2011, February). A 65 nm dual-band 3-steam 802.11n MIMO WLAN SoC. In *ISSCC Digest of Technical Papers* (pp. 170-172).

28. Lee, C. P., Behzad, A., Ojo, D., Kappes, M., Au, S., Pan, M.-A., et al. (2006). A highly linear direct-conversion transmit mixer transconductance stage with local oscillation feedthrough and I/Q imbalance cancellation scheme. In *IEEE ISSCC Digest of Technical Papers* (pp. 368-369).
29. Application note APP3350. (2004). Clock jitter and phase noise conversion. Maxim Integrated. Retrieved from www.maximintegrated.com
30. Chen, Z., & Dai, F. F. (2010). Effects of LO phase and amplitude imbalances and phase noise on *M*-QAM transceiver performance. *IEEE Transactions on Industrial Electronics, 57*(5), 1505–1517.
31. Feher, K. (1995). *Wireless and digital communications; modulation & spread spectrum applications*. Upper Saddle River, NJ: Prentice-Hall PTR.
32. Behzad, A. (2008). *Wireless LAN radios—System definition to transistor design* (p. 74). Hoboken, NJ: Wiley.
33. Tanner, R., & Woodard, J. (2004). *WCDMA requirements and practical design*. Chichester: Wiley.
34. Geier, J. (2002, June 4). 802.11 MAC layer defined. http://www.wi-fiplanet.com/tutorials/article.php/1216351/80211-MAC-Layer-Defined.htm
35. He, M., Winoto, R., Gao, X., Loeb, W., Signoff, D., Lau, W., et al. (2014, February). A 40nm dual-band 3-stream 802.11a/b/g/n/ac MIMO WLAN SoC with 1.1Gb/s over-the-air throughput. In *IEEE International Solid-State Circuits Conference (ISSCC)* (pp. 350-352).
36. Chen, T. M., Chan, W. C., Lin, C. C., Hsu, J. L., Li, W. K., Wu, P. A., et al. (2013). A 2×2 MIMO 802.11 a/b/g/n/ac WLAN SoC with integrated T/R switch and on-chip PA delivering VHT80 256QAM 17.5 dBm in 55nm CMOS. In *IEEE Radio Frequency Integrated Circuits Symposium* (pp. 225-228).
37. Wu, C. H., Chen, T. M., Hong, W. K., Shen, C. H., Hsu, J. L., Tsai, J. C., et al. (2013). A 60nm WiFi/BT/GPS/FM combo connectivity SoC with integrated power amplifiers, virtual SP3T switch, and merged WiFi-BT transceiver. *IEEE Radio Frequency Integrated Circuits Symposium, 2013*, 129–132.
38. McCune, E., (2010). *Practical Digital Wireless Signals*. Cambridge University Press, New York.

Chapter 4
Energy and Bandwidth-Efficient Modulation

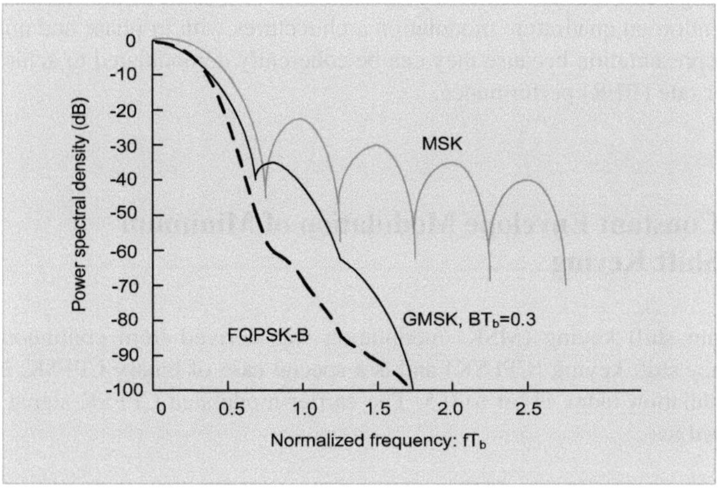

4.1 Introduction

In addition to the requirement of high spectral efficiency, energy efficiency or equivalent to battery duration is another important requirement for modulation techniques. In some applications, such as mobile handset devices, portable devices, and even satellite communication equipment, energy efficiency is crucial to achieve longer battery life or longer communication time. In these applications, the power amplifier is preferable to operate in or close to the saturation region to maximize energy efficiency or minimize DC current consumption in a saturation region. However, a saturated amplifier introduces amplitude modulation to amplitude modulation (AM-AM) and amplitude modulation to phase modulation (AM-PM)

© Springer International Publishing Switzerland 2017
W. Gao, *Energy and Bandwidth-Efficient Wireless Transmission*, Signals and Communication Technology, DOI 10.1007/978-3-319-44222-8_4

conversions into the amplified signal, which is usually the amplitude- and phase-modulated signal. If such an input signal to a power amplifier that operates in or close to a saturated condition is a non-constant envelope modulation signal, its output will be affected by the AM-AM and AM-PM conversions. As a result, a nonlinearly amplified signal at the output of the power amplifier is affected by spectrum regrowth such that its output signal cannot meet the required spectrum mask or adjacent channel power ratio (ACPR) imposed by different standards and its error vector magnitude (EVM) is degraded as well. Thus, requirements of both energy efficiency and spectral efficiency either impose constant or nearly constant envelope characteristics on the modulated signal to the power amplifier (PA).

It is desirable to choose digital modulation schemes with either constant envelope or nearly constant envelope property when considering the need for high energy and spectral efficiency transmission. There are many types of constant or nearly constant envelope modulation schemes available today to achieve a goal of high energy and spectral efficiency transmission. However, in this book, we focus our attention on quadrature modulation architectures with in-phase and quadrature (I–Q) representation because they can be coherently demodulated to achieve good bit error rate (BER) performance.

4.2 Constant Envelope Modulation of Minimum Shift Keying

Minimum shift keying (MSK) modulation was derived from continuous phase frequency shift keying (CPFSK) and is a special case of binary CPFSK, in which the modulation index is set to 0.5. The carrier-modulated CPFSK signal may be expressed as

$$s(t) = A\cos\left[2\pi f_{\mathrm{c}}t + \phi(t)\right] \tag{4.1}$$

where $\phi(t)$ represents the instant phase of the carrier and is defined as

$$\begin{aligned}\phi(t) &= k_{\mathrm{d}}\int_{-\infty}^{t} p(\tau)d\tau \\ &= k_{\mathrm{d}}\int_{-\infty}^{t}\left[\sum_{n} d_n g(\tau - nT_{\mathrm{b}})\right]d\tau\end{aligned} \tag{4.2}$$

where $p(t)$ is the instantaneous frequency deviation; d_n is an independent and identically distributed binary information sequence at the nth bit, with each element taking on equiprobable values ± 1; $g(t)$ is a square waveform pulse with amplitude 1; T_{b} is the bit duration in the case of MSK (CPFSK); and k_{d} is the constant related to the frequency deviation. Since it is often more convenient to express frequency deviation in radians per second and hertz, we further define

$$k_d = \omega_d T_b = 2\pi f_d T_b \tag{4.3}$$

where ω_d and f_d are the frequency deviation constants of the frequency modulation, expressed in radians per second per unit of $p(t)$ and in hertz per unit of $p(t)$, respectively.

The instantaneous phase $\varphi(t)$ of the carrier in the interval $nT_b \leq t \leq (n+1)T_b$ is further expressed as

$$\begin{aligned}\phi(t) &= k_d \int_{-\infty}^{nT_b} \sum_{m=-\infty}^{n-1} d_m g(\tau - mT_b)d\tau + k_d \int_{nT_b}^{t} d_n g(\tau - nT_b)d\tau \\ &= k_d \sum_{m=-\infty}^{n-1} d_m + k_d d_n (t - nT_b)/T_b \\ &= \theta_n + k_d d_n (t - nT_b)/T_b\end{aligned} \tag{4.4}$$

where θ_n, the accumulated phase for all bits up to time $(n-1)T_b$, and $g(t)$ are defined as

$$\theta_n = k_d \sum_{m=-\infty}^{n-1} d_m \tag{4.5}$$

$$g(t) = \begin{cases} 1/T_b, & 0 \leq t \leq T_b \\ 0, & \text{otherwise} \end{cases} \tag{4.6}$$

Now we need to determine k_d in (4.4). In the MSK modulation, k_d is chosen such that the contribution of an individual bit sequence $d_n = 1$ or $d_n = -1$ to the change of the phase $\phi(t)$ in the bit interval T_b is exactly equal to $\pi/2$ or $-\pi/2$. From (4.4), we have the phase value at time $t = (n+1)T_b$ for $d_n = 1$:

$$\phi(t)\big|_{t=(n+1)T_b} = \theta_n + k_d = \theta_n + \pi/2 \tag{4.7}$$

Solving (4.7), we have

$$k_d = \frac{\pi}{2} \tag{4.8}$$

From (4.3) and (4.8), we have the frequency deviation constant in hertz:

$$f_d = \frac{1}{4T_b} \tag{4.9}$$

Thus, substituting (4.8) into (4.4) and then into (4.1), we obtain the MSK-modulated signal in the interval $nT_b \leq t \leq (n+1)T_b$ in the form

$$
\begin{aligned}
s(t) &= A\cos\left[2\pi f_{\mathrm{c}}t + \phi(t)\right] \\
&= A\cos\left[2\pi f_{\mathrm{c}}t + \theta_n + \frac{\pi}{2T_{\mathrm{b}}} d_n (t - nT_{\mathrm{b}})\right] \\
&= A\cos\left[2\pi\left(f_{\mathrm{c}} + \frac{d_n}{4T_{\mathrm{b}}}\right)t - \frac{n\pi d_n}{2} + \theta_n\right] \\
&= A\cos\left[2\pi(f_{\mathrm{c}} + d_n f_{\mathrm{d}})t + \theta_n'\right]
\end{aligned}
\tag{4.10}
$$

where $f_{\mathrm{d}} = 1/4T_{\mathrm{d}}$ was substituted into (4.10), and the accumulated phase θ_n' up to time $t = nT_{\mathrm{b}}$ is simplified as

$$
\begin{aligned}
\theta_n' &= \theta_n - \frac{n\pi d_n}{2} \\
&= \frac{\pi}{2}\sum_{m=-\infty}^{n-1} d_m - \frac{n\pi d_n}{2} \\
&= \pm n\pi, \quad n = 0,\ 1,\ 2,\ \ldots
\end{aligned}
\tag{4.11}
$$

From (4.11), we see that θ_n' takes value of 0 or a positive or negative integer multiple of π. From (4.10), we know that the MSK-modulated signal has two possible instantaneous frequencies in the interval $nT_{\mathrm{b}} \le t \le (n+1)T_{\mathrm{b}}$, depending on information bit d_n:

$$
\begin{aligned}
f_1 &= f_{\mathrm{c}} + f_{\mathrm{d}}, \quad \text{for } d_n = 1 \\
f_2 &= f_{\mathrm{c}} - f_{\mathrm{d}}, \quad \text{for } d_n = -1
\end{aligned}
\tag{4.12}
$$

Therefore, the instantaneous frequency of the modulated signal is shifted between $f_{\mathrm{c}} + f_{\mathrm{d}}$ and $f_{\mathrm{c}} - f_{\mathrm{d}}$, depending on d_n. Since d_n is the random information bit, the spectrum of the MSK-modulated signal is not just two discrete frequency components.

The modulation index m is defined as

$$
m = \frac{\text{Peak-to-peak frequency deviation}}{\text{Modulation frequency}} = \frac{2f_{\mathrm{d}}}{1/T_{\mathrm{b}}}
\tag{4.13}
$$

Substituting $f_{\mathrm{d}} = 1/4T_{\mathrm{b}}$ into (4.13), we obtain the modulation index as we assumed before, or $m = 0.5$. The condition of the modulation index $m = 0.5$ is precisely required to coherently demodulate the MSK signal in the receiver because it is necessary to ensure the orthogonality of the two MSK signals with possible shift frequencies separated by $2f_{\mathrm{d}} = 1/(2T_{\mathrm{b}})$ over the bit interval of T_{b} [1]. *The modulation index* $m = 0.5$ *has the following advantages:*

- Relatively narrow main lobe and fast drop-off side-lobes
- A quadrature I–Q implementation capability to achieve a simply modulation
- Coherent detection capability to achieve a good BER

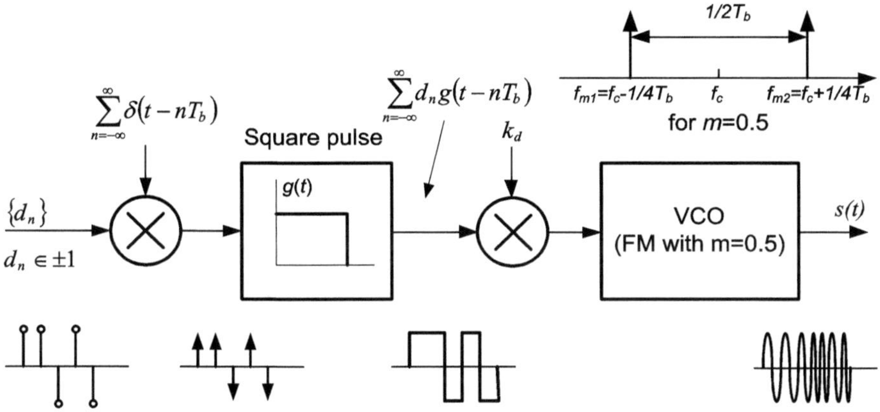

Fig. 4.1 A VCO-based MSK modulator

Figure 4.1 shows a block diagram of voltage-controlled oscillator (VCO)-based frequency modulation (FM) implementation of an MSK modulator. In such an MSK modulator, the modulation index $m=0.5$ must be kept exactly in order not to degrade the coherent demodulation performance. However, it is very difficult to keep such a value constant over temperature and other variations. Hence, an MSK modulator with a VCO structure is rarely implemented in practice, whereas the I–Q-based quadrature architecture is popularly used for the MSK modulator, which is explained in the following.

Figure 4.2 shows an illustration of the phase trajectories of MSK, where the branches are labeled with the data bits that generate the corresponding phase transition. The phase transition is calculated from (4.5) and (4.8). These phase trajectories are also called the *phase tree*. It is clear that phase change of the MSK carrier is either $\pi/2$ or $-\pi/2$ during every bit interval T_b relative to the previous phase, depending on information bit d_n. In Fig. 4.2, the initial phase $\phi(t)$ is assumed to be zero at time $t=0$.

An MSK signal can also be expressed as a form of quadrature representation in the interval $nT_\mathrm{b} \le t \le (n+1)T_\mathrm{b}$ from (4.10):

$$\begin{aligned} s(t) &= A\cos\left(d_n\frac{\pi t}{2T_\mathrm{b}}+\theta'_n\right)\cos(2\pi f_\mathrm{c}t) - A\sin\left(d_n\frac{\pi t}{2T_\mathrm{b}}+\theta'_n\right)\sin(2\pi f_\mathrm{c}t) \\ &= A\cos\theta'_n\cos\left(\frac{\pi t}{2T_\mathrm{b}}\right)\cos(2\pi f_\mathrm{c}t) - Ad_n\cos\theta'_n\sin\left(\frac{\pi t}{2T_\mathrm{b}}\right)\sin(2\pi f_\mathrm{c}t) \\ &= Aa_n\cos\left(\frac{\pi t}{2T_\mathrm{b}}\right)\cos(2\pi f_\mathrm{c}t) - Ab_n\sin\left(\frac{\pi t}{2T_\mathrm{b}}\right)\sin(2\pi f_\mathrm{c}t) \\ &= A\cos\phi(t)\cos(2\pi f_\mathrm{c}t) - A\sin\phi(t)\sin(2\pi f_\mathrm{c}t) \\ &= u_i(t)\cos(2\pi f_\mathrm{c}t) - u_q(t)\sin(2\pi f_\mathrm{c}t) \end{aligned} \tag{4.14}$$

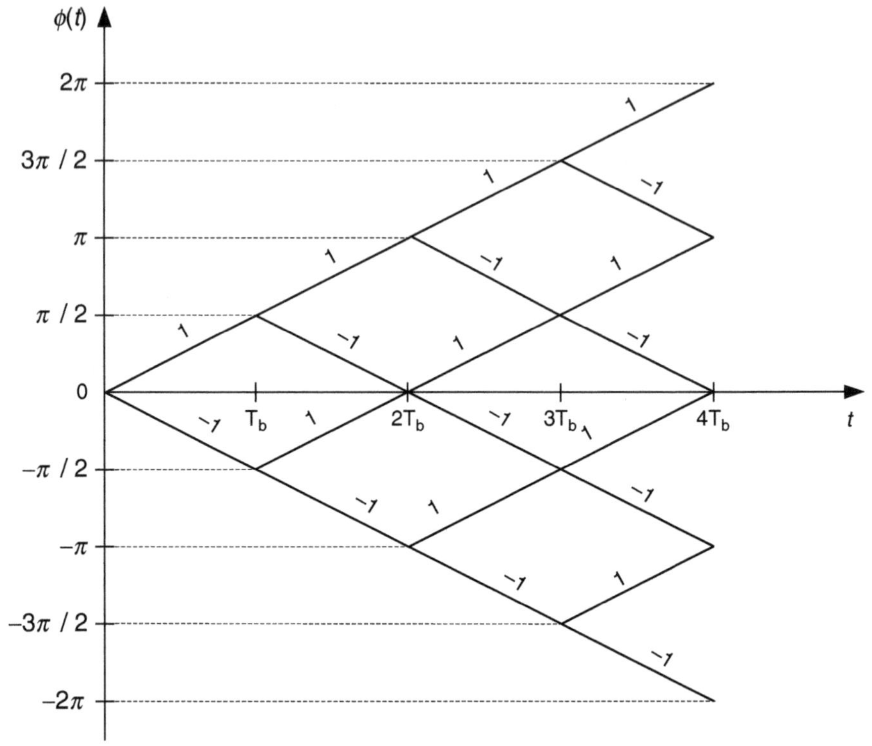

Fig. 4.2 Phase tree of MSK signal

where

$$u_i(t) = A\cos\phi(t) = Aa_nC(t), \quad u_q(t) = A\sin\phi(t) = Ab_nS(t) \tag{4.15}$$

and

$$a_n = \cos\theta'_n = \pm 1, \quad b_n = d_n\cos\theta'_n = \pm 1 \tag{4.16}$$

$$C(t) = \cos\left(\frac{\pi t}{2T_\mathrm{b}}\right), \quad S(t) = \sin\left(\frac{\pi t}{2T_\mathrm{b}}\right) \tag{4.17}$$

We can see from (4.14) that d_n is absorbed in the I branch since $\cos(d_n\pi t/2T_\mathrm{b}) = \cos(\pi t/2T_\mathrm{b})$.

Thus, the pulse shapes of the MSK-modulated signal are cosine and sinusoid waveforms with a period of $2T_\mathrm{b}$. It might be considered that the MSK of (4.14) is similar to the form of OQPSK with one-half cycle of a sinusoid. *However, it should be noted from* (4.14) *that the polarity of each pulse symbol in the I and Q channels is determined by the input bit sequence* $\mathrm{d_n}$ *and the accumulated phase up to time* $\mathrm{t = nT}_b$, *which is different from that of OQPSK*. This is because in the OQPSK signal the polarities of the I and Q symbols are determined by the even-numbered binary symbol and the odd-numbered binary symbols after a serial-to-parallel converter, respectively. We discuss the difference between them below.

To let the quadrature structure of MSK be compatible with that of OQPSK with one-half cycle of a sinusoid, Simon [2] suggested adding a differential encoder before an OQPSK modulator with one-half cycle of a sinusoid. Thus, by substituting (4.15) into (4.14) and letting $A = 1$, we have the representation of MSK in the interval $nT_b \leq t \leq (n+1)T_b$ in the form

$$s(t) = a_n C(t) \cos(2\pi f_c t) - b_n S(t) \sin(2\pi f_c t) \tag{4.18}$$

where $\{a_n\}$ and $\{b_n\}$, equivalent to those as defined in (4.16), are now the I and Q binary data sequences and are also the odd- and even-numbered sequences of a differentially encoded sequence $\{c_n\}$. The output of the differential encoder is expressed as

$$c_n = d_n c_{n-1} \tag{4.19}$$

and

$$a_n = c_{2n-1}, \quad b_n = c_{2n} \tag{4.20}$$

From (4.19) and (4.20), we can see that the binary data sequences $\{a_n\}$ and $\{b_n\}$ are obtained through the differentially encoded data sequence $\{c_n\}$ after a serial-to-parallel converter, instead of through (4.16). Figure 4.3 illustrates such a quadrature structure of an MSK modulator, which is realized by OQPSK weighted with one-half cycle of a sinusoid.

Note that the expression of MSK in (4.18) still does not exactly resemble the quadrature structure of OQPSK with one-half cycle of a sinusoid because $C(t)$ and $S(t)$ are the continuous waveforms instead of shaping pulses. To replace the continuous waveforms with the desired shaping pulse, we define a shaping pulse in the form

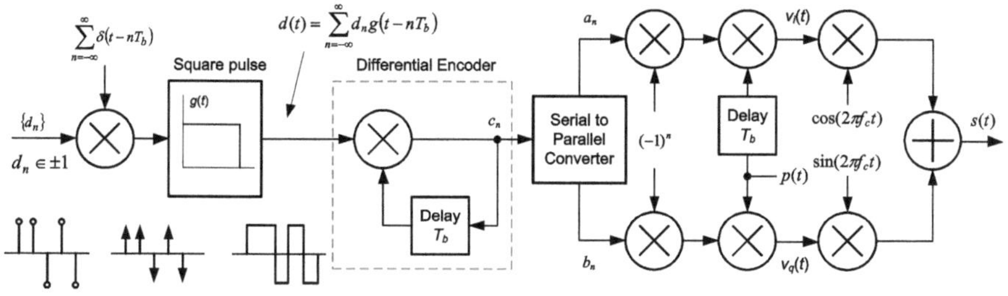

Fig. 4.3 Equivalent quadrature implementation of MSK realized by OQPSK with a half-cycle sinusoidal pulse

$$p(t) = \begin{cases} \sin\left(\frac{\pi t}{2T_\mathrm{b}}\right), & 0 \le t \le 2T_\mathrm{b} \\ 0, & \text{otherwise} \end{cases} \tag{4.21}$$

Thus, the expression of MSK in (4.18) can be rewritten as

$$s(t) = v_i(t)\cos(2\pi f_c t) - v_q(t)\sin(2\pi f_c t) \tag{4.22}$$

where

$$v_i(t) = \sum_n (-1)^n a_n p[t - (2n-1)T_\mathrm{b}] \tag{4.23}$$

$$v_q(t) = \sum_n (-1)^n b_n p(t - 2nT_\mathrm{b}) \tag{4.24}$$

The negative signs in (4.23) and (4.24) perform the alternative polarity changes of the I and Q shaping pulses at a symbol rate of $2T_\mathrm{b}$ to create the continuous waveforms expressed in (4.17). Thus, with a differential encoder before the serial-to-parallel converter, MSK can be treated as a special case of OQPSK with the pulse shape of a sinusoid lasting one-half cycle.

It can be clearly seen in Fig. 4.3 that the difference between the I–Q implementation of MSK and the conventional OQPSK with one-half cycle of a sinusoid is that a differential encoder before the serial-to-parallel converter and alternatively switching sign circuits in the I and Q channels are needed in the former. However, the latter has the same property of the power spectral density (PSD) as the former, but the latter cannot be differentially demodulated, while the former can.

To clearly compare the I–Q implementation waveforms of the equivalent quadrature MSK shown in Fig. 4.3 with ones of the conventional MSK shown in Fig. 4.1, we illustrate their waveforms in Fig. 4.4. First, we start with the input signal $d(t)$, corresponding to the sequence d_n, to the differential encoder, which is shown in Fig. 4.4a. After the differential encoder, the differentially encoded data $c(t)$ are split into the odd-numbered signal $a(t)$ on the I channel and the even-numbered signal $b(t)$ on the Q channel, as shown with the dashed-line curves in Fig. 4.4c, d. After the I–Q signals are multiplexed with properly alternative signs, and then weighted with the shape pulse $p(t)$, the final baseband signals $v_i(t)$ and $v_q(t)$ are shown with the solid-line curves in Fig. 4.3c, d. In order to compare the I–Q baseband signals generated from the quadrature MSK modulation in Fig. 4.3 with ones created from the conventional MSK modulation in Fig. 4.1, we illustrate the phase trajectory $\varphi(t)$ in Fig. 4.4e. The phase trajectory $\varphi(t)$ that corresponds to the sequence in Fig. 4.4a starts with 0 at time and changes either up $\pi/2$ for $d_n = 1$ or down $\pi/2$ for $d_n = -1$. From the phase trajectory $\varphi(t)$, the I–Q baseband signals $u_i(t)$ and $u_q(t)$ expressed by $u_i(t) = \cos\phi(t)$ and $u_q(t) = \sin\phi(t)$ under the assumption $A = 1$ in (4.15) are shown in Fig. 4.4f and g, which are identical to the I–Q baseband signals generated from the quadrature MSK modulation shown in Fig. 4.4c and d, respectively.

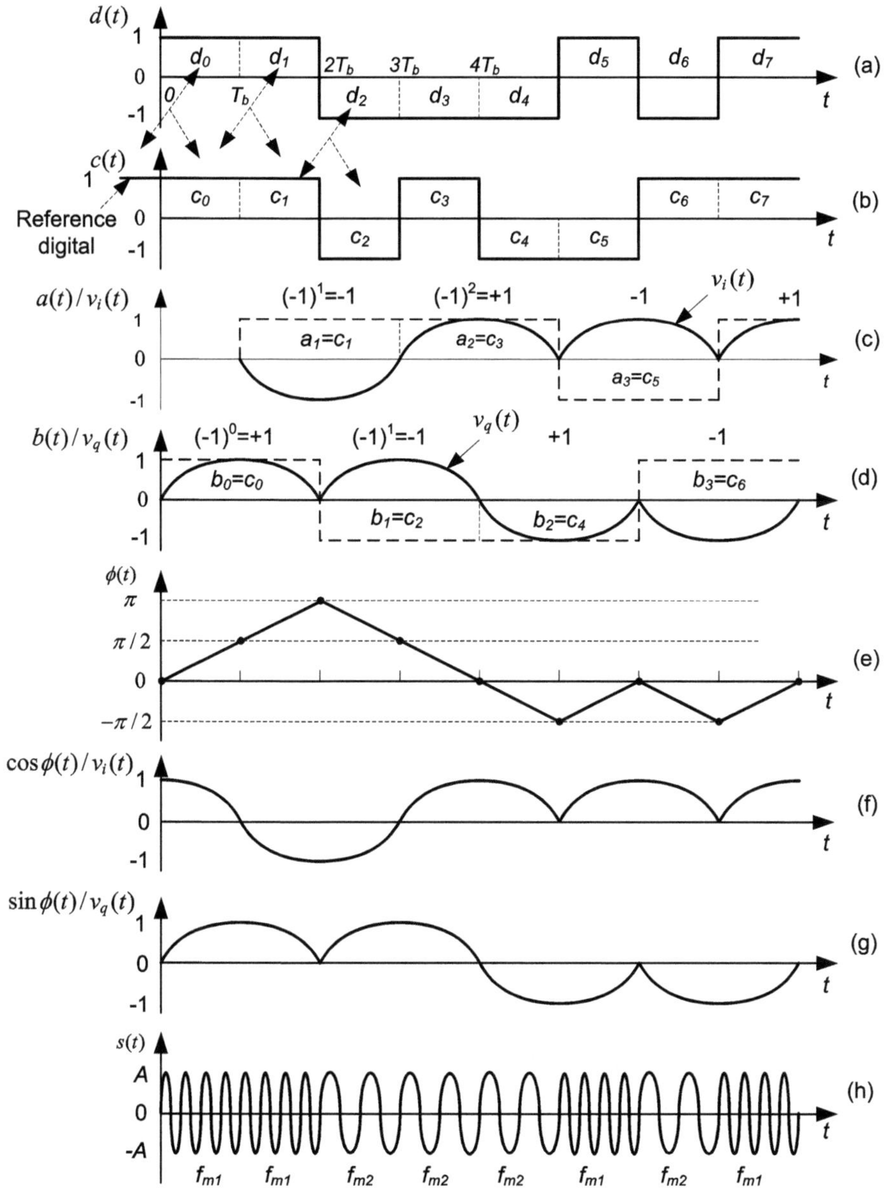

Fig. 4.4 Waveforms in Fig. 4.3: (**a**) input data, (**b**) differentially encoded data, (**c**) OQPSK even-numbered data/baseband signal in the I channel, (**d**) OQPSK odd-numbered data/baseband signal in the Q channel, (**e**) MSK instantaneous phase, (**f**) MSK baseband signal in the I channel, (**g**) MSK baseband signal in the Q channel, and (**h**) MSK-modulated signal

Hence, the equivalent quadrature MSK modulation with a differential encoder prior to a serial to parallel converter in Fig. 4.3 is identical to the conventional MSK modulation in Fig. 4.1. Finally, the modulated MSK signal switched between two frequencies of f_1 and f_2 controlled by the random sequence d_n is shown in Fig. 4.4h.

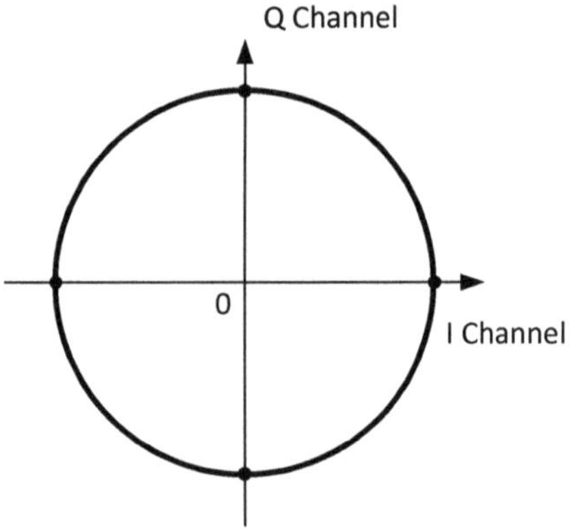

Fig. 4.5 MSK constellation with its possible phase trajectories

The constellation of the MSK signal shows a circle in Fig. 4.5. The constellation trace is represented on a X–Y plane by the magnitude and angle of the baseband vector signal, in which the X-axis stands for the I branch and the Y-axis presents the Q branch. The black dots represent all possible phase trajectory endpoints in the bit interval T_b corresponding to points in Fig. 4.4e.

The PSD of the MSK-modulated signal is identical to that of the MSK baseband signal that is expressed in (2.50) and plotted in Fig. 2.14. Even though PSD of the MSK-modulated signal has fast side-lobe roll-off compared with the unfiltered QPSK/OQPSK shown in Fig. 2.14, one of disadvantages for the MSK signal is that its spectral main lobe is 50% wider than that of QPSK/OQPSK. Furthermore, its spectral side-lobes don't drop faster than that of SQORC. These disadvantages don't satisfy the critical requirements with respect to out-of-band radiation for many wireless communication systems, such as from early SCPC satellite earth station systems to current GSM systems, and thus limit its application. To mitigate such disadvantages, a Gaussian-filtered MSK (GMSK) modulation scheme was proposed by Murota in 1981 [3], which will be introduced in the next section.

4.3 Constant Envelope Modulation of GMSK

It can be clearly seen from the phase tree of the MSK signal in Fig. 4.2 that the phase transition is not smooth at time $t = nT_b$, $n = 0,1,\ldots$ whenever two consecutive information bits are different, such as '1' to '−1' or '−1' to '1'. Such an unsmooth phase transition results in slow roll-off of spectral side-lobes. To achieve fast roll-off of spectral side-lobes, a Gaussian lowpass filter is used to suppress the high-frequency components of the NRZ data before a MSK modulator. Reasons for choosing Gaussian LPF lies in the following properties [3]:

1. Narrow main lobe and fast roll-off side-lobes
2. Lower overshooting impulse response
3. Preservation of the filter output pulse area to keep $m = 0.5$

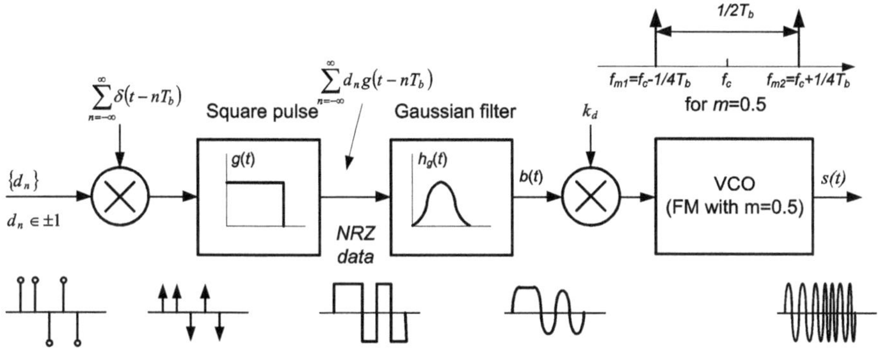

Fig. 4.6 VCO-based GMSK implementation

4.3.1 VCO-Based GMSK Modulation

Like an early MSK modulator, a GMSK modulator used to be implemented in the structure of the VCO-based FM modulation, as shown in Fig. 4.6, where the modulation index of the binary digital FM modulator was controlled to be 0.5. The input data sequence $\{d_n\}$ is passed through a square waveform pulse (or a zero-order hold) to form NRZ data, and then through a Gaussian LPF to suppress high-frequency components of the input NRZ data. The filtered NRZ signal as voltage value $b(t)$ controls VCO output frequency to form the GMSK-modulated signal. A big challenge here is to keep $m = 0.5$ as closely as possible even through temperature changes. Any inaccuracy of the modulation index m may cause intersymbol interference (ISI) when GMSK is coherently detected at the receiver. This VCO-based GMSK structure has not been used for more than two decades at least due to the problem of modulation index accuracy.

4.3.2 Quadrature Architecture of GMSK

The GMSK-modulated signal has the same expression as MSK as shown in (4.1) and (4.2). The only modification is that NRZ data must pass through a Gaussian LPF. The instant phase $\phi(t)$ in (4.2) now is

$$\begin{aligned}\phi(t) &= k_\mathrm{d}\int_{-\infty}^{t} b(\tau)d\tau \\ &= k_\mathrm{d}\int_{-\infty}^{t} p(\tau) * h_\mathrm{g}(\tau)d\tau\end{aligned} \tag{4.25}$$

where $b(t)$ is the Gaussian LPF output and $h_g(t)$ is the impulse response of the Gaussian LPF. The transfer function and impulse response of Gaussian LPF are given by

$$G(f) = \mathbf{exp}\left[-\pi\left(\frac{f}{k_g B}\right)^2\right] \tag{4.26}$$

$$h_g(t) = \Im^{-1}[G(f)] = k_g B\,\mathbf{exp}\left(-\pi k_g^2 B^2 t^2\right) \tag{4.27}$$

with

$$k_g = \sqrt{\frac{2\pi}{\mathbf{ln}\,2}} \tag{4.28}$$

where B is the -3-dB bandwidth of the Gaussian LPF. The pulse response of the Gaussian LPF is more interesting the impulse response of the Gaussian LPF because the input signal to the Gaussian LPF is actually composed of NRZ sequences instead of the impulse response sequences.

For a single square waveform pulse $g(t)$ with the amplitude 1 and time duration T_b to the input of the Gaussian LPF, the output is [4]

$$\begin{aligned} p(t) &= g(t) * h_g(t) \\ &= \int_{-\infty}^{\infty} h_g(\tau)d\tau = k_g B \int_{t}^{t+T_b} \mathbf{exp}\left(-\pi k_g^2 B^2 \tau^2\right)d\tau \\ &= \frac{1}{2}\left\{erf\left[-k_g\sqrt{B\left(t-\frac{T_b}{2}\right)}\right] + erf\left[k_g\sqrt{B\left(t+\frac{T_b}{2}\right)}\right]\right\} \end{aligned} \tag{4.29}$$

and

$$erf(t) = \frac{2}{\sqrt{\pi}}\int_0^t \mathbf{exp}\left(-\tau^2\right)d\tau \tag{4.30}$$

With

$$g(t) = \begin{cases} 1, & |t| \le T_b/2 \\ 0, & |t| > T_b/2 \end{cases} \tag{4.31}$$

Figure 4.7 shows the pulse response $p(t)$ in (4.29) with different BT_b values. The pulse response of the GLPF has a limited width in the time domain, and spans one

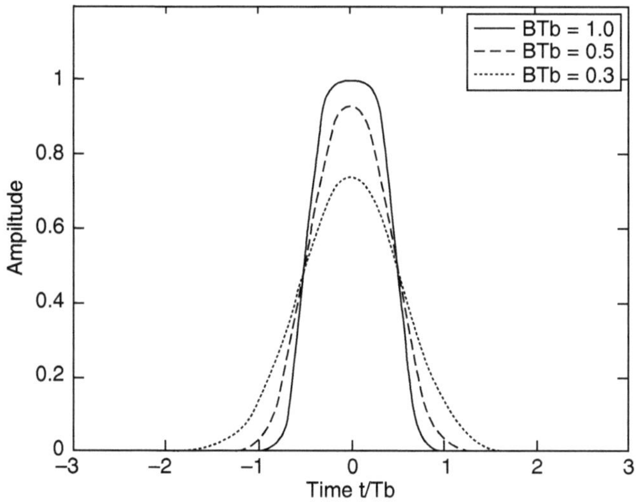

Fig. 4.7 Pulse response of Gaussian LPF with different BT_b values

T_b for $BT_b = \infty$, while it spans several bit durations for a small BT_b value. Furthermore, smaller BT_b values lead to more compact filter spectra, but at the same time the pulse response spreads over adjacent data and results in ISI.

There are two fundamental methods to implement the Gaussian LPF. The first one is to use a FIR filter to design it, whose taps can be calculated from (4.27). The number of taps is also dependent on the number of samples in the bit interval T_b. The second method is to use a lookup table (LUT) to implement it as introduced in [5]. The LUT method for the Gaussian filter is quite a bit simpler than the FIR filter, and features a straightforward implementation.

Gaussian Filter Design: In the LUT-based Gaussian filter design we first need to know how many different waveform segments at the output of the Gaussian filter are contained in the bit duration of T_b for a certain BT_b value such that we can decide how many bits at the input of the Gaussian filter are needed. For example, there are a total of 32 different segments for $BT_b = 0.3$ within one bit duration T_b (or from $t/T_b = 0.5$ to $t/T_b = 1.5$ in the eye diagram as shown in Fig. 4.8). We know the different combination of five successive input bits ($2^5 = 32$) corresponds to each segment. The relationship between the output signal segment and input data pattern is listed in Table 4.1, where only positive segments and their corresponding input data are illustrated. Negative segments are easily obtained based on the rule for positive segments, while the input data are taken *NOT* operation, or 11110 to 00001.

The number of the segments within the bit duration T_b at the output of the Gaussian filter is determined by the value of BT_b. For example, the number of the segments is 8 for $BT_b = 0.5$. In this case, three consecutive input data sequences determine one output segment at the output of the Gaussian filter from the total

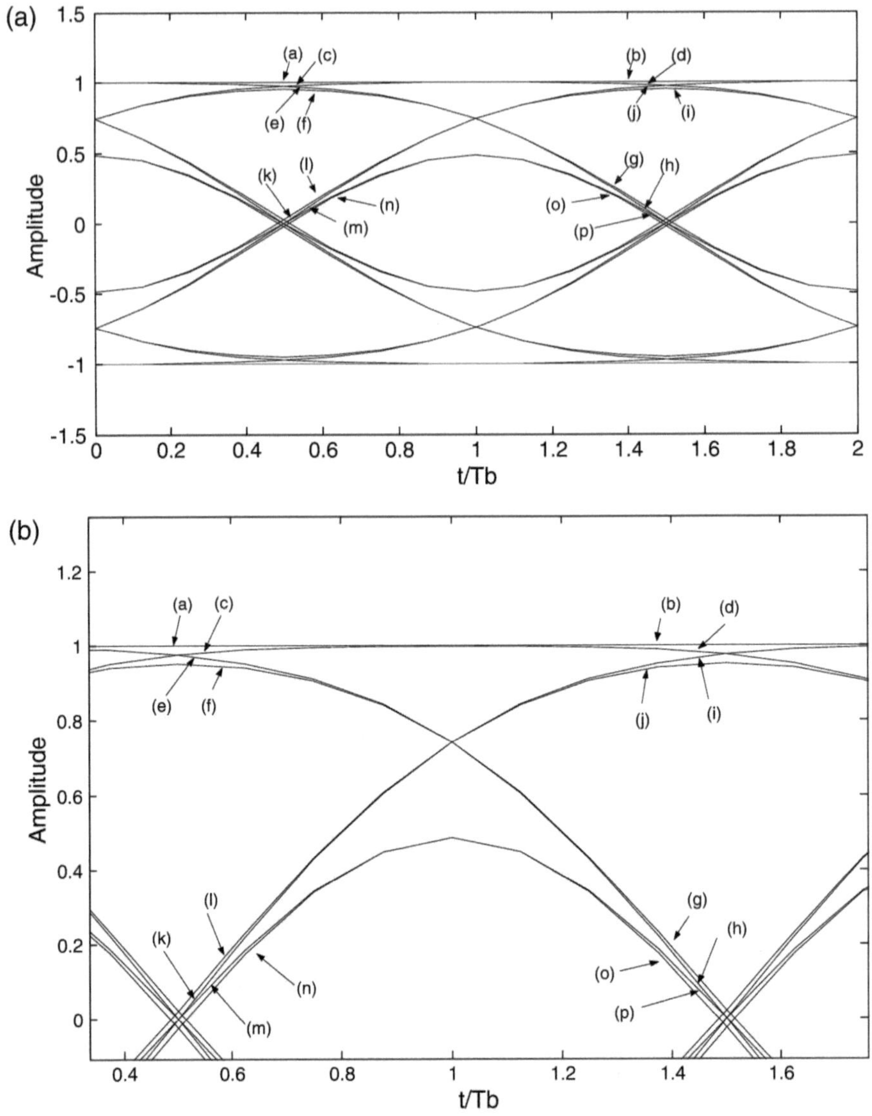

Fig. 4.8 Eye diagram at Gaussian filter output: (**a**) whole view and (**b**) detailed positive view

eight segments. Because the pulse response of the Gaussian filter is symmetrical, the number of the input sequences is always odd, or the middle sequence determines the polarity of the output segment. (For the detailed circuit design, please reference [5].) Furthermore, the number of the required segments can be reduced to half or even more if some control logics are added [5].

Table 4.1 Relationship between input data and all positive output signals for a Gaussian LPF with $BT_b = 0.3$

Waveform Index	Input data a_{n-2}, a_{n-1}, a_n, a_{n+1}, a_{n+2}	Output signal $s_i(t)$
1	11111	$s_1(t)$ [(a)–(b)][a]
2	11110	$s_2(t)$ [(a)–(d)]
3	01101	$s_3(t)$ [(f)–(g)]
4	01100	$s_4(t)$ [(f)–(h)]
5	11000	$s_5(t)$ [(e)–(h)]
6	11001	$s_6(t)$ [(e)–(g)]
7	00100	$s_7(t)$ [(n)–(o)]
8	00101	$s_8(t)$ [(n)–(p)]
9	00110	$s_9(t)$ [(k)–(i)]
10	00111	$s_{10}(t)$ [(k)–(j)]
11	01110	$s_{11}(t)$ [(c)–(d)]
12	01111	$s_{12}(t)$ [(c)–(b)]
13	10111	$s_{13}(t)$ [(l)–(j)]
14	10110	$s_{14}(t)$ [(l)–(i)]
15	10100	$s_{15}(t)$ [(m)–(o)]
16	10101	$s_{16}(t)$ [(m)–(p)]

[a][(a)–(b)] means the segment starts at the line marked by (a) and ends at the line marked by (b)

I–Q Modulation: The GMSK-modulated signal can be also expressed by (4.1), except that the phase $\phi(t)$ is substituted by (4.25) with $k_d = \pi/2$ or

$$\begin{aligned} s(t) &= A\cos\left[2\pi f_c t + \phi(t)\right] \\ &= A\cos\left[\phi(t)\right]\cos\left(2\pi f_c t\right) - A\sin\left[\phi(t)\right]\sin\left(2\pi f_c t\right) \\ &= A\cos\left[\frac{\pi}{2}\int_{-\infty}^{t} b(\tau)d\tau\right]\cos\left(2\pi f_c t\right) - A\sin\left[\frac{\pi}{2}\int_{-\infty}^{t} b(\tau)d\tau\right]\sin\left(2\pi f_c t\right) \\ &= A\cos\left[\frac{\pi}{2}\int_{-\infty}^{t} p(\tau) * h_g(\tau)d\tau\right]\cos\left(2\pi f_c t\right) - A\sin\left[\frac{\pi}{2}\int_{-\infty}^{t} p(\tau) * h_g(\tau)d\tau\right]\sin\left(2\pi f_c t\right) \\ &= u_i(t)\cos\left(2\pi f_c t\right) - u_q(t)\sin\left(2\pi f_c t\right) \end{aligned} \tag{4.32}$$

where

$$u_i(t) = A\cos\left[\frac{\pi}{2}\int_{-\infty}^{t} p(\tau) * h_g(\tau)d\tau\right] \tag{4.33}$$

$$u_q(t) = A\sin\left[\frac{\pi}{2}\int_{-\infty}^{t} p(\tau) * h_g(\tau)d\tau\right] \tag{4.34}$$

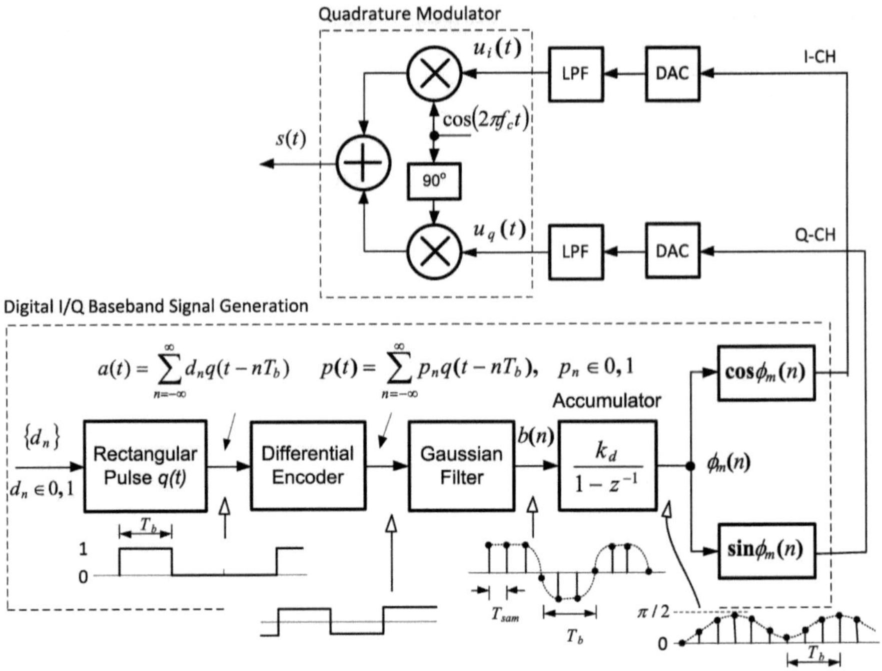

Fig. 4.9 Quadrature implementations of MSK/GMSK digital baseband I–Q signals, where Gaussian filter is bypassed for MSK

The MSK-modulated signal in (4.14) and GMSK-modulated signal in (4.32) indicate that they can be implemented in a quadrature I–Q structure as shown in Fig. 4.9. One big advantage of the quadrature implementation of GMSK is to guarantee the modulation index $m = 0.5$ exactly, which is strictly required by coherent detection.

In the GSM standard, the output data from the GSM burst is a binary $\{d_n\} \in \{0,1\}$ bit sequence. This Return to Zero (RZ) sequence is first mapped to a Non Return to Zero (NRZ) $a(t) \in \{0,1\}$ sequence, and then is differentially encoded into $\hat{a}(t)$ in the interval $(n-1)T_b \le t \le nT_b$ as follows:

$$\hat{a}\,(t - nT_b) = a(t - nT_b) \oplus a(t - (n-1)T_b) \tag{4.35}$$

where $\oplus$ denotes modulo 2 addition. To avoid an uncertain start condition, the GSM standard recommends the bit 1 is assumed to precede the burst to be processed at time $t + T_b \le t < 0$

For an analog Gaussian LPF implementation, differentially encoded sequence $\hat{a}(t)$ is mapped onto +1 and −1 symbols to form the modulation data signal $p(t)$ input to a Gaussian filter as follows:

$$p(t - nT_b) = 1 - 2\hat{a}\,(t - nT_b) \tag{4.36}$$

The procedure above is mapped a logic 1 to a symbol −1 and a logic 0 to a symbol +1, respectively. For the look-up-table (LUT) Gaussian LPF implementation as shown in Table 4.1, however, the differentially encoded sequence $\hat{a}(t)$ does not need to be mapped onto +1 and −1 symbols. In such a case, the modulation data signal $p(t)$ is given by

$$p(t - nT_b) = \hat{a}\,(t - nT_b) \tag{4.37}$$

For the GMSK modulation in the GSM standard, the output waveforms of Gaussian LPF can be generated by using the LUT method as listed in Table 4.1, according to the combination of a continuous five-bit input sequences to the Gaussian LPF with $BT_b = 0.3$. Note that a smaller BT_b value may require the combination of more input bit sequences.

A differential encoder at the transmitter is used to remove nature 180° phase ambiguity at the receiver, where a differential decoder at the receiver is employed to recover the original bit stream. Such a combination of differential encoding at the transmitter and differential decoding at the receiver results in a loss in BER performance relative to that obtained by conventional OQPSK.

The modulation data signal $p(t)$ is then passed through the Gaussian lowpass filter to form the smooth instant frequency modulation signal $b(t)$. After an integrator in the analog domain or an accumulator in the digital domain, the smooth phase modulation signal $\phi_m(t)$ is generated. The modulation phase signal $\phi_m(t)$ is then used as the argument for sine and cosine functions to create the baseband I–Q signals. Figure 4.9 also illustrates different waveforms corresponding to different circuit blocks.

To prevent the output value of the phase accumulator from being too large, we use a logic control circuit to change the unwrapped phase signal to a wrapped phase signal within the range of $\pm\pi$. Whenever the phase $\phi_m(t)$ at the accumulator output is greater than π, the accumulator extracts 2π from its current phase. On the other hand, whenever the phase is less than $-\pi$, the accumulator adds 2π to its current phase. Figure 4.10a shows both unwrapped and wrapped phase signals at the output of the accumulator, where there are 16 samples per bit interval of T_b and the phase signal starts with an initial phase of zero. Figure 4.10b illustrates the baseband I–Q signals created from the wrapped phase signal used as the argument for sine and cosine functions.

Figure 4.11 illustrates the equivalent analog I–Q waveforms corresponding to the digital waveforms on the I-CH and Q-CH branches in Fig. 4.9. The differentially encoded sequence $p(t)$ to the Gaussian LPF is processed by using (4.36) in order to change to +1 and −1 symbols as shown in Fig. 4.11d. The initial value "1" of the input sequence $a(t)$ to the differential encoder is assumed at time $-T_b \leq t \leq 0$. Due to Gaussian filtering, the phase $\phi_m(t)$ for GMSK becomes much smoother than one for MSK at the corner of the phase direction change whenever the input data change polarity as shown in Fig. 4.11e. This means that the phase change of the GMSK signal does not reach the maximum $\pm\pi/2$ phase shift at the end of the bit period whenever the input bit changes its polarity. The extent of the phase smoothing is determined by the BT_b value. The smaller the BT_b value is, the smoother the phase $\phi_m(t)$ is. It is known that the smooth phase of the GMSK signal results in its spectrum being more compact, but with more severe ISI.

Figure 4.12 shows the photographs of the eye diagrams and baseband signals of the GMSK signal used in the GSM system, where the bit rate f_b is 207.833 kbps. The GMSK baseband signals have ISI at the maximum eye-opening instants. ISI is caused by intentionally generating a constant envelope that allows the use of a nonlinear power amplifier to achieve energy efficiency. The property of the constant envelope that is characterized by the constellation is shown in Fig. 4.13. Four thick arc segments on the constellation are due to many overlapped traces shown in the range marked in (m) and (n) in Fig. 4.11f, g in a relatively short time. These thick arc segments will be shinier than any other parts if the constellation is displayed on an oscilloscope due to their staying relatively longer than any other parts in the certain time period.

Figure 4.14 shows curves of the normalized GMSK PSD related to different values of the Gaussian filter BT_b. The case of $BT_b = \infty$ corresponds to MSK. The GMSK signal, especially for smaller BT_b values, has a narrower main lobe of spectrum and faster drop-off side-lobes than that of the MSK, which leads to a great advantage of

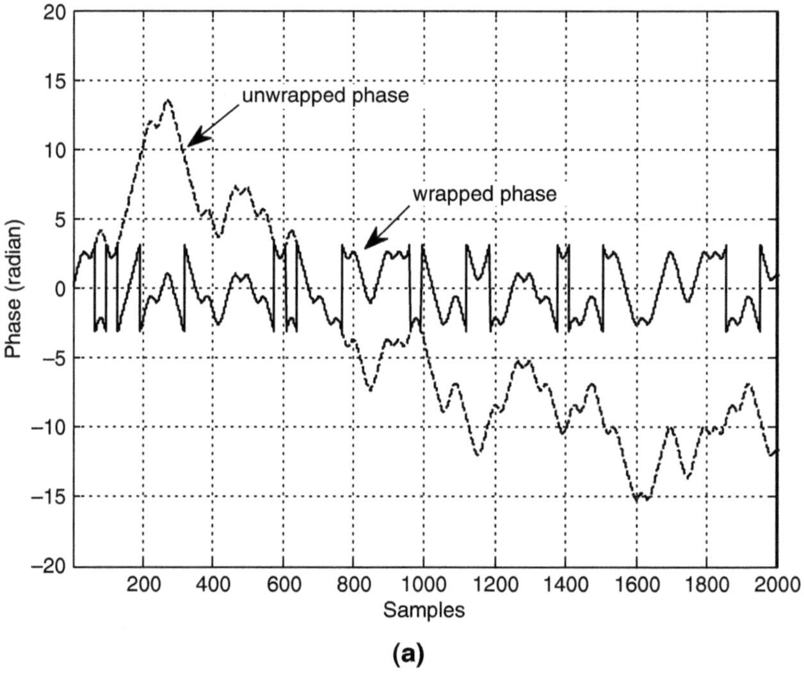

Fig. 4.10 Instant modulation phase signal and baseband I–Q signals: (**a**) Phase signal at the accumulator output, and (**b**) baseband I–Q signals, where there are 16 samples per bit interval of T_b and the phase signal starts with an initial value of zero

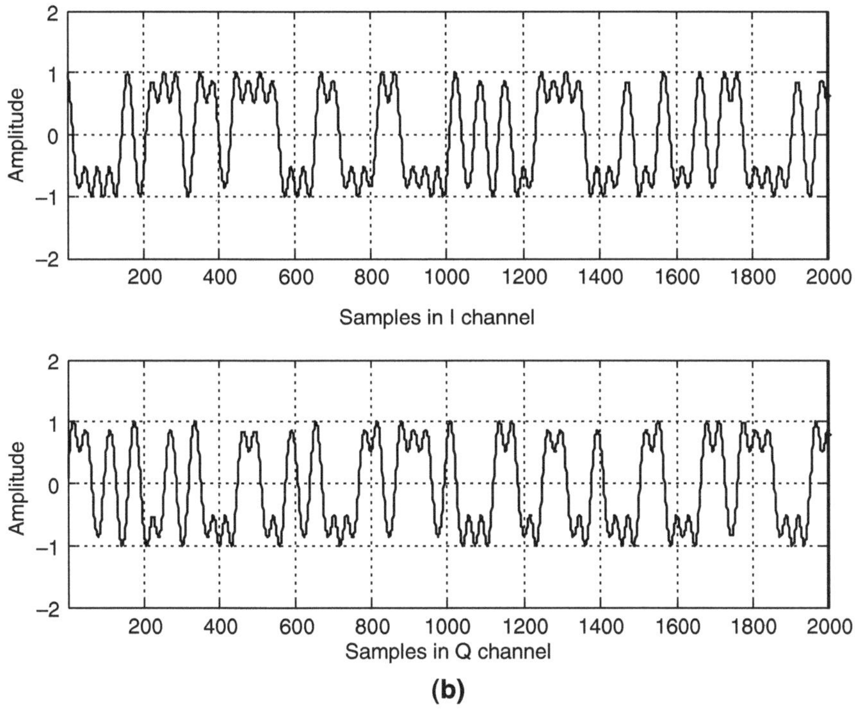

Fig. 4.10 (continued)

GMSK over MSK in the transmission bandwidth. This is major reason that MSK modulation is rarely used in today's digital communication systems. However, the smaller the BT_b value is, the worse the BER performance is due to larger ISI.

4.4 Nearly Constant Envelope Modulation of FQPSK

Previously in this chapter we mentioned that MSK and GMSK signals with constant envelope can achieve greater energy and spectral efficiency when they are transmitted over nonlinear channels compared with QPSK and OQPSK signals. The PSD of a modulated signal at the output of the power amplifier (PA) is approximately identical to its PSD at the input of the PA if the envelope of the input-modulated signal is constant. This is because AM-AM conversion and AM-PM conversion of the PA do not distort the amplified signal when a RF-modulated signal with a constant envelope is amplified by the PA.

In this section, Feher-Patented Quadrature Phase Shift Keying (FQPSK) modulation techniques [6], which have the properties of possessing either a non-constant or nearly constant envelope but achieve significant improvements in spectral efficiency and energy efficiency, are described. The family of FQPSK techniques have

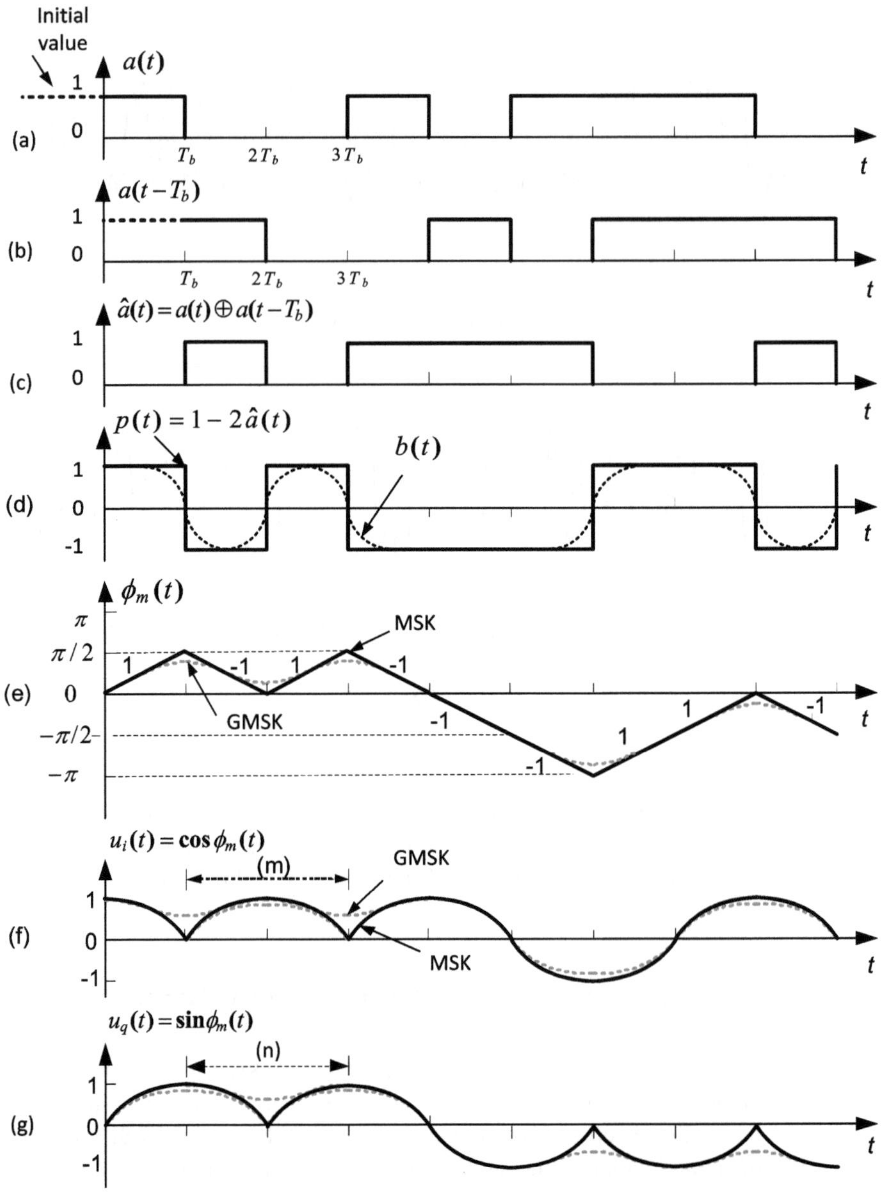

Fig. 4.11 Various waveforms in Fig. 4.9: (**a**) NRZ data, (**b**) one-bit delayed NRZ data, (**c**) differentially encoded NRZ data, (**d**) differentially encoded NRZ to +1/−1 symbol mapping and then Gaussian filtered signal (or instantaneous modulation frequency signal), (**e**) instantaneous modulation phase signal, (**f**) baseband signal in I channel (MSK in *solid-line*, GMSK in *dashed-line*), and (**g**) baseband signal in Q channel (MSK in *solid-line*, GMSK in *dashed-line*)

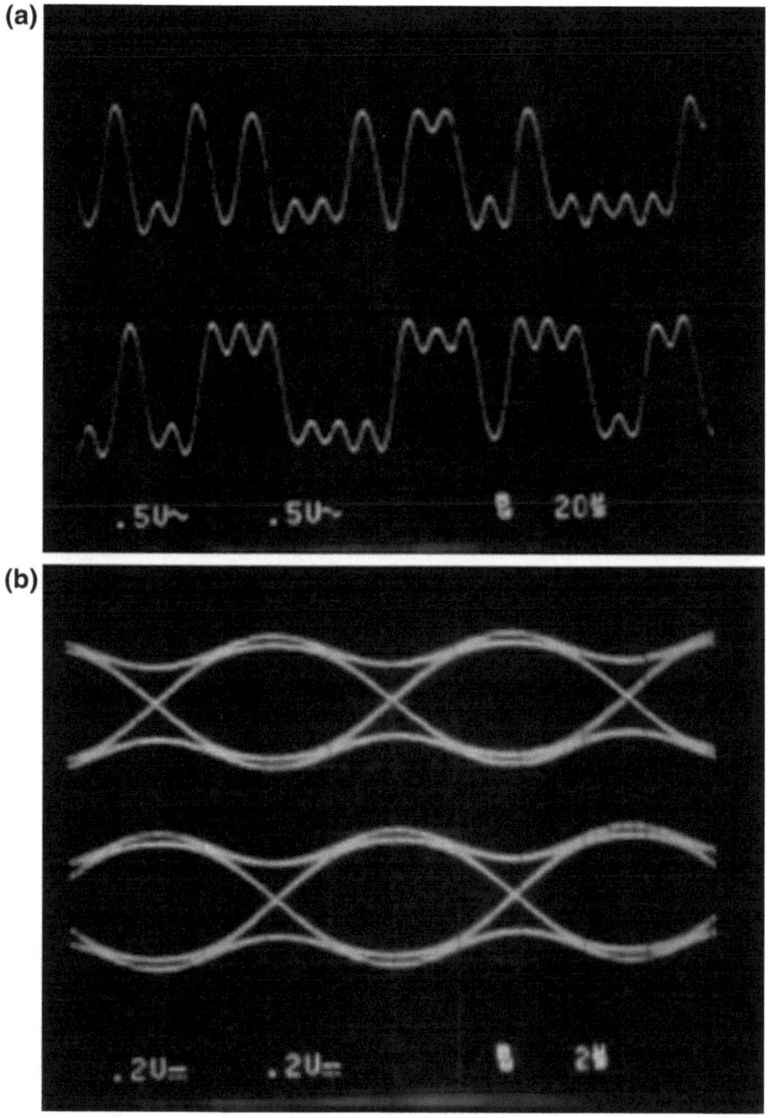

Fig. 4.12 GMSK signal with $BT_b = 0.3$ (**a**) baseband signals in I–Q branches and (**b**) eye diagrams in I–Q branches

been developing since early 1982. As of 2015, from the author's point of view, there were mainly four different versions of modulation techniques that had been studied and developed. The first version of FQPSK, which is also called intersymbol interference- and jitter-free OQPSK (IJF-OQPSK), was proposed to replace QPSK/OQPSK and MSK modulations for low-cost-power and bandwidth-efficient satellite earth stations in 1982 due to its small envelope fluctuation of 3 dB, in which

Fig. 4.13 GMSK constellation

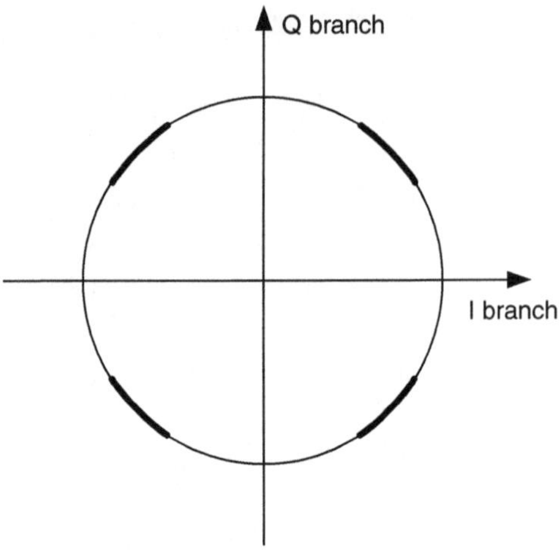

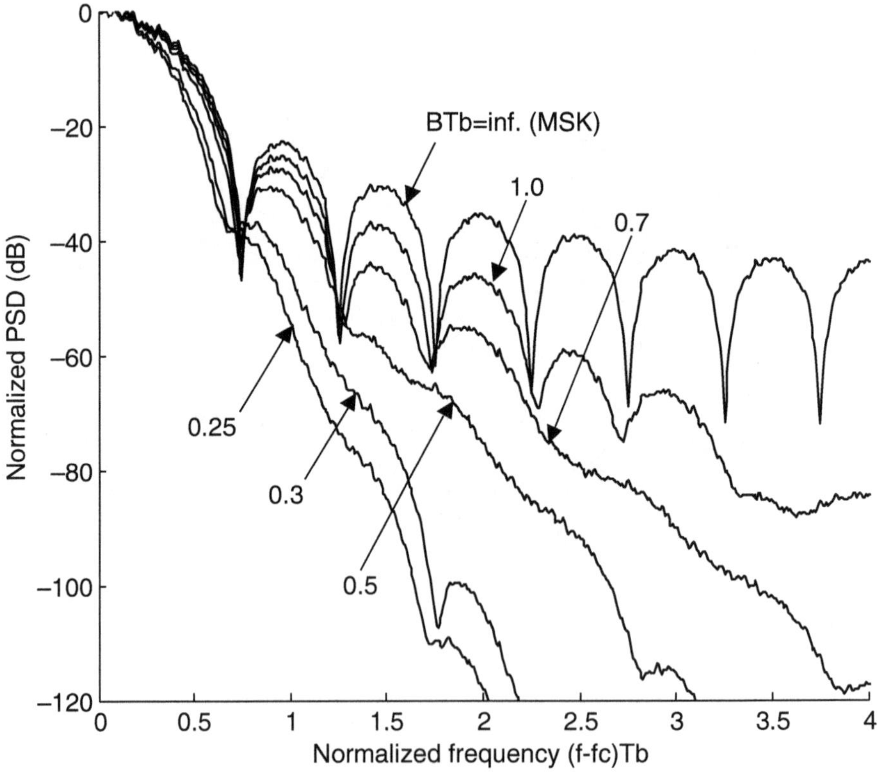

Fig. 4.14 Power spectral density of GMSK with different BT_b values

the transmission channels exhibit nonlinear characteristics or fully saturated amplifications. Besides its energy and spectral efficiency, the IJF-OQPSK signal shows BER performance that is superior to that of QPSK/OQPSK and MSK signals in an additive white Gaussian noise and adjacent channel interference environment. In order to further reduce the maximum 3-dB envelope fluctuation of IJF-OQPSK, a superposed QAM (SQAM) modulation technique was introduced in 1983 as the next version of FQPSK [7]. The maximum envelope fluctuation of the SQAM signal is reduced from 3 dB ($A = 1$ for IJF-OQPSK) to 0.7 dB ($A = 0.7$). Therefore, the SQAM signal shows further improvements over the IJF-OQPSK signal in energy, spectral efficiency, and BER performance in a nonlinear channel. At almost the same time, a cross-correlated PSK modulation technique called XPSK as the third version of FQPSK was introduced in 1983 [8] by adding the cross-correlation between the I and Q channels to obtain a nearly constant envelope. The spectral efficiency and BER performance of the XPSK signal in a nonlinear channel is almost the same as that in a linear channel due to its nearly constant envelope. The spectral efficiency of XPSK is superior to that of IJF-OQPSK, and the BER performance of XPSK is almost the same as that of the IJF-OQPSK in a nonlinear channel. It had not been further improved until 1998, when the Butterworth filtered XPSK, also named FQPSK-B as the fourth version of FQPSK, which was first studied and researched in 1996, was invented by Dr. Kamilo Feher. This more recent version of FQPSK-B or simply called FQPSK has achieved great energy and spectral efficiency in a nonlinear channel. Since then, FQPSK has been adopted as the standard modulation technique in many applications, especially for high-data-rate transmissions, where the available bandwidth is limited.

FQPSK mainly employs pulse shaping to achieve compact spectrum and cross-correlation between the I–Q channels to significantly reduce envelope fluctuation of the modulated signal and then achieve high energy and spectral efficiency through nonlinear power amplifiers. FQPSK has been demonstrated and confirmed by the extensive studies conducted by the US Department of Defense (DoD), National Aeronautics and Space Administration (NASA), and the International Consultative Committee for Space Data Systems (CCSDS) to be the most energy- and spectral-efficient systems with robust BER performance when nonlinearly amplified. In 2000, FQPSK was adopted as the standard in the Aeronautical Telemetry Standard IRIG 106 [9].

Since a nonlinear amplifier is more RF energy efficient and leads to longer battery duration, and lower cost, it is highly desirable for applications that require high transmit-energy efficiency and long battery duration such as satellite and cellular systems.

4.4.1 XPSK Modulation

In Chap. 2, we introduced the concepts and generations of IJF-OQPSK and SQAM in the FQPSK family and also demonstrated that a SQAM signal with $A = 0.8$ has the properties of smaller envelope fluctuation and higher energy and spectral

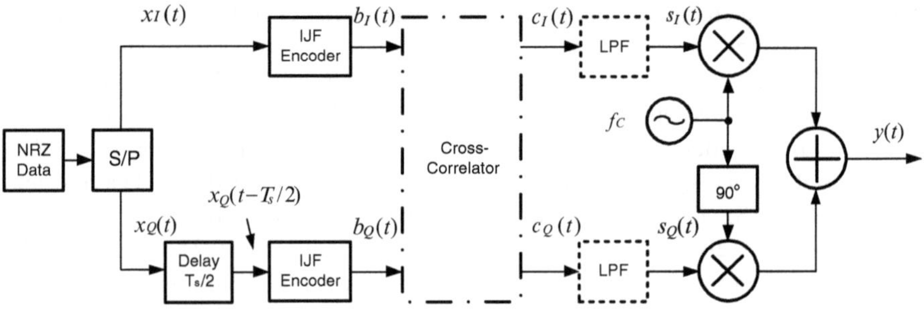

Fig. 4.15 Block diagram of an IJF-OQPSK modulator, where a cross-correlator in the dotted-dashed line and LPFs in the dotted line are excluded

efficiency than an IJF-OQKSK signal through nonlinear channels. In order to further improve constant envelope characteristics, in 1983 Sato and Feher proposed a cross-correlation operation be performed on a pair of IJF encoder outputs at every half-symbol interval between the I–Q channels, which was originally called Cross-Correlated Phase Shift Keying (XPSK) [8]. The cross-correlator is inserted at the output of the IJF encoders on the I–Q channels, as indicated by a dashed-line block in Fig. 4.15. In general, this block diagram can be used to generate the baseband waveforms of IJF-OQPSK, XPSK, and filtered XPSK (or FQPSK-B). IJF-OQPSK modulation (corresponding to all solid-line blocks) is the same as OQPSK excluding IJF-OQPSK encoders. After a serial-to-parallel (S/P) converter, the input bit non-return-to-zero (NRZ) data with the bit interval T_b are converted into the I and Q NRZ symbol data $x_I(t)$ and $x_Q(t)$ with the symbol interval of $T_s = 2T_b$:

$$x_I(t) = \sum_{n=-\infty}^{+\infty} d_{I,n} g(t - nT_s) \tag{4.38}$$

$$x_Q(t) = \sum_{n=-\infty}^{+\infty} d_{Q,n} g(t - nT_s) \tag{4.39}$$

where the pulse shaping is rectangular, or

$$g(t - nT_s) = \begin{cases} 1, & |t - nT_s| \le T_s/2 \\ 0, & |t - nT_s| > T_s/2 \end{cases} \tag{4.40}$$

and

$$d_{In} = \pm 1, \text{ with probability of } 1/2 \text{ for each}$$
$$d_{Qn} = \pm 1, \text{ with probability of } 1/2 \text{ for each}$$

The I channel data $x_I(t)$ and the half-symbol interval delayed Q channel data $x_Q(t - T_s/2)$ are then encoded into IJF baseband signals $b_I(t)$ and $b_Q(t)$, respectively, which are expressed as

$$b_I(t) = \sum_{n=-\infty}^{+\infty} b_{In}(t) \tag{4.41}$$

where

$$b_{In}(t) = \begin{cases} s_1(t - nT_s) = s_e(t - nT_s), & \text{if } d_{I,n-1} = d_{I,n} = 1 \\ s_2(t - nT_s) = -s_e(t - nT_s), & \text{if } d_{I,n-1} = d_{I,n} = -1 \\ s_3(t - nT_s) = s_o(t - nT_s), & \text{if } d_{I,n-1} = -1, d_{I,n} = 1 \\ s_4(t - nT_s) = -s_o(t - nT_s), & \text{if } d_{I,n-1} = 1, d_{I,n} = -1 \end{cases} \tag{4.42}$$

and the odd and even waveforms, $s_o(t)$ and $s_e(t)$, meet

$$\begin{aligned} s_o(t - nT_s) &= -s_o(-t + nT_s), && \text{for } |t - nT_s| < T_s/2 \\ s_e(t - nT_s) &= s_e(-t + nT_s), && \text{for } |t - nT_s| < T_s/2 \\ s_o(t - nT_s) &= s_e(t - nT_s), && \text{for } |t - nT_s| \geq T_s/2 \end{aligned} \tag{4.43}$$

and are defined by

$$\begin{aligned} s_o(t - nT_s) &= \sin \frac{\pi t}{T_s}, && \text{for } |t - nT_s| < T_s/2 \\ s_e(t - nT_s) &= 1, && \text{for } |t - nT_s| < T_s/2 \end{aligned} \tag{4.44}$$

These two fundamental waveforms are shown in Fig. 4.16 and are the same as the first and third segments in Fig. 2.15. The Q channel waveform segment $b_{Qn}(t)$ can be generated by the same mapping as $b_{In}(t)$ in (4.42), which is delayed by a half-symbol relative to $b_{In}(t)$. Figure 4.17 shows the baseband signals of

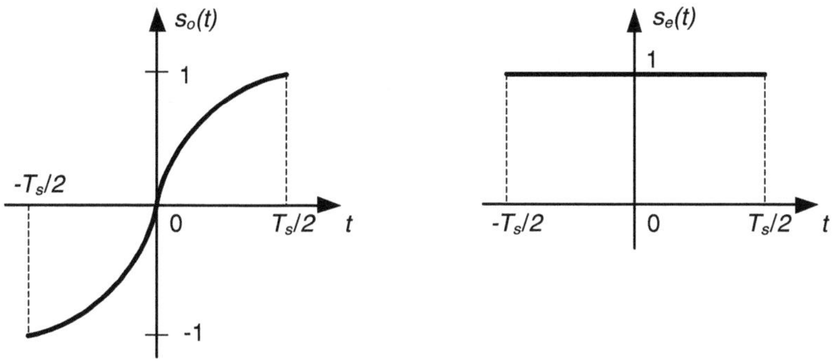

Fig. 4.16 Odd and even waveforms of $s_o(t)$ and $s_e(t)$

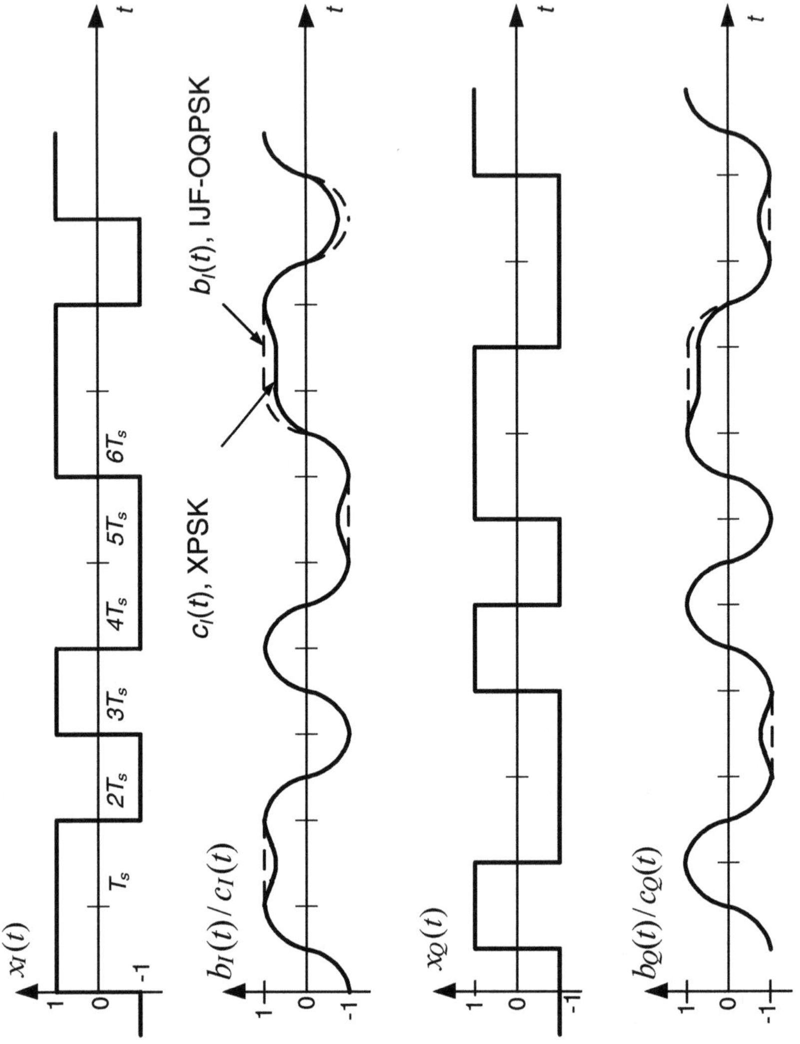

Fig. 4.17 Waveforms of IJF-QPSK and XPSK baseband signals

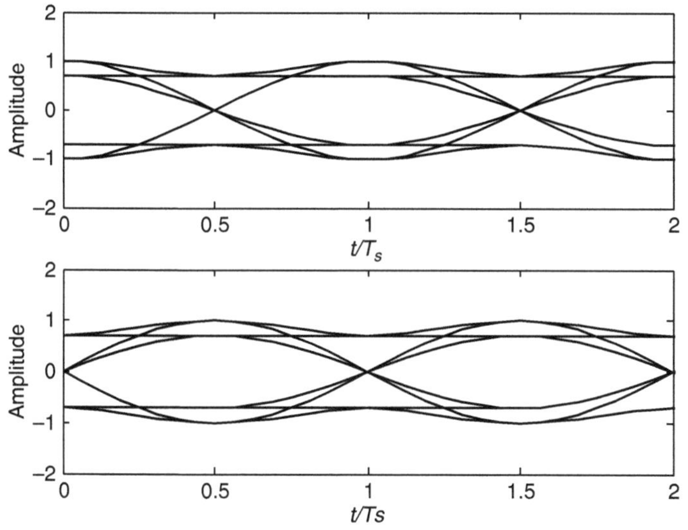

Fig. 4.18 Eye diagrams of cross-correlation FQPSK, or XPSK

IJF-OQPSK at different observation points in Fig. 4.15. The dashed-line waveforms represent the baseband signals of IJF-OQPSK.

In XPSK baseband generation, for the nth data symbol input $d_{I,n}$ in (4.38), the waveform of the I channel in the half-symbol interval at the output of the cross-correlator is dependent on not only one current data symbol $d_{I,n}$ and one previous data symbol $d_{I,\,n-1}$ on the I channel, but also one current data symbol $d_{Q,\,n}$ and two previous data symbol $d_{Q,\,n-1}$, $d_{Q,\,n-2}$ on the Q channel in order to reduce the envelope fluctuation. The same waveform segment process is applied to the baseband waveform of the Q-channel. Hence, the baseband waveform shapes on the I channel and the Q channel at the output of the correlator are correlated with each other. As a result, a nearly constant envelope modulation can be achieved. A detailed description and baseband signal generation can be found in Appendix C.

The baseband signals $c_I(t)$ and $c_Q(t)$ of the cross-correlated FQPSK or XPSK are shown in Fig. 4.17b, d (represented by the solid-line waveforms). Eye diagrams and constellation are shown in Figs. 4.18 and 4.19, respectively. As shown in Fig. 4.18, ISI at the decision instants in either the I channel or Q channel are intentionally introduced to achieve constant envelope characteristics. Mathematically, the envelope of the cross-correlated FQPSK is nearly constant, as shown in Fig. 4.19. As a result, the cross-correlated FQPSK avoids PSD regrowth when passing through a nonlinear amplification channel.

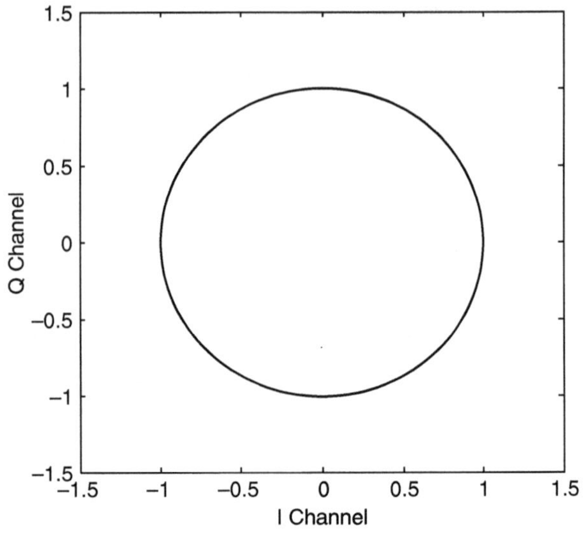

Fig. 4.19 Constellation of cross-correlation FQPSK (XPSK)

4.4.2 FQPSK-B

Even though the XPSK modulation technique was first published as early as 1983, further spectral efficiency had not been significantly improved without significant degradation of the BER until 1996 when the baseband signals of XPSK were filtered through Butterworth lowpass filters at the output of the cross-correlator, as shown in Fig. 4.15 [6]. A 3-dB bandwidth B_{3dB} of the lowpass filter is set to an appropriate value according to a multiplication of $B_{3dB}T_b$, in which T_b is the bit rate, such that the side-lobes of the PSD could roll off quickly. Due to the Butterworth filtering process, the filtered XPSK has been called FQPSK-B since then. Now FQPSK simply stands for this latest process, or FQPSK-B. With such filtering, the PSD of FQPSK-B through both linear and nonlinear channels rolls off significantly with the frequency increase compared with XPSK, while its envelope fluctuation slightly deviates from nearly constant. FQPSK-B, however, only suffers from BER degradation of 0.2 dB compared with unfiltered FQPSK.

In real applications, the baseband signals of FQPSK-B are generated in the digital domain and then are converted into the analog baseband signals through DACs. Before modulating a pair of orthogonal carrier signals, the analog baseband signals need to be passed through a lowpass filter, also called a reconstruction filter, in order to attenuate the image signals and high-order harmonic components. Thus, Butterworth lowpass filters with the cutoff frequency setting for FQPSK-B can be also used as the reconstruction filters in the I and Q channels without the need for extra lowpass filters.

Figure 4.20 illustrates the comparison of power spectral densities among FQPSK-B and other modulation formats in both linear and nonlinear channels. It is clear that the PSD of FQPSK-B is slightly affected by nonlinear amplification

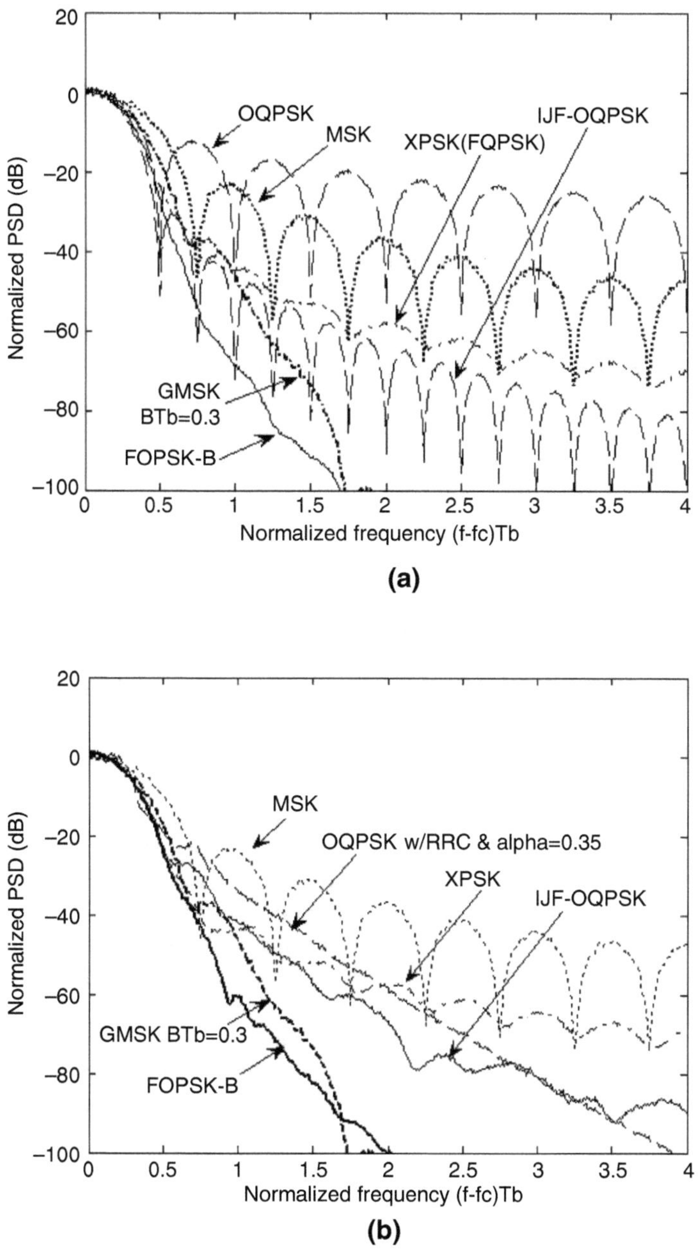

Fig. 4.20 Comparison of power spectrum density: (**a**) in a linear amplification channel and (**b**) in a nonlinear amplification channel

compared to one by linear amplification channel. FQPSK-B, however, still achieves a significant spectral advantage over the filtered OQPSK, MSK, GMSK with $BT_b = 0.3$, IJF-OQPSK, and XPSK (or unfiltered FQPSK) modulations in a nonlinear channel. Even compared with GMSK, the PSD of FQPSK-B shows a spectral advantage over GMSK with $BT_b = 0.3$ down to –90 dB.

4.5 Coherent Demodulation

In general, MSK and GMSK signals can be either coherently detected or non-coherently (or differentially) detected with the same method at the receiver. In the former detection, the MSK signal can be treated as a special case of an OQPSK signal with sinusoidal pulse shaping (or weighting). Hence, the MSK signal can be coherently detected by using the same methods as those used for OQPSK signal. In the latter case, since MSK is a type of SFSK with a modulation index of 0.5, it can also be differentially detected. The major difference between coherent detection and differential detection is that in the former a reference carrier signal in the receiver needs to phase-lock to the carrier phase of the received MSK signal, while in the latter the receiver does not need such a phase-locked reference carrier signal. In fact, in the differential detection it simply uses a delayed version of the received MSK signal as the local reference carrier to multiply the received MSK signal. Compared with the recovered carrier signal performed by a phase-locked loop (PLL), the delayed version of the received MSK signal contains not only noise, but also information data. Hence, it can be expected that the performance of differential detection for MSK should be poorer than that of coherent detection.

Since the performance of coherent detection is superior to that of differential detection and the training-sequence-aided carrier recovery required by coherent detection performs very fast in modern digital communication systems, such as GSM and WLAN, we will only introduce the coherent detection methods in this book. For differential detection, the interested reader can reference materials in [2, 10].

Considering that an equalizer may be used in conjunction with a decision-directed carrier recovery loop, we will first introduce some fundamental equalization techniques that can be used together with the decision-directed carrier recovery loop in coherent detection.

4.5.1 Adaptive Equalization

Based on its structure, an adaptive equalizer can be classified into a linear equalizer and a nonlinear equalizer. The linear equalizer is usually implemented with the transversal filter or the finite impulse response (FIR) filter, which is also called a feed-forward filter. The nonlinear equalizer that is also called a decision-feedback equalizer (DFE) has the feedback filter in addition to the feed-forward filter. Figure 4.21 illustrates a general block diagram of a decision-feedback equalizer. In the following sections, we will only introduce the linear equalizer.

4.5.1.1 Zero-Forcing Linear Equalizer

A linear equalizer is usually constructed by a feed-forward FIR filter with an order of L and adjustable coefficient length of $L+1$, as shown in Fig. 4.21. Assume the equalizer has a coefficient vector at time $t = kT$

$$\mathbf{c}_k = \begin{bmatrix} c_k(0) \\ c_k(1) \\ \vdots \\ c_k(L) \end{bmatrix} = [c_k(0) \quad c_k(1) \quad \ldots \quad c_k(L)]^T \tag{4.45}$$

The input signal vector to the equalizer is

$$\mathbf{v}_k = [v_k \quad v_{k-1} \quad \cdots \quad v_{k-L}]^T \tag{4.46}$$

The equalizer output is expressed as

$$\begin{aligned} y_k &= \sum_{l=0}^{L} c_k(l) v_{k-l} \\ &= \mathbf{v}_k^T \mathbf{c}_k = \mathbf{c}_k^T \mathbf{v}_k \end{aligned} \tag{4.47}$$

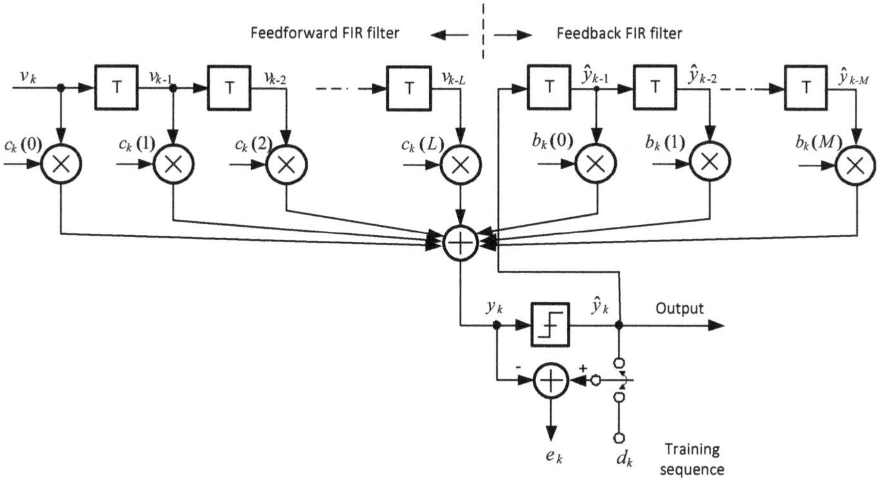

Fig. 4.21 A block diagram of an adaptive decision-feedback equalizer

The error signal at time $t = kT$ created at the output of the equalizer is

$$e_k = d_k - y_k = d_k - \mathbf{v}_k^T \mathbf{c}_K = d_k - \mathbf{c}_k^T \mathbf{v}_K \tag{4.48}$$

where d_k is the desired sequence at time $t = kT$. Theoretically, the zero-forcing algorithm completely eliminates ISI at the output of the equalizer (or the FIR filter) by using an inverse filter to the transfer function of the distorted channel, regardless of noise in the channel. The zero-forcing algorithm is obtained by adjusting the coefficients of the equalizer to force the cross-correlation between the error signal $e_k = d_k - y_k$. and the desired data signal d_k at time $t = kT$ to be zero while ignoring the ratio of the signal-to-noise at the output of the equalizer. This requires

$$E[e_k d^*{}_{k-n}] = E\left[(d_k - y_k) d^*{}_{k-n}\right] = 0, \quad n = 0, 1, \ldots, L \tag{4.49}$$

To meet the condition of (4.49), the coefficients of the equalizer are adaptively updated as follows:

$$\mathrm{c}_{k+1}(n) = \mathrm{c}_k(n) + \lambda e_k d^*{}_{k-n}, \quad n = 0, 1, \ldots, L \tag{4.50}$$

where $c_k(n)$ is the value of the n-th coefficient at time $t = kT$, and λ is the step size that controls the rate of the coefficient adjustment.

In (4.50), the desired signal d_k is usually unknown in the receiver unless the training signal is used. In practice, the detected output sequence $\hat{y}_k$ is used to replace the desired sequence d_k. Thus, (4.50) is written as

$$c_{k+1}(n) = c_k(n) + \lambda \widetilde{e}_k \hat{y}^*_{k-n}, \quad n = 0, 1, \ldots, L \tag{4.51}$$

where the error signal is $\widetilde{e}_k = \hat{y}_k - y_k$. The expression in (4.51) is a practical *zero-forcing (ZF) algorithm*. In the training-based equalizer, (4.50) is usually used during the training period if the training sequence is available. Before the training period is over, the equalized signal eye diagrams at the output of the equalizer are quite open and therefore the decisions at the output of the detector are sufficiently reliable so that the training sequence d_k can be replaced by the decision sequence $\hat{y}_k$. When the training period is over, (4.51) is switched to continue the coefficient adaptation process. Figure 4.22 illustrates the adaptive zero-forcing equalizer switched between the training mode and adaptive operation mode. A dashed-line delay block is used in the actual implementation and will be discussed in the following section.

In certain applications, (4.51) can be used at the beginning of adaption if the training sequence is not available. If the zero-forcing equalizer does not converge, the zero-forcing equalizer will restart the adaptation again with the initial coefficients until it can achieve convergence. It has been demonstrated in microwave digital communication systems that the zero-forcing equalizer has shown a superior convergence property compared to an least-mean square (LMS) equalizer at the beginning of adaption, when there is no training sequence available. When the eye

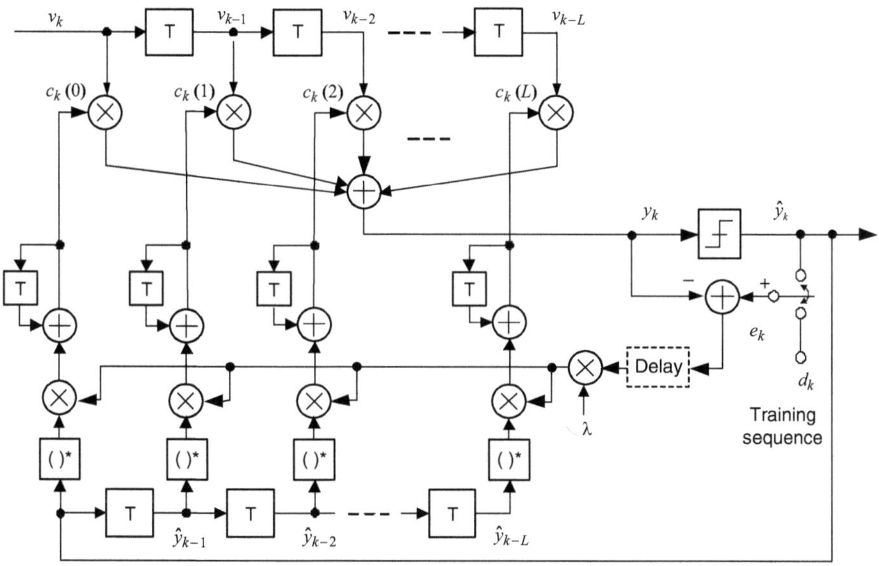

Fig. 4.22 An adaptive linear equalizer with zero-forcing algorithm

diagram or signal constellation at the output of the equalizer is quite open, the coefficient update equation in (4.51) is replaced by an equation based on an LMS algorithm. The performance of the LMS algorithm is superior to the performance of the ZF algorithm under low-SNR ratio conditions, but the LMS algorithm is identical to the ZF algorithm when the SNR is high. The LMS algorithm will be described in the following section.

4.5.1.2 Least-Mean Square Linear Equalizer

The LMS algorithm is determined by the mean square error (MSE) criterion, which minimizes the mean square value of the error signal at the output of the equalizer by adjusting the coefficients of the equalizer. It is obvious that the optimal coefficient values of the equalizer are dependent on the SNR at the input of the equalizer. In the MSE criterion, a cost function ξ is

$$\begin{aligned} \xi &= E\left[|e_k|^2\right] \\ &= E\left[e_k\left(d_k - \mathbf{c}_K^T \mathbf{v}_K\right)^*\right] \end{aligned} \tag{4.52}$$

The coefficient vector $\mathbf{c}_K$ of the equalizer should be adjusted in such a direction to minimize the mean square error function ξ. One widely used method is called the method of *steepest descent*, which leads to the LMS algorithm with the vector expression for adaptively adjusting the coefficients of the equalizer as

$$\begin{aligned}\mathbf{c_{k+1}} &= \mathbf{c_k} + \lambda\left(-\tfrac{\partial \xi}{\partial \mathbf{c}^*}\right)\Big|_{\mathrm{C}=\mathrm{C}_K} \\ &= \mathbf{c_k} + \lambda e_k \mathbf{v_k^*}\end{aligned} \tag{4.53}$$

The LMS algorithm for each individual coefficient adjustment is expressed as

$$c_{k+1}(n) = c_k(n) + \lambda \widetilde{e}_k v_{k-n}^*, \quad n = 0, 1, \ldots, L \tag{4.54}$$

where the error signal $\widetilde{e}_k = \hat{y}_k - y_k$ is used to replace $e_k = d_k - y_k$ in (4.53) in the case when the decision sequences are more reliable. Comparing (4.54) with (4.51), we can see that the difference between the LMS algorithm and the ZF algorithm is that v_k in the former is the input signal of the equalizer while $\hat{y}_k$ sin the latter is the decision signal.

Practical Implementation of the Equalizer: In practical communication systems, the received baseband signals are usually complex signals, consisting of real and imaginary parts. Thus, the coefficients of the equalizer in either the LMS algorithm or ZF algorithm are complex values. The equalizer with complex coefficients, however, can be implemented with four sub-equalizers, each with real coefficients. The real and imaginary coefficients are obtained from the real and imaginary parts of the input and output of the equalizer:

$$y_k = \mathbf{c}_k^T \mathbf{v}_k$$

or

$$\begin{aligned} y_{\mathrm{I},k} + j y_{\mathrm{Q},k} &= \left[\mathbf{c}_{\mathrm{I},k} + j\mathbf{c}_{\mathrm{Q},k}\right]^T \times \left[\mathbf{v}_{\mathrm{I},k} + j\mathbf{v}_{\mathrm{Q},k}\right] \\ &= \mathbf{c}_{0,k}^T \mathbf{v}_{\mathrm{I},k} + \mathbf{c}_{1,k}^T \mathbf{v}_{\mathrm{Q},k} + j\left[\mathbf{c}_{2,k}^T \mathbf{v}_{\mathrm{Q},k} + \mathbf{c}_{3,k}^T \mathbf{v}_{\mathrm{I},k}\right]\end{aligned} \tag{4.55}$$

In (4.55), we changed the I and Q subscripts with the numbers 0–3 representing four real coefficients. Thus, one equalizer with complex coefficients can be split into four individual equalizers with real coefficients in each. These four equalizers in practice, however, should have independent coefficients because the I–Q channels may have both amplitude and phase imbalances. Therefore, this *asymmetric* baseband equalizer architecture is used in practice instead of the conventional complex or symmetric baseband equalizer architecture.

Expressed with four independent real coefficients, the equalizer based on the ZF algorithm in (4.51) can be rewritten as

$$\begin{aligned} c_{0,k+1}(n) &= c_{0,k}(n) + \lambda \widetilde{e}_{\mathrm{I},k} \hat{y}_{\mathrm{I},k-n} \\ c_{1,k+1}(n) &= c_{1,k}(n) + \lambda \widetilde{e}_{\mathrm{I},k} \hat{y}_{\mathrm{Q},k-n} \\ c_{2,k+1}(n) &= c_{2,k}(n) + \lambda \widetilde{e}_{\mathrm{Q},k} \hat{y}_{\mathrm{Q},k-n} \\ c_{3,k+1}(n) &= c_{3,k}(n) + \lambda \widetilde{e}_{\mathrm{Q},k} \hat{y}_{\mathrm{I},k-n}, \quad k = 0, 1, \ldots, L \end{aligned} \tag{4.56}$$

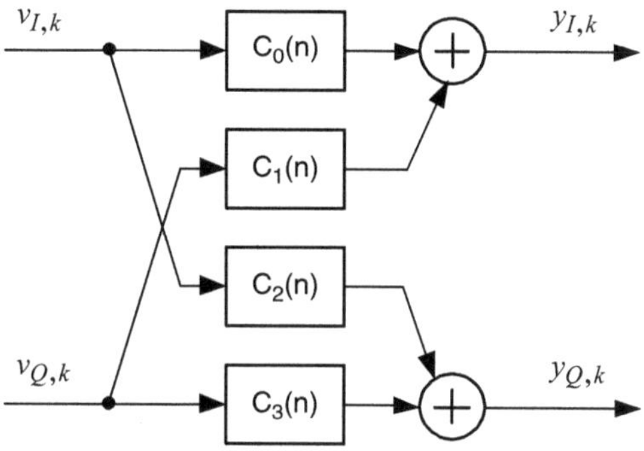

Fig. 4.23 Structure of an equalizer with four-group real coefficients in time domain

Similar to the ZF algorithm, the LMS algorithm in (4.54) can be rewritten as

$$\begin{aligned} c_{0,k+1}(n) &= c_{0,k}(n) + \lambda\tilde{e}_{\mathrm{I},k}v_{\mathrm{I},k-n} \\ c_{1,k+1}(n) &= c_{1,k}(n) + \lambda\tilde{e}_{\mathrm{I},k}v_{\mathrm{Q},k-n} \\ c_{2,k+1}(n) &= c_{2,k}(n) + \lambda\tilde{e}_{\mathrm{Q},k}v_{\mathrm{Q},k-n} \\ c_{3,k+1}(n) &= c_{3,k}(n) + \lambda\tilde{e}_{\mathrm{Q},k}v_{\mathrm{I},k-n}, \quad n = 0, 1, \ldots, L \end{aligned} \tag{4.57}$$

The real and image output signals in (4.55) are

$$\begin{aligned} y_{\mathrm{I},k} &= \sum_{n=0}^{L} c_{0,k}(n) \times v_{\mathrm{I},k-n} + \sum_{n=0}^{L} c_{1,k}(n) \times v_{\mathrm{Q},k-n} \\ y_{\mathrm{Q},k} &= \sum_{n=0}^{L} c_{2,k}(n) \times v_{\mathrm{I},k-n} + \sum_{n=0}^{L} c_{3,k}(n) \times v_{\mathrm{Q},k-n} \end{aligned} \tag{4.58}$$

Figure 4.23 shows a block diagram of a finite impulse response (FIR) filter or equalizer, in which one FIR filter with the complex coefficients is split into four identical FIR filters. Each filter has independent and real coefficients that can be updated by the ZF or LMS algorithm, depending on actual applications. Four identical FIR filters with independent coefficients are capable of cancelling the gain and phase imbalance errors on the I–Q branches at the receiver besides multipath fading compensation.

Sign Simplification: In practice, sign information obtained in the delayed input signal of v_k can be used to replace its actual value in order to simplify some calculations. In such a simplification, the LMS algorithm in (4.57) becomes

$$\begin{aligned} c_{0,k+1}(n) &= c_{0,k}(n) + \lambda\tilde{e}_{\mathrm{I},k}\mathrm{sign}[v_{\mathrm{I},k-n}] \\ c_{1,k+1}(n) &= c_{1,k}(n) + \lambda\tilde{e}_{\mathrm{I},k}\mathrm{sign}[v_{\mathrm{Q},k-n}] \\ c_{2,k+1}(n) &= c_{2,k}(n) + \lambda\tilde{e}_{\mathrm{Q},k}\mathrm{sign}[v_{\mathrm{Q},k-n}] \\ c_{3,k+1}(n) &= c_{3,k}(n) + \lambda\tilde{e}_{\mathrm{Q},k}\mathrm{sign}[v_{\mathrm{I},k-n}], \quad n = 0,\ 1,\ \ldots,\ L \end{aligned} \tag{4.59}$$

where sign(x) is defined as

$$\text{sign}(x) = \begin{cases} 1, & \text{for } x \geq 0 \\ -1, & \text{for } x < 0 \end{cases} \tag{4.60}$$

With such a simplification, it is clear that the equalizer can be simply implemented without such a multiplication with the error signal for each of the four updated equations in (4.59). Of course, the error signal may also be replaced with its sign information for further simplification, but the step size value should be reduced.

Latency in Multiplication and Addition Operations: In a practical implementation, each multiplication or addition causes one delay unit because all mathematical operations are updated at each clock period. Furthermore, the addition operation after the multiplication with more than two inputs should be decomposed into several adders each with only two inputs. Hence, an adder with more inputs results in more stages of the addition operation after the decomposition.

Figure 4.24 shows the latency caused by actual multiplication and addition operation in the five-coefficient equalization implementation. The total latency

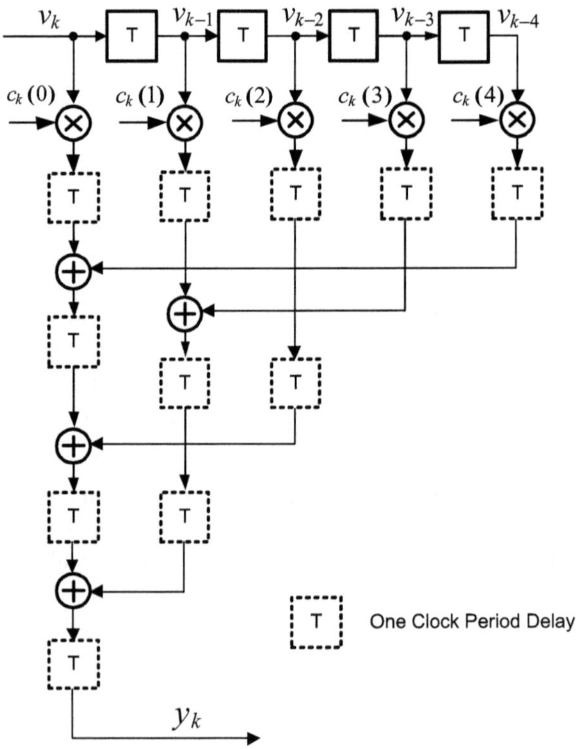

Fig. 4.24 A block diagram of a practical FIR filter branch implementation

for the five-coefficient equalizer is a four-clock period time of 4 T. Therefore the error signal also needs to be delayed by 4 T before being used for the coefficient adaption. For an equalizer with five coefficients, the amount of delay on the error signal path in Fig. 4.24 is a four-clock period of 4 T. Corresponding to the four-clock period of delay, the I and Q error signals in (4.56) and (4.57) should be replaced by $\widetilde{e}_{\mathrm{I},k-4}$ and $\widetilde{e}_{\mathrm{Q},k-4}$ to ensure that the coefficients at the next time instant are updated from ones at the current time instant and the associated error signal at the correct time instant. The latency in actual multiplication and addition operations clearly indicates why an an extra delay block in Fig. 4.24 is needed.

4.5.1.3 Blind Equalizer Based on Constant Modulus Algorithm

One problem that the adaptive equalizer updated with the LMS algorithm faces is the need for training sequences. In trained equalization techniques, a known training sequence is transmitted with information data to the receiver for the purpose of initially adjusting the equalizer coefficients. Once the eye diagram at the output of the equalizer is open before the training sequence is over, we may switch the equalizer from a training mode to a decision-directed mode to form the error signal. However, the training-sequence-based equalizer may not be practical in some applications, such as aeronautical communication systems and multipoint communication networks, where the receivers synchronize to the received signal and adjust the coefficients of the equalizers without having a known training sequence available. Therefore blind (no training sequence) equalization techniques represent an attractive alternative for these applications.

One of the broadly defined classes of adaptive blind equalization algorithms developed over the last 40 years is the group of *steepest descent based algorithms*, which rely solely on the equalized output signal and a priori statistical knowledge of the transmitted symbol constellation. These algorithms include the Sato's algorithm [11] and Godard's algorithm [12] or the constant modulus algorithm (CMA) [13]. The CMA for complex two-dimensional data communication systems is the most widely referenced blind equalization technique in both industry and academia due to its simplicity and ease of implementation with digital signal processing (DSP) chips. The CMA is well suited for use with a constant envelope modulation signal at sampling points because it has a constant modulus at the channel input. It can also be used for non-constant envelope modulation signals, like *M-ary* QAM, at a low convergence rate. Furthermore, unlike the LMS algorithm, the CMA approach has the advantage of allowing the equalizer to be adapted independent of the carrier recovery because the CMA cost function used to derive the CMA is insensitive to the phase of the equalizer output. This advantage makes the CMA more powerful in opening the constellation at the equalizer output than the LMS without the need to care about carrier frequency offset and phase error. The carrier recovery can be carried out after the equalizer.

The CMA blind approach minimizes a cost function, whose minimum is equivalent to minimizing the MSE for the LMS case. The cost function of the CMA depends on the output of the equalizer and some a priori knowledge of the statistics of the transmitted symbol constellation. A general cost function proposed by Godard [12] is of the form

$$\mathrm{CF}_p(k) = \frac{1}{4} E\left[\left(|y_k|^p - R_p\right)^2\right] \tag{4.61}$$

where $E[\bullet]$ indicates statistical expectation, y_k is the output of the equalizer at time $t = kT$, p is a positive integer, and R_p is a positive real constant depending on the transmitted signal constellation points and is given by

$$R_p = \frac{E\left[|a_k|^{2p}\right]}{E[|a_k|^p]} \tag{4.62}$$

where a_k is the complex symbol information at time $t = kT$ at the transmitter. It was demonstrated by Godard [12] that a relatively simple algorithm and fast convergence speed can be obtained in the case of $p = 2$ compared with $p = 1$. For QPSK, R_p can be normalized to be one, and thus the cost function in (4.61) will be close to zero at the decision instants due to the constant envelope of the QPSK modulation signal at the middle instant points of the baseband signals at the transmitter. For a high-order QAM modulation, R_p is a constant value that projects all of the constellation points onto the same circle [13].

For the case of $p = 2$, minimization of $\mathrm{CF}_2(k)$ with respect to the equalizer coefficients by using a stochastic gradient method results in the coefficient update equation

$$\begin{aligned} c_{k+1}(n) &= c_k(n) + \lambda\left(-\frac{\partial \mathrm{CF}_2(k)}{\partial c}\right)\bigg|_{c=c_k} \\ &= c_k(n) + \lambda e_{ck} v_{k-n}^*, \quad n = 0,\ 1,\ \ldots,\ L \end{aligned} \tag{4.63}$$

with

$$e_{ck} = y_k\left(R_2 - |y_k|^2\right) \tag{4.64}$$

where λ is the step size used to control the rate of the coefficient adjustment. Note that the coefficient update equation of (4.63) is independent of the carrier phase. Hence, the CMA-based blind equalizer has a particular advantage in allowing the equalizer to be adapted independent of the carrier recovery. The carrier phase tracking can be performed in a decision-directed mode after the equalizer, which will be described in Sect. 4.5.2.5.

4.5.2 Coherent Detection

In coherent detection, a receiver should synchronize the phase of the local reference carrier with the phase of the received signal carrier and also synchronize the phase of the local timing sequence with the phase of the recovered symbol sequence.

A fundamental MSK receiver based on the coherent detection principle is shown in Fig. 4.25 [2, 14]. Without loss of generality, the received signal $r(t)$ to the input of the receiver in Fig. 4.25 is a modulated signal at either a radio frequency (RF) or an intermediate frequency (IF). In the latter case, it is down-converted from the RF modulated signal through a mixer. In either case, the received signal $r(t)$ is expressed by the transmitted signal $s(t)$ that has the same expression as one in (4.14) plus white Gaussian noise $n(t)$. Then, the received signal $r(t)$ is coherently demodulated with a pair of orthogonal local signals $C(t)\cos(2\pi f_c t + \hat{\varphi})$ and $-S(t)\sin(2\pi f_c t + \hat{\varphi})$ on the I–Q channels, respectively. where $C(t)$ and $S(t)$ are two half-cycle sinusoidal pulse shapes given in (4.17), $\hat{\varphi}$ is the carrier phase estimate for the carrier phase $\varphi(t)$ of the received signal $r(t)$.

In an ideal carrier recovery case, the carrier phase difference between the transmitter and receiver is $\Delta\varphi = \hat{\varphi} - \varphi = 0$. The coherently demodulated baseband I–Q signals are passed through integrate and dump (I&D) circuits that perform correlation detection or matched filtering to achieve optimum coherent detection. The outputs of the correlator are sampled by the recovered symbol clock signal at the bit rate or 1/2 symbol rate clock, alternatively. After each decision on the I–Q channels, the I channel sequences are combined with the Q channel sequences through a combiner, which consists of two XOR gates performing a kind of differential decoding before producing the recovered data $\{\hat{d}_n\}$.

Compared with the optimum detection for MSK described above, the optimum coherent detections for GMSK and FQPSK are more complicated. They need more I&D branches corresponding to the different waveform segments to perform

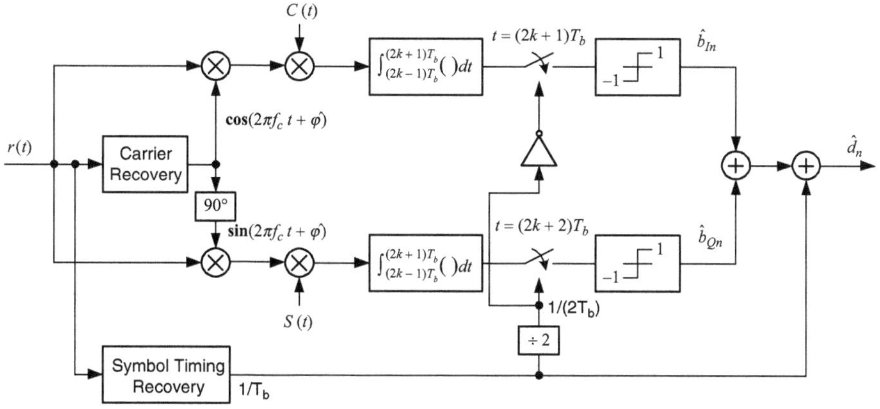

Fig. 4.25 A optimum coherent receiver of MSK

correlation detections on both I–Q channels and employ the Viterbi Algorithm (VA) to minimize the BER. In optimum coherent detection for FQPSK, FQPSK can be considered as a trellis-coded modulation due to its cross-correlation and inherent memory. The interested reader can read [2] for details.

In practice, it is preferable for the receiver to have low current consumption and low cost as high priorities the conditions that do not significantly degrade the performance, especially for mobile handset and portable communication devices. In this section, we focus on traditional lowpass filtering (LPF) detection, simply called a LPF detection, for both GMSK and FQPSK, which is implemented by using lowpass filters to replace I&D circuits in Fig. 4.25. This type of LPF detection is usually used to detect QPSK and OQPSK signals and has been demonstrated to coherently detect MSK, GMSK, and FQPSK signals with slight performance degradation relative to the optimum detection of GMSK and FQPSK.

It is clear in coherent detection that the carrier recovery and symbol timing recovery play key roles in recovering the transmitted data. Hence we discuss some practical options for implementing carrier synchronization methods. Generally, there are two basic approaches to performing carrier synchronization at the receiver. One is that a transmitter transmits a special signal called a pilot signal or a training sequence together with the information-bearing signal. This pilot signal can be inserted either at a certain frequency band or at a certain time slot depending on applications. This pilot signal allows the receiver to extract and then track the received carrier frequency so that its local oscillator can quickly synchronize to the carrier frequency and phase of the received signal because most phase- and amplitude-modulated signals don't contain a carrier component in their spectrum due to the carrier's being suppressed. However, such pilot-approach–based carrier synchronization has the distinct disadvantage that the pilot signal occupies certain frequency resources or time slots without carrying any information. The second approach is to derive the carrier phase synchronization directly from the received signal by means of some nonlinear methods, such as squaring loop carrier recovery for MSK [14].

In the following sections, we will introduce two major types of carrier synchronization techniques: reverse modulation or remodulation-loop–based carrier synchronization and Costas-loop–based carrier synchronization. Both of them can be used for the optimum detection and LPF detection.

4.5.2.1 Reverse Modulation Carrier Recovery

Generally, carrier recovery is performed at either an intermediate frequency (IF) domain after down-conversion of the RF signal or all the way down to baseband domain due to advanced DSP techniques. Hence, we focus our discussion on the carrier recovery techniques in the IF and BB domains in the following sections. For simplicity's sake, we still use the symbol f_c to represent for the carrier frequency regardless of RF and IF signals.

Before starting our discussion, we need to distinguish reverse modulation carrier recovery from remodulation carrier recovery. In reverse modulation carrier recovery, the recovered data are used to reversely modulate *the received IF-modulated signal* in order to remove the modulation data and obtain a pure carrier component as the input to the phase detector of the PLL. In remodulation carrier recovery, the recovered data are employed to remodulate *the recovered carrier signal* as the input to the phase detector of the PLL. In the former case, two input signals to the phase detector are both pure carrier signals, while in the latter case they are both modulated signals.

The reverse-modulation–based phase-locked loop is a popular carrier recovery method that is very suitable for frame-based time-division multiplexing (TDM) transmission. This method can achieve fast carrier recovery if a short length of training sequences is inserted at the beginning of each frame. Of course, the reverse modulation or remodulation loop can also be applied to non-training-sequence–based carrier synchronization. It would be better for us to start with the reverse-modulation-loop–based carrier recovery for a BPSK signal. A block diagram of the reverse-modulation-loop–based carrier recovery for an intermediate frequency (IF) BPSK signal $r(t)$ is shown in Fig. 4.26, which may be expressed as

$$\begin{aligned} r(t) &= s(t) + n(t) \\ &= A(t)\cos(2\pi f_c t + \varphi) + n(t) \end{aligned} \tag{4.65}$$

where $s(t)$ is the BPSK modulated signal at the IF, $A(t)$ is the modulation waveform carrying ± 1 data sequence, $\varphi(t)$ is the carrier phase, and $n(t)$ is a realization of a zero-mean Gaussian noise with double-sided power spectral density $N_0/2$ over the bandwidth of $s(t)$. For simplicity's sake, noise is assumed to be zero in the received signal.

First of all, the PLL is initially phase-locked to a local reference carrier signal whose frequency set to be close to the carrier frequency f_c of the input IF signal $r(t)$ in the absence of the received signal. When the input IF signal $r(t)$ is detected in the

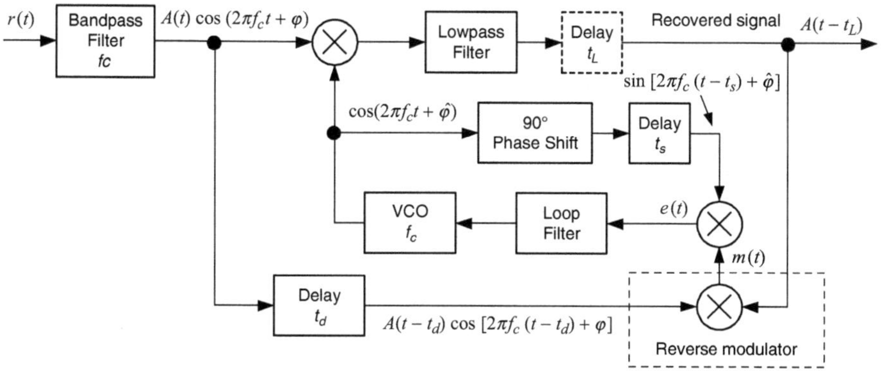

Fig. 4.26 Reverse modulation carrier recovery for BPSK signal

receiver, the PLL is immediately switched to the input IF signal to extract the carrier component from the input-modulated IF signal. In order to recover the carrier component from the IF-modulated signal, the delayed IF-modulated signal is reversely modulated by the recovered data to remove the modulation data through the reverse modulator. The local carrier signal is phase-locked to the input frequency f_c shortly if a training sequence is inserted at the beginning of each frame. Otherwise, it would take a little bit more time for the local carrier signal to be phase-locked to the input frequency f_c.

During the carrier synchronization process, the IF BPSK signal is coherently demodulated with the recovered carrier signal from the PLL and the information waveform $A(t)$ is recovered with the delay t_L at the output of the lowpass filter where the delay t_L is caused by the lowpass filter. In turn, the recovered baseband waveform is used to reversely modulate the delayed IF BPSK signal with time t_d in order to remove the modulation signal $A(t)$ from the IF signal. Thus, if time is aligned well, or $t_d = t_L = t_s$, in the reverse modulation procedure, the input signal to the phase detector is

$$m(t) = A^2(t - t_d)\cos\left[2\pi f_c(t - t_d) + \varphi\right] \tag{4.66}$$

Since the amplitude $A^2(t - t_d)$ is always positive, the sign information or ± 1 contained in $A(t)$ is removed or the phase modulation is removed completely from the delayed IF BPSK signal. Hence, the input signal $m(t)$ to the phase detector has a pure carrier component at the frequency of f_c and is then used to drive the PLL.

Now we introduce reverse-modulation-loop–based carrier recovery for a GMSK signal, as illustrated in Fig. 4.27. The bandpass filtered IF signal $r(t)$ is expressed as

$$r(t) = s(t) + n(t) \tag{4.67}$$

where $s(t)$ is the received GMSK signal at the IF and $n(t)$ is the bandpass Gaussian noise. The signal $s(t)$ is given as

$$\begin{aligned} s(t) &= A\cos\left[2\pi f_c t + \phi(t) + \varphi\right] \\ &= u_i(t)\cos(2\pi f_c t + \varphi) - u_q(t)\sin(2\pi f_c t + \varphi) \end{aligned} \tag{4.68}$$

where $u_i(t)$ and $u_q(t)$ expressed in (4.33) and (4.34) are the baseband signals on the I channel and Q channel, respectively, and φ, the phase constant, is created through a transmission channel and is used here for the purpose of carrier recovery. The bandpass noise is

$$n(t) = n_c(t)\cos(2\pi f_c t + \varphi) - n_s(t)\sin(2\pi f_c t + \varphi) \tag{4.69}$$

where $n_c(t)$ and $n_s(t)$ are the in-phase and quadrature components, respectively, of the Gaussian noise and are assumed to be statistically independent.

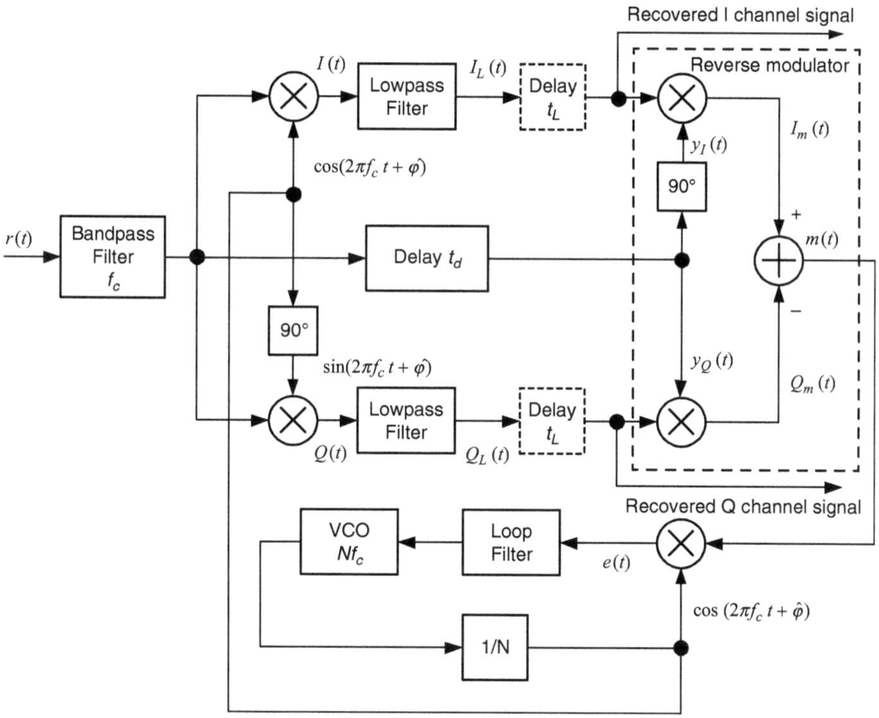

Fig. 4.27 Reverse modulation carrier recovery for QPSK/OQPSK/MSK/GMSK

After passing through the bandpass filter without causing any intersymbol interference, $r(t)$ is multiplied by $\cos(2\pi f_c t + \hat{\varphi})$ and $\sin(2\pi f_c t + \hat{\varphi})$, which are the estimated orthogonal carrier signals from the VCO, and $\hat{\varphi}$ is the estimated carrier phase. The double frequency components due to the multiplication process are removed by the lowpass filters following the mixers and the lowpass filtered baseband signals are

$$I_{\mathrm{L}}(t) = \frac{1}{2}[u_i(t) + n_{\mathrm{c}}(t)]\cos(\hat{\varphi} - \varphi) - \frac{1}{2}[u_q(t) + n_{\mathrm{s}}(t)]\sin(\hat{\varphi} - \varphi) \qquad (4.70)$$

$$Q_{\mathrm{L}}(t) = \frac{1}{2}[u_q(t) + n_{\mathrm{s}}(t)]\cos(\hat{\varphi} - \varphi) - \frac{1}{2}[u_i(t) + n_{\mathrm{c}}(t)]\sin(\hat{\varphi} - \varphi) \qquad (4.71)$$

The delayed and 90° phase-shifted IF signal as the input to the I channel mixer of a reverse modulator is

$$\begin{aligned} y_I(t) = {} & u_i(t - t_{\mathrm{d}})\mathbf{sin}[2\pi f_{\mathrm{c}}(t - t_{\mathrm{d}}) + \varphi] + u_q(t - t_{\mathrm{d}})\mathbf{cos}[2\pi f_{\mathrm{c}}(t - t_{\mathrm{d}}) + \varphi] \\ & + n_{\mathrm{c}}(t - t_{\mathrm{d}})\sin[2\pi f_{\mathrm{c}}(t - t_{\mathrm{d}}) + \varphi] + n_{\mathrm{s}}(t - t_{\mathrm{d}})\cos[2\pi f_{\mathrm{c}}(t - t_{\mathrm{d}}) + \varphi] \end{aligned} \qquad (4.72)$$

Similar to the I channel, the delayed IF signal as the input to the Q channel mixer of a reverse modulator is

$$\begin{aligned} y_Q(t) &= r(t - t_d) \\ &= u_I(t - t_d)\cos[2\pi f_c(t - t_d) + \varphi] - u_Q(t - t_d)\sin[2\pi f_c(t - t_d) + \varphi] \\ &\quad + n_c(t - t_d)\cos[2\pi f_c(t - t_d) + \varphi] - n_s(t - t_d)\sin[2\pi f_c(t - t_d) + \varphi] \end{aligned} \tag{4.73}$$

If the phase error $\Delta\varphi = \hat{\varphi} - \varphi = 0$ in (4.70) and (4.71), then the filtered baseband signals with the delay t_L become

$$I_L(t - t_L) = \frac{1}{2}[u_i(t - t_L) + n_c(t - t_L)] \tag{4.74}$$

$$Q_L(t - t_L) = \frac{1}{2}[u_q(t - t_L) + n_s(t - t_L)] \tag{4.75}$$

If the delay t_d has compensated for the delay t_L of the lowpass filters, or $t_d = t_L$, and $t' = t - t_d = t - t_L$ is met from (4.72) to (4.75), two products at the mixer's outputs of the reverse modulator are

$$\begin{aligned} I_m(t) &= y_I(t) \times I_L(t - t_L) \\ &= \frac{1}{2}\left[u_i^2(t') + 2u_i(t')n_c(t') + n_c^2(t')\right]\sin(2\pi f_c t' + \varphi) \\ &\quad + \frac{1}{2}\left[u_i(t')u_q(t') + u_i(t')n_s(t') + u_q(t')n_c(t') + n_c(t')n_s(t')\right]\cos(2\pi f_c t' + \varphi) \end{aligned} \tag{4.76}$$

$$\begin{aligned} Q_m(t) &= y_Q(t) \times Q_L(t - t_L) \\ &= \frac{1}{2}\left[u_i(t')u_q(t') + u_q(t')n_c(t') + u_i(t')n_s(t') + n_c(t')n_s(t')\right]\cos(2\pi f_c t' + \varphi) \\ &\quad - \frac{1}{2}\left[u_q^2(t') + 2u_q(t')n_s(t') + n_s^2(t')\right]\sin(2\pi f_c t' + \varphi) \end{aligned} \tag{4.77}$$

The output of the sum is

$$\begin{aligned} m(t) &= I_m(t) - Q_m(t) \\ &= \frac{1}{2}\left[u_i^2(t') + u_q^2(t') + 2u_i(t')n_c(t') + 2u_q(t')n_s(t') + n_c^2(t') + n_s^2(t')\right] \\ &\quad \times \sin(2\pi f_c t' + \varphi) \end{aligned} \tag{4.78}$$

Since the modulation is a cyclostationary stochastic process [1], the expected value of $m(t)$ is

$$E[m(t)] = \frac{1}{2}E\left[u_i^2(t') + u_q^2(t')\right] \sin\left(2\pi f_c t' + \varphi\right) \\ + \frac{1}{2}E\left[n_c^2(t') + n_s^2(t')\right] \sin\left(2\pi f_c t' + \varphi\right) \tag{4.79}$$

From the equation above, we can see that a pure carrier component at the frequency f_c appears in addition to bandpass white noise. Such a carrier component can be acquired by the phase-locked loop (PLL).

The reverse modulation carrier recovery technique has been hardware-implemented to coherently demodulate both GMSK and FQPSK signals with a data rate of 270.833 kbps, in which the reverse modulator is digitally implemented to accurately remove the modulation data from the IF input-modulated signal, as shown in Fig. 4.28 [15]. This digital reverse modulator can replace the analog reverse modulator represented by the dashed-line block in Fig. 4.27, including a delay t_d block. In the digital reverse modulator, a delay with time t_d and four-state phase shifter can be precisely implemented at a clock signal CLK. The recovered data on the I–Q channels have a total of four combinations after the hard limiters in each half-symbol (or a bit) duration due to an offset QPSK receiver structure with each combination corresponding to one of four-state phase shifter outputs. Thus, the output of the multiplexer controlled by the recovered data is the removed modulation carrier signal at the IF frequency.

Figure 4.29a shows a relatively pure carrier component at the frequency of 1.25 MHz at the output of the reverse modulator. The power level of the carrier component is about 30 dB above the noise level. From the carrier component purity point of view, the reverse modulation carrier recovery loop is not only better than the fourth power loop where the pure carrier component at the frequency of $4f_c$ is

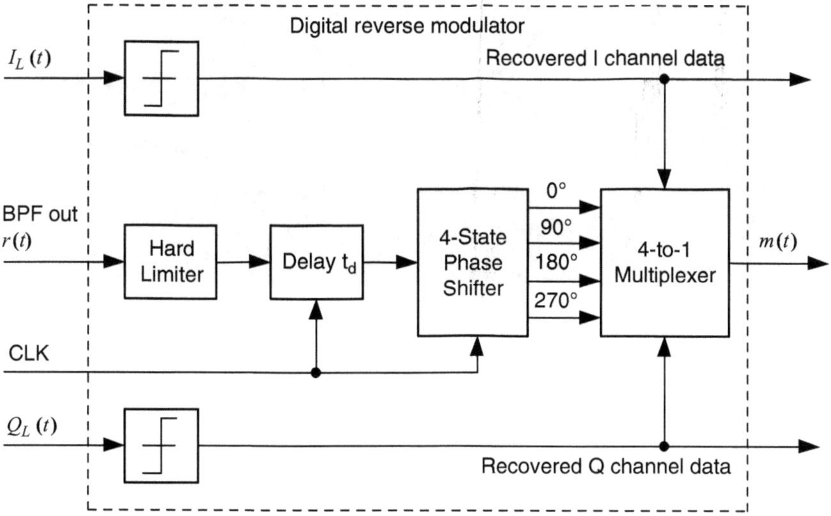

Fig. 4.28 FPGA implementation of digital reverse modulator for GMSK and FQPSK

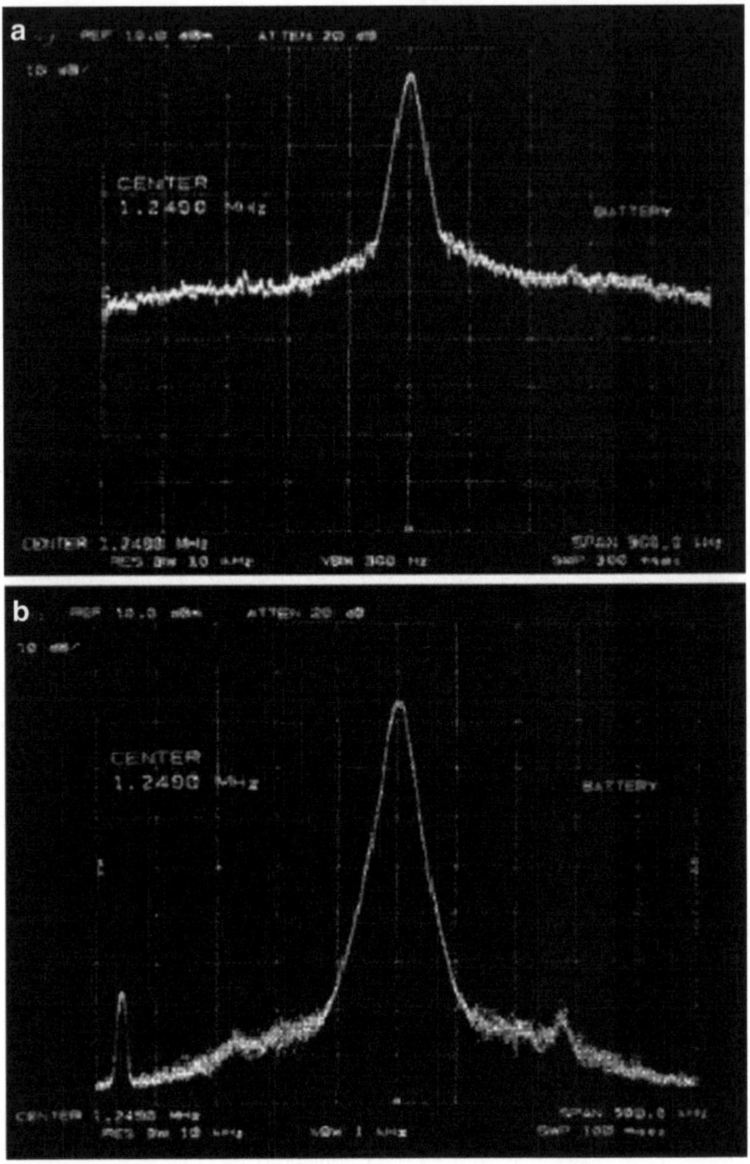

Fig. 4.29 Carrier component spectrum at an IF of 1.25 MHz: (**a**) measured at PLL input and (**b**) measured at PLL output

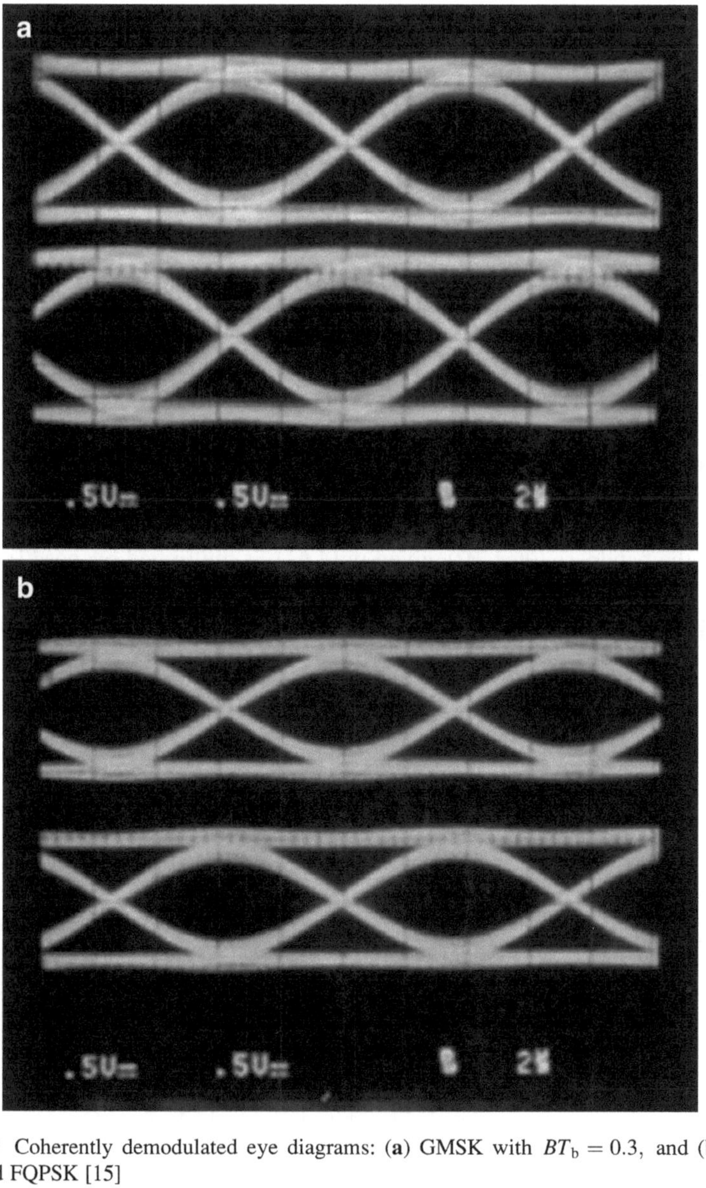

Fig. 4.30 Coherently demodulated eye diagrams: (**a**) GMSK with $BT_b = 0.3$, and (**b**) cross-correlated FQPSK [15]

attenuated if there is amplitude imbalance [16], but also better than the pilot-tone–aided carrier recovery where additional power is consumed by pilot-tone signal [17]. The phase noise of the synchronized carrier component at the output of the PLL is shown in Fig. 4.29b. It is clear that the phase noise is further improved by another 30 dB compared with the phase noise at the input of the PLL.

Figure 4.30 illustrates eye diagram patterns at the output of the lowpass filters in the receiver after coherent demodulation when GMSK and FQPSK signals are passed through nonlinear amplification channels, respectively. A 3-dB bandwidth of the bandpass filter of the receiver before the coherent detection is equal to 200 kHz, i.e., the normalized equivalent bandwidth $B_w T_b = 0.74$, which is close to the optimum parameter of 0.63 reported in [3].

4.5.2.2 Remodulation Carrier Recovery

Unlike the reverse modulation loop in Fig. 4.27, the remodulation loop [18, 19] uses the recovered I–Q baseband signals to phase-remodulate a pair of quadrature carrier signals from the PLL so that both inputs to the phase detector are the modulated signals as shown in Fig. 4.31 rather than the un-modulated (or carrier) signals. Here the recovered baseband I–Q signals are hard-limited before remodulating the recovered carrier signal because the error signal $e(t)$ is only dependent on the phase difference between the received phase-modulated signal and the phase-remodulated signal. In the following, we shall derive the phase error signal at the phase detector output.

After the bandpass filter, $r(t)$, given in (4.67), is multiplied by $\cos(2\pi f_c t + \hat{\varphi})$ and $\sin(2\pi f_c t + \hat{\varphi})$, which are the estimated orthogonal carrier signals from the VCO and $\hat{\varphi}$ is the estimated carrier phase. The output baseband signals on the I–Q channels after coherent demodulation and lowpass filtering are expressed by

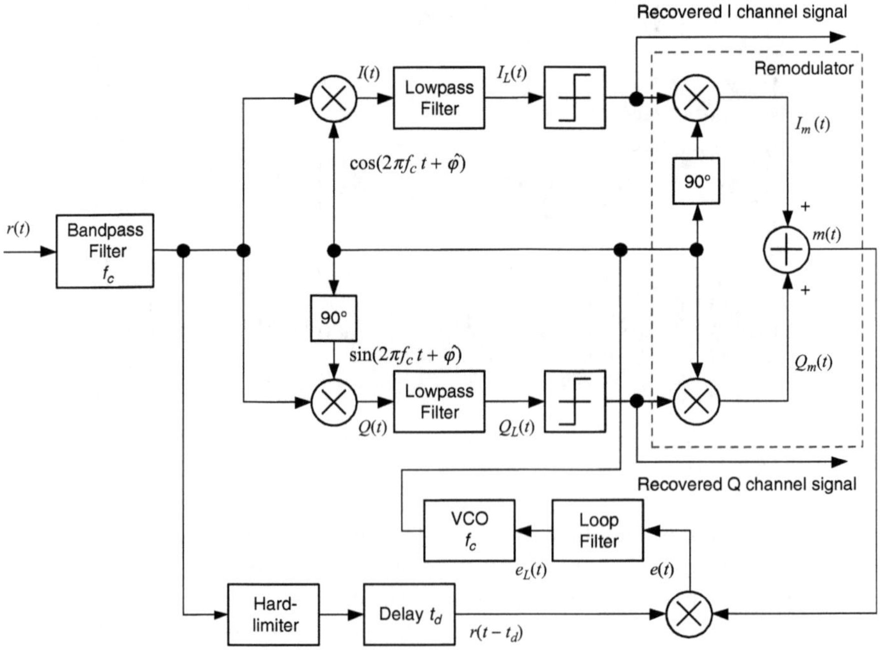

Fig. 4.31 Remodulation carrier recovery for QPSK and OQPSK, where inputs to the phase detector are both modulated signals

$$I_{\mathrm{L}}(t) = \frac{1}{2}\big[u_i(t)\mathbf{cos}\,(\varphi - \hat{\varphi}\,) + u_q(t)\mathbf{sin}\,(\varphi - \hat{\varphi}\,) \\ + n_{\mathrm{c}}(t)\mathbf{cos}\,(\varphi - \hat{\varphi}\,) + n_{\mathrm{s}}(t)\mathbf{sin}\,(\varphi - \hat{\varphi}\,)\big] \tag{4.80}$$

$$Q_{\mathrm{L}}(t) = \frac{1}{2}\big[u_q(t)\mathbf{cos}\,(\varphi - \hat{\varphi}\,) - u_i(t)\mathbf{sin}\,(\varphi - \hat{\varphi}\,) \\ + n_{\mathrm{s}}(t)\mathbf{cos}\,(\varphi - \hat{\varphi}\,) - n_{\mathrm{c}}(t)\mathbf{sin}\,(\varphi - \hat{\varphi}\,)\big] \tag{4.81}$$

It is assumed in (4.80) and (4.81) that any distortions caused by the lowpass filters on the filtered baseband signals and Gaussian noise are ignored and the phase difference $\varphi - \hat{\varphi}$ is very small. The recovered data of sgn[$I_{\mathrm{L}}(t)$] and sgn[$Q_{\mathrm{L}}(t)$] at the outputs of hard-limiters remodulate the recovered quadrature carrier signals and the summed signal $m(t)$ at the remodulator output is then expressed by

$$\begin{aligned} m(t) &= Q_m(t) + I_m(t) \\ &= \mathrm{sgn}[Q_{\mathrm{L}}(t)]\cos\,(2\pi f_{\mathrm{c}}t + \hat{\varphi}\,) + \mathrm{sgn}[I_{\mathrm{L}}(t)]\sin\,(2\pi f_{\mathrm{c}}t + \hat{\varphi}\,) \\ &= \mathrm{sgn}\big[u_q(t)\big]\cos\,(2\pi f_{\mathrm{c}}t + \hat{\varphi}\,) + \mathrm{sgn}[u_i(t)]\sin\,(2\pi f_{\mathrm{c}}t + \hat{\varphi}\,) \end{aligned} \tag{4.82}$$

The remodulated signal $m(t)$ is now used as the reference signal input to the phase detector. Another input to the phase detector is a delayed version of the received signal $r(t)$. The delay time t_{d} is used to compensate for the time delay t_{L} that is caused by the lowpass filters and is not shown in Fig. 4.31. When the delay compensation is perfect, the lowpass filtered phase error after the phase detector is

$$e(t) = r(t) \times m(t) \tag{4.83}$$

It can be noted in (4.83) that the time delays t_d and t_{L} were omitted for perfect delay compensation. After the double-frequency terms are filtered out and noise is ignored, the phase error at the output of the loop filter is

$$e_{\mathrm{L}}(t) = \frac{1}{2}\,\{-u_i(t)\mathrm{sgn}[u_i(t)]\sin\,(\varphi - \hat{\varphi}\,) - u_q(t)\mathrm{sgn}\big[u_q(t)\big]\sin\,(\varphi - \hat{\varphi}\,) \\ + u_i(t)\mathrm{sgn}\big[u_q(t)\big]\cos\,(\varphi - \hat{\varphi}\,) - u_q(t)\mathrm{sgn}[u_i(t)]\cos\,(\varphi - \hat{\varphi}\,)\} \tag{4.84}$$

With a perfect carrier phase estimate, or $\varphi - \hat{\varphi} = 0$, the phase error is rewritten as

$$e_{\mathrm{L}}(t) = \frac{1}{2}\{u_i(t)\mathrm{sgn}\big[u_q(t)\big] - u_q(t)\mathrm{sgn}[u_i(t)]\} \tag{4.85}$$

For OPSK/OQPSK signals, we have $\mathrm{sgn}[u_i(t)] = \pm 1$ and $\mathrm{sgn}\big[u_q(t)\big] = \pm 1$. Thus, the error signal $e_{\mathrm{L}}(t)$ is very small if $u_i(t)$ is close to $u_q(t)$.

A 100-Mbit/s prototype MSK modem with an optimum receiver using the remodulation loop for satellite communications was proposed in [20] and shown in Fig. 4.32. The received MSK signal is first down-converted into the IF signal $r(t)$, as given in (4.67), where the MSK signal is written as

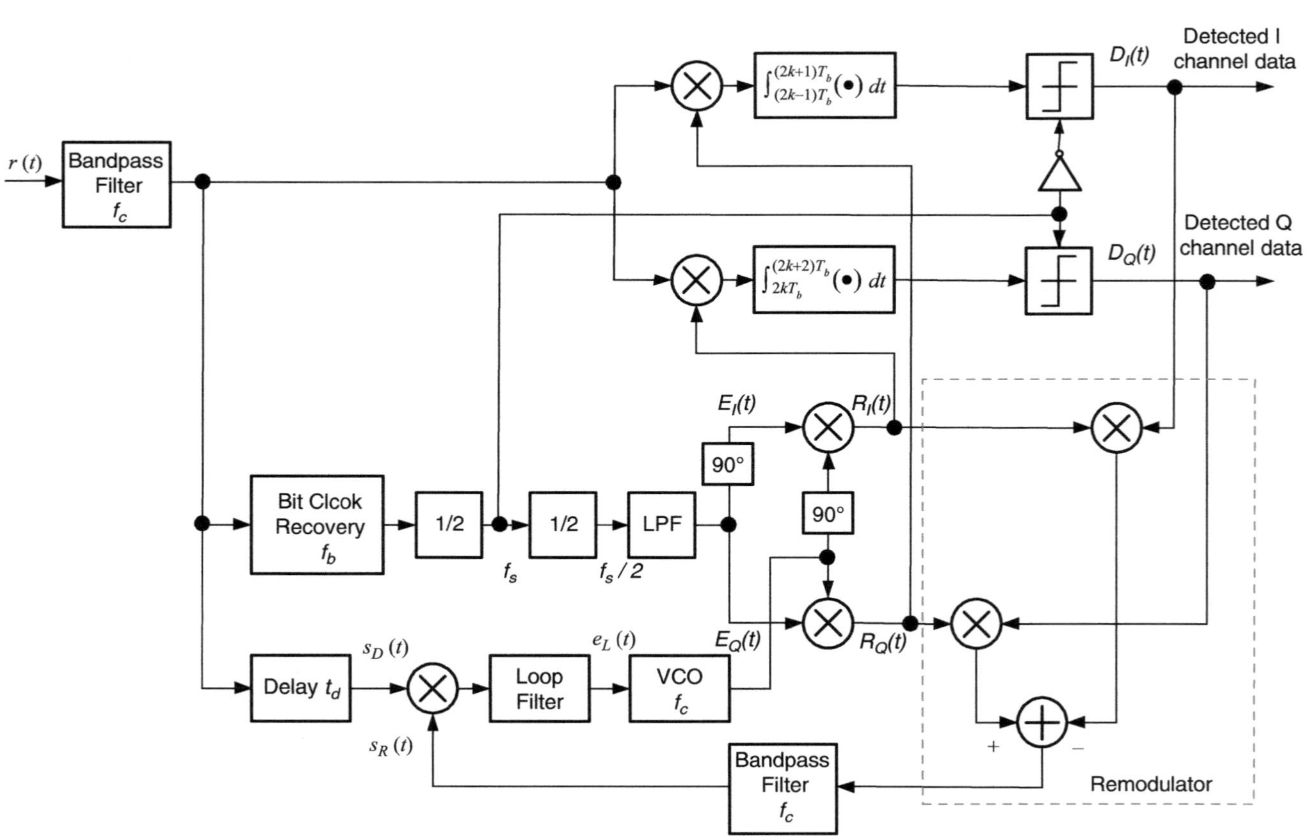

Fig. 4.32 Block diagram of an optimal detection receiver for MSK signal [20]

$$s(t) = D_{\mathrm{I}}(t)C_{\mathrm{I}}(t)\cos(2\pi f_{\mathrm{c}}t + \varphi) + D_{\mathrm{Q}}(t)C_{\mathrm{Q}}(t)\sin(2\pi f_{\mathrm{c}}t + \varphi) \tag{4.86}$$

The baseband signals $C_{\mathrm{I}}(t), C_{\mathrm{Q}}(t), D_{\mathrm{I}}(t)$, and $D_{\mathrm{Q}}(t)$ are defined as

$$D_{\mathrm{I}}(t) = \sum_{n=-\infty}^{\infty} a_n, \quad D_{\mathrm{Q}}(t) = \sum_{n=-\infty}^{\infty} b_n \tag{4.87}$$

$$C_{\mathrm{I}}(t) = \cos\left(\frac{\pi t}{2T_{\mathrm{b}}}\right), \quad C_{\mathrm{Q}}(t) = \sin\left(\frac{\pi t}{2T_{\mathrm{b}}}\right) \tag{4.88}$$

Equation (4.86) is identical to (4.14) except for the plus sign before the second term and phase constant. In (4.87), a_n and b_n represent the differentially encoded sequences in the I–Q channels and are expressed in (4.20).

In Fig. 4.32, it is assumed that the bit clock signal with the frequencyf_{b} is recovered. The symbol clock signal with the frequency f_{s} is obtained by passing through a divider by two and used to detect the recovered baseband I–Q signals in the samplers. For the MSK signal, one of the inputs to the samplers is phased-shifted by 180° for the offset sampling purposes. The sinusoidal signal and its 90° phase-shifted signal with the frequency $f_{\mathrm{s}}/2$, represented by $E_{\mathrm{I}}(t)$ and $E_{\mathrm{Q}}(t)$, are generated by further dividing the symbol clock signal by two. These signals are expressed by

$$E_{\mathrm{I}}(t) = \mathbf{cos}\left(\frac{\pi t}{2T_{\mathrm{b}}} + \theta\right), \; E_{\mathrm{Q}}(t) = \mathbf{sin}\left(\frac{\pi t}{2T_{\mathrm{b}}} + \theta\right) \tag{4.89}$$

where θ represents an initial phase status with one value of $0, \pi, \pi/2$, or $3\pi/2$, depending on the initial conditions of the divide-by-two circuits. These two signals will be used as the matched pulses to correlate with the received MSK signals so that they have the same polarity in the case $\theta = 0$. For other initial phase values of θ, the relationships between the polarities of the half-cycle sinusoidal pulses in the transmitter and that of the matched pulses in the receiver are listed in Table 1 in [20].

The regenerated quadrature carrier signals $R_{\mathrm{I}}(t)$ and $R_{\mathrm{Q}}(t)$ in Fig. 4.32 are written as

$$R_{\mathrm{I}}(t) = E_{\mathrm{I}}(t)\mathbf{cos}(2\pi f_{\mathrm{c}}t + \hat{\varphi}) = \mathbf{cos}\left(\frac{\pi t}{2T_{\mathrm{b}}} + \theta\right)\mathbf{cos}(2\pi f_{\mathrm{c}}t + \hat{\varphi}) \tag{4.90}$$

$$R_{\mathrm{Q}}(t) = E_{\mathrm{Q}}(t)\mathbf{sin}(2\pi f_{\mathrm{c}}t + \hat{\varphi}) = \mathbf{sin}\left(\frac{\pi t}{2T_{\mathrm{b}}} + \theta\right)\mathbf{sin}(2\pi f_{\mathrm{c}}t + \hat{\varphi}) \tag{4.91}$$

where $\hat{\varphi}$ is the estimated carrier phase. The received MSK-modulated signal is coherently demodulated by multiplying $R_{\mathrm{I}}(t)$ and $R_{\mathrm{Q}}(t)$ in the I channel and Q channel, respectively, so that two correlators (or matched filters) correlate the received signal with the two quadrature carrier signals $R_{\mathrm{I}}(t)$ and $R_{\mathrm{Q}}(t)$

Meanwhile, $R_I(t)$ and $R_Q(t)$ are remodulated by detected data $D_I(t)$ and $D_Q(t)$ in order to obtain another remodulated MSK signal at the input of the phase detector in the same baseband I–Q polarities as that of modulation. Taking the delay time t_d caused by the lowpass filters into account, the remodulated signal $s_R(t)$ is given by

$$\begin{aligned} s_R(t) &= D_Q(t-t_d)E_Q(t-t_d)\cos\left(2\pi f_c t+\hat{\varphi}\right) \\ &\quad - D_I(t-t_d)E_I(t-t_d)\sin\left(2\pi f_c t+\hat{\varphi}\right) \end{aligned} \tag{4.92}$$

A delay block with t_d needs to be inserted on the received signal path at the input of the phase detector in order to compensate for the delay t_d. The received signal $s_D(t)$ with the delay t_d at the phase detector after ignoring noise $n(t)$ is

$$\begin{aligned} s_D(t) &= r(t-t_d) = s(t-t_d) \\ &= D_I(t-t_d)C_I(t-t_d)\cos\left(2\pi f_c t+\varphi\right) \\ &\quad + D_Q(t-t_d)C_Q(t-t_d)\sin\left(2\pi f_c t+\varphi\right) \end{aligned} \tag{4.93}$$

Note that the extra phase corresponding to delay time t_d is simply taken account into φ. If the initial phase θ is equal to zero in (4.89), then $E_I(t) = C_I(t)$ and $E_Q(t) = C_Q(t)$. With these relationships above, the phase error after the loop filter is

$$e(t) = \frac{1}{2}\left(D_I^2E_I^2 + D_Q^2E_Q^2\right)\sin\left(\varphi-\hat{\varphi}\right) = \frac{1}{2}\sin\left(\varphi-\hat{\varphi}\right) \tag{4.94}$$

where the expression of time $t - t_d$ is omitted for simplicity, $D_I = D_Q = \pm 1$ and $E_I^2 + E_Q^2 = 1$ are used.

It is reported that although stochastically equivalent, the quadrature remodulation loop has been shown to exhibit a somewhat faster acquisition time when compared to a conventional quadrature Costas loop. Even though both inputs to the phase detector can be either un-modulated signals in the reverse modulation loop or modulated signals in remodulation-loop–based carrier recovery, the former is more popular than the later. This is especially true for the case when the pilot carrier signal is inserted at the beginning of data sequences because the carrier loop performs without the need for the recovered data.

Optimum detection or matched filter detection receiver with an integrate-and-dump (I&D) circuit achieves the best performance for the MSK signal. This detection receiver, however, requires a very wideband transmission channel, and the I&D circuit is difficult to create at a high bit rate [20]. In practice, wideband transmission channels result in less spectrum efficiency and include the penalty of high cost. Furthermore, in wireless handsets and other portable devices, the goal of achieving low power consumption and low cost along with acceptable BER performance is a higher priority than achieving the optimum BER performance at the price of high power consumption and high cost.

As mentioned previously in this Chapter, another type of the detection receivers performed by substituting a lowpass filter for an I&D, shown in Fig. 4.32, has the

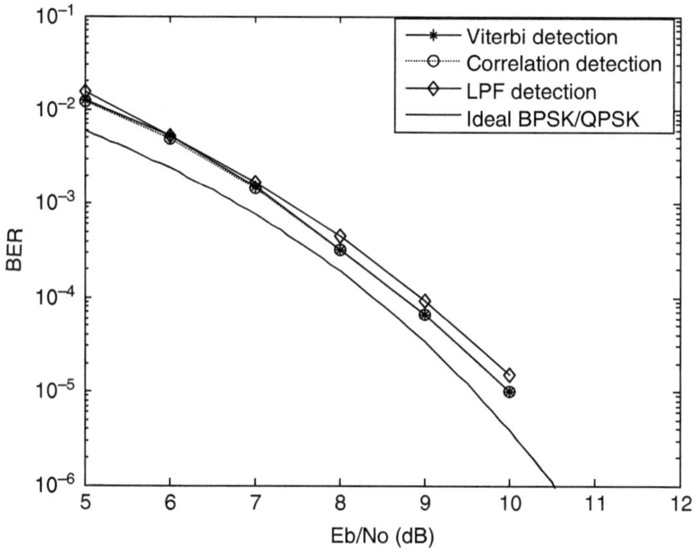

Fig. 4.33 BER performance of coherent detection MSK

advantages of simple implementation and low cost in a bandlimited channel and only leads to slight BER degradation compared to the optimum receiver. The receiver with the LPF instead of the I&D is widely used for QPSK and OQPSK detections, as illustrated in Figs. 4.27 and 4.31. Such receivers with LPFs have been experimentally demonstrated to coherently demodulate MSK, GMSK, and FQPSK signals in [3, 15].

To evaluate the performance of the MSK system with different detection methods, we use SIMULINK to simulate the BER performance of coherent detection for MSK, as shown in Fig. 4.33. A total of three different detection methods are used to evaluate the MSK performance in a white Gaussian noise (WGN) channel. The first BER curve stands for the optimum receiver using matched filter detection with the Viterbi decoder, called *Viterbi detection,* while the second one represents the same optimum receiver as the first one, except without the Viterbi decoder; this is called *correlation detection.* It can be seen that they have no differences because there is no the encoder at the transmitter. The third curve is obtained from the receiver with the LPF and sample detection circuit and is called *LPF detection.* In LPF detection, the Gaussian lowpass filters with normalized bandwidth of $BT_b = 0.63$ are used in the I–Q channels, where B is the –3-dB bandwidth of Gaussian LPF and T_b is the bit duration,. This normalized –3-dB bandwidth of $BT_b = 0.63$ for MSK was demonstrated to be optimal in achieving the best BER in a Gaussian channel [3].

The E_b/N_o degradation for the receiver with the LPF detection is only about 0.2–0.3 dB at BER $= 10^{-4}$ compared to the optimum detection receiver, whose BER performance degrades about 0.5 dB compared to ideal QPSK. However, the optimum receiver achieves its performance in an infinite-bandwidth condition, which is impossible in practice. Therefore, the receiver with the Gaussian LPF or other lowpass filters would be more practical, especially in mobile and portable RF IC designs.

A variety of different types of optimum receivers have been proposed for coherent optimum detection of GMSK [21–23]. In [21] the optimum receivers are based on the representation of the binary continuous phase modulation (CPM) signal as a sum of phase-shifted amplitude-modulation pulse (AMP) streams, in which the number of such AMP streams is dependent on the value of BT_b in the modulation. Such a decomposition of this representation for CPM using the form of a superposition of AMP streams developed by Laurent [24] helped Kaleh [21] simplify the implementation of coherent receivers for CPM signals, especially for GMSK.

In [24], the baseband signal of GMSK can be expressed as a superposition of 2^{L-1} AMP streams or the equivalent pulses $\{h_k(t)\}$, where L is used to present the duration LT_b for the pulses $\{h_k(t)\}$ and is determined by the value of BT_b. For a GMSK signal with $BT_b \geq 0.25$, the value of $L = 4$ is adequate to represent such a GMSK signal. Thus, GMSK signal needs $2^{L-1} = 8$ different pulses $\{h_k(t); k = 0, 1, \ldots, 7\}$ in its expression. In this case, the optimum receiver that minimizes the message error probability employs a bank of eight matched filters $\{h_k(-t); k = 0, 1, \ldots, 7\}$ corresponding to each of the $2^{L-1} = 8$ pulses and a Viterbi algorithm (VA) decoder. The number of states in the trellis diagram characterizing the VA is $2^L = 16$.

The complexity of the optimum receiver is directly proportional to the number of states 2^L. Using approximate signals composed of a smaller number of AMP streams, Kaleh [21] proposed a simplified Viterbi receiver, which is composed of only two matched filters, or $h_0(-t)$ and $h_1(-t)$, and results in a four-state VA and achieves suboptimum performance. This suboptimum or nearly optimum receiver has a performance degradation of less than 0.24 dB compared with the optimum receiver. The optimum filter is obtained by inserting a Wiener filter in the receiver after the simplified matched filters and before the threshold detector based on the minimum mean square error criterion (MMSE). The purpose of adding such a Wiener filter (or an equalizer taking the form of a FIR filter) is to reduce noise and ISI caused by adjacent symbols. Figure 4.34 shows the eye diagram of GMSK with $BT_b = 0.25$ at the output of the optimum filter with the Wiener filter coefficients of 11. For more detailed results, the interested reader can reference [21].

As pointed out in [21], the nearly optimum receiver consisting of two matched filters and a four-state VA for GMSK with $BT_b = 0.25$ degrades less than 0.24 dB compared with the optimum receiver that is constructed by combining the matched filter and Wiener filter, and less than 0.7 dB + 0.24 dB = 0.94 dB compared with the optimum detection for MSK, respectively. In fact, these two matched filters in the nearly optimum receiver can be replaced with Gaussian lowpass filters with $B_L T_b = 0.63$, where B_L is the 3-dB down bandwidth of the Gaussian LPF and is equivalent to the half of the 3-dB down bandwidth B_i of the Gaussian BPF in [3]. By

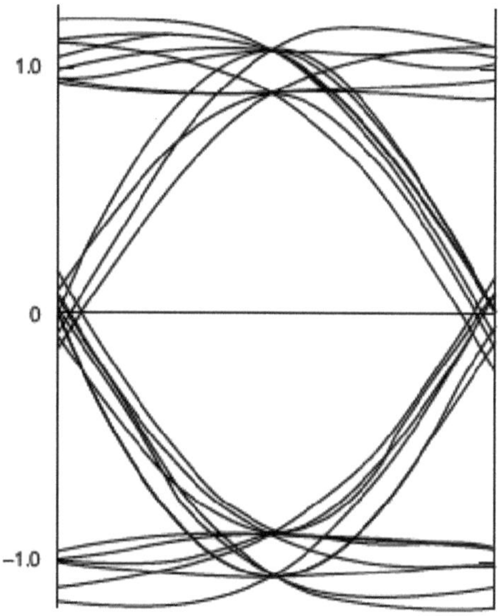

Fig. 4.34 Eye diagram at the output of the optimum receiver filter for GMSK with $BT_b = 0.25$, where the optimal filter is formed by the combination of the matched and Wiener filters [21]

using the fourth-order Gaussian filters in the receiver, Murota [3] experimentally tested the static BER performance of GMSK and showed that the BER performance of GMSK with $BT_b = 0.25$ degrades by 1.0 dB relative to MSK, which is very close to 0.94 dB in [21] above. The eye diagram of GMSK at the output of the fourth-order Gaussian LPF is illustrated in Fig. 4.35a, where the ISI is slightly larger than the ISI in Fig. 4.34 at the sampling instants. To reduce the ISI caused by the non-constant group delay property of the analog Gaussian LPF, we illustrate the eye diagram of GMSK with $BT_b = 0.25$ at the output of an approximate eighth-order LPF including the group delay compensation in Fig. 4.35b, which was used in a commercial GSM transceiver chip. It is seen from Fig. 4.35b that the eye diagram becomes symmetrical with group delay compensation.

Figure 4.36a shows the experimental test BER curves of coherent detection GMSK systems in a stationary AWGN environment where the normalized –3-dB bandwidth of the pre-detection Gaussian BPF is $B_i T_b \cong 0.63$ [3]. The parameter B_i is the –3-dB bandwidth of the Gaussian BPF and is equal to twice the –3-dB bandwidth of the Gaussian *equivalent* LPF, or $B_i = 2B_L$. This condition of $B_i T_b \cong 0.63$ is nearly optimum for these values of $BT_b = 0.25$ and ∞, as illustrated in Fig. 4.36b.

4.5.2.3 Analog Costas Loop

Another widely used method for generating a proper phase-locked carrier for a double-sideband suppressed carrier signal is a Costas loop, which is insensitive to the presence of data modulation. The Costas loop was invented by John P. Costas in 1956 [25]. Unlike the reverse modulation and remodulation loops, an error signal in

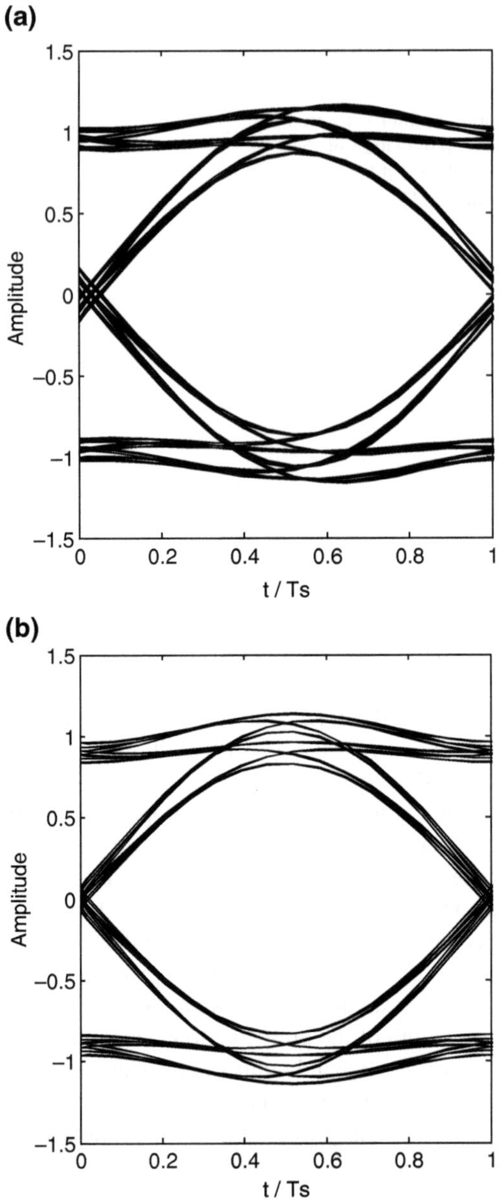

Fig. 4.35 Eye diagrams at the output of the nearly optimum receiver filter for GMSK with $BT_b = 0.25$. (**a**) a Gaussian fourth-order LPF and (**b**) an eighth-order LPF, including group delay compensation, is used and $T_s = 2T_b$ is the symbol interval

the Costa loop is generated from the recovered baseband data on the I–Q branches. Figure 4.37 shows a Costas loop used for synchronizing the received BPSK-modulated signal. Again we consider that a suppressed carrier signal of BPSK plus noise is as follows:

$$r(t) = A(t)\cos(2\pi f_c t + \varphi) + n(t) \tag{4.95}$$

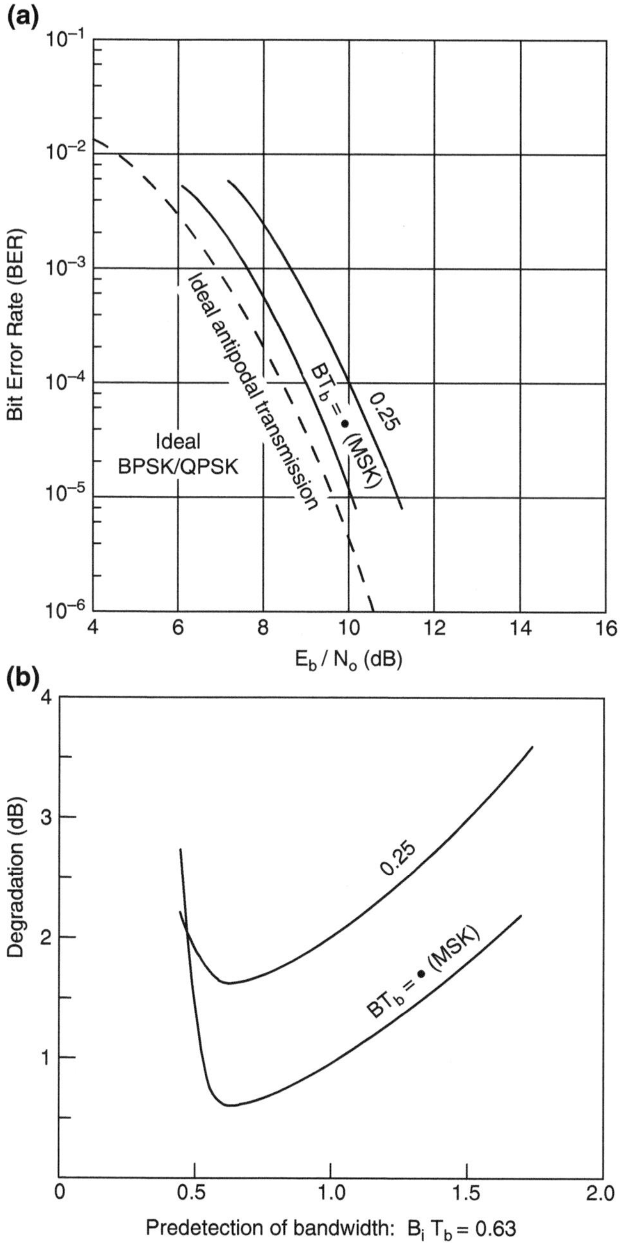

Fig. 4.36 Performance of MSK and GMSK in a stationary AWGN environment: (**a**) BER versus $E_b/N_0$and (**b**) degradation of required E_b/N_0 to maintain BER $= 10^{-3}$. Redrawn from [3]

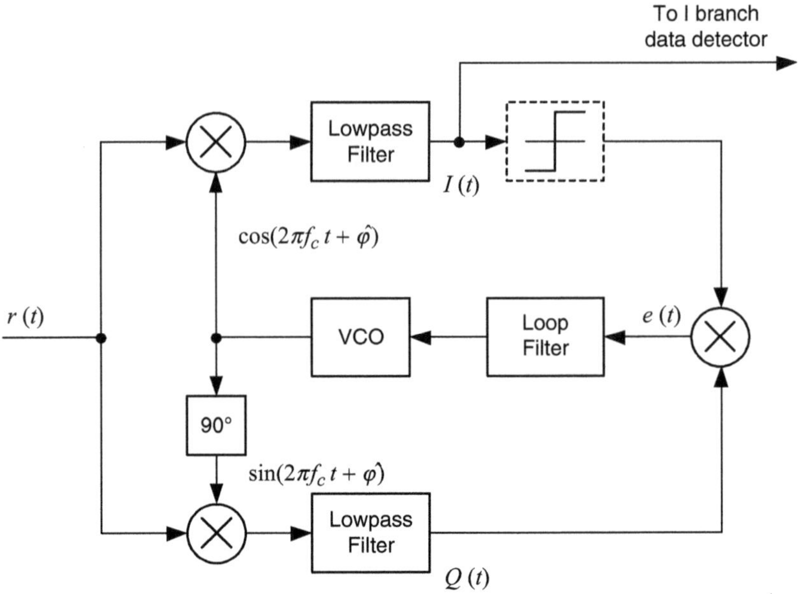

Fig. 4.37 Costas loop for BPSK with hard-limiter in-phase branch, also called a modified version [27]

where $A(t)$ carriers the binary bit information, φ is the unknown phase of the carrier signal, and $n(t)$ is the white Gaussian noise given in (4.69). The received signal $r(t)$ is multiplied by the local carrier replicas $\cos(2\pi f_c t + \hat{\varphi})$ on the I branch and $\sin(2\pi f_c t + \hat{\varphi})$ on the Q branch, respectively. After the double-frequency components are eliminated by the lowpass filters following the multiplications, the outputs of the lowpass filters are

$$I(t) = \frac{1}{2}[A(t) + n_c(t)]\mathbf{cos}(\Delta\varphi) + \frac{1}{2}n_s(t)\mathbf{sin}(\Delta\varphi) \tag{4.96}$$

$$Q(t) = \frac{1}{2}[A(t) + n_c(t)]\mathbf{sin}(\Delta\varphi) - \frac{1}{2}n_s(t)\mathbf{cos}(\Delta\varphi) \tag{4.97}$$

where the phase error $\Delta\varphi = \varphi - \hat{\varphi}$. An error signal is then generated by multiplying the output of the I-branch lowpass filter with the output of the Q branch lowpass filter.

Thus,

$$\begin{aligned} e(t) = &\frac{1}{8}\left\{[A(t) + n_i(t)]^2 - n_q^2(t)\right\}\sin(2\Delta\varphi) \\ &-\frac{1}{4}n_q(t)[A(t) + n_i(t)]\cos(2\Delta\varphi) \end{aligned} \tag{4.98}$$

The error signal is then filtered by the loop filter and the filtered output voltage drives the VCO in such a way that the phase of the local carrier replica reaches the phase of the received carrier step by step. If noise is ignored, the error signal in (4.98) is

$$e(t) = \frac{1}{8}A^2(t)\mathbf{sin}\,(2\Delta\varphi) \tag{4.99}$$

where the bit information transition determined by *A*(*t*) is removed. The error *e*(*t*) closely approaches zero when the loop is phase-clocked, depending on the SNR in the received signal.

A modified version of the Costas loop requiring even less hardware is illustrated in Fig. 4.37 when the hard-limiter as indicated by the dashed-line block is included [26, 27]. Using this modified Costas loop, the multiplier of the phase detector can be replaced with a simple chopper multiplier. In this case, the error *e*(*t*) ignoring noise is given as

$$e(t) = \frac{1}{2}A(t)\mathbf{sin}\,(\Delta\varphi) \times \text{sign}\left[\frac{1}{2}A(t)\mathbf{cos}\,(\Delta\varphi)\right] \tag{4.100}$$

It can be seen clearly from (4.100) that the error signal is proportional to the desired term $|A(t)|\,\sin(\Delta\varphi)$, regardless of the bit information transition of *A*(*t*).

The Costas loop used for a four-phase modulated signal like QPSK is different from that used for binary modulation as shown above. Figure 4.38 shows a block diagram of the Costas loop for a four-phase modulation [28]. In a four-phase transmission, the modulated signal is expressed a

$$s(t) = u_i(t)\cos\,(2\pi f_c t + \varphi) - u_q(t)\sin\,(2\pi f_c t + \varphi) \tag{4.101}$$

where $u_i(t)$ and $u_q(t)$ are independent baseband waveforms in the I and Q branches, respectively. The received input signal *r*(*t*) is approximately equal to the transmitted signal *s*(*t*) after neglecting AWGN. The error *e*(*t*) is generated as

$$\begin{aligned} e(t) = {} & \tfrac{1}{2}\left[u_i(t)\sin\,(\Delta\varphi) + u_q(t)\cos\,(\Delta\varphi)\right] sign\left\{\tfrac{1}{2}\left[u_i(t)\cos\,(\Delta\varphi) - u_q(t)\sin\,(\Delta\varphi)\right]\right\} \\ & - \tfrac{1}{2}\left[u_i(t)\cos\,(\Delta\varphi) - u_q(t)\sin\,(\Delta\varphi)\right] sign\left\{\tfrac{1}{2}\left[u_i(t)\sin\,(\Delta\varphi) + u_q(t)\cos\,(\Delta\varphi)\right]\right\} \end{aligned} \tag{4.102}$$

where sign() stands for the hard-limiting operation. For a specific case where the baseband waveforms are unfiltered or rectangular and the bandwidth of the transmission channel is unlimited, the average error signal e_{ave} is calculated as [28]

$$e_{\text{ave}} = \begin{cases} \sin\,(\Delta\varphi), & -45^\circ < \Delta\varphi \le 45^\circ \\ -\cos\,(\Delta\varphi), & 45^\circ < \Delta\varphi \le 135^\circ \\ -\sin\,(\Delta\varphi), & 135^\circ < \Delta\varphi \le 225^\circ \\ \cos\,(\Delta\varphi), & 225^\circ < \Delta\varphi \le 315^\circ \end{cases} \tag{4.103}$$

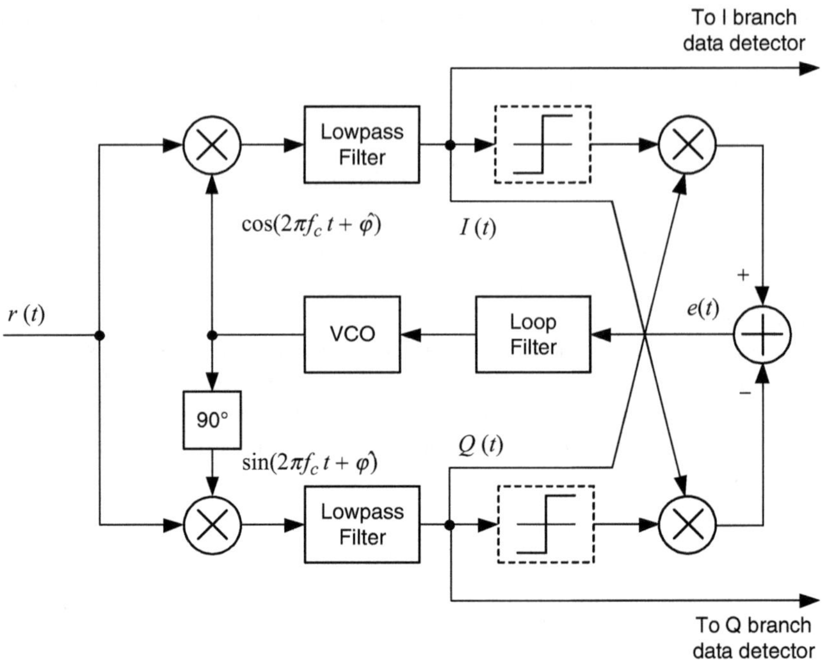

Fig. 4.38 Costas loop for QPSK [28]

The characteristic of the average error signal expressed in (4.103) is illustrated by the solid curve of Fig. 4.39. It approximates to a sawtooth pattern.

The baseband waveforms, however, are not rectangular and the bandwidth is limited as well in practical applications. For instance, the baseband signal during the symbol interval is generated by overlapping several adjacent filtered pulses in raised-cosine pulse shaping. If one-half cycle of a sinusoidal waveform is used to replace the rectangular pulse in the I–Q branches, the sharp peaks of the sawtooth curve are rounded off as shown by the dashed-line curve in Fig. 4.39.

A stable lock can occur at any of the four different phases: −90, 0, 90, and 180 [28]. An inherent fourfold ambiguity can be resolved by the differential encoding in the transmitter.

4.5.2.4 Digital Costas Loop

Digital carrier recovery is widely used in most communication systems due to its flexibility, precision, and robustness. Chung [27] derived and analyzed the linear PLL model in 1993. A first-order loop filter is used because zero steady-state phase error and frequency error can be achieved if the DC gain of the loop filter is infinite and the frequency offset is constant.

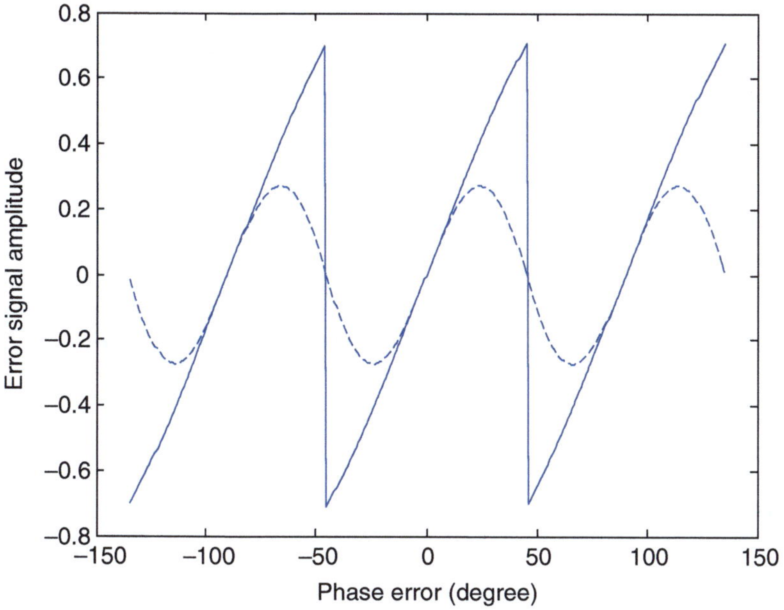

Fig. 4.39 Phase detector characteristics of a Costas loop for a QPSK modulation signal, where the solid curve depicts the case of a rectangular pulse while the dashed curve depicts one-half cycle of a sinusoid

A second-order PLL system with a first-order loop filter $F(s)$ and a voltage controlled oscillator (VCO) $N(s)$ is shown in Fig. 4.40a. Their transfer functions are given by

$$F(s) = \frac{1}{s}\frac{\tau_2 s + 1}{\tau_1} \tag{4.104}$$

$$N(s) = \frac{K_v}{s} \tag{4.105}$$

where $\tau_1 = R_1C$ and $\tau_2 = R_2C$ are time constants in an active loop filter and K_v is gain of the VCO. By using the bilinear transformation, the digital filter and NCO (digital VCO) are

$$F(z) = \frac{(K_1 + K_2) - K_1 z^{-1}}{1 - z^{-1}} \tag{4.106}$$

$$N(z) = \frac{K_v z^{-1}}{1 - z^{-1}} \tag{4.107}$$

where the coefficients $K_1 = \tau_2/\tau_1 - T/(2\tau_1)$, $K_2 = T/\tau_1$, and T is the sampling interval. The digital loop filter is shown in Fig. 4.41.

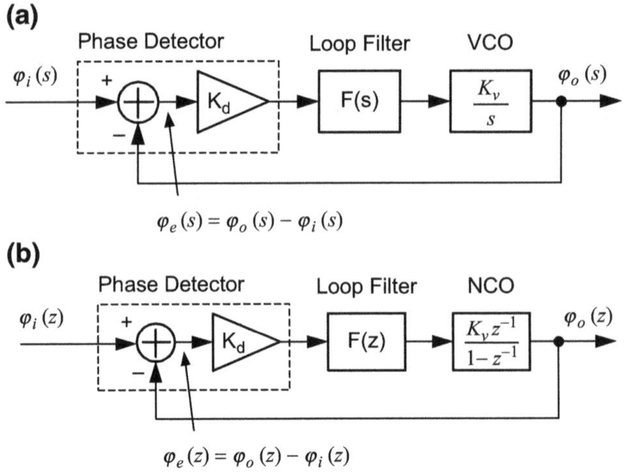

Fig. 4.40 Block diagram of a second-order PLL: (**a**) basic analog phase-locked loop (PLL) and (**b**) digital phase-locked loop

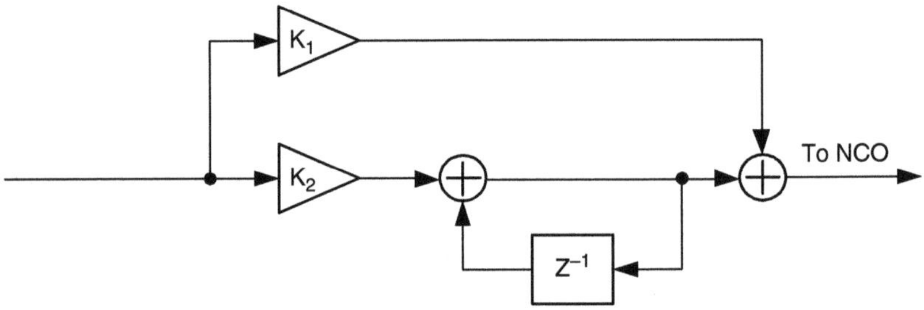

Fig. 4.41 A digital first-order loop filter

The transfer function of a linearized analog PLL model is [27]

$$H(s) = \frac{\varphi_o(s)}{\varphi_i(s)} = \frac{K_d F(s) N(s)}{1 + K_d F(s) N(s)} \tag{4.108}$$

where K_d is the gain of the phase detector. Substituting (4.104) and (4.105) into (4.108) yields the following the transfer function:

$$H(s) = \frac{2\zeta\omega_n s + \omega_n^2}{s^2 + 2\zeta\omega_n s + \omega_n^2} \tag{4.109}$$

where the natural frequency $\omega_n = \sqrt{(K_d K_v)/\tau_1}$ and the damping factor $\zeta = (\tau_2 \omega_n)/2$. Applying the bilinear transformation to (4.109) yields the digital transfer function of the PLL model given by

$$H(z) = \frac{\left[4\zeta\omega_n T + (\omega_n T)^2\right] + 2(\omega_n T)^2 z^{-1} + \left[(\omega_n T)^2 - 4\zeta\omega_n T\right] z^{-2}}{\left[4 + 4\zeta\omega_n T + (\omega_n T)^2\right] + \left[2(\omega_n T)^2 - 8\right] z^{-1} + \left[4 - 4\zeta\omega_n T + (\omega_n T)^2\right] z^{-2}} \tag{4.110}$$

Similarly, the transfer function of the digital Costas loop can be also derived from Fig. 4.40b as

$$H(z) = \frac{\varphi_o(z)}{\varphi_i(z)} = \frac{K_d F(z) N(z)}{1 + K_d F(z) N(z)} \tag{4.111}$$

By substituting $F(z)$ and $N(z)$ in (4.106) and (4.107) into (4.111), we have the following transfer function of the digital PLL model:

$$H(z) = \frac{K_d K_v (K_1 + K_2) z^{-1} - K_d K_v K_1 z^{-2}}{1 + [K_d K_v (K_1 + K_2) - 2] z^{-1} + (1 - K_d K_v K_1) z^{-2}} \tag{4.112}$$

Comparing (4.112) with (4.110) yields the following equations:

$$K_1 = \frac{8\zeta\omega_n T}{K_d K_v \left[4 + 4\zeta\omega_n T + (\omega_n T)^2\right]} \tag{4.113}$$

$$K_2 = \frac{4(\omega_n T)^2}{K_d K_v \left[4 + 4\zeta\omega_n T + (\omega_n T)^2\right]} \tag{4.114}$$

The sampling interval T is usually equal to the symbol interval in the case of QPSK modulation or the bit interval in the case of BPSK modulation. If an integrate and dump (I&D) circuit is used rather than a lowpass filter after the multiplier, the sampling interval T at the I&D output is naturally equal to the symbol interval because the I&D performs both lowpass filtering and decimation functions [27].

The natural frequency for an active lag-lead loop filter of the second-order PLL model can be found in [28] as

$$\omega_n = \frac{8\zeta B_{enb}}{4\zeta^2 + 1} \tag{4.115}$$

where B_{enb} is the noise bandwidth (one-side) of the PLL loop and has units of Hertz, despite the fact that ω_n is given in radians per second. The noise bandwidth of the PLL loop is defined as

$$B_{enb} = \int_0^\infty |H(j2\pi f)|^2 \, df \text{ (Hz)} \tag{4.116}$$

It should be noted that the noise bandwidth is different from and not equal to a 3-dB bandwidth. For the second-order PLL loop model, minimum noise bandwidth is achieved for $\zeta = 0.5$. Noise bandwidth does not exceed the minimum by more than 25% for any damping factor between 0.25 and 1.0 [28].

The noise bandwidth controls the amount of noise passed through the filter. A large noise bandwidth implies that the PLL quickly locks to the real frequency and phase in the tracking phase, but has a relatively large noise after the lock. A small noise bandwidth, on the other hand, indicates that the PLL takes more time to lock to the real frequency and phase, but has less noise after the lock. Therefore, the PLL may have two kinds of noise bandwidths, used for pull-in and tracking states, respectively.

Like other transfer functions, the transfer function $H(s)$ in (4.109) has a well-defined 3-dB bandwidth, labeled by $\omega_{3\mathrm{dB}}$. Generally, there is very little interesting in $\omega_{3\mathrm{dB}}$ of a PLL, but its relation to ω_{n} is provided here as a comparison with a familiar concept of bandwidth and is given by [28]

$$\omega_{3\mathrm{dB}} = \omega_{\mathrm{n}} \left[2\zeta^2 + 1 + \sqrt{\left(2\zeta^2 + 1\right)^2 + 1} \right]^{1/2} \tag{4.117}$$

Substituting (4.115) into (4.117) gives the relationship between $\omega_{3\mathrm{dB}}$ and B_{enb} as

$$\omega_{3\mathrm{dB}} = \frac{8\zeta B_{\mathrm{enb}}}{4\zeta^2 + 1} \left[2\zeta^2 + 1 + \sqrt{\left(2\zeta^2 + 1\right)^2 + 1} \right]^{1/2} \tag{4.118}$$

In the case of $\zeta = 0.707$, it is found from (4.117) and (4.118) that $\omega_{3\mathrm{dB}} \approx 2\omega_{\mathrm{n}}$ and $f_{3\mathrm{dB}} = \omega_{3\mathrm{dB}}/(2\pi) \approx 0.62 B_{\mathrm{enb}}$.

The damping factor controls not only how fast the PLL reaches its settle point, but also how much overshoot the PLL could tolerate. The tradeoff between overshoot and settling time should be taken into account. In most applications, the damping factor is chosen to be 0.707.

Digital frequency synthesis can be implemented by using a numerically controlled oscillator (NCO) for precise carrier replication in the receiver. Basically, one replica carrier cycle is completed each time the NCO overflows. A block diagram of the carrier loop NCO and its sine and cosine look-up table (LUT) functions are illustrated in Fig. 4.42. The system clock frequency f_{sam} or sampling frequency is usually set to $N_{\mathrm{sam}} \times f_{\mathrm{NCO}}$, where N_{sam} is the number of samples per carrier (or NCO) cycle and f_{NCO} is the frequency of the center NCO operation (or carrier). The phase offset step is set to $2\pi/N_{\mathrm{sam}}$. Then, the NCO output waveforms are controlled by the system clock at the frequency of f_{sam} and shown in Fig. 4.43.

In Fig. 4.43a, the NCO output phase accumulates by an offset step at every clock cycle until it reaches 2π after every N_{sam} cycles. Then sine and cosine values are extracted from the sine and cosine LUT. In this example, $N_{\mathrm{sam}} = 8$ and $f_{\mathrm{sam}} = 8 \times f_{\mathrm{NCO}}$ are assumed, where f_{NCO} is the frequency of the center NCO operation.

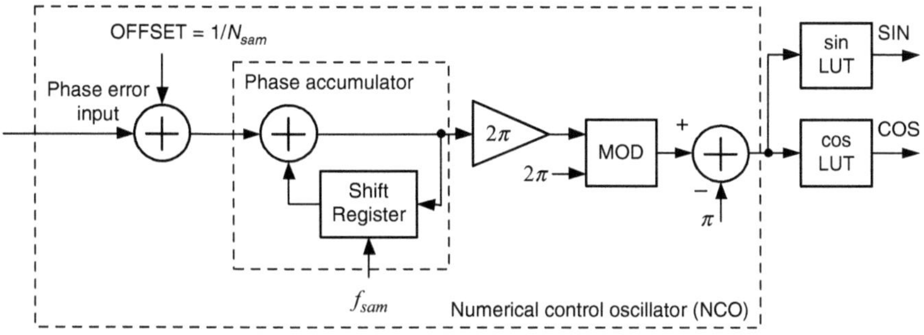

Fig. 4.42 Block diagram of numerical controlled oscillator

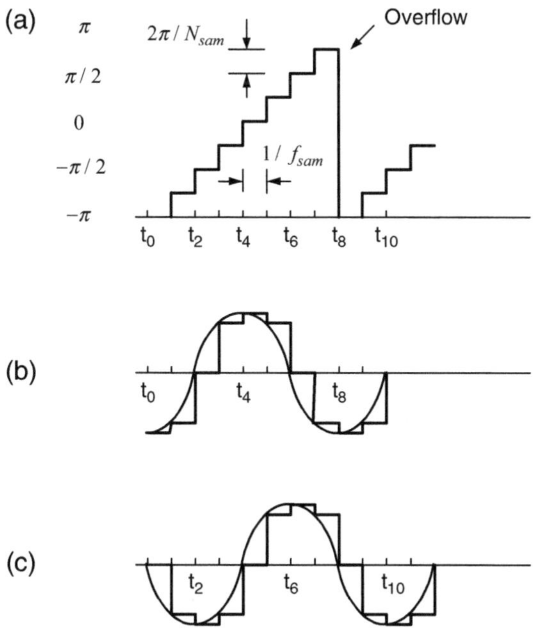

Fig. 4.43 Numerical controlled oscillator waveforms: (**a**) NCO phase waveform, (**b**) COS LUT output, and (**c**) SIN LUT output

When the input error signal is zero, the NCO accumulated phase reaches 2π or generates exactly one cycle of sinusoidal waveform in time of N_{sam} cycles. However, when the input error signal is greater than zero, the NCO accumulation speed gets higher. Then the accumulated phase reaches 2π in a time less than N_{sam} cycles, which corresponds to a higher frequency than the frequency of f_{NCO}. Conversely, when the input error signal is less than zero, the NCO accumulation speed gets lower. Then a lower frequency than the frequency of f_{NCO} is generated. Consequently, the NCO operation frequency will be controlled by the input error signal with a center frequency of f_{NCO}. The NCO gain K_{v} is set to a $2\pi\times$ NCO input scaling factor of $(1/N_{\text{sam}})$, which finally is equal to $K_{\text{v}} = 2\pi/N_{\text{sam}}$ [27].

Next, the phase detection gain K_d is needed to solve the equation. The phase error signal in the modified Costas loop for BPSK is relatively easy to get; it is expressed in (4.100) and is rewritten here:

$$e(n) = \frac{|A(n)|}{2}\mathbf{sin}\,(\Delta\varphi) \tag{4.119}$$

where $A(n)$ is the pulse-shaping baseband binary signal at time $t = nT_s$, $n = 0, 1, 2, \ldots$, and T_s is the symbol duration.

Then, in the digital modified Costas loop of Fig. 4.40b, the phase detector gain K_d is $A_{dec}/2$, where A_{dec} is the average positive decision values of $A(n)$.

In the QPSK case, the error signal in (4.103) can be approximately expressed as

$$e(n) \approx \begin{cases} |u_q(n)|\mathbf{sin}\,(\Delta\varphi), & \text{if } u_i(n) \times u_q(n) \geq 0 \\ |u_i(n)|\mathbf{sin}\,(\Delta\varphi), & \text{if } u_i(n) \times u_q(n) < 0 \end{cases} \tag{4.120}$$

Because the averagely sampled absolute values at the symbol instants in the I branch are equal to those in the Q branch, the phase error signal in (4.120) can be simply expressed as

$$e(n) = A_{\text{ave}}\mathbf{sin}\,(\Delta\varphi) \tag{4.121}$$

Then, the phase detector gain K_d is A_{ave}, where A_{ave} is the averagely sampled absolute values of $u_i(n)$ and $u_q(n)$ at the symbol instants. The phase detector gain for BPSK is half that for QPSK.

For a BPSK signal case, a Costas loop phase discriminator is usually used to generate the precise phase error signal. From (4.96) and (4.97), the phase different signal $\Delta\varphi$ when the noise is ignored for a moment can be found as

$$\frac{Q(n)}{I(n)} = \frac{\frac{1}{2}A(n)\sin\,(\Delta\varphi)}{\frac{1}{2}A(n)\cos\,(\Delta\varphi)} = \tan\,(\Delta\varphi) \tag{4.122}$$

$$\Delta\varphi = \tan^{-1}\left(\frac{Q(n)}{I(n)}\right) \tag{4.123}$$

Now, the phase error $\Delta\varphi$ is precisely expressed in (4.123) in terms of the I sample $I(n)$ and Q sample $Q(n)$ regardless of the noise, where the samples are generated at every symbol instant. It can be seen that the phase error is minimized when the sampled value in the Q branch is zero and the sampled value in the I branch is at its maximum. Table 4.2 summarizes the most commonly used Costas PLL discriminators and their characteristics.

A graphical comparison of these discriminator algorithms is made in the absence of noise and the results are plotted in Fig. 4.44. These phase discriminator outputs are calculated using the actual expressions of the Costas PLL different discriminators for a BPSK signal rather than those given in Table 4.2. For example, the curve of $Q(n) \times I(n)$ is computed using (4.96) and (4.97) with the baseband amplitude $A(n) = 1$ or -1 at time $t = nT_s$.

Table 4.2 Different types of Costas PLL discriminators for BPSK signals

Discriminator operation	Error signal proportional to	Phase detector gain K_d[a]	Loop characteristics
$Q(n) \times I(n)$	$A^2(n)\sin(2\Delta\varphi)$	$\frac{A^2(n)}{4}$	Classic Costas discriminator Phase detector output proportional to I- or Q-squared amplitude
$Q(n) \times \text{sign}[I(n)]$	$A(n)$[b] $\sin(\Delta\varphi) \times \text{sign}[A(n)\cos(\Delta\varphi)]$	$\frac{A(n)}{2}$	Modified Costas discriminator Phase detector output proportional to I or Q amplitude
$\mathbf{tan}^{-1}[Q(n)/I(n)]$	$\Delta\varphi$	1	Precise Costas digital discriminator Phase detector output independent of I and Q amplitude

[a]It is obtained when the received BPSK signal is $A(t)\cos(2\pi f_c t + \varphi)$ at the demodulator input and the amplitude of the carrier replica is unit

[b]$A(n)$ is the average sample value prior to a decision device when +1 is transmitted at the transmitter

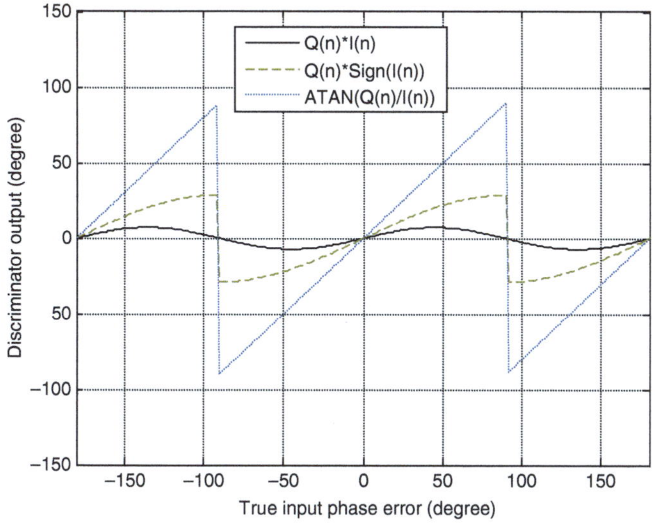

Fig. 4.44 Comparison of Costas PLL discriminators

The slope of the $Q(n) \times I(n)$ curve near zero degrees is much lower than that of the ATAN[$Q(n)/I(n)$] curve because the phase detector gain is small and the phase detector output of the ATAN[$Q(n)/I(n)$] is independent of the amplitude values of the I and Q branches. It can also be seen that the ATAN[$Q(n)/I(n)$] type of Costas discriminator has a linear property over half of the input phase error range of $\pm 90°$. Furthermore, it can be seen that all Costas PLL discriminator outputs reach zero when the real phase error is 0 and $\pm 180°$. This tells us that the Costas loop has an $180°$ phase ambiguity, which can be solved using a differential encoder in the transmitter.

Design Example 4.1 Design a digital loop filter of the second-order Costas PLL to track a real GPS carrier signal with a frequency offset of 50 Hz, in which the rate of the C/A spreading code is 1.023 Mchips/s. In this example, a low-IF digital signal with a center frequency of 9.548 MHz is the input signal to the Costas carrier loop. The modified Costas loop is chosen due to its simplified implementation. The parameters of the Costas loop are given as follows:

Noise bandwidth:	$B_{\text{enb}} = 80\,\text{Hz}$
Dumping factor:	$\zeta = 0.707$
Phase detector gain:	$K_{\text{d}} = A/2$ with $A = 0.5$
VCO gain:	$K_{\text{v}} = 2\pi \times \text{NCO input scaling factor}\,(1/4) = \pi/2$
Chip rate:	$R_{\text{chip}} = 1.023\,\text{Mchips/s}$
Sampling interval:	$T = 1/R_{\text{chip}}$

Solution Substituting the corresponding parameters above into (4.115) yields the natural frequency of ω_{n}. Then, substituting ω_{n} with other necessary parameters above into (4.113) and (4.114) yields the loop filter coefficients of $K_1 = 5.31e - 4$, and $K_2 = 5.54e - 8$, respectively. With these parameters solved, the amplitude response of the PLL transfer function in (4.112) is sketched in Fig. 4.45.

From (4.118), a 3-dB frequency for this case is calculated as $f_{3\text{dB}} \approx 0.62B_{\text{enb}} = 0.62 \times 80 = 50\,\text{Hz}$, which is the same as the value read on the curve of Fig. 4.45.

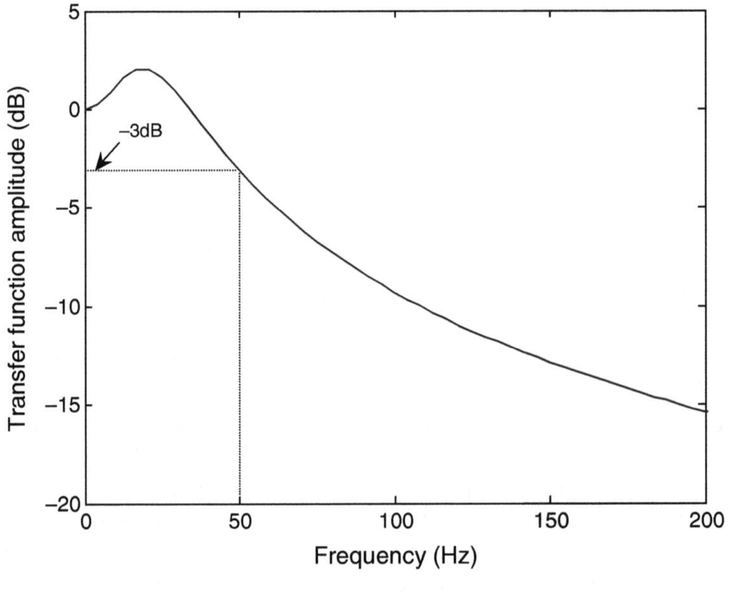

Fig. 4.45 Frequency response of a second-order Costas PLL transfer function

4.5.2.5 Decision-Directed Carrier Recovery

In modern digital communication systems, carrier recovery can be simply performed at the baseband domain rather than at the IF domain as described in the previous sections. In this case, the RF received signal is directly down-converted to the I–Q baseband signals with a pair of orthogonal local oscillator signals at a fixed frequency. Or alternatively, the received RF signal is first down-converted to the quadrature low-IF signals for the appropriate process and then to the I–Q baseband signals both with quadrature local RF and IF oscillator signals at the fixed frequency, respectively.

Consider a noiseless RF received signal with frequency offset and phase jitter as shown below:

$$y(t) = \mathbf{Re}\left\{e^{j(\omega_c t+\theta(t))}\left[u_i(t) + ju_q(t)\right]\right\} \tag{4.124}$$

and

$$u_i(t) = \sum_{n=-\infty}^{\infty} A_{\mathrm{I}n} g(t - nT_s) \tag{4.125}$$

$$u_q(t) = \sum_{n=-\infty}^{\infty} A_{\mathrm{Q}n} g(t - nT_s) \tag{4.126}$$

where $A_{\mathrm{I}n} = \pm 1$ and $A_{\mathrm{Q}n} = \pm 1$ are random sequences for QPSK on the I–Q channels, $g(t)$ stands for the transmit pulse shape, ω_c is the carrier frequency, $\theta(t) = \omega_o t + \theta_o$ models the frequency offset of ω_o and constant phase θ_o, and $u_i(t)$ and $u_q(t)$ are the I–Q baseband signals. The received signal in (4.124) is down-converted to the complex baseband signal with the local quadrature carrier signals

$$e^{-j(\omega_c t+\varphi(t))} \tag{4.127}$$

where $\varphi(t)$ is the phase of the local oscillator signal at the receiver. After down-conversion and lowpass filtering through a SRRC filter, which is matched to the transmitter-side SRRC filter to achieve the minimal ISI, the complex baseband signal is

$$r(t) = e^{j\phi(t)} \sum_{n=-\infty}^{\infty} (A_{\mathrm{I}n} + jA_{\mathrm{Q}n}) p(t - nT_s) \tag{4.128}$$

where $\phi(t) = \theta(t) - \varphi(t)$ is the phase difference between the transmitter and receiver, including the frequency offset, and $p(t)$ is the pulse representing the response of the receiving SRRC filter to its input pulse. The complex baseband signals sampled at the symbol rate $t = KT_s$ are

$$r(kT_{\mathrm{s}}) = r_k = e^{j\phi_k} \sum_{n=-\infty}^{\infty} (A_{\mathrm{I}n} + jA_{\mathrm{Q}n}) p_{k-n} \tag{4.129}$$

where ϕ_k and p_k are samples of $\phi(t)$ and $p(t)$, respectively. If a transmission channel is ideal without causing any distortion, then the pulse shape should satisfy the Nyquist criterion without ISI

$$p_k = \begin{cases} 1, & k = 0 \\ 0, & k \neq 0 \end{cases} \tag{4.130}$$

and consequently

$$r_k = e^{j\phi_k}(A_{\mathrm{I}k} + jA_{\mathrm{Q}k}) = e^{j\phi_k} A_k \tag{4.131}$$

If the transmission channel is not ideal, the pulse shape does not satisfy ISI-free condition. In this case, if the carrier recovery joins together with the adaptive baseband equalizer, then the equalized pulse shape should approximately satisfy the Nyquist criterion.

The sampled baseband signal r_k is multiplied with the estimated carrier phase $e^{-j\hat{\phi}_k}$ from a digital complex VCO as shown in Fig. 4.46, and then the recovered baseband signal is obtained:

$$q_k = r_k e^{-j\hat{\phi}_k} = e^{j(\phi_k - \hat{\phi}_k)} A_k \tag{4.132}$$

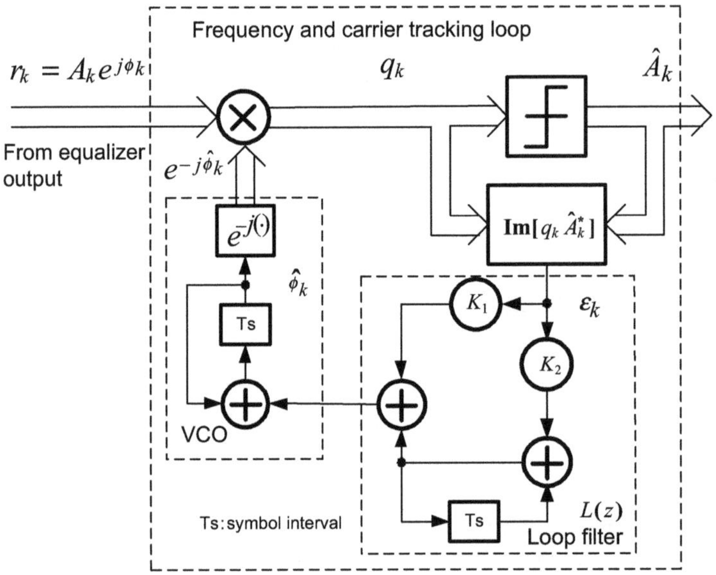

Fig. 4.46 A second-order carrier recovery loop with frequency offset correction

From (4.132), the phase error can be expressed as [29]

$$\mathbf{sin}\left(\varepsilon_k\right) = \mathbf{sin}\left(\phi_k - \hat{\phi}_k\right) = \frac{\mathbf{Im}\left(q_k A_k^*\right)}{\left|A_k\right|^2} \tag{4.133}$$

Considering that the phase error is usually small and the denominator can be omitted to avoid the division operation, we can further simplify (4.133) as

$$\varepsilon_k = \mathbf{Im}\left(q_k A_k^*\right) \tag{4.134}$$

Actually, the complex information symbols A_k that are transmitted from the transmitter are not known in the receiver unless some training sequences are inserted into the information sequences during a short period of time. In a decision-directed carrier-tracking loop, the decision values $\hat{A}_k$ can be used to replace A_k. Thus, the error signal can be expressed as

$$\varepsilon_k = \mathbf{Im}\left(q_k \hat{A}_k^*\right) \tag{4.135}$$

The error above is then filtered by a lowpass filter to remove out-of-band noise. When K_2 is equal to zero and K_1 is not equal to zero, or the transfer function of the loop filter $L(z) = K_1$, the carrier recovery is a first-order PLL loop and is able to correct the phase error. When both K_1 and K_2 are not equal to zero, corresponding to the second-order PLL loop, the carrier recovery loop is capable of tracking the frequency offset and phase jitter. By properly choosing the parameters of K_1 and K_2, which are mainly determined by the noise bandwidth of the PLL and the damping factor as well, the PLL is able to perfectly track some frequency offset and phase jitter. From the design example in the previous section, it can be seen that the noise bandwidth of the PLL is determined by the maximum frequency offset and should be set to be larger than the maximum frequency offset that the PLL is designed to track. After the frequency offset is acquired, the noise bandwidth can be reduced by adjusting K_1 and K_2 in order to achieve low noise in the tracking state.

The error signal in (4.135) can be expressed in another form. Let e_k be the difference between the output and the input of the decision or $e_k = \hat{A}_k - q_k$; then the error signal in (4.135) has the following property:

$$\mathbf{Im}\left\{q_k (e_k)^*\right\} = \mathbf{Im}\left\{q_k \hat{A}_k^*\right\} \tag{4.136}$$

In the derivative of (4.136), $\mathbf{Im}\left(q_k q_k^*\right) = 0$ is used. Substituting (4.136) into (4.135), we have another form of the error signal;

$$\varepsilon_k = \mathbf{Im}\left\{q_k \left(\hat{A}_k - q_k\right)^*\right\} \tag{4.137}$$

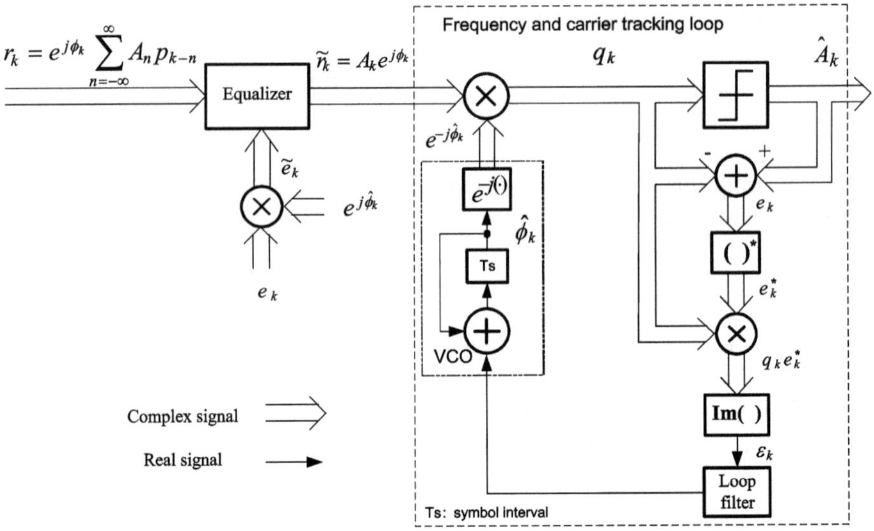

Fig. 4.47 Block diagram of a typical baseband equalizer followed by a carrier recovery loop

If a first-order carrier phase tracking loop $(L(z) = K_1)$ is used, the adaptive phase algorithm is expressed as

$$\hat{\phi}_{k+1} = \hat{\phi}_k + K_1 \varepsilon_k \tag{4.138}$$

Figure 4.47 illustrates a block diagram of an adaptive baseband equalizer in conjunction with a decision-directed carrier recover loop for QPSK/OQPSK-type signals, including GMSK and FQPSK signals. The equalizer is adaptively updated by using a least-mean square (LMS) algorithm. In this block diagram, the phase error ε_k is performed based on (4.137) instead of (4.135) even though they are equivalent. The reason for using (4.137) is that the error signal can be shared by both theequalizer and carrier recovery loop in Fig. 4.47. The principle of the adaptive equalizer was introduced in Sect. 4.5.1 and the vector coefficient $\mathbf{c}_{k+1}$ of the equalizer with LMS is adaptively adjusted at time $t = (k+1)T_s$ as

$$\mathbf{c}_{k+1} = \mathbf{c}_k + \lambda e_k \mathbf{r}_k^* \tag{4.139}$$

where $e_k = \hat{A}_k - q_k$ is the error signal, $\mathbf{r}_k = \begin{bmatrix} r_k & r_{k+1} & \ldots & r_{k-L} \end{bmatrix}^{\mathrm{T}}$ is the transpose of the input signal vector to the equalizer with a coefficient number of $L+1$, and λ is the step size that controls the rate of the coefficient adjustment.

Note that in Fig. 4.47 the error signal e_k to the equalizer adaptation should be phase de-rotated back by multiplying the complex carrier reference $\mathbf{exp}(j\hat{\phi}_k)$ because the equalizer is prior to the carrier recovery loop. The term *derotation* means that the equalizer is phase-blind updated to open the signal's constellation at the output of the equalizer regardless of the phase error. Prior to a decision circuit, the equalized signal is multiplied with the complex carrier reference $\mathbf{exp}(-j\hat{\phi}_k)$ from the carrier recovery loop to correct the phase error. For a QPSK-type signal, the decision for the complex signal q_k is made at the same time $t = kT_s$ between the I–Q branches, while for an OQPSK-type signal like GMSK and FQPSK signals, the decision is alternatively made at time $t = kT_s/2$ between the I–Q branches due to the offset property of a half-symbol time interval $T_s/2$ between the I–Q signals. Because of such an offset property, a fractionally spaced equalizer, or tap-spacing of $T_s/2$, is used to replace the symbol-spaced equalizer used for QPSK. The fractionally spaced equalizer has an advantage over the symbol-spaced equalizer of no sensitivity to timing phase and can achieve superior performance in most cases, such as high SNR and low BER ratios in the receivers.

Another type of the equalization technique described above can be also applied to adopt this equalizer in conjunction with a decision-directed carrier recovery loop. For example, a blind equalizer using a CMA can be used to replace the LMS-based equalizer in Fig. 4.47. Unlike the LMS algorithm, the error signal used in the coefficient adaptation of the blind equalizer is generated from the constellation of the signal modulation format. Thus, the blind equalizer operates independently of the carrier recovery loop.

Figure 4.48 shows a relatively detailed block diagram for Fig. 4.47, and is actually close to an implementation approach. One complex vector coefficient of the equalizer is split to four real independent vector coefficients that are capable of correcting the gain and phase imbalances on the I–Q branches in the receiver. The recursive coefficient equations of four real vector coefficients are also shown in Fig. 4.48. The de-rotated complex error signal is split to real and imaginary parts, as illustrated by the dashed rectangle representing the error signal de-rotator at the bottom of Fig. 4.48. The real and imaginary error signals are used to update the coefficients of the LMS-based equalizer. Each vector coefficient length is $L + 1$, corresponding to L shift registers and is updated for the n-th coefficient at time $t = (k + 1)T_s/2$ based on the previous coefficient and error at time $t = kT_s/2$, where $T_s = 2T_b$ is the symbol interval and is equal to twice bit interval $2T_b$ for an OQPSK-type signal, like GMSK and FQPSK. All blocks operate at a bit rate of $1/T_b$ except that error signals at the symbol rate of $1/T_s = 1/(2T_b)$, but toggled at the bit rate of $1/T_b$ between the I–Q channels due to an offset property of the OQPSK-type signal.

To see adaptive processing through joint equalization and carrier recovery in two dimensional digital communication systems, we simulate the performance of a GMSK signal with $BT_b = 0.3$. The GMSK signal with the GSM standard data rate of 270.833 kbits/s through a Gaussian channel with the frequency offset of $\Delta fT_b = 0.02$ (or $\Delta f = 5.4\,\text{kHz}$) due to the frequency difference between a source frequency at the transmitter and a reference frequency at the receiver. Figure 4.49 shows the constellation and eye diagrams of the received GMSK signal at different locations as shown in Fig. 4.48. The adaptive equalizer with the LMS algorithm is used in the simulation.

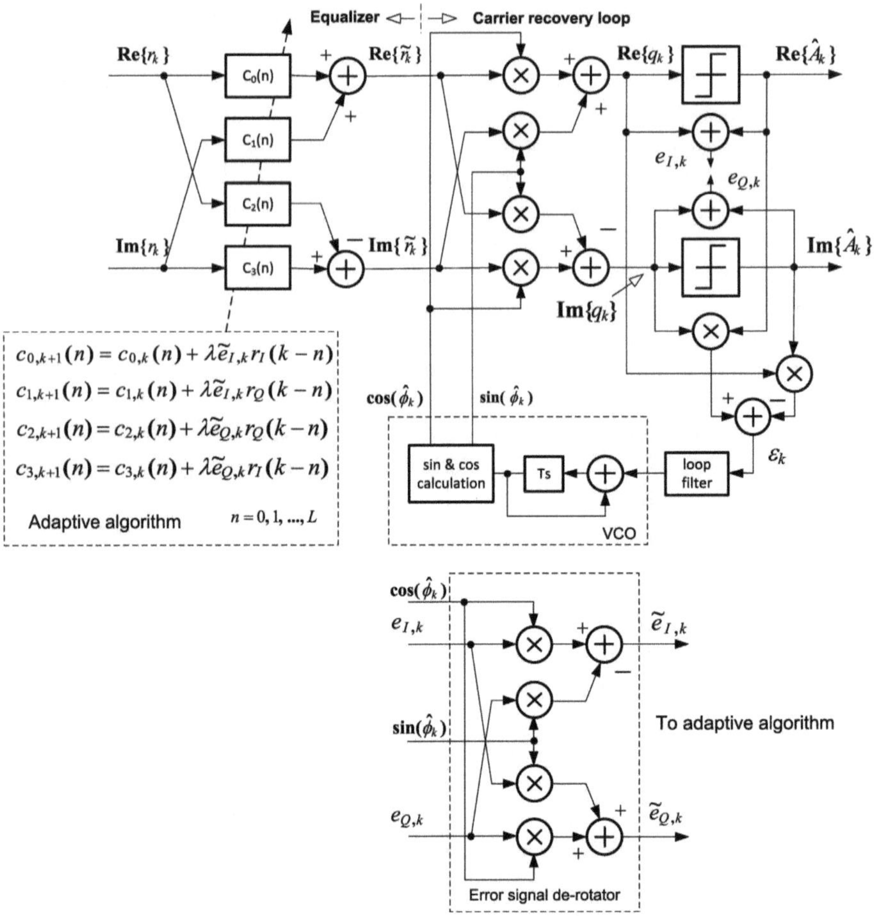

Fig. 4.48 Block diagram of a practical baseband equalizer with LMS algorithm followed by a carrier recover loop

Figure 4.49a shows the constellation of the complex signal $\tilde{r}_k$ at the output of the equalizer using the LMS algorithm. The constellation rotates with the frequency offset of $\Delta f T_b = 0.02$ and mainly consists of two circles because of inherent ISI feature of GMSK signal at the maximum eye opening instants.

Figure 4.49b illustrates the eye diagrams of the baseband I–Q signals of $\mathbf{Re}\{q_k\}$ and $\mathbf{Im}\{q_k\}$ at the output of the carrier recovery loop, where the frequency offset and phase error are completely corrected before the decision blocks. The decisions are made at the maximum eye-opening instants on the I–Q branches, alternatively.

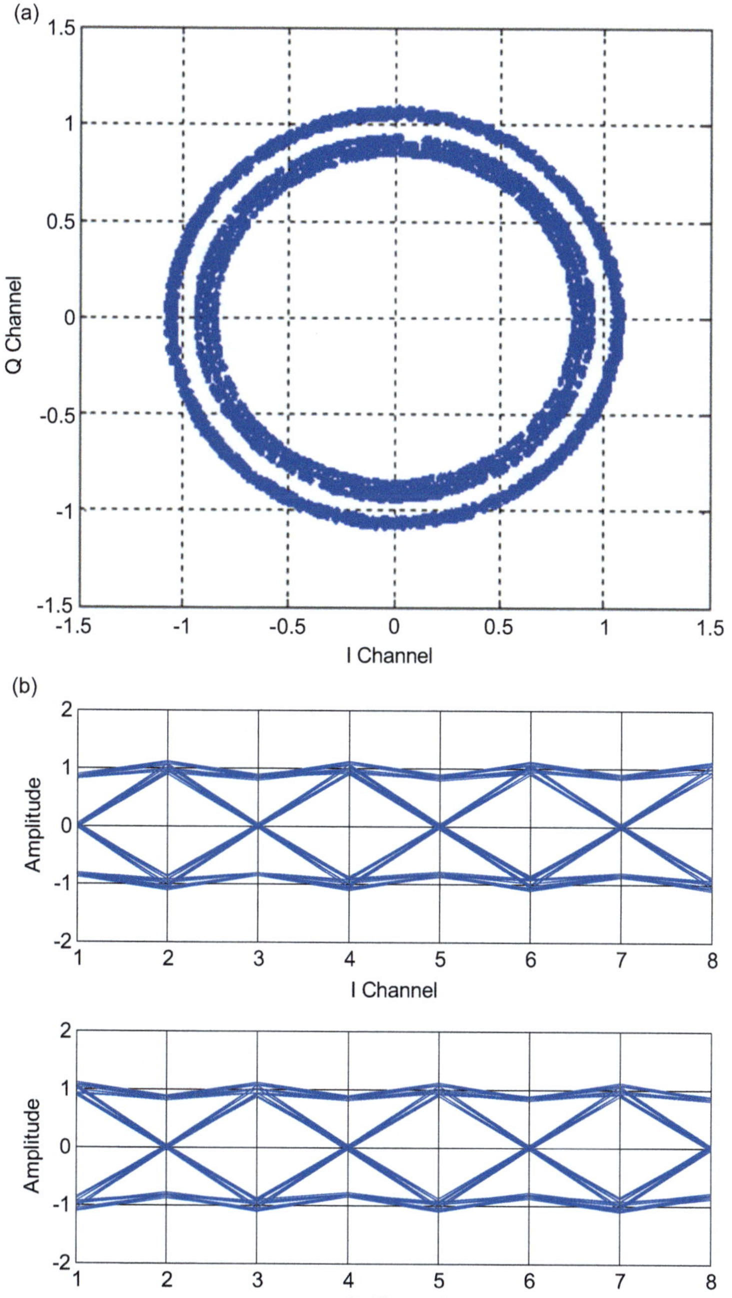

Fig. 4.49 Constellation and eye diagram of GMSK with $BT_b = 0.3$ at different points in Fig. 4.47 under high SNR and wide bandwidth conditions: (**a**) complex signal $\tilde{r}_k$ at the output of the equalizer with LMS and (**b**) eye diagram of complex signal q_k at the output of the carrier recovery loop

4.6 RF Transmitter Architectures for GMSK

In user equipment (UE), a RF transmitter transfers information-bearing baseband signals into the RF signal by means of modulation, up-conversion, filtering, and power amplification in order to send out the RF signal through a small antenna. Among these function blocks, modulation plays a very important role in enhancing spectral efficiency (or narrow bandwidth occupancy) and energy efficiency (long battery duration) of the UE transmitter system. Very large scale integrated (VLSI) circuit technology provides tremendous momentum to miniaturize the transmitter in a RF transceiver at low cost. Usually, typical requirements for the transmitters are high spectral and energy efficiency, low PSD regrowth, and low EVM. In general, GMSK transmitter architectures can be divided into two categories: *mixer-based frequency up-conversion* and *phase-locked-loop (PLL)–based frequency up-conversion*. Which category to use to implement the transmitter is mainly dependent on practical applications. Wise selection of the transmitter architecture can bring a high-quality transmitter with the best performance to its application. In this section, we introduce the main implementation structures of GMSK modulation that are usually adopted for the GSM application.

4.6.1 System Specifications of Quad-Band GSM Transmitter

The quad-band GSM transmitters targeted for GSM 850, E-GSM 900, DCS 1800, and PCS 1900 applications use GMSK modulation with $BT_b = 0.3$ to achieve energy- and spectrum-efficient transmission. In this transmission information data with a bit rate of 270.833 kbps modulates the RF carrier signal to perform the frequency transfer so that the RF-modulated signal can be effectively emitted through the air within a desired bandwidth of 200 kHz. The quad-band frequency bands and channel arrangement are depicted in Fig. 4.50.

Another important specification is the output RF modulation spectrum mask summarized in Tables 4.3, 4.4, and 4.5 [30]. The most stringent requirement for all frequency bands is –60 dBc at the frequency offset of 400 kHz. Furthermore, the levels of the phase noise are specified to be –112 dBc/Hz at the frequency offset 400 kHz and –162 dBc/Hz at the frequency offset 20 MHz from the carrier signal frequency, respectively.

The next important specification for the transmitter is the modulation accuracy, which is actually defined by the phase error of the transmitted waveform. A RMS phase error of 5° and a maximum peak phase error of 15° are specified when the system typically has 6-dBm output power delivered into a 50-Ω load. The phase error is defined as the difference between the phase trajectory of the transmitted waveform and the phase trajectory of the theoretical transmitted waveform. Since the larger RMS phase error degrades the system performance, the RMS phase error is the most stringent and important specification for GMSK modulation in the transmitter.

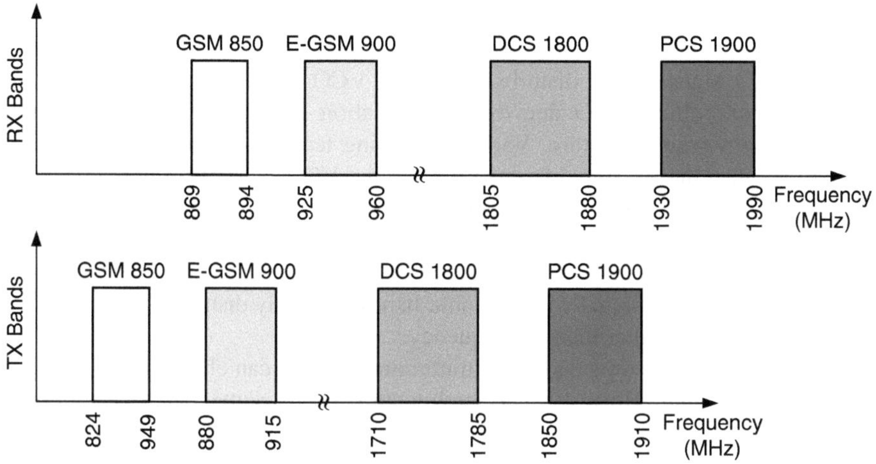

Fig. 4.50 Quad-band frequency bands of GSM mobile station

Table 4.3 Spectrum mask specifications of GSM 805 and E-GSM 900 mobile stations

Frequency Offset (kHz)	200	250	400	≥600 <1800	≥1800 <3000	≥3000 <6000	≥6000
Maximum Level (dBc)	−30	−33	−60	−60	−63	−65	−71

Table 4.4 Spectrum mask specifications of a PCS 1900 mobile station

Frequency Offset (kHz)	200	250	400	≥600 <1800	≥1800 <6000	≥6000
Maximum Level (dBc)	−30	−33	−60	−60	−65	−73

Table 4.5 Spectrum mask specifications of PCS 1900 mobile station

Frequency Offset (kHz)	200	250	400	≥600 <1200	≥1200 <1800	≥1800 <6000	≥6000
Maximum Level (dBc)	−30	−33	−60	−60	−60	−65	−73

Note: the measurement bandwidth and video bandwidth of 30 kHz should be used in the above measurements

4.6.2 *Mixer-Based Frequency Up-Conversion*

The most simple and common type of transmitter architecture used in the past and present is mixer-based. Mixer-based frequency up-conversion architecture can be implemented in the form of a direct up-conversion transmitter, two-stage up-conversion transmitter, and harmonic-rejection transmitter [31].

Due to its simplicity, the direct up-conversion transmitter architecture is very attractive for high-level-integration applications. In the case of the generation of quadrature LO signals by means of dividing the VCO signal by 2, however, the second harmonic output of the power-amplified RF signal with the same frequency as the VCO signal often disturbs the clean VCO spectrum through injection, disturbing or pulling VCO due to finite isolation in a quadrature-modulation–based up-conversion structure. Various shielding techniques have been proposed either to isolate the RF signal from disturbing the VCO or to shift the frequency of the RF signal from the VCO frequency with an offset by using the second LO source in order to minimize the VCO-disturbing phenomenon [32]. Here the word *disturbing* instead of *pulling* is used due to the fact that the second-order harmonic of the RF-modulated signal with a double-bandwidth only disturbs the phase noise of the VCO signal rather than its frequency.

The two-stage up-conversion transmitter architecture can eliminate the problem of VCO disturbing. However, this architecture needs some bandpass filters to suppress the harmonics or remove the unwanted sideband signals. Implementing these filters in the RF transceiver increases cost and die-area as well as complexity. Even though the harmonic rejection transmitter architecture may eliminate the need for the IF and the RF bandpass filters to be used in the two-stage transmission, rejections of these harmonics and sideband signals are still limited due to the amplitude and the phase imbalance errors on the I–Q branches. Hence, this kind of transmitter still requires an RF filter to sufficiently reject the transmit noise falling into other channels, such as the receive band, which would degrade the receiver performance. For example, the GSM standard requires that the level of the transmit noise at a 20-MHz offset away from the carrier frequency with the range from 880 to 915 MHz should be below –162 dBc/Hz, which is within the receive band. Furthermore, when using a mixer-based frequency up-conversion transmitter it is very hard to achieve such a low level of phase noise for the GSM systems without extra bandpass filters because active mixer circuits usually generate relatively high noise levels. In addition, it is not realistic to add an extra BPF after the PA to filter out the out-of-channel phase noise generated by the mixer-based quadrature modulator at the RF band because such a BPF with a narrow bandwidth is very complicated to be designed at the RF band. Hence, a GMSK transmitter is generally implemented by using the architecture of the phase-locked-loop–based frequency up-conversion in the GSM systems, which will be introduced in the following two sections.

4.6.3 Phase-Locked Loop-Based Frequency Up-Conversion

Quadrature-based MSK or GMSK modulators usually require at least one IF BPF or RF BPF before a power amplifier (PA) to suppress harmonics. For example, in a high level of integration modulator, an external SAW filter is inserted before the PA. This external SAW filter causes extra cost as well as additional power losses in applications. In addition, a quadrature architecture modulator also suffers I–Q amplitude and phase imbalances and carrier leakage, which in turn result in poor sideband and carrier suppressions of the modulated RF signal.

GSM systems actually have very stringent requirements on far-out suppression, in which the transmitted power spectral density must be lower than –60 dBc at a frequency of 400 kHz away from the carrier frequency. In addition, a very low phase noise below –162 dBc/Hz at a frequency of 20 MHz offset from the transmission carrier is required in order to avoid interfering the received signal in the receive band. To achieve such high requirements on low sideband suppression and phase noise without using any bandpass filter, one great method is to exploit the merit of the inherent narrow filtering property of a phase-looked loop (PLL) to reduce the harmonics and phase noise generated during the up-conversion process [33], which will be discussed later in this section. Generally, this kind of PLL-based frequency up-conversion can be accomplished in at least three different structures used for frequency up-conversion:

(a) Open-loop structure
(b) Closed-loop structure
(c) Offset (or translation) loop structure.

4.6.3.1 Open-Loop–Based PLL

The PLL open-loop–based frequency up-conversion technique is a relatively simple method to achieve frequency modulation. It consists of a basic PLL with the ability to open the loop during data transmission period and to close the loop to have the VCO frequency lock to the desired frequency. One of designs is realized with the LMX3162 RF transceiver chip manufactured by National Semiconductor Corp [34] and used commercially for a GFSK system, as shown in Fig. 4.51. First, the PLL loop is closed so that the PLL locks to the reference frequency of f_{REF}, or the output frequency of the VCO is equal to $N \times f_{REF}$, which is 1.2 GHz and then is doubled to achieve the 2.4-GHz Industrial, Scientific, and Medical (ISM) band. During data

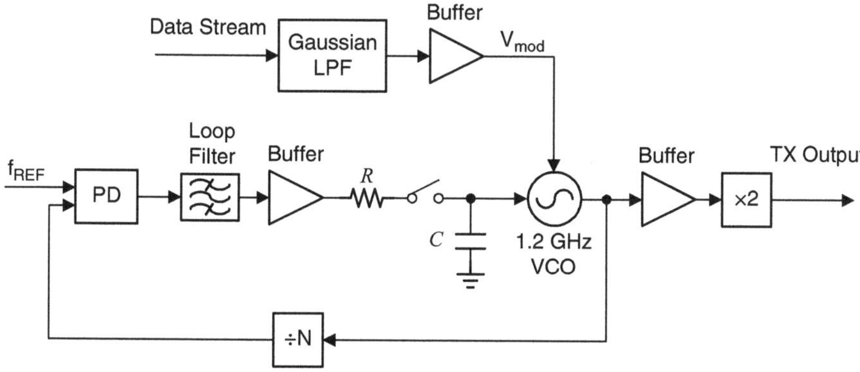

Fig. 4.51 Open-loop architecture

transmission, the loop is opened and data is applied to the tuning pin of the VCO such that the Gaussian-filtered data stream can directly modulate the VCO. By properly setting the modulation voltage V_{mod}, the maximum frequency deviation $\pm f_{\text{d}}$ of the VCO output is ± 175 kHz for Bluetooth standard because the modulation index is equal to $2f_{\text{d}}/f_{\text{b}} = 350/1000 = 0.35$, and ± 250 kHz for the DECT standard. In order to achieve low frequency draft, a large capacitor is connected at the input of the VCO to discharge and charge during the loop opening and closing periods. After the transmission, the data path is disabled again and the loop is closed to let the VCO lock to the reference frequency again.

This architecture allows a significant reduction in components; no mixers are required since the VCO performs the frequency translation, and no high-Q, low-noise, low-distortion bandpass filters are required at IF and RF. Only one DAC may be required to produce the Gaussian filtered signal as the modulation signal to the input of the VCO (not drawn in Fig. 4.51) if the precise modulation signal as the control voltage to the VCO is needed. With such architecture, significant power savings are achieved [35] compared with the mixer-based designs. However, two main drawbacks of this architecture are that the carrier frequency drifts without phase tracking during the period of open-loop modulation due to the drafting of the control voltage to the VCO and that the modulation index is inaccurate because it is difficult to accurately maintain the frequency deviation f_{d}. The first drawback limits its use to frame-based burst transmission, where the frame time will not last long. Thus, the PLL is able to periodically close the loop to prevent the tuned VCO frequency from drifting out of its allowable range. The second drawback renders it unsuitable for use with MSK and GMSK modulations, which require a modulation index of exactly 0.5.

Therefore, the open-loop architecture is mainly suitable for GFSK modulation because it is unnecessary for the GFSK modulation to keep an accurate modulation index. In the Bluetooth system, the modulation index will be between 0.28 and 0.35, while in the Digital European Cordless Telephone (DECT) system, it will be without the restriction of 0.5, or between 0.35 and 0.70. In contrast, coherent detection on the GMSK signal requires the exactly accurate modulation index value of 0.5.

Although there are numerous disadvantages in this scheme, the open-loop architecture has been used in some applications with more relaxed specifications such as DECT. For this reason, we only introduce this architecture simply here.

4.6.3.2 Modulated Fractional-N Synthesizer

Unlike the open-loop modulation architecture associated with the PLL, a modulated fractional-*N* synthesizer performs the frequency modulation through the PLL in the closed-loop format. The key point lies in using a fractional-*N* synthesizer to perform the frequency modulation of the VCO through appropriate control of a frequency synthesizer that sets the VCO frequency and yields the simplest transmitter solution. One of the fractional-*N* synthesizer's key advantages is its ability to use a high reference frequency along with fine resolution in the frequency steps. This property is not achievable with the conventional integer-*N* synthesizer. Since the VCO frequency change takes place in the closed-loop format during modulation, the problem of frequency drift during modulation is eliminated. For more detailed information regarding basics principles of fractional-*N* synthesizer, the interested reader can reference [36, 37].

Since the fractional-*N* frequency synthesizer has fine resolution in the frequency steps, Riley [38] proposed to add the Gaussian filtered modulation information in the loop through the input of the delta-sigma modulator to realize the RF GMSK-modulated signal at the output of the transmitter. This implementation achieves a significant performance improvement over the open-loop modulation based architecture. The modulated fractional-*N* synthesizer, however, is only suitable to low data rate information due to the fact that the PLL functions lowpass filtering. For high data rate information like 270.833 kbits/s GMSK signal, the performance of the modulated GMSK signal through this fractional-*N* synthesizer such as phase accuracy degrades because modulation information in high frequency range is filtered out by the loop lowpass filter of the PLL. In order to prevent information in high frequency from loss, the PLL needs to have a wider loop bandwidth. On the other hand, the PLL with wider bandwidth has worse phase noise as well as big quantization noise produced by the delta-sigma modulator. Therefore, this topology has a natural conflict between the required low phase noise and actually high data rate, which results in such a way that we could not have both fish and bear foot. Hence, designers are encouraged to find new methods to achieve a new balance between high data rate and low phase noise.

The obstacles of achieving high data rate modulation mentioned above are significantly mitigated if the bandwidth of the modulation signal is allowed to exceed the bandwidth of the PLL without filtering out high frequency components of the modulation signal. One method to achieve such a goal is that the high data rate modulation is achievable by inserting a pre-distortion filter before the delta-sigma modulator [39–41] as shown in Fig. 4.52. The basic idea for this compensation is to boost the high frequency components of the modulation signal, which are attenuated by the PLL bandwidth later so that the bandwidth of the overall PLL from the input of the compensator to the output of the PLL is extended after such compensation. The fact is that modulation baseband signals suited to the modulated fractional-*N* synthesizer should have a constant envelope feature to allow direct modulation on a VCO embedded inside a PLL.

Figure 4.53 depicts an equivalent baseband model of the proposed compensation method as in Fig. 4.52 in the frequency domain, where the Gaussian lowpass filter and the pre-distortion filter are both digital, and the equivalent baseband filter of the PLL is analog. To extend the modulation bandwidth, the transfer function of the pre-distortion is the inverse of the PLL transfer function. By doing so the overall transfer function allows the data rate to exceed the PLL bandwidth.

However, it is a difficult design task to approach the inverse of the PLL transfer function in the digital domain to match the analog transfer function of the PLL [42]. This is because the transfer function of the PLL has many unknowns

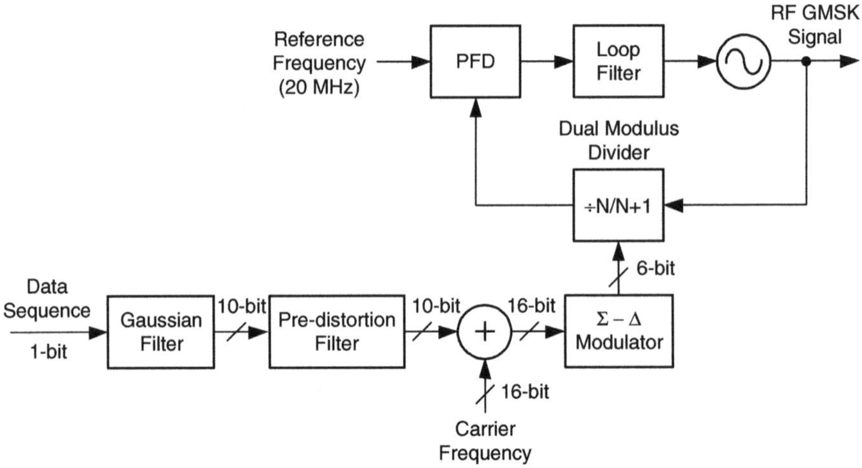

Fig. 4.52 Modulated fractional-*N* synthesizer. Redrawn from [40]

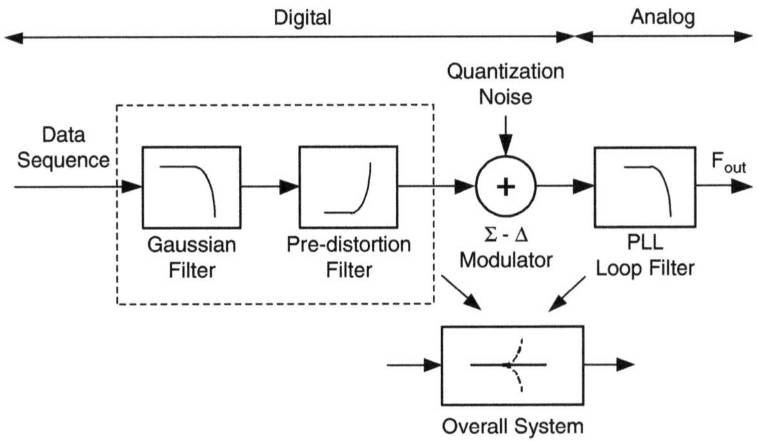

Fig. 4.53 Simplified compensation block diagram in equivalent baseband domain. Redrawn from [40]

such as open-loop gain and the pole/zero frequencies. Moreover, these parameters are sensitive to process and temperature variations. All these uncertain parameters indicate that expensive factory calibrations are often necessary in production [43, 44].

In order to not physically extend the bandwidth of the closed-loop PLL transfer function due to phase noise problem the compensation filter or pre-distortion filter, $H_C(s)$, needs to be inserted before the delta-sigma modulator. Its amplitude response is the inverse of the PLL transfer function $H_{PL}(s)$, or

$$H_C(s) = \frac{1}{H_{PL}(s)} \tag{4.140}$$

Thus, the overall transfer function seen by the Gaussian filtered data is flat within a relative wide bandwidth at the least. If the pre-distortion filter matches the inverse of the PLL transfer function well, this method is equivalent to the traditional FM modulator based VCO with the well-controlled modulation index of 0.5.

One of the tough requirements of this approach is that the pre-distortion filter and PLL transfer function must have complementary transfer function. Otherwise, the phase error of the GMSK-modulated signal would be relatively big if the pre-distortion filter does not match the inverse of the PLL transfer function well. Good matching requires well control of the parameters of the PLL transfer function, such as the phase detector gain, VCO gain, and loop filter RC values. To achieve good matching, the calibration for each of these characteristics is needed.

Mathematically, the transfer function of the pre-distortion filter can be realized by inversing the transfer function of the PLL. This requires that the transfer function of the PLL should have its zero to be the left range of S-plan in order to have a stable pre-distortion filter. Since the parameters of the PLL transfer function are usually unknown and may suffer change due to temperature change, and furthermore in practice, the pre-distortion filter is implemented digitally and the PLL transfer function is of the form of analog, close matching between them is impossible without calibration. Therefore, the behavior of the compensated PLL transfer function needs to be analyzed due to possible mismatches in these parameters.

Due to the inverse relationship between the pre-distortion filter and the PLL as given in (4.140), the shape of the amplitude response of the closed-loop PLL affects the compensation implementation and accuracy. After studying the effects of a type-1 loop and type-2 loop on the modulated fractional-N synthesizer, Lee [42] utilized the type1 loop in the implementation of the modulated fractional-N synthesizer based on the following main advantages:

(a) By choosing the damping factor of 0.707, a type-1 loop has a maximally flat Butterworth frequency response. This PLL is relatively easy to be implemented in its inverse form.
(b) Mismatches in the charge pump currents associated with type-2 loop are not present in the type-1 loop. The sample-and-hold phase detector in the type 1-loop can be designed with high linearity to minimize noise folding to lower offset frequencies that reappear as spurious tones in the output spectrum.
(c) The implementation of the pre-distortion filter is relatively easier for the type-1 loop since the closed-loop PLL has a maximally flat Butterworth response.
(d) A type-1 loop does not face a jitter peaking problem that a type-2 loop faces since the type-1 loop filter consists of poles only [42, 45].

A linearized model for the modulated fractional-N synthesizer is shown in Fig. 4.54 [42]. The phase detector is modeled as a subtractor with a gain K_p. This phase detector gain that depends on the absolute values of resistance and

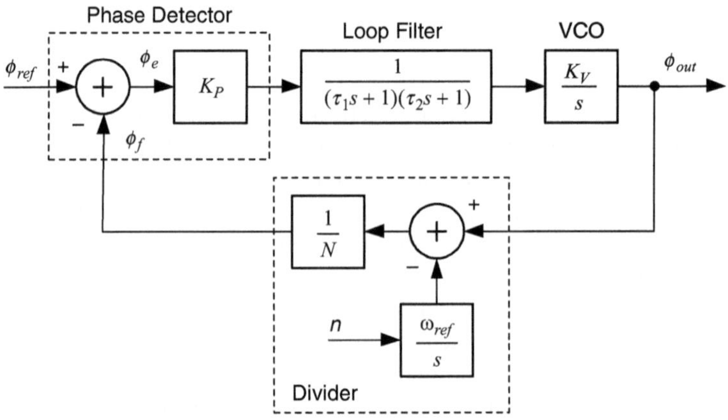

Fig. 4.54 Simplified block diagram of a modulated fractional-*N* synthesizer. Referenced from [42]

capacitance is process dependent. Therefore, any gain variation on K_p needs to be calibrated using an automatic K_p tuning circuit. The loop filter consisting of one dominant pole and one out-of-band pole is used to simply form a type-1 PLL. Here, the loop filter with two poles instead of three is used for the sake of simplicity. In the actual implementation, the first pole is embedded in the phase detector through switched capacitor techniques and its location is mainly determined by the ratio of capacitances, which can be controlled accurately and does not depend on process. The second out-of-band pole is used to attenuate the quantization noise from the sigma-delta modulator and also minimize the reference feedthrough. The VCO is modeled as an integrator with a gain K_V, which is also process dependent. Therefore, it must also be calibrated. The divider is linearized about its operating point where N is the nominal divide value and n is the variation about the operating point.

From the description of all parameters in the PLL, we know that the closed-loop transfer function of the PLL is well defined with just two unknown parameters; namely the gain K_p of the phase detector and the gain K_V of the VCO. Any deviations of these two parameters from the desired values need to be calibrated using automatic tuning circuits.

The complete PLL open-loop gain with type-1 PLL is given by

$$\begin{aligned} L(s) &= K_p\left(\frac{1}{(\tau_1 s+1)(\tau_2 s+1)}\right)\left(\frac{K_v}{s}\right)\left(\frac{1}{N}\right) \\ &= K\left(\frac{1}{s(\tau_1 s+1)(\tau_2 s+1)}\right) \end{aligned} \tag{4.141}$$

where the total gain constant $K = \left(K_p K_v\right)/N$. The transfer function of the closed-loop PLL is given by

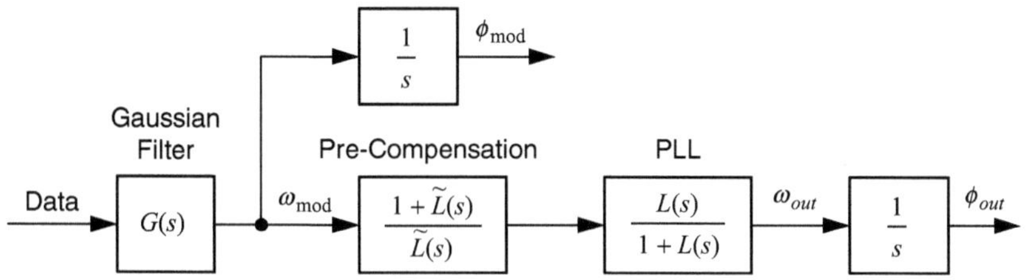

Fig. 4.55 Simplified compensation model. Redrawn from [42]

$$\begin{aligned} H_{\mathrm{PL}}(s) &= \frac{L(s)}{1+L(s)} \\ &= \frac{K}{s(\tau_1 s+1)(\tau_2 s+1)+K} \end{aligned} \tag{4.142}$$

As mentioned previously, the transfer function $H_{\mathrm{C}}(s)$ of the pre-distortion filter is the inverse of the transfer function $H_{\mathrm{PL}}(s)$ of the PLL closed-loop and is given by

$$\begin{aligned} H_{\mathrm{C}}(s) &= \frac{1}{\widetilde{H}_{\mathrm{PL}}(s)} = \frac{1+\widetilde{L}(s)}{\widetilde{L}(s)} \\ &= \frac{s(\widetilde{\tau}_1 s+1)(\widetilde{\tau}_2 s+1)+\widetilde{K}}{\widetilde{K}} \end{aligned} \tag{4.143}$$

where par denotes approximation due to impossible complementary. A simplified block diagram of the cascade of the Gaussian filter, the pre-distortion filter, and the PLL is shown in Fig. 4.55.

It is easy to find the transfer function from the output of the Gaussian filter to the modulation part of the instantaneous output as given by

$$\begin{aligned} H_{\mathrm{mod}}(s) &= \frac{\omega_{\mathrm{out}}(s)}{\omega_{\mathbf{mod}}(s)} = H_{\mathrm{C}}(s) \times H_{\mathrm{PL}}(s) \\ &= \frac{\frac{1}{\widetilde{K}}s(\widetilde{\tau}_1 s+1)(\widetilde{\tau}_2 s+1)+1}{\frac{1}{K}s(\tau_1 s+1)(\tau_2 s+1)+1} \end{aligned} \tag{4.144}$$

It can be seen from the equation above that the modulation transfer function is determined by how close the estimated parameters of the gain constant $\widetilde{K}$, time constants $\widetilde{\tau}_1$ and $\widetilde{\tau}_2$ are to the actual parameters K, τ_1 and τ_2, where K is dependent on K_{P} and K_{V}, τ_1 and τ_2 are determined by R and C values of the loop filter. Figure 4.56 shows the amplitude response of the modulation transfer function versus the mismatch in value of K when $\widetilde{\tau}_1 = \tau_1$, $\widetilde{\tau}_2 = \tau_2$, where the first pole is located at 200 kHz and the second pole is located at 4.6 MHz.

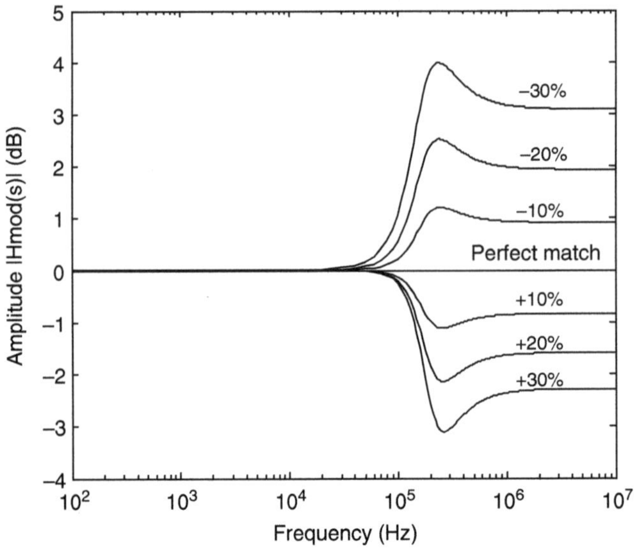

Fig. 4.56 Frequency response of modulation transfer function with mismatch in K

It can be seen that the amplitude of the modulation transfer function is almost constant before the frequency of about 50 kHz and then reaches its peak value at the first pole with the frequency of 200 kHz for different variations of K. The second pole at the frequency of 4.6 MHz has no effect on the amplitude of the modulation transfer function at low frequencies.

Figure 4.57 shows amplitude response of the modulation transfer function versus different variations of $\tau_1 = R1C1$ values related to the first pole. Compared with the impact of gain K on the amplitude, the values of $R1$ and $C1$ have less effect on the amplitude.

The most important parameter that we are concerned is the phase error, which determines the modulation accuracy, or EVM. The phase error between the desired output phase $\phi_{\mathbf{mod}}$ and the actual output phase ϕ_{out} due to mismatch is calculated as in [42]:

$$\phi_e(s) = \phi_{\mathbf{mod}}(s) - \phi_{\text{out}}(s) = [1 - H_{\mathbf{mod}}(s)]\phi_{\mathbf{mod}}(s) \tag{4.145}$$

Substituting (4.144) into (4.145) gives

$$\phi_e(s) = \left(\frac{\left(\frac{1}{K} - \frac{1}{\widetilde{K}}\right) s(\tau_1 s + 1)(\tau_2 s + 1)}{1 + \frac{1}{K} s(\tau_1 s + 1)(\tau_2 s + 1)} \right) \phi_{\mathbf{mod}}(s) \tag{4.146}$$

where $\widetilde{\tau}_1 = \tau_1$ and $\widetilde{\tau}_2 = \tau_2$ are assumed.

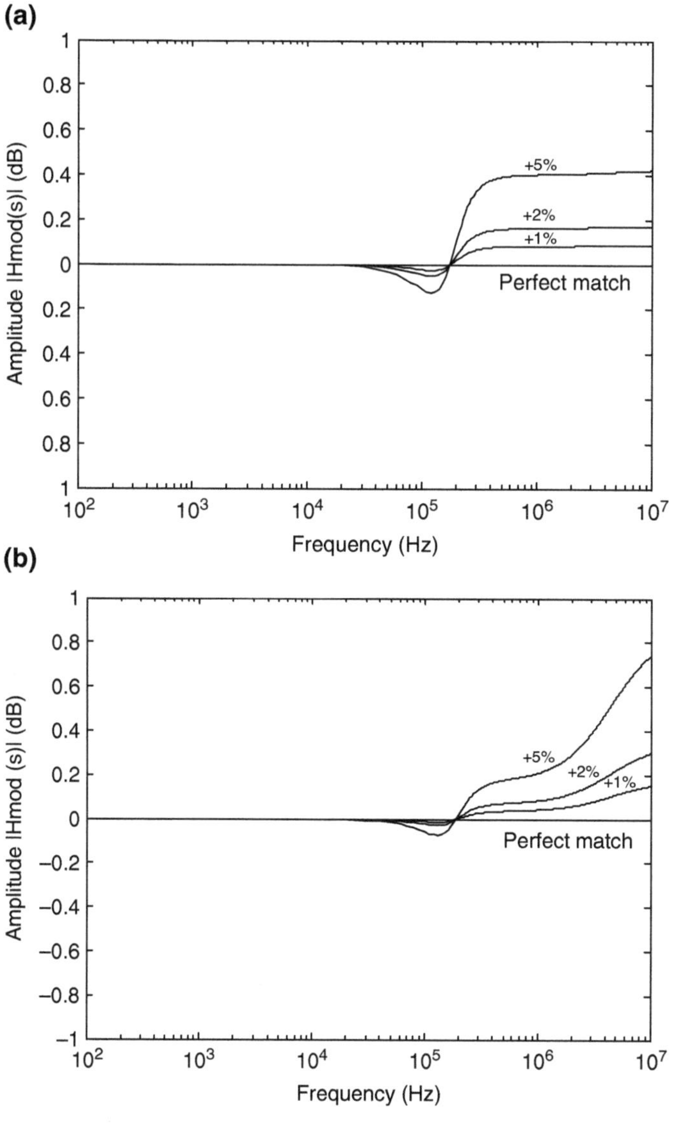

Fig. 4.57 Frequency response of modulation transfer function with mismatch in $R1C1$: (**a**) Mismatch in $R1$ and (**b**) mismatch in $C1$

The term $\left(\frac{1}{K} - \frac{1}{\widetilde{K}}\right)$ can be further written as follows:

$$\begin{aligned} \left(\frac{1}{K} - \frac{1}{\widetilde{K}}\right) &= \frac{1}{K}\left(\frac{\widetilde{K} - K}{\widetilde{K}}\right) \\ &= \frac{1}{K}\left(\frac{\delta}{1+\delta}\right) \end{aligned} \tag{4.147}$$

where the estimated gain constant $\widetilde{K} = (1+\delta)K$ and δ is gain error. Substituting (4.147) into (4.146) and assuming $\delta \ll 1$, the phase error is rewritten as

$$\phi_e(s) = \left(\frac{\delta}{K}\right)\left(\frac{s(\tau_1 s+1)(\tau_2 s+1)}{1+\frac{1}{K}s(\tau_1 s+1)(\tau_2 s+1)}\right)\phi_{\mathbf{mod}}(s) \qquad (4.148)$$

If gain matching is pretty well, or $\delta = 0$, the phase error is zero. Otherwise, the phase error is dependent on the gain error δ or mismatching extent if the locations of all poles (or all zeros if the PLL loop filter has them) are estimated precisely. It is actually impossible to precisely estimate the locations of all poles and zeros of the closed-loop PLL transfer function. Therefore, calibration for mismatches between the pre-distortion filter and the closed-loop PLL transfer function is necessary.

As mentioned previously, the transfer function of the closed-loop PLL is only well defined with just two unknown parameters of K_P and K_V after the cut-off frequency of the first pole is well controlled through the use of switched capacitor techniques [42]. Then, any deviations of the phase detector gain of K_P—which depends on the absolute value of the current source and the sampling capacitor—from the desired value are first calibrated out using an automatic tuning circuit. Then the estimate of the VCO gain K_V, and in turn the estimate of the loop gain of the closed-loop PLL, is performed through measuring the amplitude of the single-tone–modulated signal after demodulating it from the RF band to the baseband signal by using a first-order frequency discriminator. Finally, the measured amplitude value is used as the input address to an LUT to update the parameters of the pre-distortion filter and to ensure the best compensation [42]. The test results in [42] showed that the RMS phase error for the low-band was 2.6°, and the spectrum at a 400-kHz offset was –67 dBc with a resolution bandwidth of 30 kHz.

Another compensation (or calibration) solution is to apply a frequency step signal to the input of the sigma-delta modulator in the time domain and then to measure the integrated phase error to estimate the K value, which is directly proportional to the product of K_V and K_P. In this solution, a type-2 PLL is used, where the loop filter has one zero and three poles. Figure 4.58 illustrates a block diagram of the modulated fractional-N synthesizer with the digital calibration circuits. The upper portion is the modulated fractional-N synthesizer and the lower portion is the phase error measurement circuit, which consists of a delta-sigma frequency discriminator and a lowpass filter. The delta-sigma frequency discriminator realized by a sampled-data delta-sigma modulator is used to measure the phase error trace between the RF signal divided by $N/N+1$ and the reference clock signal in the time domain. Since different loop gain K values depend on different phase error traces, the compensation algorithm circuit compares the measured phase error trace with the stored phase error curves with time, which are related to different known transfer functions of the closed-loop PLL, to find the best-fit curve with the smallest phase error. Then corresponding coefficients of the pre-distortion filter related to the smallest phase error are read out and loaded to the pre-distortion filter. With this solution, hardware-measured RMS

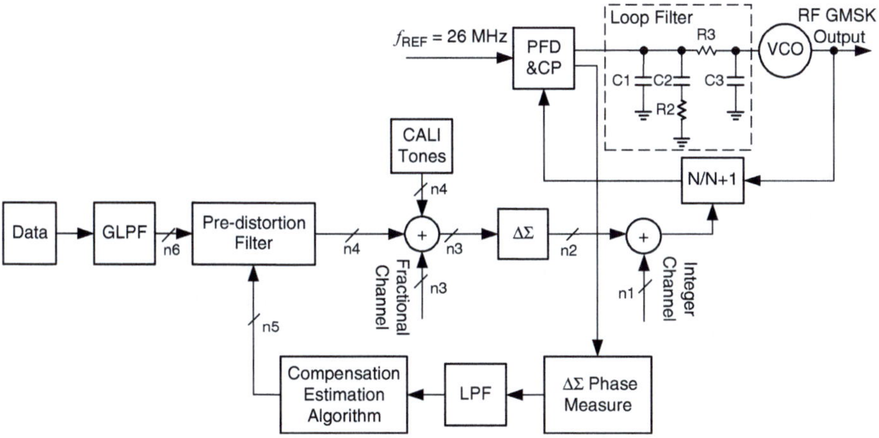

Fig. 4.58 Block diagram of the modulated fractional-*N* synthesizer with the pre-distortion filter and calibration circuit

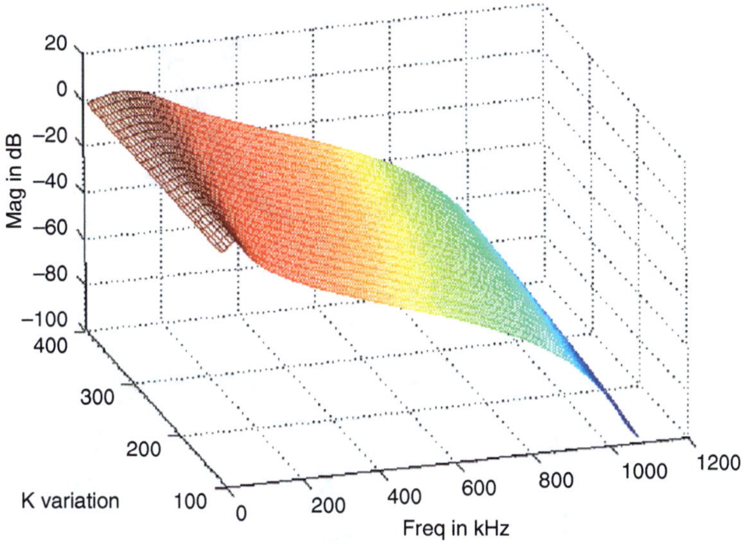

Fig. 4.59 Amplitude response of the closed-loop PLL versus *K* variation with parameters of $C1 = 66$ pF, $C2 = 2134$ pF, $R2 = 2.17$ kΩ, $C3 = 300$ pF, $R3 = 2.049$ kΩ, $K_P = 36 \sim 130$ μA/2π, $K_V = 45$ MHz/V, $N = 15$, FREF = 26 MHz, where $K = K_P K_V / N$ and the loop filter consists of a PLL type 2 (Courtesy Shaolin Li)

phase error from 1.6 to 2.0° is achieved. More accurate compensation can be achieved if capacitors with ±5% tolerance are used in the loop filter. A total of 21 pre-stored curves in the LUT are used.

Figure 4.59 illustrates the amplitude response of the closed-loop PLL versus *K* variation by using MATLAB simulation, where the pre-distortion filter is bypassed. It can be seen that the amplitude response of the closed-loop PLL has a narrow loop bandwidth and a large peak for small K values, while it has a wide loop bandwidth and a smooth peak for large K values.

With this calibration solution, we simulated the demodulated eye diagrams from the modulated fractional-N synthesizer, as shown in Fig. 4.60. Figure 4.60a displays the eye diagrams without the pre-distortion filter, where the parameter K_P is within the range from 36 to 130 μA/2π. In this case, the RMS phase error is 6.2°, which is above the standard defined value of 5°. After the pre-distortion filter, the RMS phase error is reduced to 1.3°, as shown in Fig. 4.60b. It is obvious that the modulated fractional-N synthesizer could not be used to realize GMSK modulation in the GSM standard without the pre-distortion filter or another compensation method.

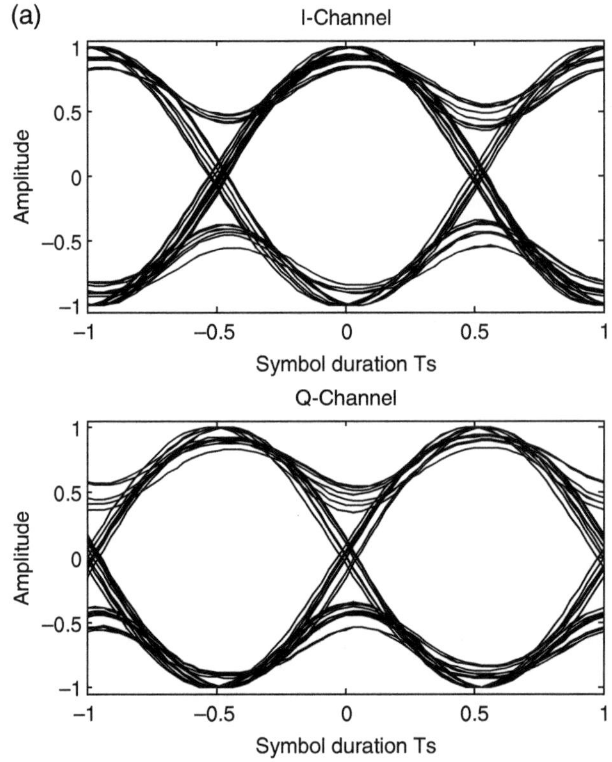

Fig. 4.60 Eye diagrams in the transmitter for type 2 filter: (**a**) phase error = 6.2° without the pre-distortion filter and (**b**) phase error = 1.3° with the pre-distortion filter

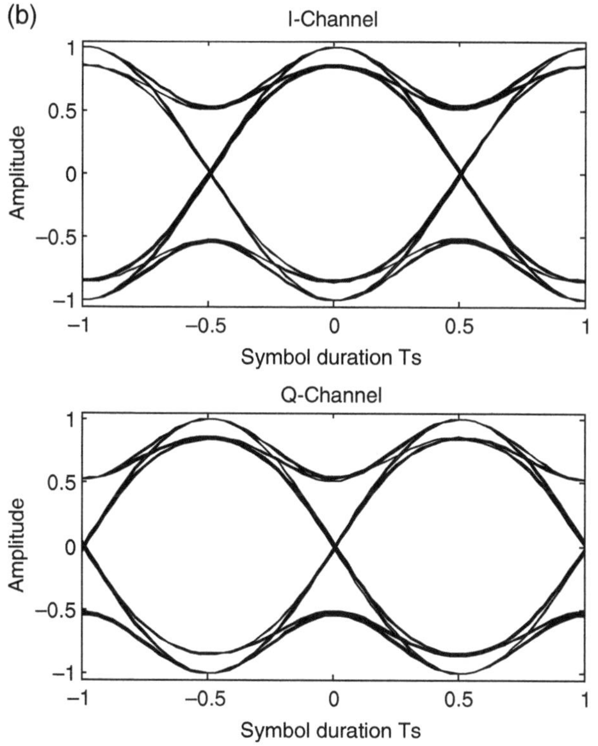

Fig. 4.60 (continued)

4.6.3.3 Offset Phased-Locked Loop

Offset or translation PLL-based modulation has been widely developed over more than two decades for the GSM transceiver due to its superior performance, specifically the sufficient suppression level of the noise transmitted in the GSM receive band, the small phase noise, and fast PLL setting time [46–52]. A sufficient suppression level of the noise at the output of the PA falling in the receive band eliminates the need for the duplexer and the transmit SAW filter. The low level of phase noise is required to meet the RMS phase error of less than 5° defined by the GSM standard. The fast setting time can reduce the transmitter's current consumption during the transmit mode. Figure 4.61 presents a block diagram of the quadrature modulation of an IF carrier by means of a frequency translation loop or an offset phased-locked loop. The modulator-based offset phased-locked loop consists mainly of a down-conversion mixer, a quadrature IF modulator, a phase detector, a loop filter, and a VCO. All filters and VCO are built inside the RF transceiver chip. The modulated RF signal at the RF VCO output with the frequency f_{RF} is first down-converted to an intermediate frequency (IF) signal having the frequency f_{IF} with a local oscillator through a mixer. The output IF signal is then filtered through

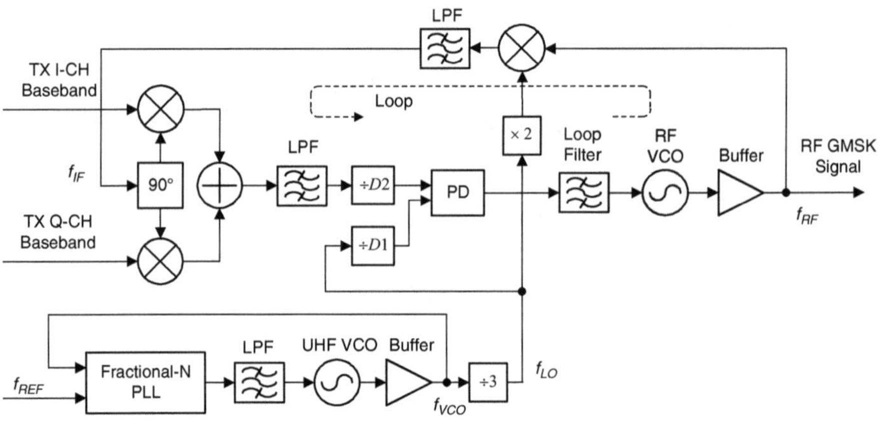

Fig. 4.61 Block diagram of OPLL transmitter for a GSM system with a quadrature modulator inside-offset PLL. Referenced from [52]

a bandpass or lowpass filter to remove high-frequency harmonics. Then, the IF signal that actually is a modulated signal is modulated by the I–Q baseband signals, lowpass filtered, divided by *D*2, and finally phase-compared with the local reference signal after being divided by *D*1.

The local reference signal as another input to the phase detector is obtained from the UHF VCO signal after the divided-by-three and then the divided-by-D1 circuits. The phase difference then drives the charge pump and is filtered by the loop filter. The filtered error signal drives the RF VCO so that the phase difference between inputs of the phase detector reaches zero in steady state. Thus, the IF modulated signal is translated into the RF modulated signal, which is available at the RF VCO output. Now it is easy to understand that the *offset* here means that the center frequency f_{RF} of the RF VCO is offset from the doubled local oscillator frequency f_{LO} by f_{IF}, or $f_{RF} = 2 \times f_{LO} - f_{IF}$.

For example, during the GSM TX mode, if the output frequency of UHF VCO is $f_{VCO} = 1425$ MHz, the frequency of the LO output is $f_{LO} = f_{VCO}/3 = 475$ MHz. The LO frequency is further divided by $D1 = 10$ as the reference signal with the frequency of 47.5 MHz to the phase detector. Another input signal with the carrier frequency of 47.5 MHz to the phase detector is obtained by dividing the GMSK-modulated IF signal with the carrier frequency of 95 MHz with $D2 = 2$. After the PLL is in its locked state, the RF GMSK signal at the RF VCO output is down-converted with a doubled LO signal with the frequency of 950 MHz. The difference frequency component of the down-converted signal is used as the IF carrier signal with a frequency of 95 MHz for the I–Q modulator after passing through the on-chip lowpass filter (LPF) following the mixer. This LPF attenuates the unwanted harmonics as well as the unwanted side-lobes. Hence, the carrier frequency of the RF GMSK signal is $950 - 95 = 855$ MHz. *D*1 and *D*2 can be programmed independently to make the device particularly suitable for dual-band applications. The LO frequency planning plays an important role in designing the multi-band GSM transceivers because it strongly influences the performance of the transceiver [53]. For example, some combinations result in poor power spectral density (PSD) and spurious performance if the frequency f_{LO}, $D1$ and $D2$ are not chosen appropriately.

A key feature of operations is that the PLL cancels the phase shift of the IF modulated signal at the input of the phase detector by generating an equal and opposite phase shift at the RF VCO output [47]. As a result, the RF VCO follows the inverse phase of the I–Q baseband signals of the modulator. Thus, the RF modulated signal with an equal and opposite phase shift is generated at the RF VCO output relative to the phase shift of the IF modulated signal at the input of the phase detector. Swapping I–Q signals can reverse the polarity of the modulation. Finally, the RF modulated signal has low phase noise and suppressed harmonics.

In Fig. 4.61, the I–Q modulator is in the feedback path of RF PLL. The phase noise property of the RF modulated signal is determined by the PLL. The PLL must be fast enough to follow the modulation signal, whose data rate is 270.833 kbps, but it must also be slow enough to suppress the high frequency phase noise [47]. Therefore, the trade-off between phase accuracy and phase noise suppression is dependent of the closed-loop bandwidth. In general, a loop bandwidth two to four times bigger than the modulation data rate is acceptable in GSM applications. For example, the loop bandwidth values of 500 kHz and 800 kHz were used in [47] and [54], respectively. These values of the bandwidth give an acceptable RMS phase error and phase noise suppression at the frequency offset of 20 MHz from the carrier. The phase detector is frequency sensitive and its frequency range is typically between 50 and 150 MHz. In addition, the frequency range at the input of the phase detector can be up to about 400 MHz [50].

In contrast, the I–Q modulator can be inserted outside the feedback loop of the RF PLL as well. In [47] and [33], the modulator is placed at the reference input, as shown in Fig. 4.62. In this case, an IF continuous waveform (CW) carrier is

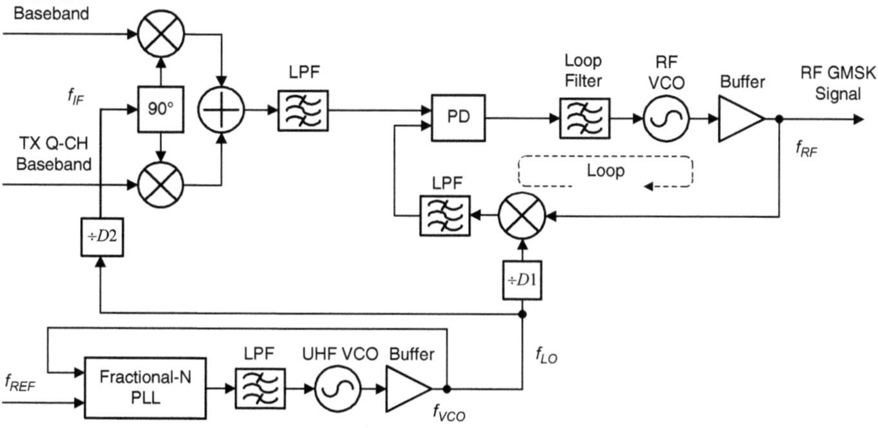

Fig. 4.62 Block diagram of OPLL transmitter with a quadrature modulator outside-offset PLL

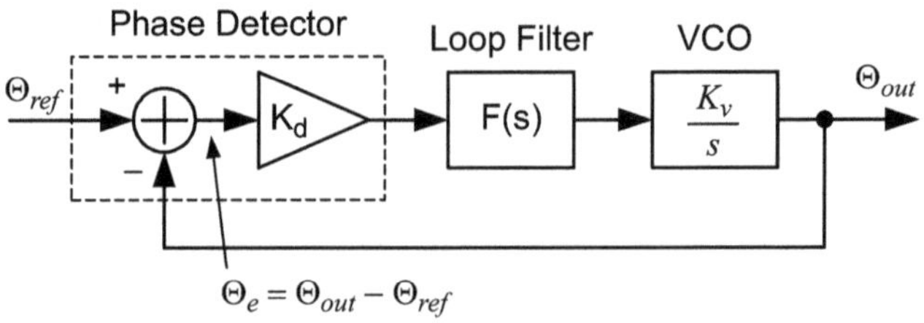

Fig. 4.63 Linear model of an OPLL in the s-domain

modulated by the baseband I–Q signals to form the IF modulated reference signal, which is used as one of the input signals of the phase detector. The other input is the output of the down-converter mixer after lowpass filtering at the frequency of $f_{\mathrm{MIX}} = f_{\mathrm{LO}}/D1 - f_{\mathrm{RF}}$. The phase detector output is an error current whose value is proportional to the phase difference between the feedback signal f_{MIX} and the reference IF signal f_{IF}. The error current signal is lowpass filtered to form an error voltage to further drive the RF VCO. Thus, the RF VCO output is frequency-modulated with the GMSK baseband signal. In the locked status, the frequency of the down-converted signal is equal to that of the reference IF-modulated signal. The center frequency of the RF VCO output is offset from the LO frequency by f_{IF} or $f_{\mathrm{RF}} = f_{\mathrm{LO}}/D1 - f_{\mathrm{IF}}$.

The performance is equivalent wherever the I–Q modulator is replaced, whether inside the PLL or outside the PLL. It should be noted that the major difference between the conventional PLL and the offset PLL (OPLL) is that the frequency modulation of the reference input is *reproduced* at the output of the OPLL without scaling [46]. Precisely speaking, the offset PLL-based transmitter performs *both frequency up-conversion and narrow bandpass filtering* after the I–Q modulation. On the other hand, an RF transmitter performed by an I–Q modulator with one-stage direct up-conversion is very difficult to meet the stringent requirements of both low phase noise and suppressed harmonics in the GSM standard.

The closed-loop transfer function $H_{\mathrm{CL}}(s)$ of the offset PLL can be derived from Fig. 4.63:

$$H_{\mathrm{CL}}(s) = \frac{\Theta_{\mathrm{out}}(s)}{\Theta_{\mathrm{ref}}(s)} = \frac{K_{\mathrm{d}} K_{\mathrm{v}} F(s)}{s + K_{\mathrm{d}} K_{\mathrm{v}} F(s)} \tag{4.149}$$

where K_{d} is the phase comparator constant in A/rad, K_{v} is the gain of the VCO in rad × s^{-1}/V, and $F(s)$ is the transfer function of the loop filter.

In general, the loop filter is a third-order lowpass filter, so that the closed-loop transfer function is the fourth-order. This selection is chosen based on sufficient suppression level of the transmitter noise. Figure 4.64 illustrates an example of the magnitude response of the closed-loop function $H_{\mathrm{CL}}(s)$. The flat 0-dB region is defined as the loop bandwidth for the case of the normalized frequency response.

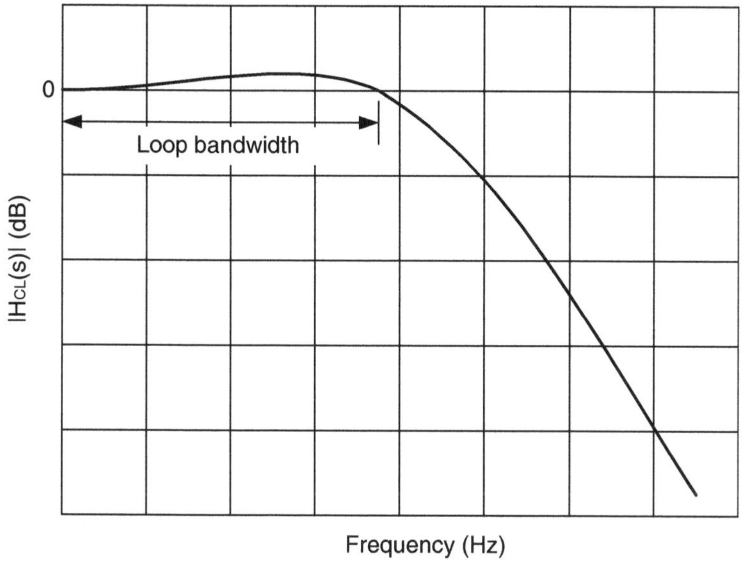

Fig. 4.64 Amplitude response of the closed-loop transfer function

The input phase variation within the loop bandwidth at the input of the phase detector is reproduced at its output, and the input phase variation outside the loop bandwidth is suppressed by the PLL. Consequently, the OPLL performs the function of a bandpass filter whose bandwidth equals twice the loop bandwidth. The trade-off between the transmit noise level and the phase error must be considered in the design of the OPLL because they are both related to the loop bandwidth [46]. For example, the narrower bandwidth increases the suppression level of the transmit noise but also increases the RMS phase error because the input phase variation at the input of the phase detector is not completely reproduced at the output of the phase detector.

The trade-off between the RMS phase error and the 20 MHz offset noise was simulated and measured in [50] and [46], respectively. In order to achieve RMS phase error of less than 2° and 20-MHz offset noise of less than –165 dBc/Hz in [46], the optimum bandwidth must be in the range from 0.6 to 2.6 MHz. In practice, the bandwidth around 1 MHz can yield acceptable performance.

So far we have introduced the offset or translation PLL-based GMSK-modulation scheme. Now we shall show some curves of the measured GMSK signal from a commercial GSM chip. The first curve is the PSD of GMSK, as shown in Fig. 4.65, where the measured PSD is about −68 dBc at a frequency offset of 400 kHz on the OPLL-based GMSK transmitter output in the E-GSM 900 band and meets the most critical specification value of −60 dBc at the frequency offset of 400 kHz.

The transmission format of the GSM system is TDM burst transmission. One frame consists of eight time slots or bursts. Each time slot has a total of 156.25 bits, of which the first three are tail bits, followed by 58 encrypted bits, 26 training

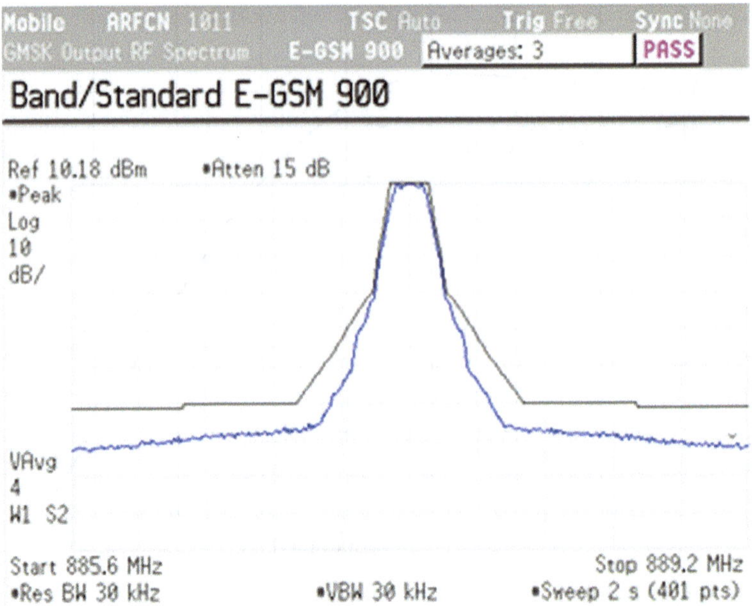

Fig. 4.65 Measured power spectrum density (PSD) at transmitter output from a commercial GSM transceiver chip, where the PSD is located at the center frequency of 891 MHz with a span of 3.6 MHz or the range from 885.6 to 889.2 MHz

sequence bits, another 58 encrypted bits, another 3 tail bits, and finally 8.25 guard period (bits). Hence, one time slot lasts $156.25/270833 \approx 577$ μs. Eight time slots l form a TDMA frame, which lasts about 4.62 ms.

The modulation accuracy of the GMSK-modulated signal is defined by the phase accuracy of the-modulated signal and is specified in the following way:

For any 148-bit subsequence of the 511-bit pseudo-random sequence, the phase error trajectory on the useful part of the burst shall be measured by computing the difference between the phase of the transmitted signal and the phase of the theoretical modulated signal. The RMS phase error shall not be greater than 5° with a maximum peak deviation during the useful part of the burst less than 20°.

Figure 4.66 illustrates the transmitted RF signal power level versus time within one-burst duration for GMSK modulation, where the transmitted power level must be within the time mask to qualify the specification. To measure the RMS phase error of the transmitted GMSK signal, the transmitted signal shall be demodulated to the I and Q baseband signals in the receiver. Figure 4.67 illustrates the measured PSD of the transmitted GMSK signal and the RMS phase error for one burst slot, where the RMS phase error is 1.42°.

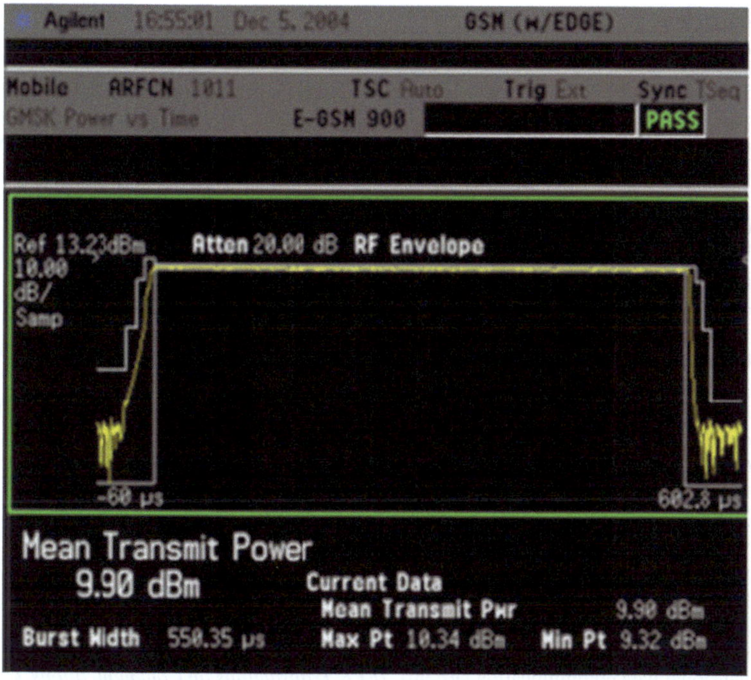

Fig. 4.66 Transmitted GMSK signal power level versus time in one time slot duration for one commercial RF transceiver and baseband chips. (Rise time = 4 bits, fall time = 2 bits)

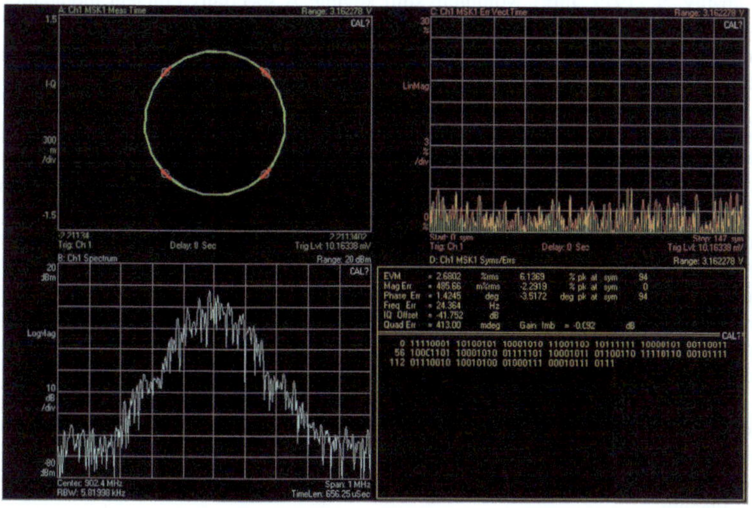

Fig. 4.67 Measured constellation and PSD of RF offset loop-based I–Q GMSK signal for one commercial RF transceiver and baseband chips, where the RMS phase error is 1.42°

References

1. Proakis, J. G. (1995). *Digital communications* (3rd ed.). New York, NY: McGraw-Hill Inc.
2. Simon, M. K. (2001). *Bandwidth-efficient digital modulation with application to deep-space communications* (Deep-space communications and navigation series). New York, NY: John Wiley & Sons Inc.
3. Murota, K., & Hirade, K. (1981, July). GMSK modulation for digital mobile radio telephony. *IEEE Transactions on Communications, 29*(7), 1044–1050.
4. Feher, K. (1995). *Wireless and digital communications; modulation & spread spectrum applications*. Upper Saddle River, NJ: Prentice-Hall PTR.
5. Gao, W., Soderstrand, M., & Feher, K (1995, May). Gaussian filter screens TDMA and frequency-hopping spread-spectrum signals. *Microwave & RF* (pp. 17–20).
6. Feher, K., & Kato, S. U.S. patents: 4,567,602; 4,339,724; 4,644,565; 5,784,402; 5,491,457. Canadian patents: 1,211,517; 1,130,871; 1,265,851.
7. Seo, J. S., & Feher, K. (1985, May). SQAM: A new superposed QAM modem technique. *Transactions on Communications, COM-33*(3), 296–300.
8. Kato, S., & Feher, K (1983, May). XPSK: A new cross-correlated phase shift keying modulation technique. *IEEE Transactions on Communications, COM-31*(5), 701–707.
9. Telemetry Group. (2004, May). Telemetry Standards, IRIG Standard 106-04.
10. Simon, M. K., & Wang, C. C. (1984, November). Differential detection of Gaussian MSK in a mobile radio environment. *IEEE Transactions on Vehicular Technology, VT-33*(4), 307–320.
11. Sato, Y. (1975, June). A method of self-recovering equalization for multilevel amplitude modulation systems. *IEEE Transactions on Communications, COM-23*, 679–682.
12. Godard, D. N. (1980, November). Self-recovering equalization and carrier tracking in two dimensional data communication systems. *IEEE Transactions on Communications, COM-28*, 1867–1875.
13. Treichler, J. R., et al. (1983, April). A new approach to multipath correction of constant modulus signals. *IEEE Transactions on Acoustics, Speech, and Signal Processing, ASSP-31* (2), 459–472.
14. Pasupathy, S. (1979, July). Minimum shift keying: A spectrally efficient modulation. *IEEE Communications Magazine* (pp. 14–22).
15. Gao, W., & Feher, K. (1996, March). All digital reverse modulation architecture based carrier recovery implementation for GMSK and compatible FQPSK. *IEEE Transaction on Broadcasting, 42*(1), 55–62.
16. Spilker, J. J., Jr. (1977). *Digital communication by satellite* (pp. 31–312). Englewood Cliffs, NJ: Prentice-Hall, Inc.
17. Cavers, J. (1991, May). Performance of tone calibration with frequency offset and imperfect pilot filter. *IEEE Transactions on Vehicular Technology, VT-40*, 426–434.
18. Jain, P. K. (2004, December). Regenerate coherent carriers from PSK signals. *Microwaves & RF* (pp. 52–68).
19. Weber, C. L., & Alem, W. K. (1980, December). Demod-remod coherent tracking receiver for QPSK and SQPSK. *IEEE Transactions on Communications, COM-28*(12), 1945–1954.
20. Morihiro, Y., Nakajima, S., & Furuya, N. (1979, October). A 100 Mbit/s prototype MSK modem for satellite communications. *IEEE Transactions on Communications, COM-27*(10), 1512–1518.
21. Kaleh, G. K. (1989, December). Simple coherent receivers for partial response continuous phase modulation. *IEEE Journal on Selected Areas in Communications, 7*(9), 1427–1436.
22. Anderson, J. B., Aulin, T., & Sundberg, C. E. (1986). *Digital phase modulation*. New York, NY: Plenum.
23. Liu, G. L. (1998, October). Threshold detection performance of GMSK signal with $BT_b = 0.5$. *MILCOM'98 Conference Proceedings, 2*, 515–519.

24. Laurent, P. A. (1986, February). Exact and approximate construction of digital phase modulations by superposition of amplitude modulated pulse. *IEEE Transactions on Communications, COM-34*(2), 150–160.
25. Costas, J. P. (1956). Synchronous communications. *Proceedings of the IRE, 44*, 1713–1718.
26. Holmes, J. K. (1982). *Coherent spread spectrum systems*. New York, NY: John Wiley & Sons Inc.
27. Chung, B. Y., et al. (1993, September). Performance analysis of an all-digital BPSK direct-sequence spread-spectrum IF receiver architecture. *IEEE Journal on Selected Areas in Communications, 11*(7), 1096–1107.
28. Gardner, F. M. (1979). *Phase lock techniques*. New York, NY: Jon Wiley & Sons Inc.
29. Lee, E. A., & Messerschmitt, D. G. (1994). *Digital communication*. Norwell, MA: Kluwer Academic Publishers.
30. GSM 05.05 version 8.5.1 Release 1999. *Digital Cellular Telecommunications Systems (Phase 2+); Radio Transmission and Reception*. ETSI EN 300 910 V8.5.1 (2000-11).
31. Weldon, J. A., Narayanaswami, R. S., Rudell, J. C., Lin, L., Otsuka, M., & Dedieu, S. (2001, December). A 1.75 GHz highly integrated narrow-band CMOS transmitter with harmonic-rejection mixers. *IEEE Journal of Solid-State Circuits, 36*(12), 2003–2015.
32. Stetzler, T. D., Post, I. G., Havens, J. H., & Koyama, M. (1995, December). A 2.7-4.5V single chip GSM transceiver RF integrated circuit. *IEEE Journal of Solid-State Circuits, 30*, 1421–1429.
33. Tham, J. I., et al. (March 1999). A 2.7V 900 MHz/1.9 GHz dual band transceiver IC for digital wireless communication. *IEEE Journal of Solid-State Circuit, 34*(3), 286–291.
34. LMX3162 data sheet, National Semiconductor Corporation, January 2000.
35. Heinen, S., Beyer, S., & Fenk, J. (1995, February). A 3.0V 2 GHz transmitter IC for digital radio communication with integrated VCO's. *Proceedings of the IEEE International Solid-State Circuits Conference* (pp. 150–151).
36. Razavi, B. (2003). *RF microelectronics*. Taiwan: Pearson Education.
37. Goldberg, B. (1999, June). Analog and digital fractional-*N* PLL frequency synthesis: A survey and update. *Applied Microwave & Wireless* (pp. 32–42).
38. Riley, T. A. D., & Copeland, M. A. (1994, May). A simplified continuous phase modulator. *IEEE Transactions on Circuits and Systems II: Analog and Digital Signal Processing, 41*(5), 321–328.
39. Vandegraff, J. J. (1989, September 12). Phase locked frequency synthesizer with single input wideband modulation systems. US Patent 4,866,404.
40. Perrott, M. H., Tewksbury, T. L., & Sodini, C. G. (1997, December). A 27-mW CMOS fractional-N synthesizer using digital compensation for 2.5 Mbit/s GFSK modulation. *IEEE Journal of Solid-State Circuits, 32*(12), 2048–2060.
41. Bax, W. T., & Copeland, M. A. (2001, August). A GMSK modulator using a $\Delta\sum$ frequency discriminator based synthesizer. *IEEE Journal of Solid-State Circuits, 36*(8), 1218–1227.
42. Lee, S. T. (2003). *Quad-band global system for mobile communications complementary metal-oxide-semiconductor transmitter*. Doctor's dissertation, University of Washington.
43. McMahil, D. R., & Sodini, C. G. (2001). Automatic calibration of modulated $\sum-\Delta$ frequency synthesizers. *Symposium* on *VLSI Circuits Digest of Technical Papers* (pp. 51–54).
44. McMahill, D. R., & Sodini, C. G. (2002, January). A 2.5 Mb/s GFSK 5 Mb/s 4-FSK automatically calibrated $\sum-\Delta$ frequency synthesizer. *IEEE Journal of Solid-State Circuits, 37*(1), 18–26.
45. Lee, T. H., & Bulzacchelli, J. F. (1992, December). A 155 MHz clock recovery delay- and phase-locked loop. *IEEE Journal of Solid-State Circuits, 27*(12), 1736–1746.
46. Yamawaki, T., Kokubo, M., Irie, K., Matsui, H., Hori, K., Endou, T., et al. (1997, December). A 2.7V GSM RF transceiver IC. *IEEE Journal of Solid-State Circuit, 32*(12), 2089–2096.
47. Imine, G., Herzinger, S., Schmidtz, R., Kubetzko, D., & Fenk, J. (1998, February). An up-conversion loop transmitter IC for digital mobile telephones. *ISSCC Digest of Technical Papers* (pp. 364–365).

48. Tham, J. L., Margarit, M. A., Pregardier, B., Hull, C. D., Magoon, R., & Carr, F. (1999, March). A 2.7V 900 MHz/1.9 GHz dual-band transceiver IC for digital wireless communication. *IEEE Journal of Solid-State Circuit, 34*(3), 286–291.
49. Molnar, A., Magoon, R., Zachan, J., Hatcher, G., & Rhee, W. (2002, February). A single-chip quad-band (850/900/1800/1900 MHz) direct conversion GSM/GPRS RF transceiver with integrated VCOs and fractional-N synthesizer. *ISSCC Digest of Technical Papers* (pp. 184–185).
50. Song, E., Koo, Y., Jung, Y.-J., Lee, D.-H., Chu, S., & Chae, S.-I. (2005, May). A 0.25 μm CMOS Quad-Band GSM RF transceiver using an efficient LO frequency plan. *IEEE Journal of Solid-State Circuits, 40*(5), 1094–1106.
51. Durrant, M., & Nitschke, A. (2005, May). Design considerations for an ultra-compact GSM radio solution. *RF Design* (pp. 46–54).
52. Data sheet (2001, December 3). *CX74017 RF transceiver for multi-band GSM/GPRS/EDGE applications,* Conexant.
53. Strange, J., & Atkinson, S. (2000, June). A direct conversion transceiver for multi-band GSM application. *Proceedings of IEEE RFIC Symposium* (pp. 25–28).
54. Cipriani, S., Carpineto, L., Bisanti, B., Hogervorst, I. R., Puccio, G., & Mouralis, N. (2002). Fully integrated zero IF transceiver for GPRS/GSM/DCS/PCS application. *ESSCIRC 2002* (pp. 439–442).

Chapter 5
Linearization Techniques for RF Power Amplifiers

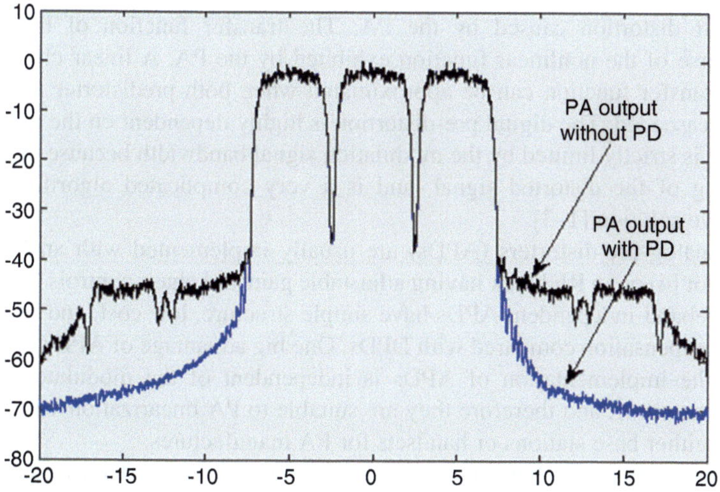

5.1 Introduction

Orthogonal frequency division multiplexing (OFDM) with high-order modulation formats of *M-ary* QAM adopted by many wireless standards are widely used in modern wireless communication systems due to its advantages over single-carrier schemes in robustly combating multipath fading with simple equalization filters and over classical FDM schemes in achieving spectrally efficient transmission. OFDM signals naturally behave with non-constant envelope characteristics and have a very high peak-to-average power ratio (PAPR). The RF modulated signal with a high PAPR highly requires that RF transmitter power amplifiers (PAs) operate with a

© Springer International Publishing Switzerland 2017
W. Gao, *Energy and Bandwidth-Efficient Wireless Transmission*, Signals and Communication Technology, DOI 10.1007/978-3-319-44222-8_5

large backoff power from their P1dB compression point to perform linearly, resulting, however, in low energy efficiency. If its back-off power is not enough, or if it operates close to its P1dB point, the PA causes spectral regrowth, which leads to adjacent channel interference. Meanwhile, it also causes in-band signal error vector magnitude (EVM) degradation, which in turn degrades the bit error rate (BER) performance at the receiver. It is commonly known that achieving both highly spectral efficiency and energy efficiency is contradictory in wireless communications. To obtain high energy efficiency to reduce severe distortion caused by PA non-linearity, pre-distortion or linearization techniques have been widely used in applications. Generally, predistortion is classified as digital predistortion (DPD) and RF analog predistortion (APD), according to its operation domain at either a digital or analog domain.

Digital predistorters (DPDs) are often implemented in either a *look-up table* (LUT) or a *digital signal processor* (DSP) in the digital domain, which can accurately generate the predistorted baseband I–Q signals to compensate for the nonlinear distortion caused by the PA. The transfer function of the DPD is the inverse of the nonlinear function exhibited by the PA. A linear characteristic of the transfer function can be approximated when both predistorter and PA are serially cascaded. The digital pre-distortion is highly dependent on the modulation formats, is strictly limited by the modulation signal bandwidth because of the band spreading of the distorted signal, and is a very complicated algorithm and an expensive solution [1–3].

RF analog pre-distorters (APDs) are usually implemented with an RF vector modulator [4] or an RF block having adjustable gain and phase controls [5] before a PA. Baseband-independent APDs have simple structure, low cost, and wide bandwidth compensation compared with DPDs. One big advantage of APDs over DPDs is that the implementation of APDs is independent of the modulator and is a standalone mode, and therefore they are suitable to PA linearization with applications in either base stations or handsets for PA manufactures.

In this chapter, a Volterra model of a PA with a memory effect, which approximates PA nonlinearity characterization, is described first. Then, a precise and simple Volterra model of a pre-distorter, which is used to approximate the inverse of a nonlinear PA, is presented. Next, the structure of a vector modulator used as an APD is illustrated in greater detail. Finally, a RF APD system-on-chip (SoC) chipset that has been commercially used in cellular base stations is introduced as the example of a study of APD applications.

5.2 Memory Model of Power Amplifiers

Memory effects are a general phenomenon of typical RF PAs. The memory effects of RF PAs can arise from multiple sources, including bias circuit effects, self-heating, and trapping effects. They are mainly characterized by both amplitude-to-amplitude modulation (AM-AM) conversion and amplitude-to-phase modulation (AM-PM) conversion; AM-AM is generated mainly by the voltage drop across the loss resistance of the drain inductor and AM-PM is mainly created by the variations of the drain source capacitance as a function of drain voltage. One of the obvious distortions affected by the memory effects is asymmetric sidebands of RF power spectral density (PSD), which may violate the required specification on one side and have more margins on another side. Therefore, from a hardware design point of view, the memory effects of RF PAs should be minimized. On the other hand, it is necessary for circuit designers to precisely set up a PA model by considering its nonlinearity and memory effects.

The Volterra series is a generalization of the Taylor series and has been widely used for modeling PAs with mild nonlinearities and memory effects. One serious drawback of the Volterra model is the large number of coefficients that need to be extracted based on the amount of the collected input and output data of the PA. Therefore, it is very complicated, and sometimes it may be impossible.

In practical applications, the Volterra series must be simplified by avoiding summation over an infinite number of terms. A sufficiently accurate model, depending on an actual application, can be obtained by using a finite number of nonlinear terms and a few memory terms, which characterize the order of the nonlinear model and the amount of memory, respectively. With this kind of simplification, the number of coefficients of the Volterra model can be dramatically reduced and the structure of this model can be significantly simplified.

There are some simplified approaches to approximating the Volterra model with memory, such as the truncated Volterra series presented by Zhu et al. [6, 7], the Wiener model proposed by Clark et al. [8], and the memory polynomial (MP) model proposed by Kim and Konstantinous [9]. Most of the approaches are based on an analysis of the physical characteristics of PAs or structure modification of the original polynomials. It has been shown that the MP model has low complexity in its hardware implementation and easy extraction from the captured input and output data of the PA [10, 11].

In addition, some efforts have been made to adaptively prune the Volterra models based on captured input and output signals of the PA, where small kernels were considered negligible and therefore removed [12, 13]. Even these methods based on adaption can update the coefficients of the Volterra model with time to avoid any changes caused by supply voltage variation, temperature change, agility, and other factors, but they also increase the complexity and power consumption in the hardware implementation.

A complex baseband Volterra series with odd-order kernels only in a discrete time domain can be expressed as [6]

$$\widetilde{y}(n) = \sum_{m=0}^{M} h_1(m)z(n-m) + \sum_{m_1=0}^{M}\sum_{m_2=m_1}^{M}\sum_{m_3=m_2}^{M} h_3(m_1,m_2,m_3)z(n-m_1)z(n-m_2)z^*(n-m_3)$$
$$+\sum_{m_1=0}^{M}\sum_{m_2=m_1}^{M}\sum_{m_3=m_2}^{M}\sum_{m_4=m_3}^{M}\sum_{m_5=m_4}^{M} h_5(m_1,m_2,m_3,m_4,m_5)\prod_{j}^{4} z(n-m_j)z^*(n-m_5)+\ldots \tag{5.1}$$

where $h_i(m_1, m_2, \cdots, m_i)$ is the ith-order Volterra kernel, M is the highest memory unit, the symbol * represents the conjugate transpose, and $\widetilde{y}(n)$ is the estimate value of $y(n)$ at the PA output. If $h_i(m_1, m_2, \cdots, m_i)$ is equal to 0, except along the diagonal $m_1 = m_2 = \cdots = m_i$, then (5.1) can be simplified by

$$\widetilde{y}(n) = \sum_{m=0}^{M}\left[h_1(m)z(n-m) + h_3(m,m,m)z(n-m)\left|z(n-m)\right|^2+\cdots\right.$$
$$\left.+\,h_{2n+1}(m,m,\cdots,m)z(n-m)\left|z(n-m)\right|^{2n}+\cdots\right],\ n=0,1,\cdots \tag{5.2}$$

Equation (5.2) is referred to as an MP. In practice, the finite-order $N = 2n + 1$ is used to truncate the model shown above.

A relatively simple approximation to the Volterra model is the MP model proposed by Kim and Konstantinous [9], which captures both memory effects and nonlinear behavior of a PA, and is suitable to modeling a PA with the wideband input signal. Similar to the Volterra model of a PA, the MP model of the digital pre-distortion (DPD) is described as an approximate inverse of a PA. Two advantages of MPs are their simple implementation and the good approximation they yield.

More generally, a baseband memory polynomial model of a PA with input $z(n)$ and output $y(n)$ as given in (5.2) can be expressed as

$$\widetilde{y}(n) = \sum_{k=1}^{K}\sum_{m=0}^{M} b_{km}z(n-m)\left|z(n-m)\right|^{k-1} \tag{5.3}$$

where K is the order of nonlinearity, M is the maximum delay unit, b_{km} is the kernel charactering the nonlinearity of the system, and $\widetilde{y}(n)$ is the approximate output $y(n)$ of the PA. If M is equal to 0, or memoryless system, then (5.3) reduces to

$$\widetilde{y}(n) = \sum_{k=1}^{K} b_k z(n)\left|z(n)\right|^{k-1} \tag{5.4}$$

The model in (5.4) has a good approximation for a PA with a narrowband signal as its input. However, most PAs have memory effects on the output signals. Therefore, the memoryless polynomial in (5.4) does not approximate a model well for an actual PA, especially for a PA operating in a wideband frequency domain. In contrast, (5.3) has a better approximation than (5.4) and presents both a nonlinearity property and a memory effect, which means that the output of the PA is not only a function of the current input, but also a function of the past inputs and their power terms. Based on practical PA memory modeling simulations, one memory delay unit ranging from 1 to 3 ns is a good approximation.

An important advantage of MP models, including memory and memoryless polynomials, is that their expressions of (5.3) and (5.4) are linear combinations of the basic functions. Thus, the least squares (LS) estimation method can be applied to identify the coefficients of the model. In practice, the LS estimator is more commonly used due to its easy implementation and acceptable performance. The criteria are that it be unbiased and have minimum variance. The LS estimator uses variance as a measure of performance to minimize the error between the estimate and the true value, which is expressed by

$$E(\mathbf{b}) = \sum_{n=0}^{N-1} |e(n)|^2 = \sum_{n=0}^{N-1} |y(n) - \tilde{y}(n)|^2 \tag{5.5}$$

where N is the length of the captured input and output data, and $\mathbf{b}$ is the vector of coefficients of b_{km} in (5.3) or b_k in (5.4). The minimum error, corresponding to the optimal vector of $\mathbf{b}_{\text{opt}}$ value, is dependent on the highest orders of nonlinearity and the maximum delay unit of memory effect. If the process is ergodic and stationary, the LS error (LSE) estimator approaches the minimum mean squared error (MMSE) estimator as the length of the captured data grows, in which the MMSE is defined as

$$E(\mathbf{b}_{\text{opt}}) = \mathbf{min}\{E[|y(n) - \tilde{y}(n)|^2]\} \tag{5.6}$$

where $E[\cdot]$ is the expectation, and $\mathbf{b}_{\text{opt}}$ is the optimal vector that leads to the MMSE. The PA model and algorithm accuracy actually depends on the selected nonlinear polynomial order, the captured data length, and the MATLAB computation accuracy. The modeling accuracy for the PA is usually evaluated by using the normalized mean square error (NMSE), defined as

$$\text{NMSE}_{\text{PA}}[\text{dB}] = 10\,\mathbf{log}_{10}\left(\frac{\sum_{n=1}^{N} |y(n) - \tilde{y}(n)|^2}{\sum_{n=1}^{N} |y(n)|^2}\right) \tag{5.7}$$

where $y(n)$ is the measured data at the output of PA, and $\tilde{y}(n)$ is the modeled data at the output of Volterra model.

The coefficients b_{km} in (5.3) can be solved by using the LS method. By defining a new variable

$$u_{km}(n) = z(n-m)\left|z(n-m)\right|^{k-1} \tag{5.8}$$

then (5.3) can be written in matrix form as

$$\widetilde{\mathbf{y}} = \mathbf{U}\mathbf{b} \tag{5.9}$$

where $\widetilde{y}$ is the vector of the output, **b** is the vector of the coefficients to be determined, and **U** is a matrix filled with all input series' values. In details, they are expressed as

$$\begin{aligned} \mathbf{z} &= [z(0), \ldots, z(N-1)]^{\mathrm{T}} \\ \mathbf{b} &= [b_{10}, \ldots, b_{K0}, \ldots, b_{1M}, \ldots, b_{KM}]^{\mathrm{T}} \\ \mathbf{U} &= [\mathbf{u}_{10}, \ldots, \mathbf{u}_{K0}, \ldots, \mathbf{u}_{1M}, \ldots, \mathbf{u}_{KM}] \\ \mathbf{u}_{km} &= [u_{km}(0), \ldots, u_{km}(N-1)]^{\mathrm{T}} \end{aligned} \tag{5.10}$$

Consequently, the LS solution for (5.9) is given by

$$\mathbf{b} = \left(\mathbf{U}^{\mathrm{H}}\mathbf{U}\right)^{-1}\mathbf{U}^{\mathrm{H}}\widetilde{\mathbf{y}} \tag{5.11}$$

where $(\cdot)^{\mathrm{H}}$ denotes the Hermitian transpose operator, $(\cdot)^{-1}$ stands for the inverse operator, and the hat indicates an estimator.

Design Example 5.1 As an example, a memory polynomial model is used to approximate a nonlinear system which is to obey a Wiener-Hammerstein (W-H) model [8, 10], i.e., a linear time-invariant (LTI) system with memory (or a TX Butterworth filter including the pulse-sharping filter) followed by a memoryless nonlinearity and then by another LTI system with memory (or a RX Butterworth filter), as shown in Fig. 5.1. Such a configuration is usually used to simulate satellite communication channels in which the PA of the satellite transponder is driven at near saturation by the PSK-modulated signals to achieve maximum energy efficiency.

The transfer functions of two LTI blocks are assumed to be expressed as

$$H_{\mathrm{F}}(z) = \frac{1+0.4z^{-2}}{1-0.2z^{-1}}, H_{\mathrm{E}}(z) = \frac{1-0.1z^{-1}}{1-0.4z^{-1}} \tag{5.12}$$

In the discrete time domain, the expression for the input and output of the memoryless nonlinear block is given by

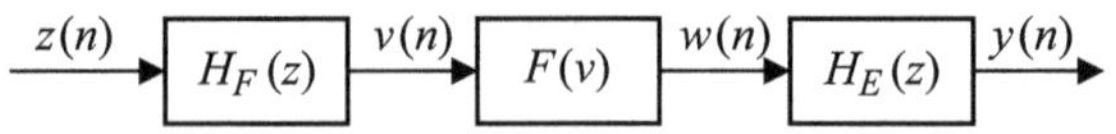

Fig. 5.1 A Wiener-Hammerstein block diagram

$$w(n) = \sum_{\substack{k=1 \\ k\,\mathrm{odd}}}^{K} b_k v(n)|v(n)|^{k-1} \tag{5.13}$$

where the coefficients actually extracted from a real Class AB PA are

$$\begin{aligned} b_1 &= 1.0108 + j0.0858 \\ b_3 &= 0.0879 - j0.1583 \\ b_5 &= -1.0992 - j0.8891 \end{aligned} \tag{5.14}$$

The complex gain of the memoryless block $F(v)$ is obtained as in (5.13):

$$F(v) = b_1 + b_3|v(n)|^2 + b_5|v(n)|^4 \tag{5.15}$$

It can be seen that the complex gain is dependent on the magnitude of the complex baseband envelope.

Solution To collect the output data of $y(n)$, the baseband input $z(n)$ to the Wiener-Hammerstein (W-H) model is a three-carrier WCDMA signal with each having a 5-MHz bandwidth. Around 10,000 I–Q samples used as the output data $y(n)$ at the output of the W-H model are collected at a sampling rate of 61.44 MHz. The coefficients of the memory polynomial model with fifth odd-order $K=5$ and maximum memory unit $M=2$ can be extracted by substituting both input and output data into (5.11), each having 10,000 data samples. The memory polynomial model used here is expressed by

$$\tilde{y}(n) = \sum_{\substack{k=1 \\ k\,\mathrm{odd}}}^{5} \sum_{m=0}^{2} b_{km} z(n-m)|z(n-m)|^{k-1} \tag{5.16}$$

The coefficients are

$$\begin{aligned} &b_{10} = -0.7199 + j0.06523, b_{30} = 58.6535 + j6.4971, b_{50} = -942.6704 - j151.8989 \\ &b_{11} = 3.3960 + j0.2959, b_{31} = -117.0267 - j13.7877, b_{51} = 1869.1053 + j282.9810 \\ &b_{12} = -0.0255 + j0.0059, b_{32} = 59.2177 + j6.0675, b_{52} = -956.1663 - j154.2885 \end{aligned} \tag{5.17}$$

Figures 5.2a, b show the simulated AM-AM and AM-PM characteristics for the output $y(n)$ of W-H system and the output $\tilde{y}(n)$ of the memory polynomial model. It can be seen from the curves that AM-AM and AM-PM curves of the memory polynomial model are identical to those of the W-H system well due to their complete overlap. Such accurate behavioral modeling of nonlinear PA based on memory polynomial series can be also seen in Fig. 5.3 where the PSD of the

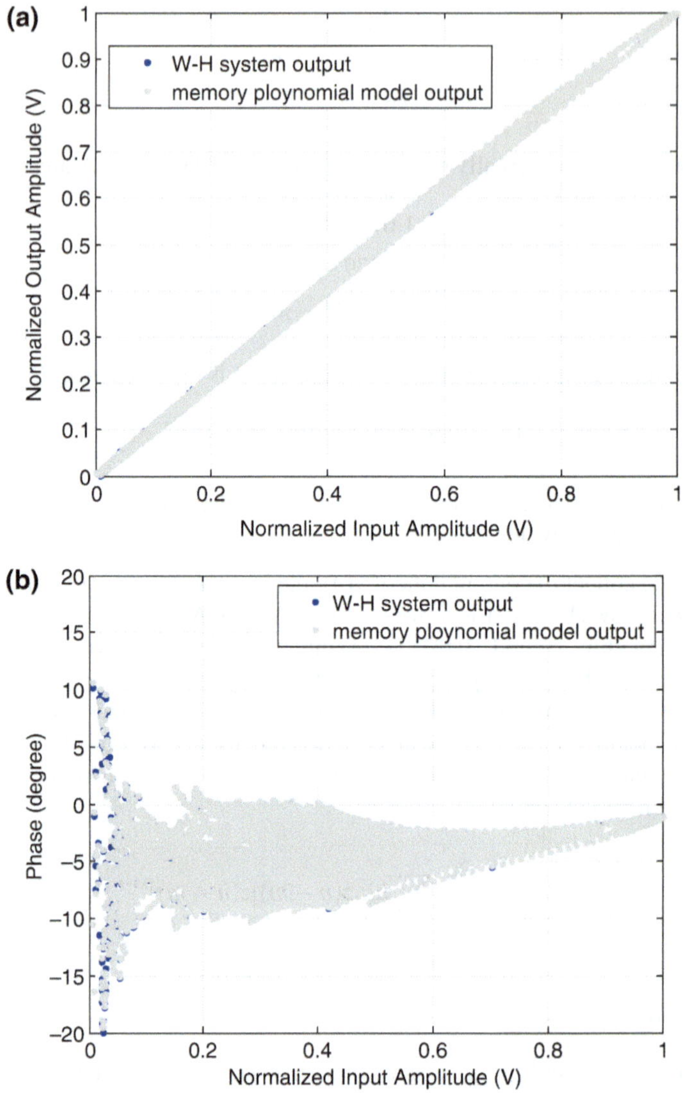

Fig. 5.2 AM-AM and AM-PM characteristics of the outputs $y(n)$ and $\widetilde{y}(n)$ of W-H model and memory polynomial model, respectively: (**a**) AM-AM and (**b**) AM-PM characteristics

curve (c) matches PSD of the curve (b) in the frequency domain well, down to −70 dB relative to the peak level.

The best method to evaluate modeling accuracy is to calculate NMSE using (5.7). Thus, an appropriate order number with satisfactory performance can be obtained through NMSE, as shown in Fig. 5.4, where the targeted PA characteristic is the W-H model expressed in (5.12)–(5.14). It can be seen from Fig. 5.4 that the memory polynomial model with odd-order only has a similar NMSE value as one with even-odd order when the odd-order number is beyond 5, but a simple

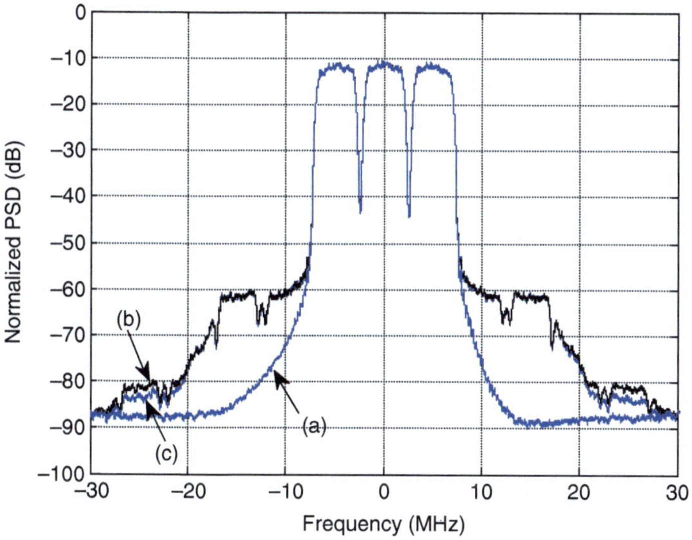

Fig. 5.3 Power spectral density of three-WCDMA signals at baseband: (**a**) input, (**b**) output of W-H system, and (**c**) output of memory polynomial model

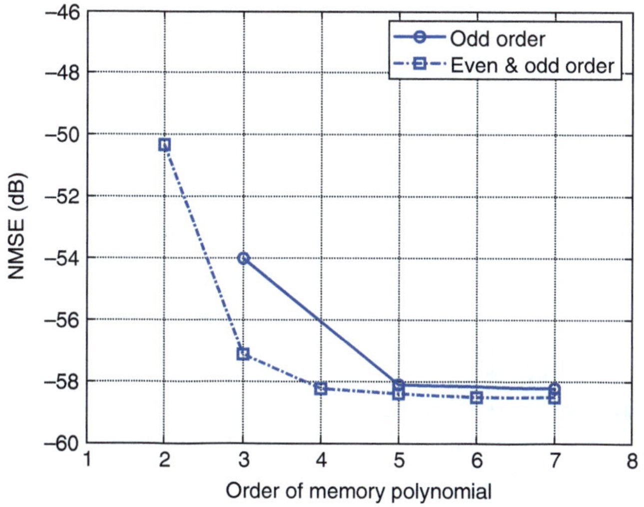

Fig. 5.4 NMSE versus order of memory polynomial model with maximum delay unit $M = 2$

expression. NMSE with order number 1 is not included here because it corresponds to a linear model rather than a nonlinear model.

In this section, the memory polynomial series expressed in (5.3) has been used to model the behavioral characteristics of a nonlinear PA based on the amount of the collected input and output data from the PA through the LS algorithm (5.11).

Actually, an example introduced above can be applied to modeling the behavioral characteristics of a practical PA. Specifically, attention should be first paid to aligning the collected output data of the PA with the collected input data of the PA accurately. Then the LS algorithm can be used to calculate the coefficients of the memory polynomial model. Try different number of even-odd order and odd-order, and different number of memory delay unit to find available numbers for fitting practical applications based on the calculated NMSE value. From a hardware-implementation point of view, simplicity and satisfactory performance is always preferred.

In the following section, we will introduce some procedures for extracting the coefficients of the Volterra polynomial by means of the LS algorithm based on the measured data from the input and output of a practical power amplifier.

5.3 Behavioral Modeling of a Practical Power Amplifier

In this section, we present the nonlinear characteristics of a practical power amplifier (PA) and coefficient extractions from the measured input and output data of the power amplifier. From these extracted coefficients, an accurately behavioral model with memory effects for this practical power amplifier is approximated using the Volterra polynomial model.

A test-bench setup used to collect data from the input and output of the PA is shown in Fig. 5.5, where the PA from Skyworks Solutions is referred to as a device under test (DUT). A MXG ES4438C Agilent signal generator (SG) is used to generate an 802.11a OFDM signal sent as a stimulating signal with a 20-MHz bandwidth to the PA, which is biased as a Class-AB model, is operated at an RF frequency of 5 GHz, and has a P1dB output compression point of 27 dBm with

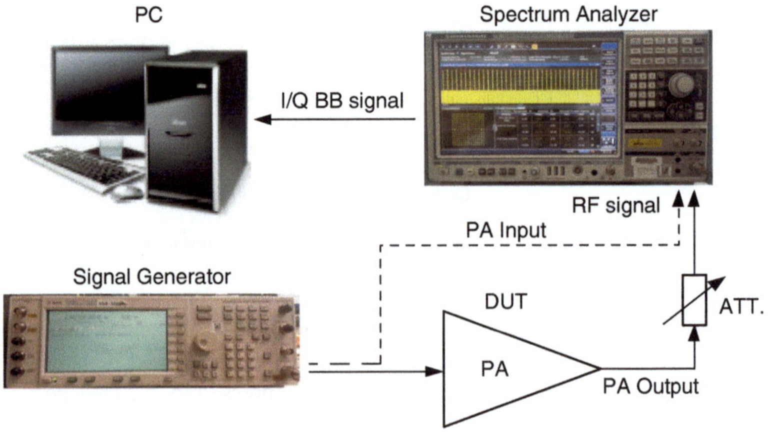

Fig. 5.5 Experimental setup for extracting the coefficients of the Volterra Polynomial

measurement. An R&S FSW spectrum analyzer (SA) is employed to down-convert the RF modulated signal to the analog baseband I–Q signals and then to recover the transmitted baseband I–Q signals in digital formats after correcting frequency offsets and then performing carrier phase, frame, and symbol timing synchronizations. A personal computer (PC) is used to import the digital baseband I–Q signals from the I–Q output interface of the SA. The PC needs to collect two groups of data: one group data is obtained when the output of the DUT is connected to the input of the SA, while one another group is acquired when the output of the SG is linked to the input of the SA. The former corresponds to the output signal data of the PA and the latter corresponds to the input signal data of the PA. With input and output data, coefficients of the Volterra polynomial that model an actual PA can be extracted by using the LS algorithm expressed in (5.11).

In data collection experiment, a frame-based 802.11a OFDM signal as complex I–Q data with a modulation format of 64-QAM and a bandwidth of 20 MHz is downloaded from an N7617B Signal Studio of KEYSIGHT Technologies to an MXG ES4438C Agilent Signal Generator, where each frame is identical and is continuously repeated. The frame-based OFDM baseband I–Q signals then modulate a pair of quadrature carrier signals to generate a RF-modulated signal at 5500 MHz or the channel 100, which has about a 9.5-dB peak-to-average power ratio with measurement. In order to extract the coefficients of the Volterra polynomial used to approximate to the PA behavioral modeling, we need to collect the input and output data of the PA through the I–Q output interface of the SA, both when the output of the SG is connected to the SA and then when the output of the PA is connected to the SA. The collected data signals need to be interpolated and aligned in MATLAB before the PA coefficient extraction. These procedures are described below:

1. ***Data Collection From PA Input***
   When the output of the SG is connected to the SA, it is treated as the input signal of the PA. To collect the complex I–Q data from the SA, the RF output signal of the SG is first down-converted to the baseband I–Q signals and then the digital baseband I–Q signals after carrier phase and symbol timing synchronizations are sent to the PC through the I–Q output interface of the SA at a sampling rate of 40 MHz, i.e., four times higher than the baseband bandwidth of about 8.5 MHz For the frame-based WLAN OFDM signals, one complete frame data should be collected at least.
2. ***Data Collection From PA Output***
   In order to characterize nonlinearity of the PA, the average output power of the PA is set to a 4-dB back-off from its P1dB compression point. As a rule of thumb, a PA it is preferred to operate back-off from the P1dB compression point by a PAPR value in order to avoid causing nonlinear distortions. Here, PA operation with a 4-dB back-off implies that the PA operates in near-saturated region. The output of the PA is then connected to the input of the SA through a 20 dB attenuator to prevent SA saturation. After down-conversion and carrier phase and symbol timing synchronizations are performed inside the SA, the digital baseband I–Q signals are collected by the PC through the I–Q output interface of the SA.

3. ***Data Interpolation and Alignment***

   As described above, data are collected from both the input and output of the PA at a sampling rate of 40 MHz, which corresponds to the interval of 25 ns between two consecutive samples. It has been found from actually modeling approximation that a delay unit of less than 5 ns can achieve a good approximation to the memory effects of the PA. If the interval of 2.5 ns between two consecutive samples is chosen, interpolated data samples, corresponding to a sampling rate of 400 MHz, can be obtained in MATLAB by interpolating the original sampled data at the sampling rate of 40 MHz with a factor of 10.

   After interpolation, the input and output data of the PA need to be aligned before the PA coefficient extraction. Alignment can be performed with one frame by searching for a cross-correlation peak between the interpolated input and output data. Due to pseudo-random characteristics of the frame data, the cross-correlation peak is unique and occurs significantly.

4. ***Coefficient Extraction***

   After alignment, the coefficients of the Volterra polynomial can be extracted by substituting the aligned input and output data into (5.11) with proper arrangements in (5.10). In practice, the coefficient extraction is performed in MATLAB.

   Around 7000 I–Q samples in one frame time duration were collected at a sampling rate of 40 MHz. After interpolation with a factor of 10, 70,000 interpolated I–Q samples were obtained and about 30,000 interpolated samples were used for coefficient extraction. A small number of samples less than 30,000 can be also used for coefficient extraction. In order to accurately model the nonlinear characteristics of the PA, the nonlinearity order K in the Volterra model (5.3) was set to 5 and the maximum memory unit M to 2. By substituting the interpolated input and output data into (5.11), coefficients of the Volterra polynomial can be extracted. Then, by substituting these coefficients and interpolated input data into (5.3), we can obtain the baseband I–Q data at the output of the Volterra model that approximates to the collected and interpolated baseband I–Q data at the output of the practical or actual PA. Figure 5.6 shows the AM-AM and AM-PM curves of both measured and extracted group data. It can be seen that data at the output of the memory polynomial model match data at the output of the actual PA well on both AM-AM and AM-PM curves.

   The modeling accuracy for this case can be evaluated with the NMSE value that is calculated in (5.7) and is equal to -42.77 dB. Table 5.1 lists some NMSE values with nonlinear order numbers up to 7 and with a memory delay unit $M = 2$. It shows that the NMSE value decreases very slowly when the nonlinearity order K is greater than or equal to 4. Hence, the Volterra polynomial with $K = 4$ and $M = 2$ is an accurate modeling approximation to the AM-AM and AM-PM characteristics of this actual PA.

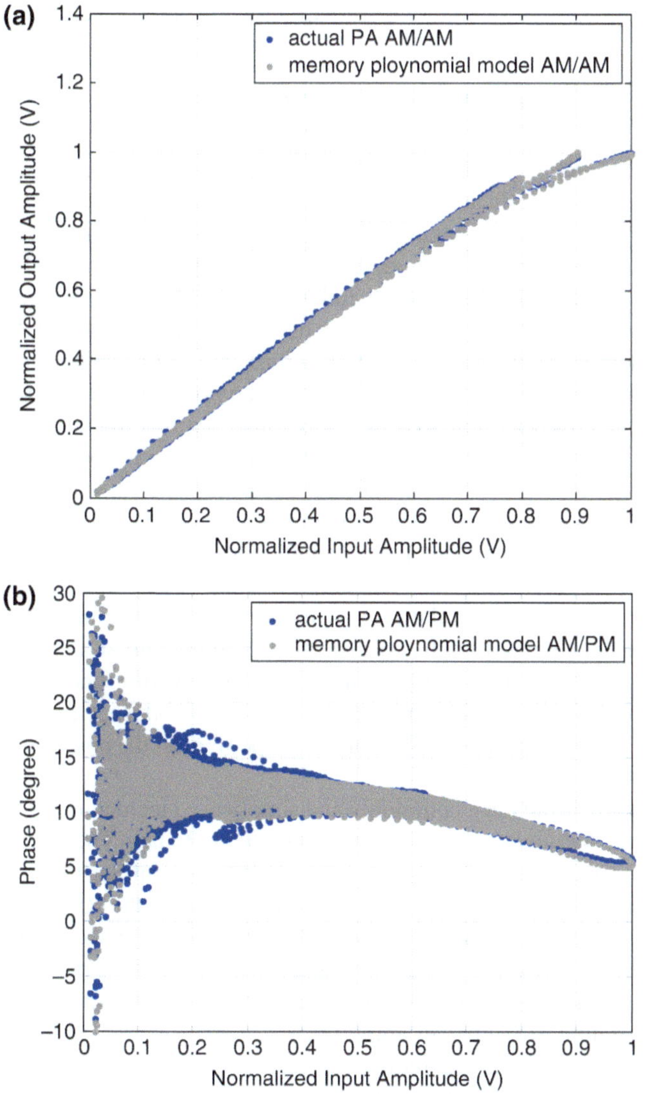

Fig. 5.6 Normalized AM-AM and AM-PM characteristics of a practical PA for 802.11a OFDM transmission and a Volterra polynomial modeling with fifth even-odd order $K = 5$ and maximum memory unit $M = 2$: (**a**) AM-AM characteristic and (**b**) AM-PM characteristic

Table 5.1 NMSE value of a PA model versus nonlinearity order K at maximum delay unit $M = 2$

Nonlinearity order K	Maximum delay unit M	NMSE (dB)
2	2	−33.65
3	2	−39.12
4	2	−42.43
5	2	−42.77
6	2	−43.22
7	2	−43.27

5.4 Power Amplifier Linearization

A power amplifier is the last active stage of the transmitter before a RF signal is transmitted. Most specifications for the transmission part are defined at the output of the power amplifier. In general, these specifications for the transmitter are more stringent than those for the receiver because any violations would affect the receptions of other users, such as spectral emission, out-of-band emission, and adjacent channel power ratio (ACPR). Considering that nonlinear distortion of the transmitter is usually dominated by a PA, all specifications for the transmitter are usually met with some amount of margin before the power amplifier and after a RF modulator. Therefore, the power amplifier plays a very important role in the transmitter.

When transmitting constant-envelope–modulated signals, power amplifiers can achieve energy efficiency without causing spectrum regrowth when operating at the saturation region or close to the saturation region. This is because both the AM-AM and AM-PM characteristics of power amplifiers do not distort the amplified RF signals due to their constant envelope features. For the transmissions of non-constant envelope modulation signals, on the other hand, one way for the power amplifier to avoid nonlinear distortion is to operate at a certain amount of back-off from its P1 dB compression point, which in turn decreases the energy efficiency of the power amplifier. In order to achieve both spectrum efficiency and energy efficiency for such non-constant envelope signals, the power amplifier has to be linearized.

Predistortion is the most commonly and simply used technique for linearizing an amplifier and has existed for many decades until today. The basic concept of pre-distortion is to insert a nonlinear pre-distortion module or block before either the baseband modulation signal at the digital domain or the RF modulated signal prior to the power amplifier at the analog domain. The nonlinear module generates intermodulation distortion (IMD) products that are in anti-phase with the IMD products generated by the power amplifier, hence, reducing out-of-channel emissions at the output of the power amplifier.

In general, predistorter techniques can be classified as three kinds: the baseband predistorter, the cuber (or third-order) predistorter, and the RF envelope predistorter. The first kind of predistorter generates the predistorted baseband I–Q signals in the digital baseband domain. In the second technique, the RF input signal is split into a delay path and an error generation path. A pair of diodes in the error generation path of the nonlinear module keeps the distorted signals after canceling the fundamental signal. Then, the distorted signals with the controllable magnitudes and phases are recombined with the original RF signal via the delay path at the end of the predistorter. Finally, the spurious signals are cancelled after the combiner [4]. The third predistortion technique creates two second-order nonlinear functions that interpolate the inverse AM-AM and AM-PM nonlinearities of the power amplifier by using the envelope of the modulating signal [14]. These two second-order nonlinear functions are used to generate the pre-distorted RF signal prior to the power amplifier. Furthermore, its coefficients can be adaptively updated by using some forms of adaptive algorithms since the power amplifier's characteristics tend to drift with time, such as changes in temperature, supply voltage variations, aging of devices, and switching among bands or channels.

The nonlinearity of the PA primarily results in degradations of the adjacent channel power ratio (ACPR) and error vector magnitude (EVM) at the transmitter. There are several types of linearization techniques to minimize the nonlinear effects on ACPR and EVM in literature, such as feedback linearization, feed-forward linearization and predistortion linearization. Due to its low cost and simple hardware implementation, the predistortion technique is introduced in this chapter.

5.4.1 Digital Baseband Pre-distortion

A general way to compensate for the nonlinear distortions of a PA by means of a pre-distortion technique is to insert a pre-distortion block with a transfer function of F in the signal path prior to the PA's having a transfer function of G, as shown in Fig. 5.7. A linear amplification Q can be achieved when these two blocks are cascaded together if F is the inverse of G or $F = G^{-1}$ under the assumption that G is invertible. It can be also be expressed as

$$\begin{aligned} y(n) &= g\left(|z(n)|^2\right)z(n) \\ &= g\left(\left|f\left(|x(n)|^2\right)x(n)\right|^2\right) \times f\left(|x(n)|^2\right)x(n) \end{aligned} \tag{5.18}$$

where f and g stand for gains of the pre-distorter and power amplifier in time domain, respectively. The PA is linearized when the cascaded gain q meets the following condition:

$$q = g\left(\left|f\left(|x(n)|^2\right)x(n)\right|^2\right) \times f\left(|x(n)|^2\right) \approx \text{constant} \tag{5.19}$$

Since it is very complicated to directly solve an inverse function of a PA with memory effect, the parameters of a digital pre-distorter (DPD) are iteratively estimated through either a closed-loop adaptation mechanism or an indirect learning structure for DPD identification. Either one avoids obtaining the transfer function of the PA first, and then solves its inverse function for the DPD.

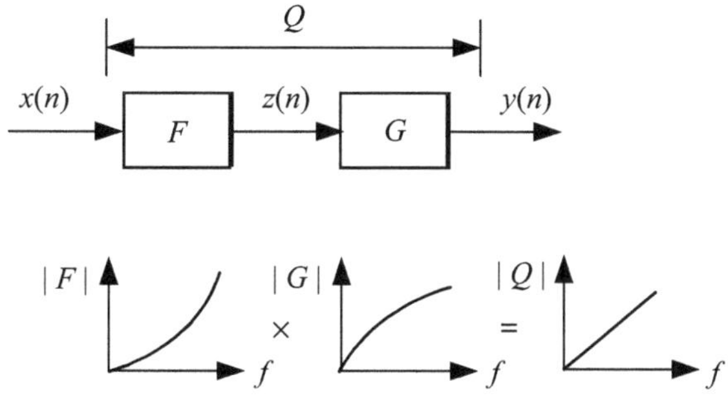

Fig. 5.7 Block diagram of a pre-distortion system

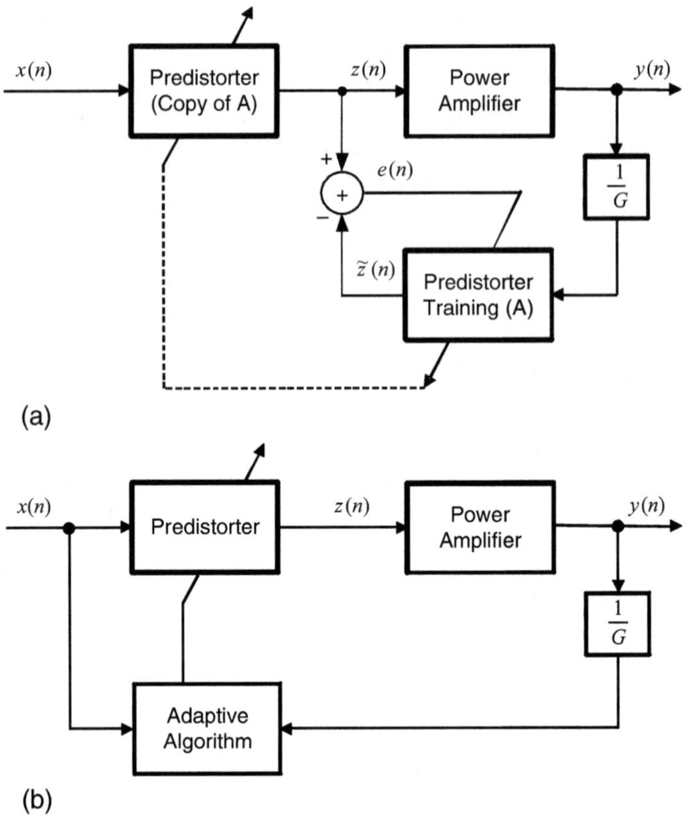

Fig. 5.8 Block diagram of equivalent baseband for a PA with a pre-distorter: (**a**) indirect learning structure, (**b**) direct learning structure

Indirect learning [10, 15, 16] and direct learning [17, 18] structures for pre-distortion identification are illustrated in Fig. 5.8. The indirect learning method first identifies the coefficients of the pre-distorter training block on the feedback path by minimizing the error signal of $e(n)$ in Fig. 5.8a, and then copies these coefficients to the pre-distorter on the feed-forward path. On the contrary, the direct learning strategy in Fig. 5.8b directly adjusts the coefficients of the pre-distorter in the feed-forward path based on the input of the pre-distorter and the output of the power amplifier. Both architectures have similar compensation principles, but they also have different features. One advantage of the indirect learning structure is that the system operation is very stable and its performance is relatively good because it is an open-loop system and the coefficients are only updated once they converge. The updating rate can be very slow depending on the variations of the power amplifier, such as the changes of temperature and supply voltage. Hence, the indirect learning configuration shall be introduced in this section, while the direct learning structure shall be presented in the application section of this chapter.

In Fig. 5.8a, a block of pre-distorter training (A) on the feedback path has the input of $y(n)/G$, where G is the linear gain of the PA, and the output of $\tilde{z}(n)$. The actual pre-distorter prior to the PA is an exact copy of the pre-distorter training block (A) on the feedback path; its input is $x(n)$ and its output is $z(n)$. In an ideal case, the output of the PA would be equal to $y(n) = Gx(n)$, corresponding to $z(n) = \tilde{z}(n)$, which leads to the error signal $e(n) = 0$.

A basic procedure for obtaining the parameters of pre-distorter training A is to collect a number of data samples of the input $y(n)/G$ of the predistorter training block and the input z(n) of the PA that is used to approximate the output $\tilde{z}(n)$ of the predistorter training block after convergence; then the parameters can be solved through the least-squared solution offline, which can be used as the initial parameters or coefficients of the pre-distorter. The system can run in open loop—i.e., the pre-distorter training block is shut down—if the PA characteristics do not change rapidly with time due to power supply voltage variation, temperature drift, aging, and other factors, which have long time constants. If the PA characteristics change too much with time, an adaptive algorithm may be used to update the coefficients of pre-distorter when the feedback loop is closed.

Similar to the memory polynomial expression of the PA model in (5.3), the memory polynomial of the pre-distorter can be also expressed as

$$z(n) = \sum_{k=1}^{K}\sum_{m=0}^{M} a_{km}x(n-m)|x(n-m)|^{k-1} \tag{5.20}$$

where M is the largest delay unit representing the memory effect, and K is the highest nonlinear order. In most practical pre-distortion designs, the compensation for the nonlinearity of the PA is well approximated by using odd-order nonlinearities only. Then, the expression (5.20) can be written as

$$z(n) = \sum_{k=2l+1}^{K}\sum_{m=0}^{M} a_{km}x(n-m)|x(n-m)|^{k-1}, l = 0, 1, 2, \ldots \tag{5.21}$$

where K is an odd number. Actually, if even-order nonlinearity items in (5.20) are included, the spectral regrowth can be further reduced slightly. Therefore, from the hardware simplicity point of view, it is very worthwhile to use odd-order nonlinearity items for the pre-distorter.

From the collected input and output data of the PA offline, a transfer function of the pre-distorter on the feedback path can be estimated through a LS algorithm. The output signal $y(n)$ of the PA is first attenuated by a linear gain G of the PA to avoid excessive input to the pre-distorter training block. Defining the input to the pre-distorter training as a new signal [10] yields

$$u_{kq}(n) = \frac{y(n-q)}{G}\left|\frac{y(n-q)}{G}\right|^{k-1} \tag{5.22}$$

At convergence, $\tilde{z}(n) \approx z(n)$ and the output of the pre-distorter training block can be expressed by

$$\mathbf{z} = \mathbf{U}\mathbf{a} \tag{5.23}$$

where $\mathbf{z} = \left[z(0), \ldots, z(N-1)^{\mathrm{T}}\right], \mathbf{U} = \left[\mathbf{u}_{10}, \ldots, \mathbf{u}_{K0}, \ldots, \mathbf{u}_{1Q}, \ldots, \mathbf{u}_{KQ}\right], \mathbf{u}_{kq} = \left[u_{kq}(0), \ldots, u_{kq}(N-1)\right]^{\mathrm{T}}$, and $\mathbf{a} = \left[a_{10}, \ldots, a_{K0}, \ldots a_{1Q}, \ldots, a_{KQ}\right]^{\mathrm{T}}$. Solving the LS solution for (5.23), the coefficients of the pre-distorter training block are

$$\mathbf{a} = \left(\mathbf{U}^{\mathbf{H}}U\right)^{-1}U^{\mathrm{H}}z \tag{5.24}$$

where $(\bullet)^{\mathrm{H}}$ stands for complex conjugate transpose.

The coefficients of the pre-distorter obtained through (5.24) offline are copied to the pre-distorter on the feed-forward path as the initial coefficients when the system is powered on. When needed because of variations of power supply voltage and temperature, the coefficients can be updated by the adaptive algorithm [9].

Similar to the NMSE used for evaluating PA modeling accuracy defined in (5.7), the compensation accuracy for the PD is usually evaluated using the NMSE defined as

$$\mathrm{NMSE_{PD}}\,[\mathrm{dB}] = 10\,\mathbf{log}_{10}\left(\frac{\sum_{n=1}^{N}\left|z(n) - \tilde{z}(n)\right|^2}{\sum_{n=1}^{N}\left|z(n)\right|^2}\right) \tag{5.25}$$

where $z(n)$ is the input data to the PA and $\tilde{z}(n)$ is the output data from the pre-distorter shown in Fig. 5.8a. If $\tilde{z}(n)$ is close to $z(n)$, then $\mathrm{NMSE_{PD}}$ becomes very small, which implies that the PD accurately compensates for the nonlinear distortions of the PA.

The digital pre-distorters (DPDs) are implemented in the digital baseband domain before the digital to analog converter (DAC). To see how the DPD compensates for the nonlinear distortions, we assume the behavioral model of the targeted PA is the same as one expressed in (5.16), which is extracted from the W-H model given in (5.12)–(5.15). Figure 5.9 shows the performance of the pre-distorter with a nonlinear order $K = 5$ and delay unit $M = 2$ of the memory polynomial series. It can be clearly seen that the memory polynomial PD is able to suppress most of the spectral regrowth as indicated by the curves (b) and (c), where the PD with odd-order nonlinearity terms has only about 3-dB degradation compared with the one with even-odd order nonlinearity terms, but it has fewer coefficients. From a simplicity point of view, the lower-cost implementation would be preferred.

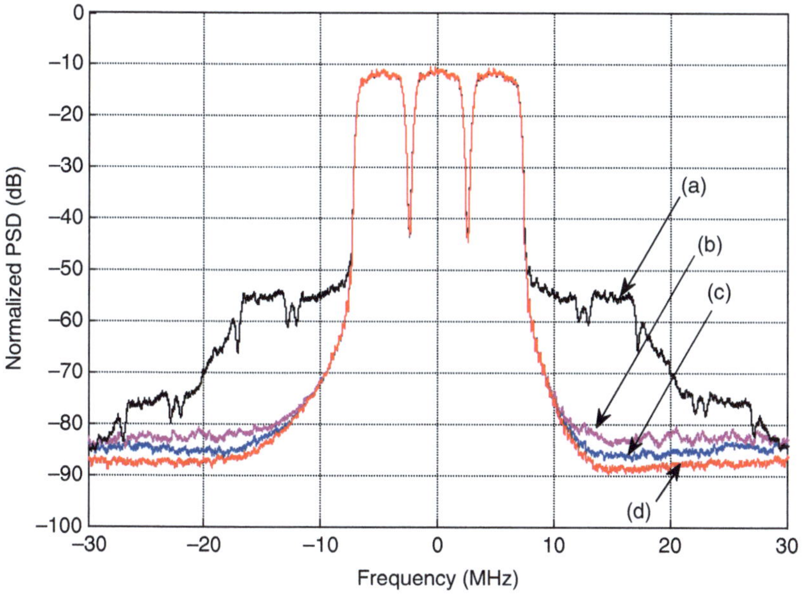

Fig. 5.9 Power spectral density of a three-carrier WCDMA signal at baseband: (**a**) PA output without PD, (**b**) PA output with fifth odd order PD, (**c**) PA output with fifth even-odd order PD, and (**d**) PA input

Table 5.2 Compensation NMSE versus maximum memory delay M at even-odd order $K = 5$

Memory effect	Maximum delay unit M	NMSE (dB)
Memoryless	0	−29.6
Memory	1	−50.1
Memory	2	−50.7

Table 5.2 illustrates $NMSE_{PD}$ results versus different the maximum delay number M. $NMSE_{PD}$ is not much improved after M is equal to or greater than 1. This tendency is similar to that shown in Fig. 5.10. The differences in spectral regrowth at the upper and lower adjacent channels between $M = 0$ corresponding to the curve (b) and $M = 2$ corresponding to the curve (d) are about 10 dB because the memory effect of the PA model used here is not severe. Otherwise, the memoryless DPD would be incapable of suppressing the spectral regrowth, especially for wideband PA applications with significant memory effects, such as LTE and 802.11 Wi-Fi standards.

For $K = 5$ and $M = 2$, the memory polynomial model of DPD used in Fig. 5.10 is expressed as

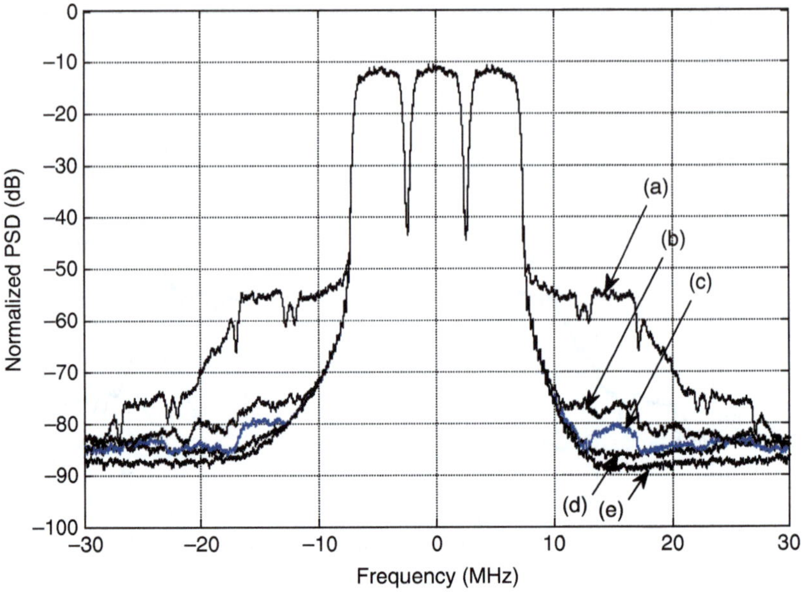

Fig. 5.10 Power spectral density of a three-carrier WCDMA signal at baseband: (**a**) PA output without PD, (**b**) PA output with PD having $M=0$, (**c**) PA output with PD having $M=1$, (**d**) PA output with PD having $M=2$, and (**e**) PA input. All PD models have even-odd order $K=5$

$$
\begin{aligned}
z(n) &= \sum_{k=1}^{5}\sum_{m=0}^{2} a_{km}x(n-m)\left|x(n-m)\right|^{k-1} \\
&= a_{10}x(n) + a_{11}x(n-1) + a_{12}x(n-2) \\
&\quad + a_{20}x(n)\left|x(n)\right| + a_{21}x(n-1)\left|x(n-1)\right| + a_{22}x(n-2)\left|x(n-2)\right| \\
&\quad + a_{30}x(n)\left|x(n)\right|^2 + a_{31}x(n-1)\left|x(n-1)\right|^2 + a_{32}x(n-2)\left|x(n-2)\right|^2 \\
&\quad + a_{40}x(n)\left|x(n)\right|^3 + a_{41}x(n-1)\left|x(n-1)\right|^3 + a_{42}x(n-2)\left|x(n-2)\right|^3 \\
&\quad + a_{50}x(n)\left|x(n)\right|^4 + a_{51}x(n-1)\left|x(n-1)\right|^4 + a_{52}x(n-2)\left|x(n-2)\right|^4
\end{aligned}
\tag{5.26}
$$

where the coefficients are

$$
\begin{aligned}
&a_{10} = 1.8517 + j0.0976, a_{20} = 15.0286 + j2.5592, a_{30} = -225.4617 - j28.4730, \\
&a_{40} = 1090.1275 + j130.1822, a_{50} = -1719.0851 - j254.8413 \\
&a_{11} = -0.4903 + j0.0874, a_{21} = -29.9729 - j5.1971, a_{31} = 452.3464 + j53.8019, \\
&a_{41} = -2198.3903 - j239.4420, a_{51} = 3516.1881 + j424.4039 \\
&a_{12} = -0.3743 - j0.1007, a_{22} = 15.1299 + j2.6554, a_{32} = -229.2100 - j26.4797, \\
&a_{42} = 1116.9208 + j115.4541, a_{52} = -1796.8289 - j193.8795
\end{aligned}
\tag{5.27}
$$

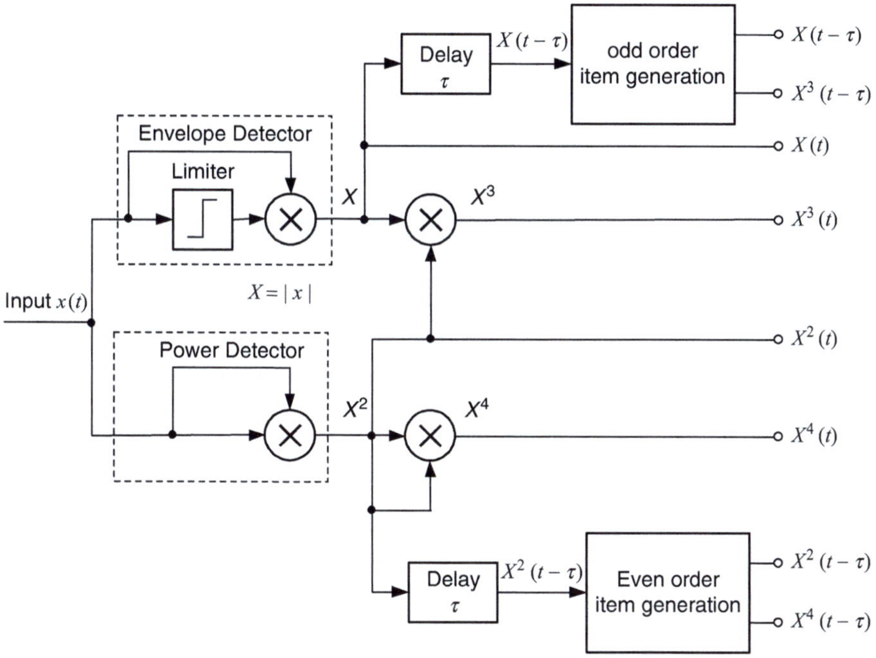

Fig. 5.11 Block diagram of high-order nonlinear term generation

In (5.26), the output $z(n)$ of the pre-distorter consists of even-odd order nonlinear items with both memory and memoryless polynomial series. These nonlinear items can be implemented with the structure as shown in Fig. 5.11. This nonlinear term generation can be used either for the digital pre-distortion (DPD) or for the RF analog pre-distortion (APD). In the latter case, the baseband signal $x(t)$ is recovered from the RF input signal to the PA after down-conversion. Even- and odd-order item generators following the delay blocks generate the delayed even- and odd-order components, respectively. These delay components are used to compensate for the memory effects of the power amplifier.

To validate the effective compensation for the memory effects of the PA, Fig. 5.12 illustrates the AM-AM and AM-PM plots of the PA without DPD and with DPD, where the PA obeys the W-H model given in (5.16)–(5.17) and the coefficients of the PD are expressed in (5.27). It can be clearly seen that both nonlinear characteristics and memory effects of the PA have been significantly compensated. For a comparison between the memory DPD and memoryless DPD, the AM-AM and AM-PM plots of the same PA with and without memoryless DPD are shown in Fig. 5.13, where memory effects still remain although the gain and phase nonlinear features have been mostly compensated.

Design Example 5.2 As we have introduced four procedures regarding to the coefficient extraction of the PA behavioral modeling from a practical PA measurement in Sect. 5.3, now we need to design a pre-distorter to compensate for the PA nonlinear distortions by using the nonlinear order $K = 5$ and the maximum delay unit $M = 2$ for the pre-distorter.

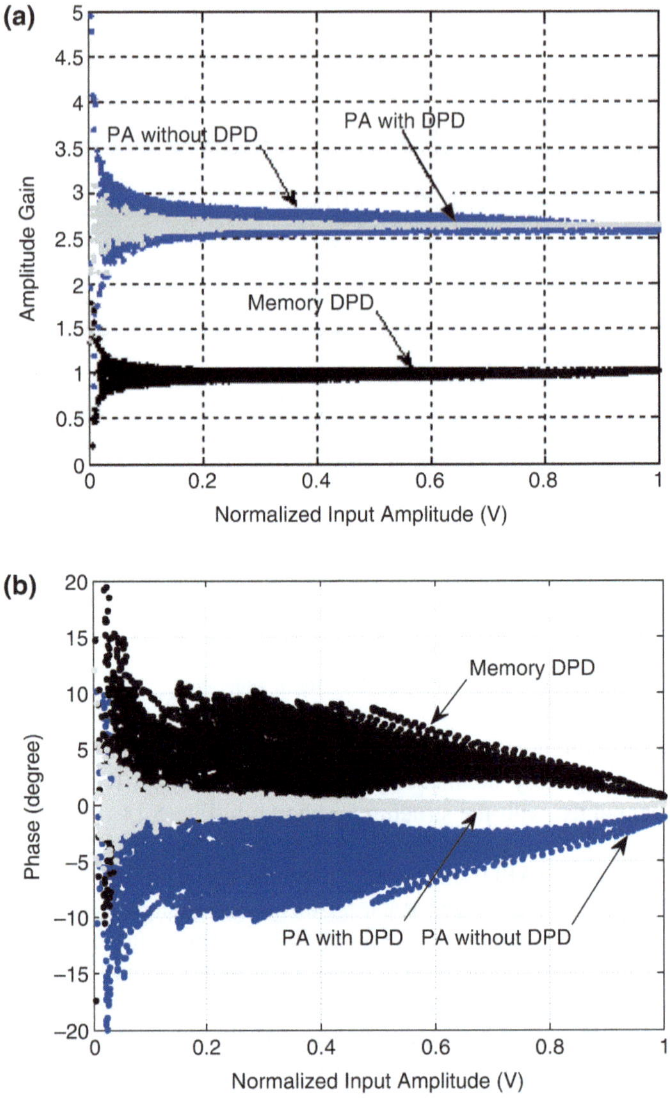

Fig. 5.12 AM-AM and AM-PM characteristics for nonlinear PA with memory effects after memory DPD compensation: (**a**) AM-AM characteristic and (**b**) AM-PM characteristic. Note: Here AM-AM is actually for AM-to-AM gain

Solution In Sect. 5.3, we extracted the coefficients of the Volterra polynomial model based on the measured data collected at the input and output of the practical PA, respectively. Thus, we have the Volterra polynomial model being used for designing a pre-distorter to compensate for the nonlinearities of this practical PA. To extract the coefficients of the pre-distorter as expressed in (5.24), we also need to collect data at the input and output of the Volterra polynomial model,

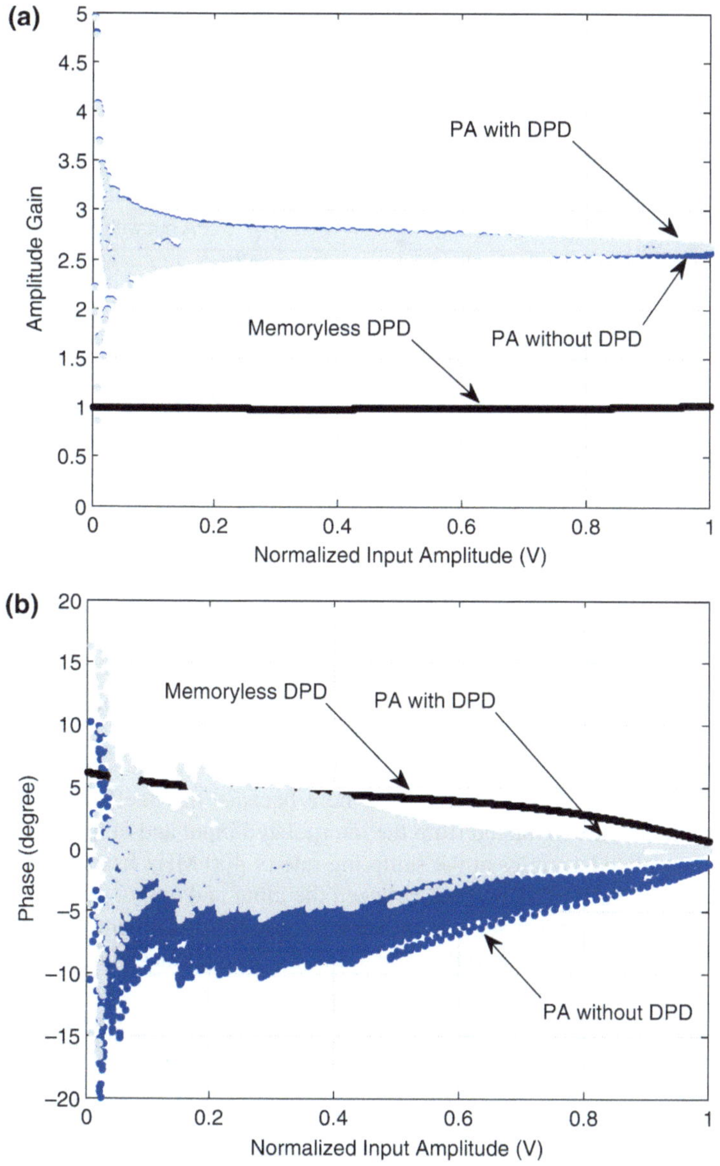

Fig. 5.13 AM-AM and AM-PM characteristics for nonlinear PA with memory effects after memoryless DPD compensation: (**a**) AM-AM characteristic, and (**b**) AM-PM characteristic. Note: Here AM-AM is actually for voltage gain

respectively. The Volterra polynomial model used above with $K = 5$ and $M = 2$ is chosen as a predistorter.

Similar to the behavioral modeling of the PA—except there is no need for interpolation here—there are also four major procedures for the coefficient

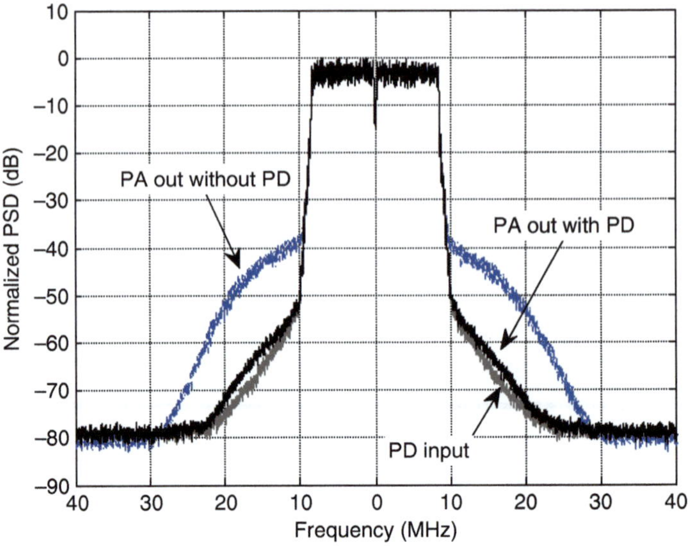

Fig. 5.14 PSD plots of an 802.11a OFDM signal at the input of the DPD, and the output of the practical PA with memory DPD and without memory DPD

extraction of the PD: *data collection from the input of the PA behavioral model, data collection from the output of the PA behavioral model, data alignment, and coefficient extraction.*

Note that interpolation is not needed here because the coefficients of the PA behavioral model are extracted from the interpolated input and output data. Around 30,000 interpolated samples at the sampling rate of 400 MHz from one frame data are collected in MATLAB, aligned between the input and output samples, and then are substituted into (5.24) for coefficient extraction. From a simplicity-of-design point of view, the nonlinear order $K = 5$ and the maximum delay unit $M = 2$ are chosen for the PD polynomial expression in (5.20).

Figure 5.14 shows the PSD plots of an 802.11a OFDM signal with a data rate of 54 Mbps that passes through a practical PA under different conditions. It is clear that the spectrum regrowth at the output of the PA without PD can be successfully compensated by at least a 15-dB reduction within the frequency offset range, either 10 to 20 MHz or -10 to -20 MHz, by inserting a pre-distorter prior to the PA. This can also be seen from Fig. 5.14, where the PSD curve of the PA with PD is very close to the PSD curve of the PD input, which indicates that the PD significantly compensates for the spectrum regrowth due to nonlinear amplification of the PA.

5.4.2 RF Analog Pre-distortion

Compared with DPD, the analog pre-distorter (APD) has advantages of simple structure, low cost, and flexible standalone operation; nor it does not affect the architecture of existing system since it is simply inserted between the RF modulator and the RF PA without direct access to a digital baseband processor. Analog pre-distorters that consist of a vector modulator were introduced to cancel the distorted harmonic components at the output of the pre-distorter by controlling the attenuator and phase shifter of the vector modulator [4, 19–21]. The APD in the system level approach is suitable for base station applications or repeater applications, where a small amount of extra power consumption due to the APD is negligible compared with the base station in these applications. A CMOS PA with a built-in RF analog pre-distorter in the circuit level approach was proposed for handset applications [5], in which the driver stage of the PA adopts a vector modulator with gain and phase of a pre-distorter controlled via a two-word look-up table (LUT) with multiple AM-AM and AM-PM curves. Another APD for the base station applications was presented to correct both nonlinearity and memory effects of the PA by using both input signal envelope and its derivative [22]. The RF input signal is pre-distorted using the vector modulator controlled with the correcting coefficients. As opposed to those the mentioned above, the RF analog pre-distorter uses the envelope of the RF-modulated signal to generate two second-order nonlinear functions that interpolate the inverse AM-AM and AM-PM nonlinearities of the power amplifier [14, 23]. Due to its features of flexible insertion and complete standalone capability between the RF modulator and power amplifier, this type of RF analog PD (APD), representing a general RF APD, will be presented in the following sections.

A simplified block diagram of the adaptive APD is shown in Fig. 5.15. The adaptive algorithm is used to update the coefficients of F_1 and F_2 for tracking

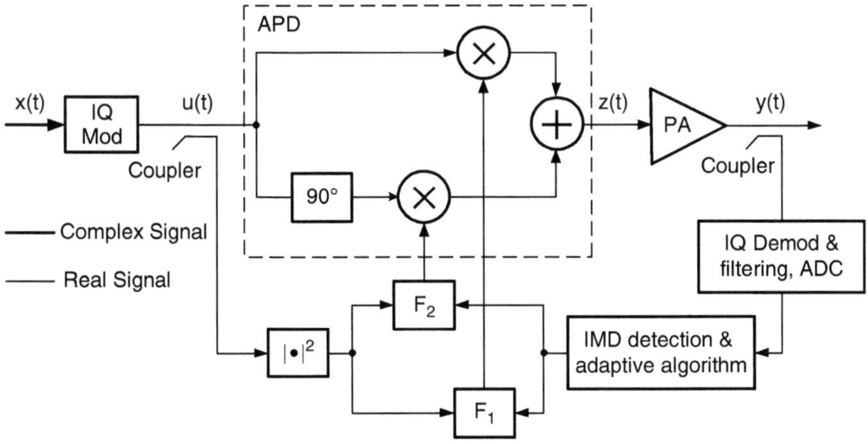

Fig. 5.15 Block diagram of an analog pre-distorter. Referenced from [14]

possible drifts of the power amplifier's characteristics with time. These drifts may be caused by temperature changes, supply voltage variations, and aging of devices.

The complex baseband signal $x(t)$ is expressed with the in-phase (I) baseband signal $x_i(t)$ and quadrature (Q) baseband signal $x_q(t)$ as

$$x(t) = x_i(t) + jx_q(t) \tag{5.28}$$

The I–Q baseband signals modulate a pair of quadrature carriers with the frequencies of $\omega_c = 2\pi f_c$, respectively, and then they are summed to form a real RF signal $u(t)$, which is expressed as

$$u(t) = x_i(t)\mathbf{cos}\,(\omega_c t) - x_q(t)\mathbf{sin}\,(\omega_c t) = \mathcal{R}e\{x(t)e^{j\omega_c t}\} \tag{5.29}$$

where $x(t)$ is referred to as the complex envelope. The RF signal is split into the I–Q paths of the pre-distorter that is made up of a complex phasor modulator [14], where a pair of the quadrature RF signals is multiplied by two nonlinear second functions F_1 and F_2 to generate the pre-distorted signals, respectively. F_1 and F_2, which interpolate the inverse AM-AM and AM-PM nonlinearities of the power amplifier, are functions of the complex envelope of the modulation signal and are represented by

$$\begin{aligned} F_1\left[|x(t)|^2\right] &= \alpha_{11} + \alpha_{13}|x(t)|^2 + \alpha_{15}|x(t)|^4 \\ F_2\left[|x(t)|^2\right] &= \alpha_{21} + \alpha_{23}|x(t)|^2 + \alpha_{25}|x(t)|^4 \end{aligned} \tag{5.30}$$

where $|x(t)|$ is the envelope of the complex baseband signal. Therefore, F_1 and F_2 are amplitude dependent of the baseband signal. With the F_1 and F_2 expressions, the complex gain $F[|x(t)|^2]$ of the pre-distorter can be expressed as

$$F\left[|x(t)|^2\right] = F_1\left[|x(t)|^2\right] + j \times F_2\left[|x(t)|^2\right] = \alpha_1 + \alpha_3|x(t)|^2 + \alpha_5|x(t)|^4 \tag{5.31}$$

where $\alpha_i = \alpha_{1i} + j\alpha_{2i}$ are complex coefficients that model the inverse AM-AM and AM-PM characteristics of the power amplifier using the 3rd order and 5th order IMD products. The reason that the third- and fifth-order IMD products are only considered here is that these IMD products are the critical unwanted frequency IMD products with relatively high amplitudes compared with any other higher odd-order IMD products. The even-order IMD products are not considered here because they are outside the frequency band of interest. When two continuous wave (CW) tones with frequencies of f_1 and f_2 are combined and input to a device under test (DUT), the frequencies of the second-, third-, and fifth-order IMD products at the DUT output are shown in Fig. 5.16, which appear at

2nd order IMD frequencies: $f_2 - f_1$ and $f_1 + f_2$
3rd order IMD frequencies: $2 \times f_1 - f_2$ and $2 \times f_2 - f_1$
5th order IMD frequencies: $3 \times f_1 - 2 \times f_2$ and $3 \times f_2 - 2 \times f_1$

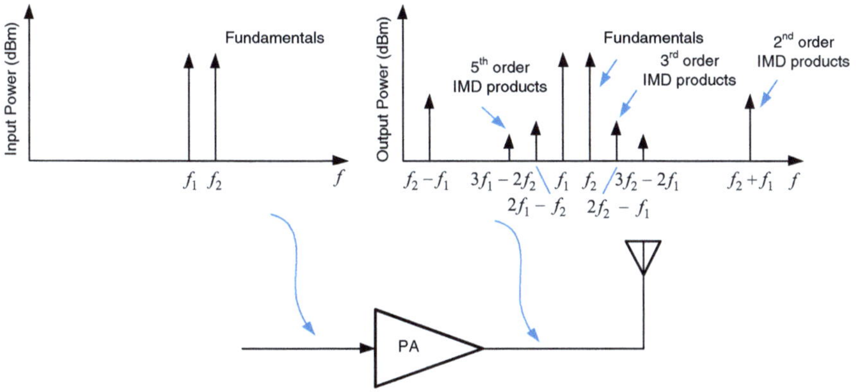

Fig. 5.16 Second-, third- and fifth-order IMD products at the output of a PA

Actually, (5.31) is the same as the gain of the DPD in (5.21) with $K = 5$ (odd-order only) and $M = 0$, which is expressed as

$$G_{\text{DPD}}\left(|x(n)|^2\right) = \frac{z(n)}{x(n)} = a_{10} + a_{30}|x(n)|^2 + \alpha_{50}|x(n)|^4 \tag{5.32}$$

The model of the APD used in [14, 23] is memoryless polynomial expression, but it can be modified to the memory polynomial model by adding the delay unit. The difference between DPD and APD is a location where the pre-distortion is performed. In the DPD, the pre-distortion is carried out in the digital domain before the DAC, while in the APD the pre-distortion is performed in the analog domain after the DAC and before the PA.

With the complex gain $F[|x(t)|^2]$ of the APD, the output of the APD can be expressed as

$$z(t) = \Re e\{\tilde{z}(t)e^{j\omega_c t}\} = \Re e\left\{x(t)F\left[|x(t)|^2\right]e^{j\omega_c t}\right\} \tag{5.33}$$

where $\tilde{z}(t)$ is the complex envelope of the real signal of $z(t)$. Hence, its equivalent complex baseband signal in (5.33) is expressed as

$$\tilde{z}(t) = x(t) \times F\left[|x(t)|^2\right] \tag{5.34}$$

With the substitution of (5.31) into (5.34), (5.34) becomes

$$\tilde{z}(t) = \alpha_1 x(t) + \alpha_3 x(t)|x(t)|^2 + \alpha_5 x(t)|x(t)|^4 \tag{5.35}$$

It can be clearly seen from (5.21) and (5.35) that (5.35) is a special case of (5.21) when the maximum odd-order $K = 5$ and maximum memory $M = 0$.

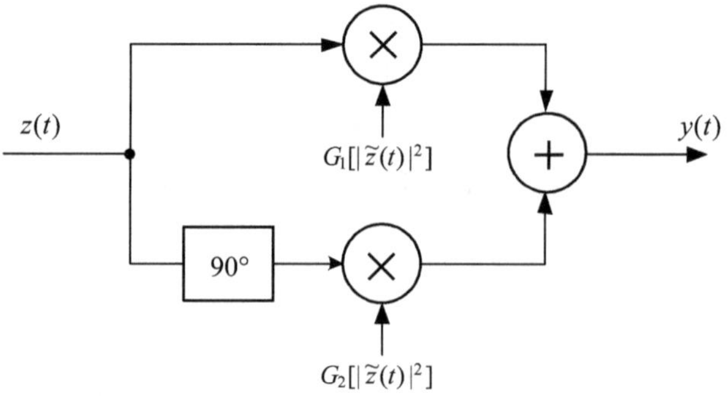

Fig. 5.17 Quadrature structure of a power amplifier. Redrawn from [14]

Considering the memory effects, (5.35) can be expressed with odd-order terms only: as

$$\tilde{z}(t) = \sum_{\substack{k=1 \\ k\,\text{odd}}}^{K} \sum_{m=0}^{M} \alpha_{km} x(t-m)|x(t-m)|^{k-1} \tag{5.36}$$

As described in the previous section, the AM-AM and AM-PM of the power amplifier can be modeled using the memory polynomial series. The quadrature model of a power amplifier can be described in a real bandpass form as illustrated in Fig. 5.17 [14], where the quadrature signal is realized by creating a 90° phase shift for the RF real signal on the quadrature path after the splitter. The complex gain is expressed as

$$G\left[|\tilde{z}(t)|^2\right] = G_1\left[|\tilde{z}(t)|^2\right] + jG_2\left[|\tilde{z}(t)|^2\right] = \beta_1 + \beta_3|\tilde{z}(t)|^2 + \beta_5|\tilde{z}(t)|^4 \tag{5.37}$$

The real and imagined parts of the complex gain in (5.37) are given by

$$\begin{aligned} G_1\left[|\tilde{z}(t)|^2\right] &= \beta_{11} + \beta_{13}|\tilde{z}(t)|^2 + \beta_{15}|\tilde{z}(t)|^4 \\ G_2\left[|\tilde{z}(t)|^2\right] &= \beta_{21} + \beta_{23}|\tilde{z}(t)|^2 + \beta_{25}|\tilde{z}(t)|^4 \end{aligned} \tag{5.38}$$

where the complex coefficients of $\beta_i = \beta_{1i} + j\beta_{2i}$ model the AM-AM and AM-PM characteristics of the power amplifier and can be extracted from the measured input and output data of the power amplifier.

With the complex gain expressions of the APD and PA, the real bandpass signal at the output of the PA becomes

$$y(t) = \Re e\{\tilde{y}(t)e^{j\omega_c t}\} = \Re e\left\{x(t)F\left[|x(t)|^2\right]G\left[|\tilde{z}(t)|^2\right]e^{j\omega_c t}\right\} \tag{5.39}$$

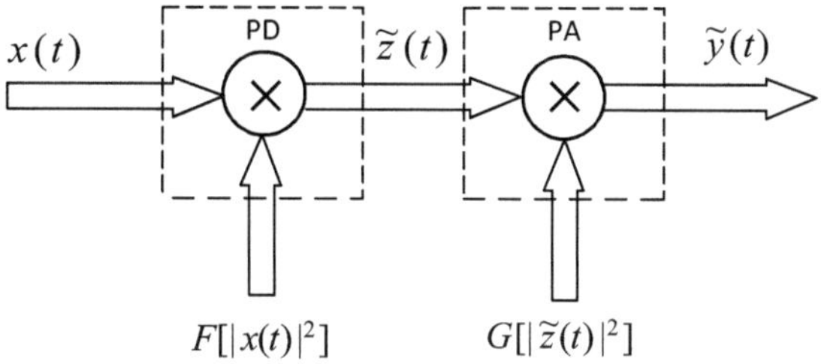

Fig. 5.18 Cascaded complex gain of pre-distorter and power amplifier. Redrawn from [14]

Its equivalent complex baseband signal in (5.39) is

$$\tilde{y}(t) = x(t)F\left[|x(t)|^2\right]G\left[|\tilde{z}(t)|^2\right] \tag{5.40}$$

It is shown in (5.40) that the complex baseband output of the power amplifier is a function of the complex baseband input, complex gain of the PD, and complex gain of the PA, where the complex gains depend on the complex envelopes of the baseband inputs of the PD and PA. A cascaded gain with a complex baseband gain form given in (5.40) is illustrated in Fig. 5.18.

Using envelope notation, the output of the power amplifier is

$$\begin{aligned}\tilde{y}(t) &= x(t)F\left[|x(t)|^2\right]G\left[|\tilde{z}(t)|^2\right] \\ &= x(t)K\left[|x(t)|^2\right]\end{aligned} \tag{5.41}$$

where the composite complex gain $K\left[|x(t)|^2\right]$ is defined as

$$\begin{aligned}K\left[|x(t)|^2\right] &= F\left[|x(t)|^2\right]G\left[|\tilde{z}(t)|^2\right] \\ &= F\left[|x(t)|^2\right]G\left[|x(t)|^2 \times \left|F\left[|x(t)|^2\right]\right|^2\right]\end{aligned} \tag{5.42}$$

Furthermore, the composite complex gain can be approximated by a power series of the form with odd-order up to fifth order

$$K\left[|x(t)|^2\right] = K_1\left[|x(t)|^2\right] + jK_2\left[|x(t)|^2\right] = \gamma_1 + \gamma_3|x(t)|^2 + \gamma_5|x(t)|^4 \tag{5.43}$$

where the in-phase and quadrature gains are expressed as

$$\begin{aligned}K_1\left[|x(t)|^2\right] &= \gamma_{11} + \gamma_{13}|x(t)|^2 + \gamma_{15}|x(t)|^4 \\ K_2\left[|x(t)|^2\right] &= \gamma_{21} + \gamma_{23}|x(t)|^2 + \gamma_{25}|x(t)|^4\end{aligned} \tag{5.44}$$

The relationship of the complex coefficients between (5.43) and (5.44) is $\gamma_i = \gamma_{1i} + j\gamma_{2i}$, which is also a function of the pre-distorter's α_i and power amplifier's β_i coefficients. Substituting (5.31), (5.37), and (5.43) into (5.42) and solving for γ_i yields [14]

$$\begin{aligned}\gamma_1 &= \alpha_1\beta_1 \\ \gamma_3 &= \alpha_3\beta_1 + \alpha_1\beta_3|\alpha_1|^2 \\ \gamma_5 &= \alpha_5\beta_1 + \alpha_3\beta_3|\alpha_1|^2 + \alpha_1\beta_5|\alpha_1|^4 + 2\alpha_1\beta_3\Re e\{\alpha_1\alpha_3^*\}\end{aligned} \quad (5.45)$$

In (5.45), the third- and fifth-order IMD products are unwanted and have to be reduced; this is determined by the coefficients of γ_3 and γ_5. There are two different approaches to minimize γ_3 and γ_5. The first one is to obtain the coefficients β_i of the power amplifier first through the LS in (5.11) based on the measured input and output data of the power amplifier, where the memory M is set to zero for memoryless nonlinearity. Then, the coefficients of α_i can be solved by using (5.45). The second approach is to directly use the LS in (5.24) to solve the coefficients α_i of the - pre-distorter based on the captured input and output data of the power amplifier. It should be noted above that when using (5.24) and (5.11) solve the coefficients of α_{i} and β_{i} in (5.45), α_{i} is the same as the predistorter coefficient of a_{k0} in (5.24) while β_{i} is the same as the power amplifier coefficient of b_{kO} in (5.11), where the largest delay M is set to zero.

5.4.3 Coefficient Adaption of Analog Pre-distortion

An initial estimate for the pre-distorter's coefficients can be obtained by using a least squares approximation method, which can be applied to the initial coefficients of either APD or DPD. The optimum coefficients can be achieved by minimizing adjacent channel emission through an adaptive algorithm. In addition, characteristics of the power amplifier drift due to aging, temperature changes, and supply voltage variations. All of these factors require the pre-distorter should be designed to have an adaptive capability.

Unlike an adaptive LMS equalizer, whose adaptive algorithm is derived by taking the derivative of the mean square error (objective) function with respect to the coefficients, an analytical expression (closed-form) for the pre-distorter is rarely available due to its nonlinear characteristics. In this case, a simple iterative algorithm called "simultaneous perturbation stochastic approximation (SPSA)" [24] for such optimization problems without available mathematical expression can be utilized to adaptively adjust the coefficients of the pre-distorter. Actually, the SPSA method is similar to Hooke and Jeeves' method, which is also known as *derivative-free* or *direct search* optimization method [29].

The essential feature of SPSA is the gradient approximation, which requires only two measurements of the objective function performed by perturbing a variable vector with upward and downward values regardless of the dimension of the optimization problems [24]. This feature is very suitable to some applications in which the mathematical objective function expressions are unavailable, and reduces the cost of optimization, especially when a large number of variables need to be optimized.

Consider the objective function $\xi(\mathbf{x})$, where the vector $\mathbf{x} = (x_1, x_2, \ldots, x_n)$ is an n-dimensional vector and presents the coefficients of the pre-distorter. The optimization problem can be translated into taking the derivative of ξ with respect to $\mathbf{x}^*$ such that it is equal to zero or $\partial\xi/\partial\mathbf{x}^* = 0$, and where the solution vector $\mathbf{x}_{\text{opt}}$ corresponds to the minimal objective function, or $\xi(\mathbf{x}_{\text{opt}}) = \xi_{\mathbf{min}}$. Actually, this is the classical formulation of local optimization for differentiable object functions. The objective function can target the integrated out-of-channel power in the frequency domain at the output of the power amplifier or the output of the equivalent complex PA in the baseband domain. For example, the out-of-channel power is integrated within the adjacent and alternative adjacent channels. It is assumed that measurements of $\xi(\mathbf{x})$ are available at various values of $\mathbf{x}$. In general, the SPSA algorithm is of a recursive form:

$$\hat{\mathbf{x}}_{\mathbf{k+1}} = \hat{\mathbf{x}}_{\mathbf{k}} - a_k \hat{\mathbf{g}}_{\mathbf{k}}(\hat{\mathbf{x}}_k) \tag{5.46}$$

where $\hat{\mathbf{g}}_{\mathbf{k}}(\hat{\mathbf{x}}_{\mathbf{k}})$ is the estimate of the gradient $\mathbf{g}(\mathbf{x}) = \partial\xi/\partial\mathbf{x}$ at the k-th iterate $\hat{\mathbf{x}}_{\mathbf{k}}$ based on measurements of the objective function, and a_k is a small positive step that usually gets smaller as k gets larger. Under appropriate conditions, the iteration in (5.46) will converge to an optimum value of $\mathbf{x}_{\text{opt}}$ that minimizes the objective function $\xi(\mathbf{x}_{\mathbf{opt}})$. Ignoring any noise, the estimated gradient $\hat{\mathbf{g}}_{\mathbf{k}}(\hat{\mathbf{x}}_{\mathbf{k}})$ for upward and downward simultaneous perturbation approximation is given by

$$\hat{\mathbf{g}}_{\mathbf{k}}(\hat{\mathbf{x}}_{\mathbf{k}}) = \frac{\xi(\hat{\mathbf{x}}_{\mathbf{k}} + c_k\Delta_k) - \xi(\hat{\mathbf{x}}_{\mathbf{k}} - c_k\Delta_k)}{2c_k} \begin{bmatrix} \Delta_{k,1}^{-1} \\ \Delta_{k,2}^{-1} \\ \vdots \\ \Delta_{k,p}^{-1} \end{bmatrix} \tag{5.47}$$

where c_k is a small positive number and becomes smaller as k increases, and $\Delta_k = (\Delta_{k,1}, \Delta_{k,2}, \ldots, \Delta_{k,p})^{\mathrm{T}}$ is the distribution of the user-specified p-dimensional random perturbation vector, in which the superscript T denotes vector transpose. In (5.46) and (5.47), the choice of the positive steps of a_k and c_k is critical to the performance of SPSA algorithm, affecting the speed of the convergence and stability. Some guidance on picking these numbers can be found in [25].

5.5 Applications

Linearization techniques of power amplification, including both analog pre-distortion (APD) and digital pre-distortion (DPD) techniques, have been widely used in modern wireless digital communication systems, especially in systems of base station PA linearization. Application requirements in these systems include wide bandwidth, high energy efficiency, multi-standard capability, and low distortion due to nonlinearity. Due to the low current consumption requirements for mobile phones and portable devices, *adaptive* and *effective* APD and DPD techniques, including down-conversion in the feedback RF path, still have a long way to go before having wide application in phones and devices even though some function-simplified and bandwidth-limited predistortion techniques have been used. In this section, one typical APD chip designed by Maxim is briefly introduced and discussed, which is suitable to applications of base station PA linearization.

5.5.1 Maxim's RF Pre-distortion Technique

Maxim's RF analog pre-distorter (RFAPD) [26, 27] is similar to DPD in the compensation for AM-AM and AM-PM distortions, IMD, and PA memory effects; both employ a feedback loop to adaptively update the coefficients of the pre-distorter for the compensation. The major difference is their circuit design and system implementation. RF APD SoC chipsets from Maxim work with RF input and output signals of the power amplifier so that they enable standalone operation without direct access to a digital modulator of the existing transmitter system. This feature is an attractive and practical solution for achieving both energy and spectral efficiencies in existing wireless communication systems. Figure 5.19

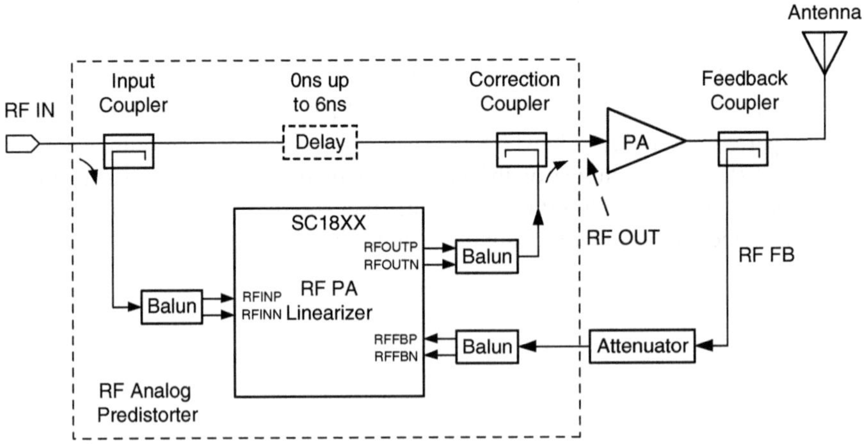

Fig. 5.19 Top-level block diagram of RF APD. Redrawn from [27]

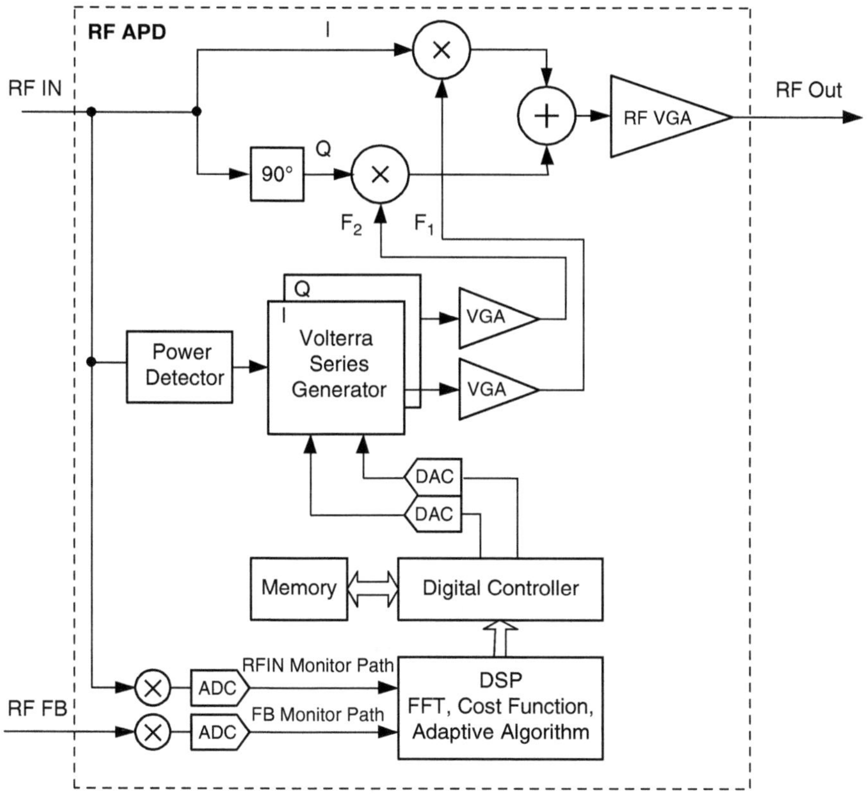

Fig. 5.20 A detailed block diagram of RF IC APD chip. Redrawn from [27]

below shows a top-level block diagram of a PA utilizing RFAPD, which is also called a RF-power amplifier linearizer (RFPAL) and was developed based on Scintera's RFPAL chipset [28]. A detailed block diagram of a RF APD IC chipset is illustrated in Fig. 5.20.

There are two input signals, the RF input and RF feedback signals, to the RF PA linearizer and one output signal, as shown in Fig. 5.19. The single-ended signals to the block of the RF APD are transferred into the differential signal through baluns so that the common mode noises are suppressed during the analog signal operations. Referenced in Fig. 5.19, the RF signal at the output of the power amplifier feeds back to the RF APD block via a feedback coupler after passing through an attenuator. The RF FB signal is first down-converted into the baseband I–Q signals, and then filtered by the lowpass filters to remove out-of-band harmonics. Then, the baseband I–Q signals are digitally processed using an FFT operation in the frequency domain after passing through the analog-to-digital converters (ADC). The IMD products in the adjacent channels and alternative adjacent channels are integrated to generate an objective function. By minimizing the objective function through the adjustments of the coefficients α_{1i} and α_{2i} in the functions of F_1 and F_2

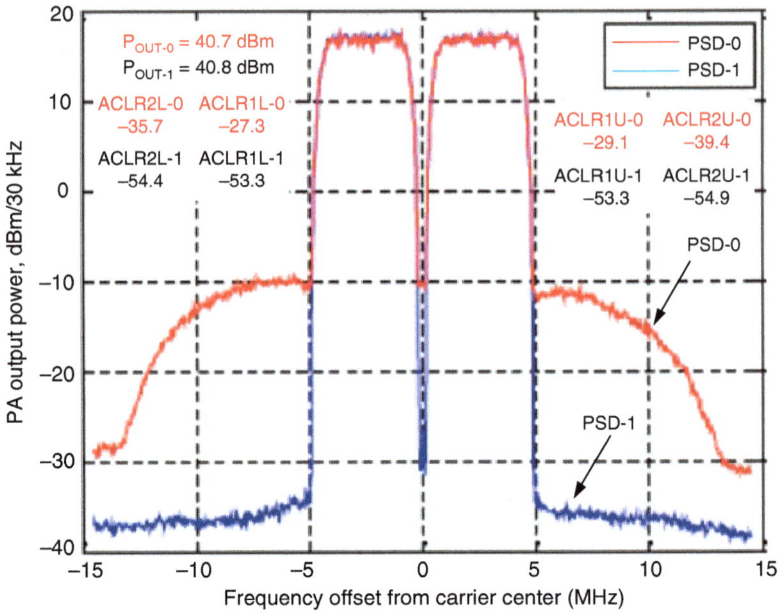

Fig. 5.21 Power spectral density of two WCDMA signals on a Doherty LDMOS PA, where PSD-0 stands for PSD without APD and PSD-1 corresponds to PSD with APD. Referenced from [28]

expressed in (5.30), the nonlinearity distortions and memory effects of the power amplifier are continuously compensated. The period of coefficient adaption can be made either faster or slower according to the environment change, such as temperature drifts, power supply variations, and PA aging. When the adaption works very slowly, the RF APD system can be treated as an open loop rather than a closed loop to ensure that the system operation is stable. F_1 and F_2 are functions of the complex envelope of the RF-modulated signal and interpolate the inverse AM-AM and AM-PM nonlinearities of the power amplifier. After passing through variable gain amplifiers as shown in Fig. 5.20, F_1 and F_2 multiply the RF in-phase and quadrature signals, respectively. Actually, this multiplication converts the even-order terms into odd-order terms to cancel the odd-order nonlinearity harmonics created by the power amplifier. The RF I–Q signals are then summed and amplified to form the pre-distorted RF signal at the input of the power amplifier.

The RF input in Fig. 5.19 is split into two parts via an input coupler. The first part goes to an orthogonal RF modulator, and another one is further split into two parts at the input of the RF APD, as shown in Fig. 5.20. In the APD, one is down-converted into the baseband I–Q signals as the reference signals relative to the RF FB path, and another one is used to generate the envelope amplitude and second-order nonlinearity items used in the Volterra series generator block. In order to compensate for the memory effects of the power amplifier, four different sets of coefficients are created based on delay terms from τ_1 to τ_4, ranging from 0 to

Table 5.3 ACLR requirements in the 3GPP WCDMA system

WCDMA system	Adjacent channel frequency offset (MHz)	ACLR limit (dB)
Base station	±5	−45
	±10	−50
User equipment	±5	−33
	±10	−43

300 ns [27] and corresponding to the maximum delay unit $M = 3$ of the polynomial memory expression in (5.36). The maximum delay unit is determined by the memory effects of an actual power amplifier and the cost and power consumption of hardware implements. These second-order nonlinearity elements and the coefficients of the memory polynomial series from the outputs of DACs compose the pre-distorter's gain series of F_1 and F_2 as given in (5.30).

Figure 5.21 shows power spectral density (PSD) curves of a two-carrier WCDMA signal at the output of the power amplifier with an APD and without an APD. A class AB power amplifier with average output power of about 41 dBm was used to evaluate the performance improvement, where PSD-0 represents PSD without APD while PSD-1 stands for PSD with APD. In the 3GPP WCDMA standard specifications, the adjacent channel leakage ratio (ACLR) requirements are listed in Table 5.3, where adjacent channels located at a ±5-MHz frequency offset from the center frequency of the desired channel are called ACLR1, while alternative adjacent channels located at a ±10-MHz frequency offset are called ACLR2.

It can clearly be seen in Fig. 5.21 that ACLR1 at the lower side is improved from −27.3 to −53.3 dB, and at the upper side from −29.1 to −53.3 dB, while ACLR2 at the lower side is improved from −36.7 to −54.4 dB and at the upper side from −39.4 to −54.9 dB under the same average output powers. Such improvements made by the RF APD can let the power amplifier have less back-off from its P1dB compression point to achieve energy efficiency. Otherwise, the power amplifier will back off more from its P1dB point in order to meet the 3GPP WCDMA ACLR specifications at the price of low energy efficiency.

As described in [27], the RF APD is suitable to small cellular base stations at power levels below 60 W, as in microcell, picocell, and enterprise femtocell applications from 470 to 2800 MHz with PAPR of up to 10 dB, and a wide range of PAs, including Class AB or Doherty amplifiers. Its power consumption is a small portion of the total system power consumption in a transmitter system of the small cell base stations, and hence it can be ignored. If the power consumption, complexity, and cost can be further reduced with open-loop–based pre-distortion by removing the complicated down-conversion, demodulation, and adaption blocks, the RF APD may find its application in user equipment, handsets, and other portable wireless devices. In some applications, such as 802.11 WALN systems, it is also possible for the RF APD to be applied if the coefficients of the APD can be initially updated through the calibration procedure during the beginning of power-on and adaptively updated by using a local receiver in a SoC transceiver chip during the receiver's idle time. Such adaption processing in these applications would not be performed often in order to reduce power consumption, depending on temperature change and power supply variation.

References

1. Sirois, J., Boumaiza, S., Helaoui, M., Brassard, G., & Ghannouchi, F. M. (2005, September). A robust modeling and design approach for dynamically load and digitally linearized Doherty amplifiers. *IEEE Transactions on Microwave Theory and Techniques, 53*(9), 2875–2883.
2. Ui, N., Sano, H., & Sano, S. (2007, June). A 80W 2-stage GaN HEMT Doherty amplifier with 50 dBc ACLR, 42% efficiency 32 dB gain with DPD for WCDMA base station. *IEEE MTT-S International Microwave Symposium* (pp. 1259–1262).
3. Kim, J., Woo, Y. Y., Moon, J., & Kim, B. (February 2008). A new wideband adaptive digital pre-distortion technique employing feedback linearization. *IEEE Transactions* on *Microwave Theory* and Techniques, *56*(2), 385–392.
4. Lee, S. Y., Lee, M. W., Kam, S. H., & Jeong, Y. H. (2010, January) Advanced design of high linearity analog pre-distortion Doherty amplifiers using spectrum analysis for WCDMA applications. *Radio and Wireless Symposium (RWS),2010 IEEE* (pp. 140–143).
5. Son, K. Y., Koo, B., & Hong, S. (2012, August). A CMOS power amplifier with a built-in RF pre-distorter for handset applications. *IEEE Transactions on Microwave Theory and Techniques, 60*(8), 2571–2581.
6. Zhu, A., Wren, M., & Brazil, T. J. (2004). An efficient Volterra based behavioral model for wideband RF power amplifiers. *IEEE MTT-S International Microwave Symposium Digest, 2*, 787–790.
7. Zhu, A., & Brazil, T. J. (2004, December). Behavioral modeling of RF power amplifiers based on pruned Volterra series. *IEEE Microwave and Wireless Components Letters, 14*(12), 563–565.
8. Clark, C. J., Chrisikos, G., Muha, M. S., Moulthrop, A. A., & Silva, C. P. (1998, December). Time-domain envelope measurement technique with application to wideband power amplifier modeling. *IEEE Transactions on Microwave Theory and Techniques, 46*, 2531–2540.
9. Kim, J., & Konstantinous, K. (2001, November). Digital pre-distortion of wideband signals based on power amplifier model with memory. *Electronics Letter, 37*(23), 1417–1418.
10. Ding, L., Zhou, G. T., Morgan, D. R., & Kenney, J. S. (2004, January). A robust digital baseband pre-distorter constructed using memory polynomials. *IEEE Transactions on Communications, 52*(1), 159–165.
11. Chen, W., Zhang, S., Liu, Y. J., Ghannouchi, F. M., Feng, Z., & Liu, Y. (2014, October). Efficient pruning technique of memory polynomial models suitable for PA behavioral modeling and digital pre-distortion. *IEEE transaction on Microwave Theory and Techniques, 62*(10), 2290–2299.
12. Zhu, A., Pedro, J. C., & Cunha, T. R. (2007, May). Pruning the Volterra series for behavioral modeling of power amplifiers using physical knowledge. *IEEE Transactions on Microwave Theory and Techniques, 55*(5), 813–821.
13. Fehri, B., & Boumaiza, S. (March 2014). Baseband equivalent Volterra series for digital pre-distortion of dual-band power amplifiers. *IEEE Transactions on Microwave Theory and Techniques, 62*(3), 700–714.
14. Costescu, F. C. (1992). *Amplifier linearization using adaptive analog pre-distortion*. Master Thesis, Simon Fraser University.
15. Morgan, D. R., Ma, Z., Kim, J., Zierdt, M. G., & Pastalan, J. (2006, October). A generalized memory polynomial model for digital pre-distortion of RF power amplifiers. *IEEE Transactions on Signal Processing, 54*(10), 3852–3860.

16. Eun, C., & Powers, E. J. (1997, January). A new Volterra pre-distorter based on the indirect learning architecture. *IEEE Transactions on Signal Processing, 45*, 223–227.
17. Choi, S., Jeong, E., & Lee, Y. H. (2007). A direct learning structure for adaptive polynomial based pre-distortion for power amplifier linearization. *2007 I.E. 65th Vehicular Technology Conference - VTC2007* (pp. 1791-1795)
18. Zhou, D., & DeBrunner, V. E. (2007, January). Novel adaptive nonlinear pre-distorters based on direct learning algorithm. *IEEE Transactions on Signal Processing, 55*(1), 120–133.
19. Lee, S. Y., Lee, Y. S., Hong, S. H., Choi, H. S., & Jeong, Y. H. (2005, February). An adaptive prediction RF power amplifier with a spectrum monitor for multicarrier WCDMA applications. *IEEE Transactions on Microwave Theory and Techniques, 53*(2), 786–793.
20. Cha, J., Yi, J., Kim, J., & Kim, B. (2004, February). Optimum design of a pre-distortion RF power amplifier for multicarrier WCDMA applications. *IEEE Transactions on Microwave Theory and Techniques, 52*(2), 655–663.
21. Yi, J., Yang, Y., Park, M., Kang, W., & Kim, B. (2000, December). Analog pre-distortion linearizer for high power RF amplifiers. *IEEE Transactions on Microwave Theory and Techniques, 48*(12), 2709–2713.
22. Braithwaite, R. N. (2010). Memory correction for a WCDMA amplifier using digital-controlled adaptive analog pre-distortion. *Radio and Wireless Symposium (RWS), 2010 IEEE* (pp. 144–147).
23. Stapleton, S., & Costescu, F. C. (1992, February). An adaptive pre-distorter for a power amplifier based on adjacent channel emissions. *IEEE Transactions on Vehicular Technology, 41*(1), 49–56.
24. Spall, J. C. (1998). An overview of the simultaneous perturbation method for efficient optimization. *Johns Hopkins Applied Technical Digest, 19*(4), 482–492.
25. Spall, J. C. (1998). Implementation of the simultaneous perturbation algorithm for stochastic optimization. *IEEE Transactions on Aerospace and Electronic Systems, 34*, 817–823.
26. Maxim Application Note. *RF Pre-distortion versus Digital Pre-distortion,* 2010. www.maximintegrated.com.
27. Quzillou, M. (2011, August 12). Linearize power amplifiers with RF pre-distortion. *Microwaves and RF*.
28. Data sheet. (2000). *RF Power Amplifier Linearizer (RFPAL).* www.scintera.com
29. Hooke, R., & Jeeves, T. A. (1961). Direct search solution of numerical and statistical problems. *Journal of the Association for Computing Machinery, 8*(2), 212–229.

Chapter 6
Transceiver I: Transmitter Architectures

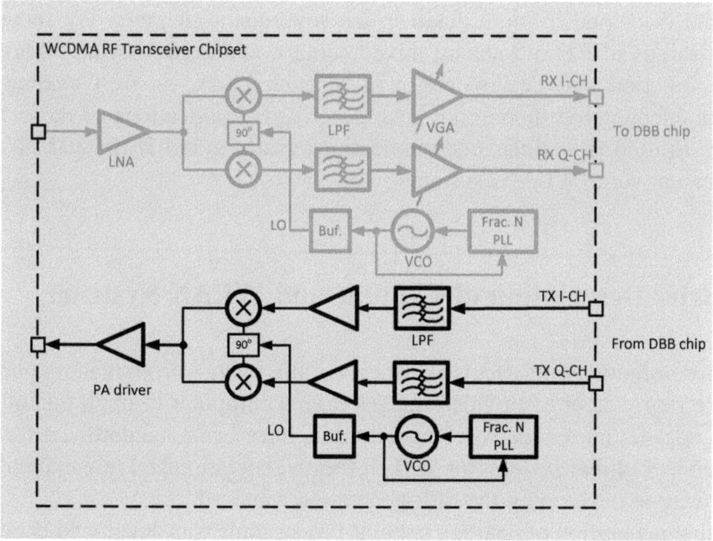

6.1 Introduction

In the previous chapters, we described the three major subsystems of modulation, demodulation, up-conversion, and power amplification in wireless communication systems. Starting with Chap. 6, we move to a top level, or the transceiver architecture that includes other functional blocks besides these three subsystems to achieve complete transmission and reception. In general, there are three types of common

© Springer International Publishing Switzerland 2017
W. Gao, *Energy and Bandwidth-Efficient Wireless Transmission*, Signals and Communication Technology, DOI 10.1007/978-3-319-44222-8_6

transmit and receive architectures available to the wireless radio frequency (RF) integrated circuit (IC) transceiver architect: *superheterodyne, low intermediate frequency* (IF), and *direct conversion*, also known as *zero-IF*. Each of these architectures has its own advantages and disadvantages, and some of the potential issues related to the particular architecture can be minimized with either careful circuit design techniques or calibration methods. We describe and analyze what advantages and disadvantages these architectures have in practical applications and what challenges RFIC designers may face in their designs, and we discuss some of these design techniques and calibration methods in more detail in the subsequent chapters.

Considering that a transmitter has different operational functions than a receiver, and that some of the potential issues are different between transmitters and receivers, we divide the transceiver architecture into two parts, or the transmitter architecture (Chap. 6) and receiver architecture (Chap. 7). Therefore, some unique and special features that the transmitter and receiver may have are introduced and discussed in more detail in the subsequent chapters. We focus on the system designs of RF and analog mixed signals, and digital signal algorithms. A detailed function and design treatment of each block, as well challenges and problems encountered therein, are also described. Applications of these architectures are limited to cellular communications systems and IEEE 802.11 wireless local area network (WLAN) systems.

6.2 Brief Description of Cellular and WLAN Systems

Wireless communication affects all aspects of life today—from making phone calls to transferring data or even video images from a computer through the internet. In the past decade, numerous wireless proposals have been standardized for various applications. Cellular systems are in their fourth generation and researchers are well on their way to developing the fifth-generation standard.

The first generation of analogy cellular FM systems was developed beginning in the 1990s and was used mainly to transfer analog voice information with a channel bandwidth of about 25 kHz. The second generation (2G) featured digital cellular systems that achieved higher capacity and improved compatibility. Digital signal processing (DSP) and digital communications techniques played a key role in the 2G because of their dramatic performance improvements and low cost and power consumption. The 2G digital cellular systems conformed to at least three standards: one for Europe and international applications ("*Global System for Mobile communication (GSM)*); one for cdmaOne (*IS*-95); and another one for North America (*IS-54* and *IS-136*). The data rate of the GSM system is 270.833 kbps and the channel bandwidth is 200 kHz. The third-generation (3G) cellular systems could transmit data at high rates—up to 3.84 Mbps—to support high-capacity messaging and used advanced time division multiple access (TDMA) and code division multiple access (CDMA). In the early 2000s, the standard for the 3G cellular systems was released to provide multi-media support along with peak data rates up to at least 200 kbps. With growing worldwide development and commercial operation in the following years, 3G systems mainly included wideband code-division multiple access (WCDMA) systems from the Universal Mobile Telecommunications System (UMTS), CDMA2000 systems from Qualcomm, and time-division synchronous CDMA (TD-SCDMA) systems from China. In the early 2010s, the Long Term Evolution Advanced (LTE Advanced) system was recognized worldwide as a fourth-generation (4G) technology able to deliver downlink speeds of 1 Gbps and 100 Mbps for stationary and mobile reception, respectively. With the mature development and successful commercial operation of 4G systems, the next generation of terrestrial mobile telecommunications (i.e., 5G) technologies have been investigated and researched so as to meet the anticipated worldwide demand in the 2020 era and beyond . These demands and needs include higher traffic volume; indoor or hotspot traffic; and spectrum, energy, and cost efficiencies. Various organizations from different countries and regions have launched research programs aimed at developing potentially key 5G technologies. Mobile and Wireless Communications Enabled for the Twenty-Twenty Information Society (METIS) is an integrated project partly funded by the European Commission and is considered the 5G flagship project. The objective of METIS is to lay the foundation for 5G mobile and wireless communications systems, whose applications are expected to begin rolling out in 2020 [1–3].

One of the major users of 3G and 4G cellular systems for high-data-rate services is the WLAN system based on IEEE 802.11a/b/g/n/ac standards. WLAN devices have been widely used to provide wireless internet access in public places and in personal homes. The current WLAN standard, or 802.11 ac, utilizes up to eight spatial streams and has a transmission channel band up to 80 MHz, which can be combined to form a 160-MHz transmission channel (option). The 802.11 ac wave 1 standard that supports single-user multiple input/multiple output (MIMO) and achieves maximum data rates up to 1.3 Gbps with three spatial streams has dominated today's WLAN product market. The 802.11ac wave 2 standard that supports multiple-user MIMO and achieves a maximum data rate up to 3.5 Gbps with eight spatial streams will be available in 2017.

Complete radio solutions are developed as RFIC transceiver chipsets mainly comprising two chipsets—an RF chip and a digital baseband (DBB) and media access control (MAC) chip. A complete RF system usually consists of several independent modules/chipsets: a radio frequency (RF) front-end module that may contain a power amplifier (PA), a low noise amplifier (LNA) and transmit/receive (T/R) switch, a RF transceiver chip, and a digital baseband (DBB) & MAC chip. Of course, a final goal of complete chipsets is to integrate the RF transceiver chip and the DBB and MAC chip into one chipset to achieve low-cost and small form factor with manageable power dissipation. Consequently, a high level of integration of radio functions becomes a necessity [4].

In the following sections, we mainly focus on the transmit system design aspects of the RFIC chipset technique, which includes the features, advantages and disadvantages that apply to different transmitter architectures with applications to cellular and Wi-Fi communications. In addition, some approaches to minimizing these disadvantages are introduced and discussed in detail.

6.3 Superheterodyne Transmitter

An RF transmitter mainly performs the functions of digital baseband modulation mapping and pulse shaping, reconstruction filtering, RF IQ modulation, frequency up-conversion, and power amplification. A majority of commercial wireless transmitters transfer the baseband spectrum to the RF spectrum in either one or two steps. In the former case, it is called *direct up-conversion.* In the latter case, it is referred to as *superheterodyne, a* system in which intermediate frequency (IF) quadrature carrier signals are first either phase-modulated or phase- and amplitude-modulated by the I and Q baseband signals, and then converted into the RF frequency with a second LO signal. Each of these transmitter structures has its own inherent strengths and weaknesses, presents many potential challenges for RF IC engineers. Some features of the superheterodyne transmitter will be described briefly below.

The superheterodyne transmitter is shown in Fig. 6.1. In this transmitter, the I and Q digital baseband (BB) signals are received from the digital baseband PHY block and passed through the DACs to generate the I and Q analog BB signals. The lowpass filters, also called reconstruction filters, with the proper cut-off frequency on the I and Q channels are used to remove image components of the digital pulse-shaping signals at the outputs of the DACs. The cutoff frequency of the lowpass filter depends on the sampling frequency of the DAC, the rate of symbol data, and the required transmit spectrum mask. The filtered I and Q baseband signals then modulate a pair of orthogonal IF carrier signals in phase, represented by LO1, to form the single-sideband–modulated signal at the IF (ω_1). The modulated IF signal is further up-converted into the RF frequency with the second local oscillator (LO2). The first bandpass filter (BPF) suppresses the harmonics of the IF signal, while the second BPF passes the wanted RF signal at the frequency of either $\omega_2 - \omega_1$ or $\omega_2 + \omega_1$ depending on the application and attenuates the harmonics of the RF signal. Note that the second BPF can be replaced with either a LPF for the RF signal at the frequency of $\omega_2 - \omega_1$ or a highpass filter (HPF) for one at the frequency of $\omega_2 + \omega_1$ respectively.

Superheterodyne transmitters allow the system designers, through proper frequency planning, to avoid the carrier feed-through (or leakage) that plagues direct up-conversion. Another advantage of the superheterodyne transmitter is that it avoids the VCO-pulling that the direct conversion transmitter may face because the PA output signal frequency and its harmonic frequencies are away from the frequency of the VCO. The VCO-pulling problem will be described in a direct-conversion transmitter in the following section. Furthermore, the superheterodyne transmitter also minimizes the performance degradation because the I–Q amplitude and phase mismatching problems are minor at the IF frequency.

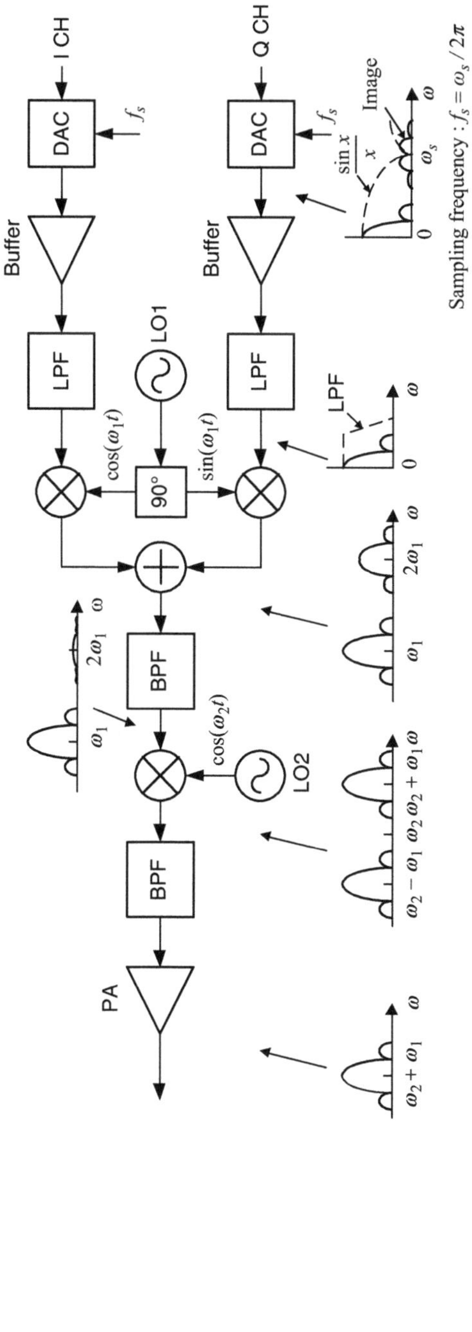

Fig. 6.1 Two-step superheterodyne transmitter

However, the main disadvantages of the superheterodyne transmitter are as follows:

- It requires an extra BPF following the second upconversion. This filter should have a high Q factor to reject the side-lobes of the RF output signal by a large certain amount, typically 50–60 dB [5]. For example, in the GSM standard specification, the spectrum sideband of the RF-modulated GMSK signal at the frequency offset of 400 kHz should not exceed –60 dBc. Without the BPF, it is not possible to achieve such a low sideband output signal in *the mixing up-conversion structure*. Such a BPF with the higher center frequency is typically a passive and is usually implemented with relatively expensive off-chip device.
- It needs the second LO source to convert the IF-modulated signal to the RF frequency. Hence, a superheterodyne transmitter usually has large size and high cost compared with a direct up-conversion transmitter. It is not suitable for RF IC transceivers, especially for multimode applications, due to the narrowband nature of the IF filter.

6.4 Direct up-Conversion Transmitter

Generally a direct up-conversion transmitter requires fewer blocks compared to a superheterodyne transmitter. From the meaning of its name, we may surmise that the direct up-conversion transmitter performs the frequency transfer from the baseband frequency to the RF frequency with one frequency mixing stage, as shown in Fig. 6.2. The filtered baseband I–Q signals directly modulate a pair of the orthogonal LO signals after DACs on the I–Q channels and then are summed to generate the single-sideband–modulated signal at the RF frequency. After passing through a lowpass filter or a bandpass filter that attenuates high-order harmonics, the RF-modulated signal is amplified via the power amplifier or power amplifier driver. It is obvious that the direct up-conversion transmitter omits the needs for the second LO source and an associated BPF after the second LO mixing, which are very important for the RF IC designers to create a low-cost transceiver with low power consumption, and a high degree of integration.

Even though the direct up-conversion transmitter has the advantages mentioned above, it may have some potential disadvantages as well:

- *Error Vector Magnitude (EVM) degradation due to the I–Q amplitude and phase imbalance or mismatching*: The I–Q amplitude imbalance is mainly caused by either gain imbalance of the I–Q baseband signals or gain imbalance of the quadrature LO carrier signals, while the I–Q phase mismatching is caused by a not exactly 90°-phase difference between quadrature LO signals.
- *The LO leakage at the RF output, also called LO feed-through (LOFT), due to the DC offsets on the IQ baseband branches*: Various wireless standards have certain requirements on the maximum amount of LO leakage allowed. For

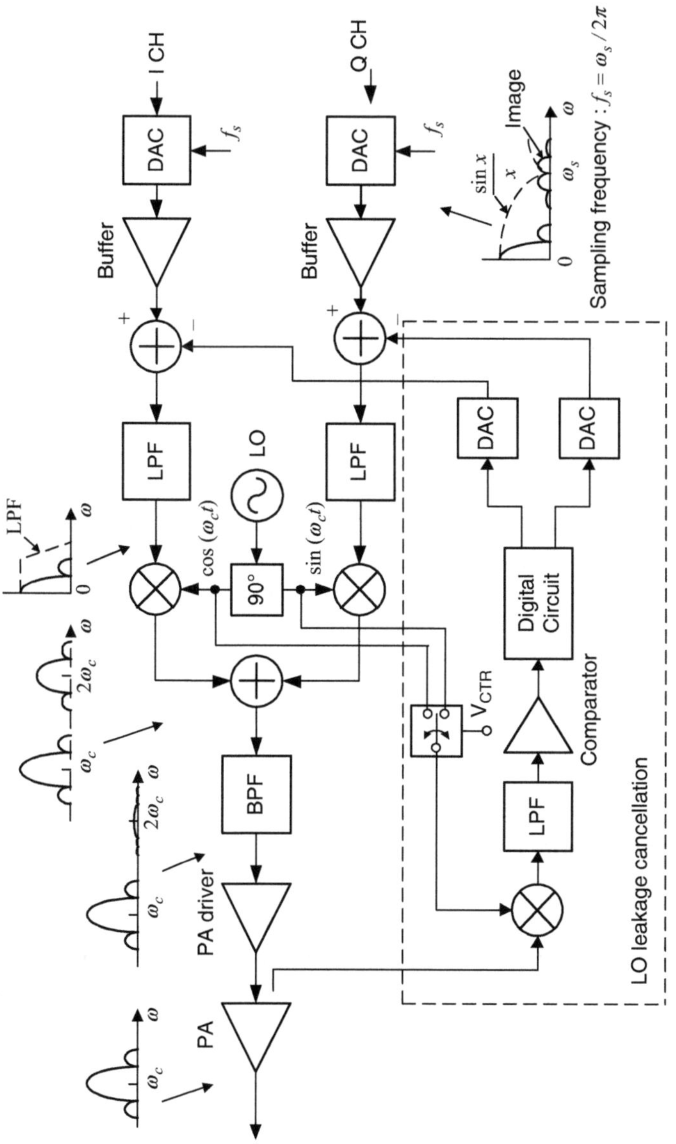

Fig. 6.2 Direct-conversion transmitter with LO leakage cancellation

example, the IEEE 802.11a standard requires that the LO leakage should not exceed -15 dB relative to overall transmitted power. In the fact, the larger LO leakage does not carry any information for the receiver, but consumes the transmitted power.

- *There is phase noise disturbance of the VCO by a strong RF signal at the output of the PA due to the fact that the VCO operates at the same frequency as the second order harmonic frequency of the RF signal*: The phase noise disturbance of the VCO is somewhat similar to the VCO-pulling except without injection locking. In this case, a pair of the orthogonal LO signals are generated by dividing the VCO frequency by 2. The second harmonic of the RF-modulated signal at the PA output could couple to the PLL loop through various paths, such as through bond wires of the package, through the substrate, and through the power supplies. The coupled second harmonic component that has the same frequency as the VCO output signal may pass through the frequency and phase detector to degrade the phase noise of the VCO after being divided by the PLL divider together with the VCO output. This phenomenon may also happen even using an external power amplifier rather than integrating a power amplifier in the RF transceiver IC chip due to finite oscillation. The phase-noise disturbance of the VCO severely degrades EVM performance.

These transmission impairments associated with the direct-conversion transmitter can be mitigated by using either calibration techniques or careful circuit designs. For example, the first two disadvantages mentioned above can be minimized by using calibration methods, while the third one can be reduced with careful printed circuit board (PCB) design and layout. Hence, the direct-conversion transmitter architectures are widely used in RF transceivers for wireless cellular and WLAN standards, especially for wideband signals, such as WCDMA signals in the 3G cellular standard, OFDM signals in the 4G (LTE Advanced) cellular standard, and OFDM signals in the 802.11 series WLAN standards.

6.5 Transmission Impairments

In this section, we will first introduce some calibration methods to minimize the I–Q imbalance errors and LO leakage and some design strategies to reduce phase-noise disturbance in detail. Then, we will discuss the performance degradations due to nonlinear distortion caused by the power amplifier and possible techniques to mitigate nonlinear distortion. It is important to note that the nonlinearity of the power amplifier dominates the performance degradations of the transmitter.

6.5.1 I–Q Gain and Phase Imbalances and DC Offsets

I–Q gain (or amplitude) and phase imbalances can happen at either a transmitter or a receiver. The I–Q gain imbalance is usually caused by the mismatches *either*

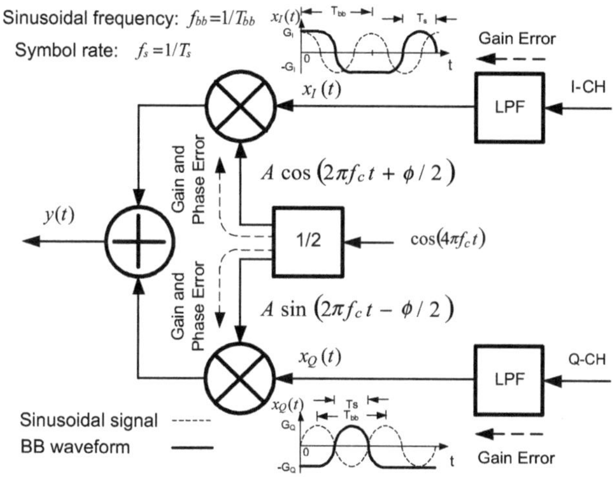

Fig. 6.3 I–Q gain and phase errors caused by possible sources in a transmitter

between the amplitudes of the I–Q baseband signal branches *or* between the amplitudes of the in-phase and quadrature (I–Q) signal of the LO paths, while the I–Q phase imbalance is caused by a non-90° phase shift between the I–Q signals of the LO only. Figure 6.3 illustrates possible major source that can generate the I–Q gain and phase imbalances at the transmitter.

The problems that the I–Q imbalance errors and DC offsets may cause at the transmitter are *intermodulation (IM) products* and *degradation of EVM*. The I–Q gain and phase imbalance errors and DC offsets cause spurious products at the output of the I–Q modulator. These distortions can interact with the amplifier nonlinearity to produce unexpected intermodulation products at the output of the amplifier. In practice, these intermodulation products are related to the regrowth of power spectral density of the RF modulated signal. The gain and phase imbalance errors and DC offsets distort the signal constellation, which results in the degradations of TX EVM and RX BER. It was demonstrated that combined impairments of the gain and phase imbalance errors of 3% and 3° and DC offset of 3% at the transmitter result in the 1-dB degradation at the BER of 10^{-5} for a 16QAM signal at the receiver [6].

To see how the I–Q imbalance affects the EVM performance, we use two types of complex baseband (BB) signals as the inputs to a quadrature modulator. The first one is a complex sinusoidal signal of $x(t) = x_I(t) + jx_Q(t) = G_I \cos(\omega_{bb}t) + jG_Q \sin(\omega_{bb}t)$ while the second one is a complex BB waveform through pulse shaping for a QPSK signal expressed generally by $x(t) = x_I(t) + jx_Q(t)$. Here ω_{bb} is the frequency of the sinusoidal signal, $x_I(t)$ and $x_Q(t)$ are real and imaginary parts of the complex signal. The sinusoidal signal frequency of f_{bb} is assumed to be equal to one half of the symbol rate of f_s, or $f_{bb} = f_s/2$. The real and imaginary waveforms of these two complex signals are plotted on the I–Q channels shown in Fig. 6.3.

In practice, the I–Q phase imbalance ϕ is caused by non-ideal 90 degree phase shift between quadrature LO signals *rather than the I-Q baseband signals*, and the gain imbalance is caused by either quadrature LO signals or I–Q baseband signals. Here, we assume that the gain imbalance is due to gain difference between the I–Q baseband signals only and the amplitudes of quadrature LO signals are the same with a gain value of A in the following derivation. Then, the modulated signal is expressed as

$$y(t) = Ax_{\mathrm{I}}(t)\mathbf{cos}\left(\omega_{\mathrm{c}}t + \frac{\phi}{2}\right) + Ax_{\mathrm{Q}}(t)\mathbf{sin}\left(\omega_{\mathrm{c}}t - \frac{\phi}{2}\right) \tag{6.1}$$

where the baseband signals in the I–Q channels are given by

$$x_{\mathrm{I}}(t) = G_{\mathrm{I}}\mathbf{cos}\,(\omega_{\mathrm{bb}}t) \approx \left(1 + \frac{\varepsilon}{2}\right)\mathbf{cos}\,(\omega_{\mathrm{bb}}t) \tag{6.2}$$

$$x_{\mathrm{Q}}(t) = G_{\mathrm{Q}}\,\mathbf{sin}\,(\omega_{\mathrm{bb}}t) \approx \left(1 - \frac{\varepsilon}{2}\right)\mathbf{sin}\,(\omega_{\mathrm{bb}}t) \tag{6.3}$$

and ε and ϕ are gain and phase errors, respectively. The gain error ε is expressed as

$$\varepsilon = \frac{G_{\mathrm{I}}}{G_{\mathrm{Q}}} - 1 \tag{6.4}$$

where G_{I} and G_{Q} are gain values of the I–Q channels, respectively. With the Taylor series first-order approximations to G_{I} and G_{Q}, we have

$$G_{\mathrm{I}} \approx 1 + \frac{\varepsilon}{2}, \quad G_{\mathrm{Q}} \approx 1 - \frac{\varepsilon}{2}, \tag{6.5}$$

The expression of (6.5) is used in (6.2) and (6.3). To obtain the constellation of the RF-modulated signal in (6.1), we need to coherently demodulate the modulated signal $y(t)$ with ideally synchronized carrier signals $\mathbf{cos}(\omega_{\mathrm{c}}t)$ and $\mathbf{sin}(\omega_{\mathrm{c}}t)$ at the receiver. After respectively multiplying $y(t)$ by these two quadrature carrier signals and lowpass-filtering the high-order harmonics, we obtain the recovered baseband I–Q signals:

$$x_{\mathrm{RI}}(t) = \frac{A}{2}\left(1 + \frac{\varepsilon}{2}\right)\mathbf{cos}\,(\omega_{\mathrm{bb}}t)\mathbf{cos}\left(\frac{\phi}{2}\right) - \frac{A}{2}\left(1 - \frac{\varepsilon}{2}\right)\mathbf{sin}\,(\omega_{\mathrm{bb}}t)\mathbf{sin}\left(\frac{\phi}{2}\right) \tag{6.6}$$

$$x_{\mathrm{RQ}}(t) = -\frac{A}{2}\left(1 + \frac{\varepsilon}{2}\right)\mathbf{cos}\,(\omega_{\mathrm{bb}}t)\mathbf{sin}\frac{\phi}{2} + \frac{A}{2}\left(1 - \frac{\varepsilon}{2}\right)\mathbf{sin}\,(\omega_{\mathrm{bb}}t)\mathbf{cos}\left(\frac{\phi}{2}\right) \tag{6.7}$$

In the ideal case, the I–Q gain and phase errors are equal to zero. The vector trace of the complex sinusoidal signal is illustrated in Fig. 6.4a by a large circle. To see

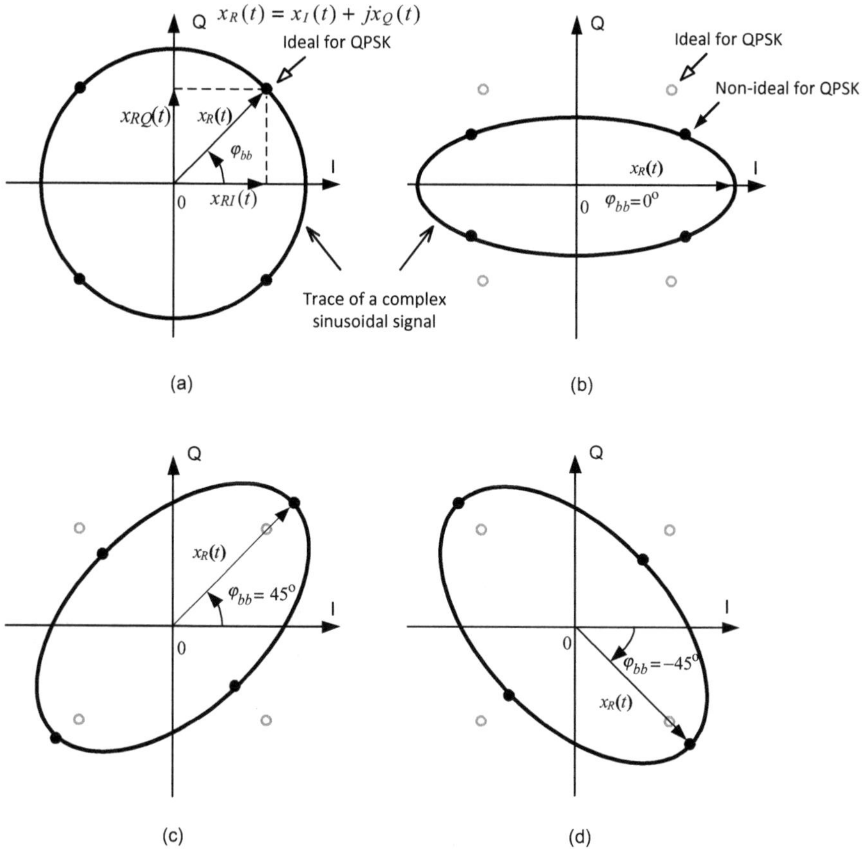

Fig. 6.4 Effect of the I–Q gain and phase imbalance on quadrature signal constellation: (**a**) ideal case, (**b**) $\phi = 0$, $\varepsilon > 0$, or $G_I > G_Q$, (**c**) $\varepsilon = 0$, or $G_I = G_Q$, $\phi < 0$, (**d**) $\varepsilon = 0$, or $G_I = G_Q$, $\phi > 0$

the effect of the I–Q gain error on the constellation vector trace, set the phase error $\phi = 0$. Equations (6.6) and (6.7) can be rewritten as

$$x_{RI}(t) = \frac{A}{2}\left(1 + \frac{\varepsilon}{2}\right)\cos\left(\omega_{bb}t\right) \tag{6.8}$$

$$x_{RQ}(t) = \frac{A}{2}\left(1 - \frac{\varepsilon}{2}\right)\sin\left(\omega_{bb}t\right) \tag{6.9}$$

From (6.4), when the I gain is larger than the Q gain, or $G_I > G_Q$ and $\varepsilon > 0$, the maximum baseband signal vector occurs at $\omega_{bb}t = \varphi_{bb} = 0$, or I-axis, as shown in Fig. 6.4b by a large ellipse. Similar to the I channel case, when the Q gain is bigger than the I gain, or $G_I < G_Q$ and $\varepsilon < 0$, the maximum baseband signal vector occurs at $\omega_b t = \varphi_{bb} = \pi/2$, or Q-axis (not shown in Fig. 6.4).

To see the effect of the I–Q phase error on the vector trace, let the gain error $\varepsilon = 0$ (6.6) and (6.7) can be rewritten as

$$x_{RI}(t) = \frac{A}{2}\, \mathbf{cos}\left(\omega_{bb}t + \frac{\phi}{2}\right) \tag{6.10}$$

$$x_{RQ}(t) = \frac{A}{2}\, \mathbf{sin}\left(\omega_{bb}t - \frac{\phi}{2}\right) \tag{6.11}$$

Equations (6.10) and (6.11) show that the maximum baseband signal vector trace occurs at $\omega_{bb}t = \varphi_{bb} = 45°$ for a negative phase error, or $\phi < 0$, while the maximum vector trace occurs at $\omega_{bb}t = \varphi_{bb} = -45°$ for a positive phase error, or $\phi > 0$, as shown in Fig. 6.4c, d by the large ellipses. It is clearly seen that the complex sinusoidal signal with either gain error or phase imbalance error causes the vector trace shape change from a perfect circle in the ideal case of zero imbalance. Next, we apply a complex-value QPSK signal; its real and imaginary parts are referred to as the baseband I–Q components of BB signals, to the input of the I–Q modulator. In the ideal case of zero imbalances, the constellation diagram of QPSK with the optimal decision values is shown in Fig. 6.4 (a) by four small filled circles. The constellation diagrams of QPSK, corresponding to other non-ideal conditions used for the complex sinusoidal signal, are illustrated in Fig. 6.4 (b), (c), and (d), respectively, by four small filled circles. It can be seen that these filled circles deviate from the ideal constellation diagram of QPSK, represented by small empty circles. As a result, BER or PER will degrade due to these deviations at the receiver. Therefore, the gain and phase imbalances need to be minimized through the calibration at the transmitter.

Quadrature Modulator Sideband Suppression: Quadrature modulators perform spectral transfer of the baseband signal by using the I–Q data streams to modulate two orthogonal carrier signals in their phases, amplitudes or both, respectively, such as QPSK and QAM formats. Important specifications for quadrature modulators are the carrier suppression and sideband suppression, which are completely correlated with DC offsets, amplitude, and phase imbalance errors on the baseband I–Q channels or the orthogonal LO signal paths. Usually DC offsets are generated on the baseband I–Q channels, while the I–Q amplitude imbalance error may be generated on either baseband I–Q channels or the orthogonal LO paths. The phase imbalance error is mainly generated in a 90° phase splitter on the LO path, which provides a pair of orthogonal carrier signals. Figure 6.5 illustrates the concept of a high sideband suppression and carrier suppression at the output of a conventional quadrature modulator, as shown in Fig. 6.3, where the baseband I–Q signals of $G_I\mathbf{cos}(\omega_{bb}t)$ and $G_Q\mathbf{sin}(\omega_{bb}t)$ are expressed in (6.2) and (6.3). In performing carrier and sideband suppression measurements, the baseband I–Q signals are single tones with the frequency of f_{bb} and orthogonal (or 90° between in–phase and quadrature paths) to each other. The baseband I–Q signals are, respectively, mixed with a pair of orthogonal carrier (LO) signals at the frequency

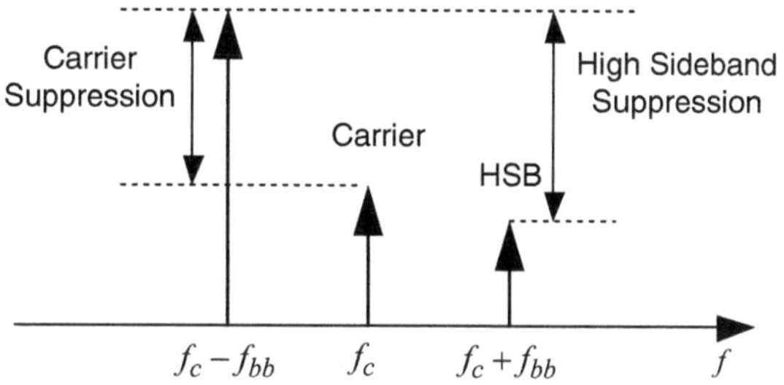

Fig. 6.5 Carrier at the frequency of f_c and high sideband suppressions for a pair of $\mathbf{cos}(\omega_{bb}t)$ and $\mathbf{sin}(\omega_{bb}t)$ baseband signals at a frequency of f_{bb}

of f_c to generate new frequency components at the outputs of the I mixer and Q-mixer at the frequencies of $f_{bb}+f_c$ and f_c-f_{bb}. In a perfect modulator, at the outputs of the I mixer and Q-mixer, the frequency components at f_c-f_{bb} have the same amplitude and phase, while the frequency components at f_c+f_{bb} have the same amplitude and an exact 180° difference in phase. After combining the outputs of the I mixer and Q-mixer, the frequency component at f_c+f_{bb} is completely cancelled, while the frequency components at f_c-f_{bb} is added with double amplitude.

If there are no DC offsets on the I–Q channels, then the carrier is completely suppressed at the output of the quadrature modulator. Similar to the carrier suppression, if the amplitudes and phases of both orthogonal carrier signals are the same and different by 90°, respectively, and the amplitudes of the baseband I–Q channels are also the same, then one of the sidebands is completely cancelled out at the output of the quadrature modulator. In practice, however, complete cancellation is never accomplished, but with careful design, achieving 40 dB of sideband cancellation is possible. Similar to the sideband suppression, the carrier suppression of –25 dBc or even more is also achievable with careful design. For higher suppression requirements, the calibration or compensation method can be used to minimize DC offsets and I–Q imbalances. To understand how compensation can be achieved, it is helpful to understand how sideband suppression and carrier suppression are related with the amplitude and phase imbalance errors.

The level of the sideband signal power can be calculated using the sideband suppression equation as follows

$$\mathrm{SBS(dBc)} = 10\,\mathbf{log}\,\frac{\gamma^2 + 1 - 2\gamma\,\mathbf{cos}\,(\phi)}{\gamma^2 + 1 + 2\gamma\,\mathbf{cos}\,(\phi)} \tag{6.12}$$

The gain ratio γ and gain imbalance ε are expressed as follows:

$$\gamma = \frac{G_I}{G_Q} \quad \text{and} \quad \varepsilon = \gamma - 1 \tag{6.13}$$

In order to plot a set of suppression contours with gain and phase errors as the axes, we can express the phase error ϕ as a function of gain error γ and sideband suppression SBS, or

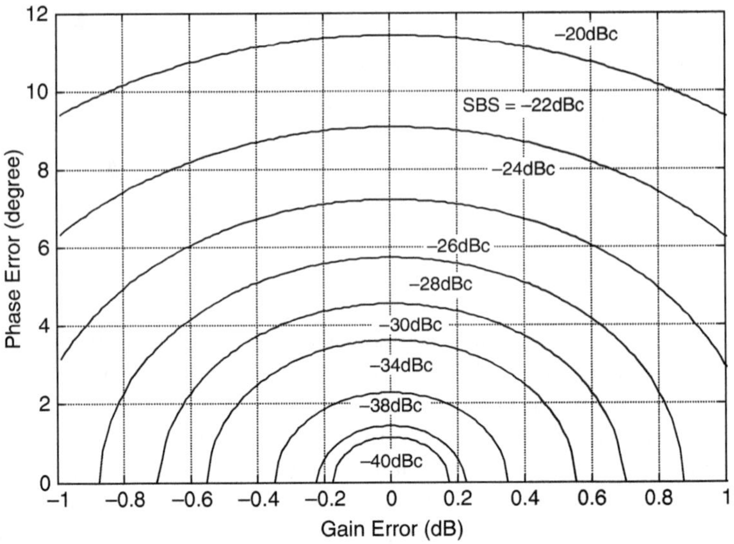

Fig. 6.6 Sideband suppression versus gain and phase imbalance errors

$$\phi = \cos^{-1}\left(\frac{\gamma^2 + 1 - \gamma^2 10^{\frac{SBS(dB)}{10}} - 10^{\frac{SBS(dB)}{10}}}{2\gamma 10^{\frac{SBS(dB)}{10}} + 2\gamma}\right) \tag{6.14}$$

Figure 6.6 illustrates different SBS values versus gain and phase imbalance errors, where SBS values versus negative phase imbalance error are not shown here because they are identical to SBS values versus positive phase imbalance error. Compared with phase imbalance adjustment, it is relatively easy to balance gain error due to no cross-adjustment between the I and Q channels. Figure 6.7 shows SBS values as a function of the phase error for different gain errors.

It should be noted that sideband suppression is also called the image rejection ratio (IRR), which quantifies the factor that describes how much the mirror signal is suppressed or removed after down-conversion or up-conversion by combining the in-phase and quadrature channels. During the down-conversion or up-conversion the mirror signal is usually folded onto the bandwidth of the wanted signal. As a result, the folded mirror signal distorts the wanted signal. Any gain and phase imbalance errors decrease such an ability to remove or suppress the mirror signal, especially in a low-IF receiver. Basically, SBS or IRR factor must be small enough in order not to distort the wanted signal.

Compensations for I–Q Imbalance and DC Offset: There are many compensation methods for minimizing the I–Q imbalances and DC offsets in literature. In this section, a compensation method of the I–Q imbalances and DC offsets by means of the FFT operation is introduced, which is very suitable for an OFDM-signal–based transceiver.

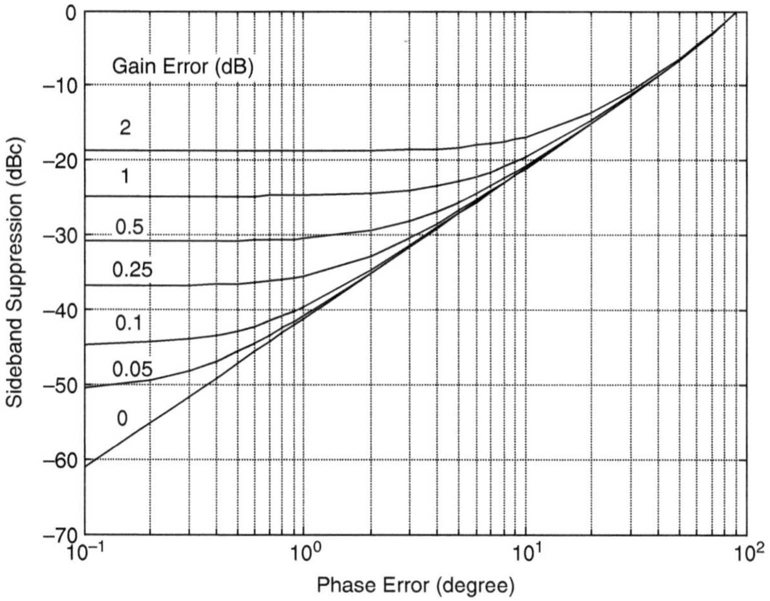

Fig. 6.7 Sideband suppression vs. phase error with different gain errors

Considering DC offsets of D_I and D_Q in the I and Q channels, we can rewrite (6.2) and (6.3) as

$$x_I(t) = \left(1 + \frac{\varepsilon}{2}\right)\cos(\omega_{bb}t) + D_I \tag{6.15}$$

$$x_Q(t) = \left(1 - \frac{\varepsilon}{2}\right)\sin(\omega_{bb}t) + D_Q \tag{6.16}$$

By substituting (6.15) and (6.16) into (6.1), and also applying the trigonometric identity to (6.1), we can write it as

$$y(t) = U(t)\cos(\omega_c t) - V(t)\sin(\omega_c t) \tag{6.17}$$

Here, $U(t)$ and $V(t)$ are expressed as

$$U(t) = x_I(t)\cos(\phi/2) + x_Q(t)\sin(\phi/2) \tag{6.18}$$

$$V(t) = x_I(t)\sin(\phi/2) + x_Q(t)\cos(\phi/2) \tag{6.19}$$

It should be pointed out that in the derivative above the amplitude of A is set to 1 and the sign before the second item in (6.1) is set to negative for the sake of simplicity.

It can be clearly seen from (6.18) and (6.19) that the phase imbalance error ϕ is equivalently transferred into the baseband domain. Thus, the gain and phase imbalance errors ε and ϕ, and DC offset errors D_I and D_Q, are enclosed in the equivalent baseband signals $U(t)$ and $V(t)$. These three kinds of errors can be detected from the envelope of the modulated signal in (6.17) by passing them through a squared device, and then a lowpass filter with DC blocking (or a series capacitor before a LPF). The envelope expression of (6.17) is given as

$$\begin{aligned}\mathrm{ENV}(t) &= U^2(t) + V^2(t)\\ &= \{[(1+\varepsilon/2)\mathbf{cos}\,(\omega_{bb}t) + D_I]\mathbf{cos}\,(\phi/2) + [(1-\varepsilon/2)\mathbf{sin}\,(\omega_{bb}t) + D_Q]\mathbf{sin}\,(\phi/2)\}^2\\ &\quad + \{[(1+\varepsilon/2)\mathbf{cos}\,(\omega_{bb}t) + D_I]\mathbf{sin}\,(\phi/2) + [(1-\varepsilon/2)\mathbf{sin}\,(\omega_{bb}t) + D_Q]\mathbf{cos}\,(\phi/2)\}^2\end{aligned} \tag{6.20}$$

For the sake of simplicity, the I–Q imbalance and DC offset errors will be treated separately in the following derivatives.

1. *I–Q Gain and Phase Imbalance Errors.*
   In this case, $D_I = D_Q = 0$. After the DC component is blocked by the lowpass filter with DC blocking, (6.20) can be written as [7]:

$$\mathrm{ENV}(t) \approx \varepsilon\,\mathbf{cos}\,(2\omega_{bb}t) + \phi\,\mathbf{sin}\,(2\omega_{bb}t) \tag{6.21}$$

   In the derivative above, gain and phase imbalance errors are assumed to be small, or $\varepsilon \ll 1$, $\phi \ll 1$. Thus, both gain and phase imbalance errors in (6.21) are related to the second harmonic frequency of the test tone. These two errors can be estimated by taking the DFT operation of (6.21), or

$$\hat{\varepsilon} = \mathbf{Re}\{\mathrm{ENV}(2\omega_{bb})\} \tag{6.22}$$

$$\hat{\phi} = \mathbf{Im}\{\mathrm{ENV}(2\omega_{bb})\} \tag{6.23}$$

   where $ENV(2\omega_{bb})$ is the discrete Fourier transform of $ENV(t)$ at $2\omega_{bb}$.
   It is convenient to use a recursive equation to approach the optimal values with a small step size λ for each iteration, or

$$\hat{\varepsilon}\,(n) = \hat{\varepsilon}\,(n-1) + \Delta\hat{\varepsilon}\,, \quad \hat{\phi}\,(n) = \hat{\phi}\,(n-1) + \Delta\hat{\phi} \tag{6.24}$$

   with

$$\Delta\hat{\varepsilon} = \lambda\,\mathbf{Re}\{\mathrm{ENV}(2\omega_{bb})\}, \quad \Delta\hat{\phi} = \lambda\,\mathbf{Im}\{\mathrm{ENV}(2\omega_{bb})\} \tag{6.25}$$

2. *DC Offset Error.*
   In this case, $\phi = 0$, and $\varepsilon = 0$. After the DC component is blocked by the lowpass filter with DC blocking, (6.20) can be written as [7]:

$$\mathrm{ENV}(t) \approx D_{\mathrm{I}} \mathbf{cos}\,(\omega_{\mathrm{bb}} t) + D_{\mathrm{Q}} \mathbf{sin}\,(\omega_{\mathrm{bb}} t) \tag{6.26}$$

DC offset errors in the I and Q channels are both related to the fundamental frequency of the test tone. DC offsets can be estimated by taking the DFT operation, or

$$\hat{D}_{\mathrm{I}} = \mathbf{Re}\{\mathrm{ENV}(\omega_{\mathrm{bb}})\} \tag{6.27}$$

$$\hat{D}_{\mathrm{Q}} = \mathbf{Im}\{\mathrm{ENV}(\omega_{\mathrm{bb}})\} \tag{6.28}$$

Similar to the estimation of the gain and phase imbalance, the recursive equations for DC offset are given by

$$\hat{D}_{\mathrm{I}}(n) = \hat{D}_{\mathrm{I}}(n-1) + \Delta\hat{D}_{\mathrm{I}}, \quad \hat{D}_{\mathrm{Q}}(n) = \hat{D}_{\mathrm{Q}}(n-1) + \Delta\hat{D}_{\mathrm{Q}} \tag{6.29}$$

with

$$\Delta\hat{D}_{\mathrm{I}} = \lambda \mathbf{Re}\{\mathrm{ENV}(\omega_{\mathrm{bb}})\}, \quad \Delta\hat{D}_{\mathrm{Q}} = \lambda \mathbf{Im}\{\mathrm{ENV}(\omega_{\mathrm{bb}})\} \tag{6.30}$$

After the gain and phase imbalance parameters $\hat{\varepsilon}$ and $\hat{\phi}$, and the DC offset parameters $\hat{D}_{\mathrm{I}}$ and $\hat{D}_{\mathrm{Q}}$ are adaptively estimated, the fundamental frequency item and second-order frequency item are minimized. Figure 6.8 shows a block diagram of the digital compensator for the I–Q gain, phase imbalance, and DC offset errors in the transmitter. In practice, the compensation for the I–Q imbalance and DC offset errors can be performed during the period of the calibration in a RF transceiver. ADC and FFT calculation blocks can be shared with a digital baseband receiver, which is capable of supporting OFDM signal receptions.

To validate the adaptive algorithm described above, we performed MATLAB simulation to adaptively compensate for these impairments, targeting the 802.11n WLAN in the 2.4 GHz band. A test tone located at the fourth subcarrier at a frequency of $f_{\mathrm{b}} = 4 \times 0.3125 = 1.25$ MHz modulates a carrier signal at a frequency of 2.412 GHz in channel 1 of the 2.4 GHz band. Figure 6.9 illustrates that adaptive compensation for the I–Q imbalance and DC offset is performed in three different stages. After the modulation, the frequency component at $f_{\mathrm{c}} + f_{\mathrm{b}}$ is partially cancelled due to I–Q imbalance error, the frequency component at $f_{\mathrm{c}} - f_{\mathrm{b}} = 2412–1.25 = 2410.75$ MHz is summed as the desired RF output signal, and the middle component at $f_{\mathrm{c}} = 2412$ MHz is a carrier leakage signal due to DC offsets, as shown in Fig. 6.9a. It can be clearly seen that at the beginning the carrier leakage located in the middle of components is about −25 dB down related to a lower sideband component in the left due to DC offsets, while an upper sideband component in the right is about −27 dB down related to the lower sideband component. After three iterations, they are reduced to −35 and −36 dB,

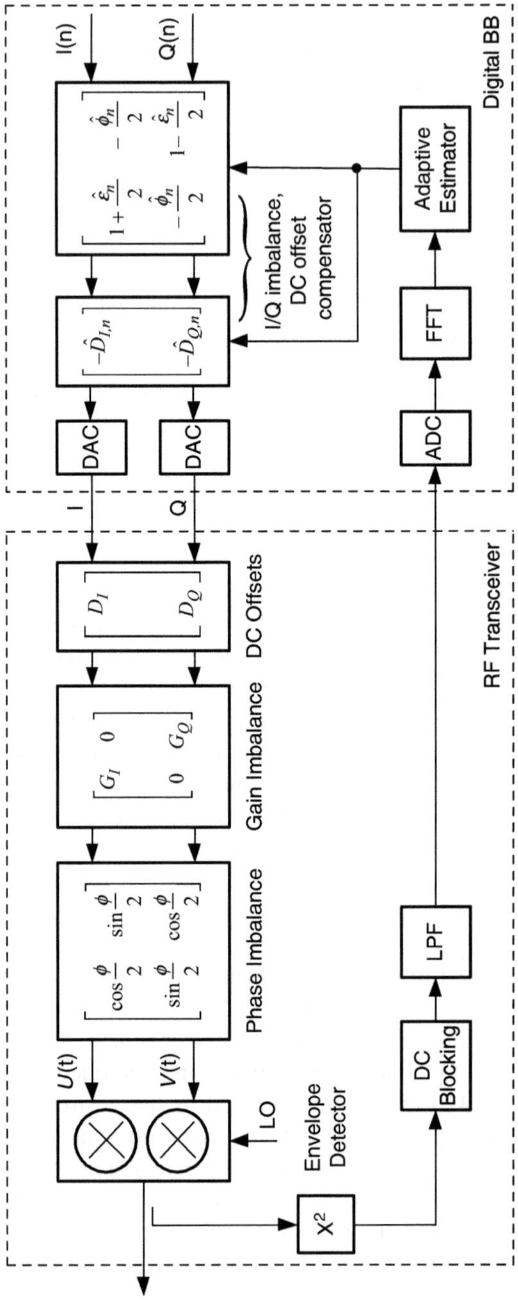

Fig. 6.8 A block diagram of I–Q imbalance and DC offset compensator in an 802.11n WLAN transmitter

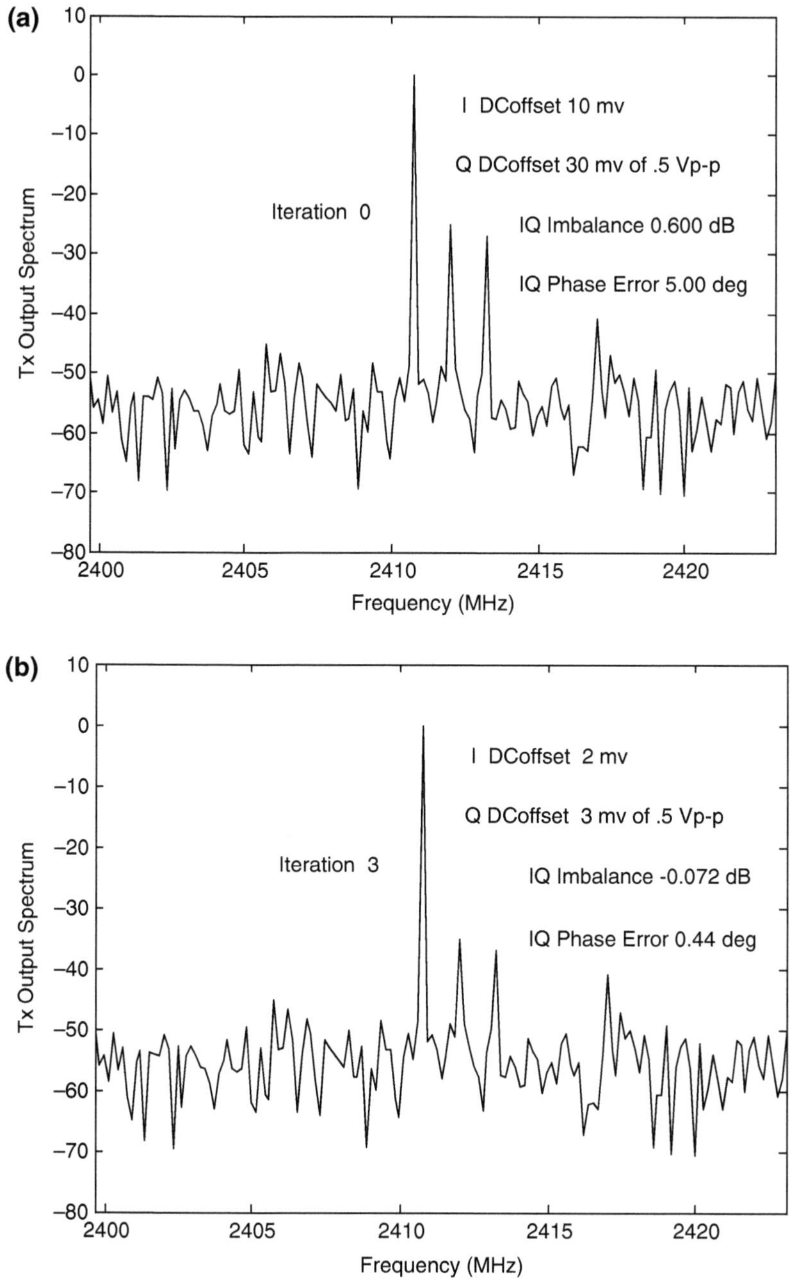

Fig. 6.9 Adaptive compensation for I–Q imbalance and DC offset errors in channel 1 at the center frequency of 2.412 MHz in the WLAN 2.4-GHz band, a test tone at a fourth subcarrier $f_b = 4 \times 312.5$ kHz = 1.25 MHz: (**a**) TX spectrum with initial errors, (**b**) TX spectrum after the 3rd iteration, and (**c**) TX spectrum after the fifth iteration

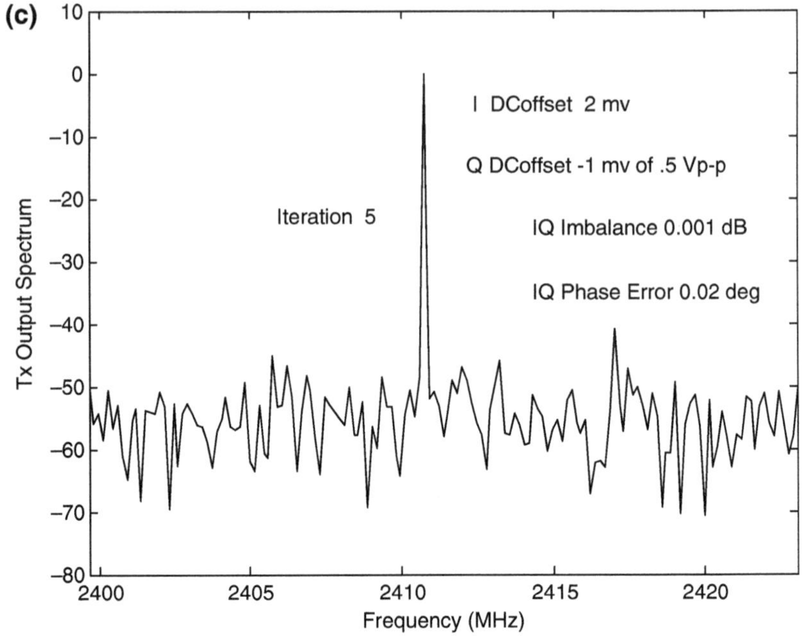

Fig. 6.9 (continued)

respectively, compared to the lower sideband component in the left, as shown in Fig. 6.9b. Finally, they are reduced to more than −45 dB after five iterations, as shown in Fig. 6.9c. Note from Fig. 6.9 that the leftmost one is the low sideband desired signal, the middle one is the carrier leakage, and the rightmost one is the high sideband suppressed signal.

6.5.2 LO Leakage

Ideally, there should be no signal at the PA output or at the RF output port when there is no data transmission. A common problem that modern communication systems face is that the local oscillator (LO) signal used as the modulator's carrier signal may leak to the RF output [8, 9]. Two major reasons for this leakage are the finite isolation between LO and RF ports and an unavoidable DC offset voltage at the input of the mixer due to mismatches and imperfections in the analog baseband components [10], including DAC offsets.

There are two main reasons to reduce LO leakage. Firstly, the LO leakage signal as interferer may drop into other channel with the same frequency band. For example, in WLAN systems, since the transmitter and receiver share the same frequency band; a TX/RX switch between the antenna and the transceiver is used to achieve transmission and reception functions in different time intervals.

The LO signal leaked into the receive chain through either the antenna or the TX/RX switch will affect the received signal when the leakage level is relatively high compared to a weakly received signal. In a WCDMA UMTS system where the transmitter and receiver operate at the same time but in different frequency bands or up-link frequency band from 1920 to 1980 MHz used for the transmission of signals from a user equipment (UE) to a base station (BS) and down-link frequency band from 2110 to 2170 MHz used for the reception of signals from a BS to a UE in the band I, the requirement for the transmitter OFF power should be less than –56 dBm [11]. If the transmitter leakage signal to the receiver chain is not too small, it may create a third-order intermodulation product with other blockers through the mixer of the receiver and drop into the desired baseband signal band. As a result, the system performance degrades.

Secondly, poor carrier suppression at the RF-modulated signal spectrum consumes the transmitted signal power such that the useful signal-to-noise ratio decreases because the carrier does not carry any information. Furthermore, the RF-modulated signal with poor carrier suppression may result in a DC offset after the down-conversion in the receiver. The IEEE 802.11WLAN standard mandates that the transmitter center frequency leakage should not exceed -15 and -20 dB relative to the overall transmitted power for transmission at a 20 MHz bandwidth and a 40 MHz bandwidth, respectively. For the carrier leakage related to such an OFDM signal, one cancellation method that was introduced for I–Q imbalance correction in the previous section can be used. Even though there are many other methods for LO leakage cancellation [9, 12], in this section another calibration method for LO leakage for a more general modulated signal, including both OFDM signals and non-OFDM signals, will be presented. This calibration method is called the synchronous detection method and its block diagram is shown in Fig. 6.2.

One key feature is to estimate the proper DC voltage in a loop back path from the PA driver output to the I–Q BB input and then to subtract the input DC offset at the I–Q BB inputs step by step until the LO leakage is minimal. The synchronous detection method uses a coherent demodulation to extract the DC voltage. In Fig. 6.2, a weak LO leakage RF signal is decoupled at the output of the PA driver as an input to a down-conversion mixer in the loop back path. Since the RF LO leakage signal consists of the RF *Sine* (quadrature) and *Cosine* (in-phase) signals, synchronous detection is performed separately through a switch controlled by V_{CTR} between the I–Q channels. First, the DC offset caused in the I channel is detected by multiplying the LO leakage signal with the synchronous *Cosine* signal. The DC offset is passed through a lowpass filter, while the second-order harmonic is filtered out. The DC offset signal is then converted into the digital signal through a voltage comparator. A binary search algorithm in the digital circuit is used to output an N-bit code word to the I channel DAC with an N-bit resolution. N depends on how small the LO leakage level needs to be. The bigger the N number, the lower the LO suppression level that can be reached.

After the LO leakage signal cancellation is done in the I channel, the LO leakage cancellation is then carried out in the Q channel; the procedure is the same as that in

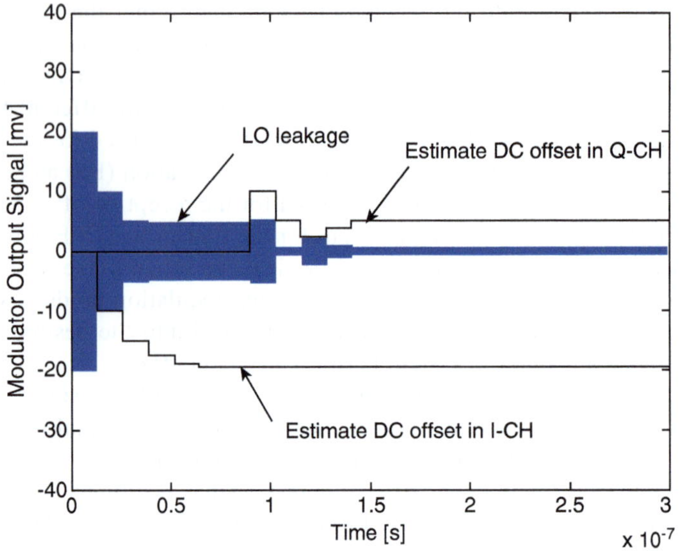

Fig. 6.10 LO amplitude is reduced from an initial value of −20 mV to a final value of 0.625 mV with 6-bit DAC; LO leakage signal at a frequency of 2.4 GHz, binary search clock frequency of 22 MHz

the I channel. An advantage of synchronous detection is that a very small LO leakage value can be reached because the method avoids measuring the small LO leakage signal level precisely as one in [9].

As an example, the principle of synchronous detection using a binary search algorithm with 6-bit logic coding is shown in Fig. 6.10, where a −20-mV DC offset voltage exists in the I channel and a 4.8-mV DC offset voltage exists in the Q channel as initial DC offsets, respectively. The calibration is performed in the I channel first until the LO leakage signal is minimized and is then carried out in the Q channel until the LO leakage signal is minimized. The solid waveform in Fig. 6.10 represents the combined carrier leakage signals at a frequency of 2.4 GHz from the I–Q channels. For an N-bit binary search, the successive approximation register (SAR) performs $(N-1)$-step operations. The residual error is usually less than or equal to the step size of ΔV.

In this case, it is assumed that the maximum peak-to-peak value is in the range from –20 to 20 mV and a 6-bit resolution DAC is used. Then, a step size $\Delta V = 40\ (\text{mV})/2^6 = 0.625$ mV is calculated. First, the LO leakage calibration is carried out in the I channel. After 5-step binary searches, the LO leakage caused by the I channel DC offset is minimized. During the calibration of the I channel DC offset, the LO leakage caused by the Q-channel DC offset cannot be reduced. In the next procedure, the LO leakage signal is cancelled by switching *Sine* signal to the cancellation circuit. After another 5-step binary searches in the Q-channel, the peak-to-peak of the LO leakage signal is reduced to −0.625 mV from the initial value 4.8 mV. The absolute residual error is equal to the step size $\Delta V = 0.625$ mV. It can be seen from Fig. 6.10 that the estimate DC offsets are close to –20 and 4.8 mV in the I and Q channels, respectively.

It should be noted that the final LO leakage signal might not be smallest among the previously reached LO leakage values, but it is less than or equal to the step size. This is because the N-bit binary search algorithm must perform $(N-1)$-time even though the LO leakage signal reaches the minimum value before $(N-1)$-time operation. It should be also pointed out that the carrier phase error between quadrature carrier signals and the synchronized quadrature signals should be as small as possible so that the detected DC offsets are close to their true values.

6.5.3 VCO Phase-Noise Disturbance

VCO phase noise can be introduced into a RF-modulated signal when a baseband signal is mixed with a local oscillator (LO) signal that is obtained from a VCO output signal to perform frequency translation from baseband to RF. Hence, the quality of VCO phase noise plays a very important role for the overall performance of both the transmitter and receiver. The LO phase-noise contribution reflects the frequency stability of the reference crystal oscillator used by the frequency synthesizer and frequency stability of the free-running voltage-controlled oscillator (VCO) used by the synthesizer's phase-locked loop (PLL) [13]. Depending on the PLL-loop bandwidth, the ideal result is that the synthesizer's output phase-noise spectral density is dominated by the generally good long-term stability of the crystal oscillator at low frequency offsets and by the generally good short-term stability of the VCO at high frequency offsets, with the in-band noise floor established by the phase detector and frequency dividers of the PLL itself [13].

The performance of the transmitter can degrade due to a VCO frequency pulling phenomenon. The VCO frequency pulling occurs when the VCO frequency changes in response to either load varying or injection pulling. In the former case, a change in impedance seen by the VCO output can induce changes in the DC voltage across junctions of the VCO's active devise. As a result, it leads to the VCO's frequency change. In the latter case, an interfering signal that is very close to the VCO's operation frequency and is of sufficient amplitude at the VCO output port can cause the VCO to shift its oscillation frequency to track the interfering frequency instead of the VCO output frequency.

The performance degradation of the transmitter can be also caused by VCO phase noise disturbance. One of major phenomena that produce the VCO phase noise disturbance is the leakage of the second-order harmonic distortion (HD2) at the PA output to the PLL due to finite isolation, especially in the case when the PA has relatively large output power.

In a direct up-conversion transmitter, the phase-noise disturbance of the VCO is the biggest potential challenge for the RFIC designers if a pair of orthogonal LO signals used for a quadrature modulator are obtained by dividing the VCO signal by two. Thus, the HD2 of the RF-modulated signal at the output of the PA has the same frequency as the VCO signal. The HD2 that has twice the bandwidth of the RF modulated signal may leak to the PLL due to finite isolation and pass through a divider with the VCO signal together. Figure 6.11 illustrates a possible mechanism of the PA leakage to the PLL. The divided HD2 component as "noise" occurs at the input of the phase and frequency detector (PFD) and then passes through a loop filter with the divided VCO signal together. The lowpass filtered phase-frequency error simply increases the phase-noise power spectral density of the VCO after the loop filter, and hence degrades the VCO phase noise and its spectral purity. As a result, the VCO phase-noise disturbance results in EVM degradation of the transmitted signal. Here, the term "VCO disturbance" is preferable to "VCO-pulling" that is commonly used in literature because there is no VCO frequency-pulling phenomenon in this case.

Figure 6.12a shows a constellation of the 802.11g DQPSK-DSSS signal measured at a RF transceiver output for the case where a strong second-order harmonic signal at a PA output affects the VCO phase noise due to insufficient isolation between the PA output and the PLL. It can be seen that the constellation degrades in the phase angle spreading rather than amplitude disturbance, as shown by the banana-shaped dark traces. The banana-shaped dark traces with a phase error of 9.6° indicate that the constellation at the sampled instants spreads along the circle due to the VCO phase noise degradation rather than PA's nonlinear distortion. The constellation diagram shown in Fig. 6.12b, however, becomes more compact at the decision points when sufficient isolation between the PA output and the VCO control loop is increased to mitigate the VCO disturbance. In Fig. 6.12b, the PA output power is the same as the one in Fig. 6.12a. Therefore, the isolation between the PA output and the PLL is a big challenge in RF transceiver designs.

In addition to increasing the isolation between the PA output signal and the PLL to reduce the effect of the PA output signal on the VCO phase disturbance, the phenomenon of VCO phase disturbance can be avoided if the RF signal frequency and its harmonic frequencies are far away from the frequency of the VCO. One effective method is to use a quadrature-structured VCO operating at two-thirds of the LO frequency, as shown in Fig. 6.13 [14, 15]. Suppose the frequency of the LO signal is represented by f_{LO}. Then the frequency of the VCO is given by $f_{\mathrm{VCO}} = (2/3)f_{\mathrm{LO}}$. The VCO signal is first split into two paths: one passes through a 90° shift network to get two quadrature signals at the frequency of f_{VCO}, while another goes to a divide-by-two circuit to obtain another two quadrature signals at one-third of the LO frequency, or $f_{\mathrm{DIV}} = (1/3)f_{\mathrm{LO}}$. These two coupled quadrature signals go to a single-sideband (SSB) circuit to obtain the quadrature upper-band LO signals and meanwhile suppress the quadrature lower-band LO signals. The detailed derivation is given by

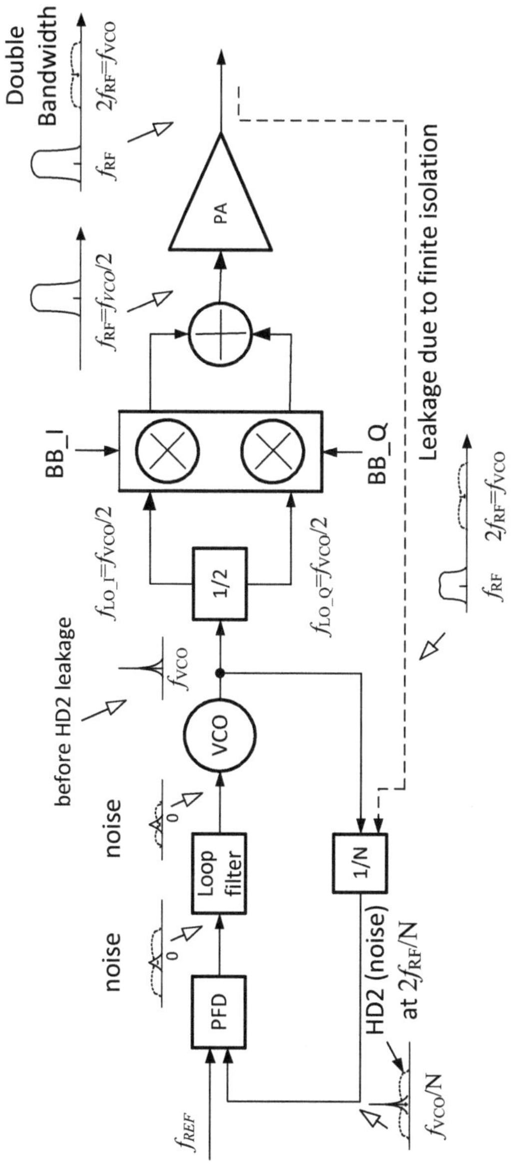

Fig. 6.11 Block diagram of PA leakage to PLL control loop

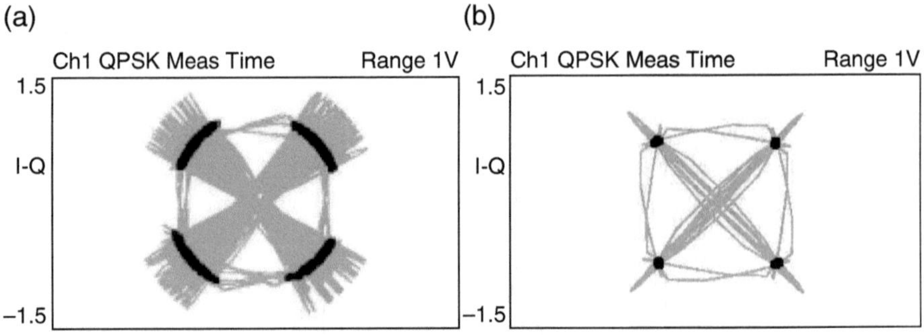

Fig. 6.12 Constellation of the 802.11g 2 Mbps DQPSK-DSSS signal due to VCO phase noise disturbance: (**a**) RMS EVM of 16.5 % and phase error of 9.4°, and (**b**) RMS EVM of 2.7 % and phase error of 1.2°

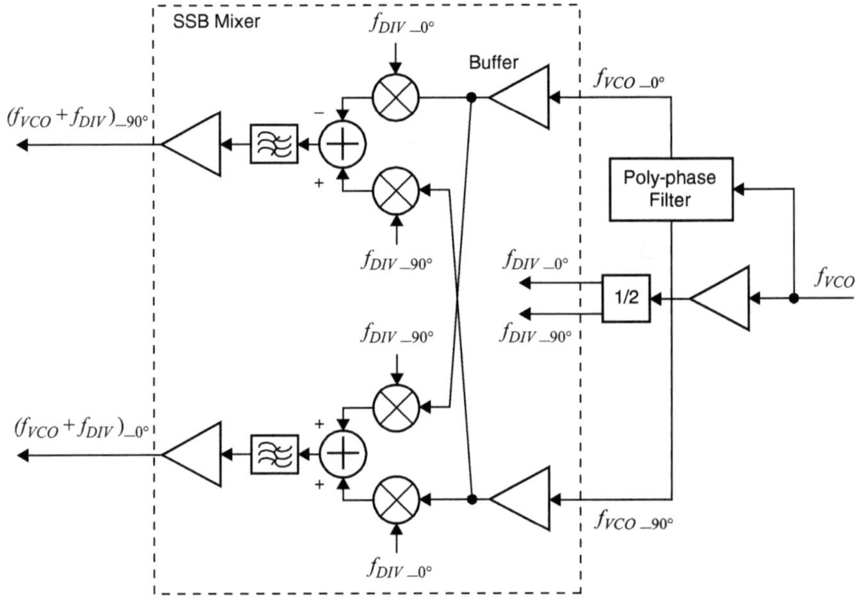

Fig. 6.13 Block diagram of single-sideband mixer

$$\begin{aligned}(f_{\rm VCO}+f_{\rm DIV})\_0^{\circ} &= 0.5\{\cos\,[2\pi(f_{\rm VCO}+f_{\rm DIV})t]+\cos\,[2\pi(f_{\rm VCO}+f_{\rm DIV})t]\}\\ &\quad -0.5\{\cos\,[2\pi(f_{\rm VCO}+f_{\rm DIV})t]-\cos\,[2\pi(f_{\rm VCO}+f_{\rm DIV})t]\}\\ &= \cos\,[2\pi(f_{\rm VCO}+f_{\rm DIV})t]\\ &= \cos\left[2\pi\left(\frac{2}{3}f_{\rm LO}+\frac{1}{3}f_{\rm LO}\right)t\right]=\cos\,(2\pi f_{\rm LO}t)\end{aligned} \tag{6.31}$$

$$
\begin{aligned}
(f_{\mathrm{VOC}} + f_{\mathrm{DIV}})\_90^\circ &= 0.5\{\mathbf{sin}\,[2\pi(f_{\mathrm{VCO}} + f_{\mathrm{DIV}})t] - \mathbf{sin}\,[2\pi(f_{\mathrm{VCO}} + f_{\mathrm{DIV}})t]\} \\
&\quad + 0.5\{\mathbf{sin}\,[2\pi(f_{\mathrm{VCO}} + f_{\mathrm{DIV}})t] + \mathbf{sin}\,[2\pi(f_{\mathrm{VCO}} + f_{\mathrm{DIV}})t]\} \\
&= \mathbf{sin}\,[2\pi(f_{\mathrm{VCO}} + f_{\mathrm{DIV}})t] \\
&= \mathbf{sin}\left[2\pi\left(\frac{2}{3}f_{\mathrm{LO}} + \frac{1}{3}f_{\mathrm{LO}}\right)t\right] = \mathbf{sin}\,(2\pi f_{\mathrm{LO}}t)
\end{aligned}
\tag{6.32}
$$

From (6.31) and (6.32), we can see that harmonic frequencies of the local oscillator do not coincide with the VCO frequency. In turn, any harmonic frequencies of the RF signal also do not coincide the VCO frequency because the LO frequency is the same as the RF frequency in the direct conversion transceiver. As a result, the VCO disturbance or pulling problem can be avoided when the PA delivers high output power. This frequency planning is very useful when the PA and the RF transceivers are integrated together in one IC chip.

In Fig. 6.13, a quadrature-structured VCO needs a 90° phase shifter. There are several ways to implement a 90° phase shifter, such as *an RC-CR network* [16], and *a Sequence Asymmetric Polyphase Network* [17]. For detailed design information, the interested reader can reference the literature.

Transient effects that happen in WLAN systems cause serious frequency pulling and can be especially difficult to isolate and identify. In the 802.11 WLAN systems, the RF power amplifier is switched on and off between transmission and reception because they share the same frequency band. When the PA is enabled before each transmission, the PA will start drawing significant current and may cause a drop in the power supply voltage or induce a ground current [13]. As a result, this transient phenomenon can affect the frequency synthesizer and introduce transient frequency drift and phase noise, which momentarily degrade the transmitted signal performance, such as EVM. Figure 6.14 shows that such frequency pulling of the synthesizer's VCO causes big frequency errors at the moment of switching on the PA, and generally settles down to its steady-state frequency within microseconds. If such a settlement takes too long and the frequency drift is too far away from its steady frequency, the transient frequency errors can degrade the received signal performance in the case that the receiver estimates the frequency offset on the first few the received preamble symbols.

Such a transient effect of switching PA on the frequency shift of the VCO can be minimized or avoided if the transmitter on is slightly delayed after the PA is powered on. Figure 6.15 shows the transient frequency drift versus time for such a case.

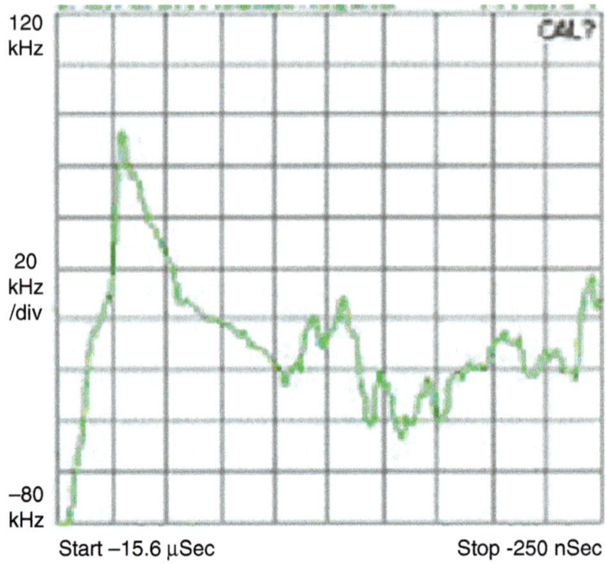

Fig. 6.14 Transient frequency drift caused by frequency pulling due to switching on PA and TX transmitter together

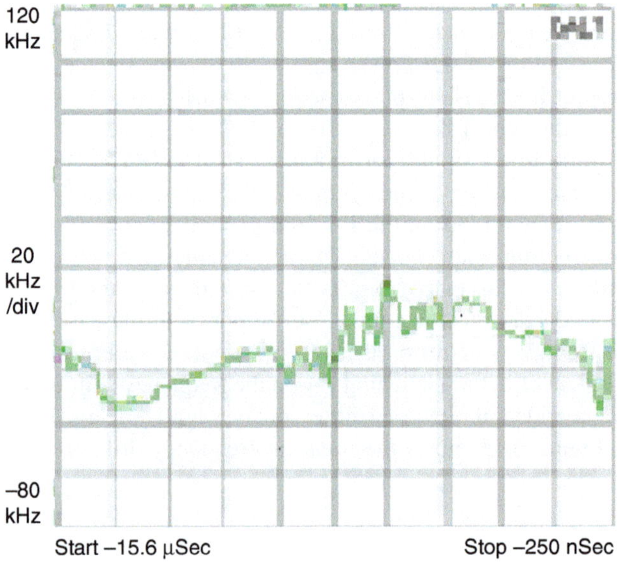

Fig. 6.15 Transient frequency versus time by switching on PA first, and then TX transmitter after 3 μs

6.5.4 *Nonlinearity of Power Amplifier*

Power amplifier is the last stage of the transmitter and delivers the required power level to an antenna. Generally, it is required for the transmitter to have a power amplifier with large linear operation range and low-voltage operation in order to avoid causing non-linearity degradation for the transmitted signal. This is because the main source of spectral re-growth is intermodulation distortion (IMD) of the modulated signal by non-linearity in the transmitted chain [18]. In many wireless systems, system designers have relied on the PA output back-off from its P1-dB compression point, which is 1 dB down from its output with the linear gain slope to ensure acceptable distortion. However, too much output back-off results in poor energy efficiency or DC-to-AC (RF) power conversion [19]. In turn, poor energy efficiency reduces talk time in handset units.

On the other hand, it is desirable for the PA to operate at a near saturation or full saturation region in order to achieve high energy efficiency, especially for wireless portable systems. The advantage of high energy efficiency is that it extends battery duration in mobile handsets and wireless local area network (WLAN)-enabled notebook PC cards. The concept and definition of power amplifiers' efficiency was discussed in Chap. 2.

However, when the power amplifier operates at saturation or the near saturation region in order to achieve high efficiency, the PA behaves in a nonlinear feature, which results in both spectral regrowth and error vector magnitude (EVM) degradation of the RF-modulated signal at the PA output. As a result, the regrowth spectrum causes severe adjacent channel interferences and EVM degradation. The nonlinearity of the PA is generally described by two functions, namely, AM-AM and AM-PM conversions. The AM-AM describes amplitude nonlinearity of the PA, while AM-PM characterizes the phase nonlinearity of the PA. In practice, most power amplifiers have significant frequency-dependent *memory effects* that are highly associated with the amplified signal wideband. The measured AM-AM and AM-PM characteristics of a commercially available and RF-integrated circuit (RFIC) power amplifier at a frequency of 5 GHz for the 802.11 a WLAN application with a 20-MHz bandwidth are illustrated in Fig. 6.16, where the input and output amplitudes of the PA are normalized to unit (Volts). The widths of AM-AM and AM-PM curves are determined by the output signals, which are historically dependent on both the current input signals and the previous input signals, i.e., memory effects, which can sometimes severely degrade system performance, especially in high-speed data transmission.

To obtain the AM-AM and AM-PM characteristics, a RF-modulated OFDM signal at the output of the PA was first down-converted to the baseband band and then coherently demodulated to the complex I–Q baseband signal with a bandwidth (a single sideband) of 10 MHz through a spectrum analyzer, or FSW from Rohde & Schwarz, and finally exported to MATLAB for the further analysis. Around 10,000 I–Q samples in one frame at the input and the output of the PA were captured at a sampling rate of 80 MHz, respectively. Time alignment and normalization were then performed between the PA input data $u(n)$ and the PA output data $y(n)$, and the AM-AM and AM-PM curves are illustrated in Fig. 6.16a, b, where the coordinates (x, y) in these plots are calculated by

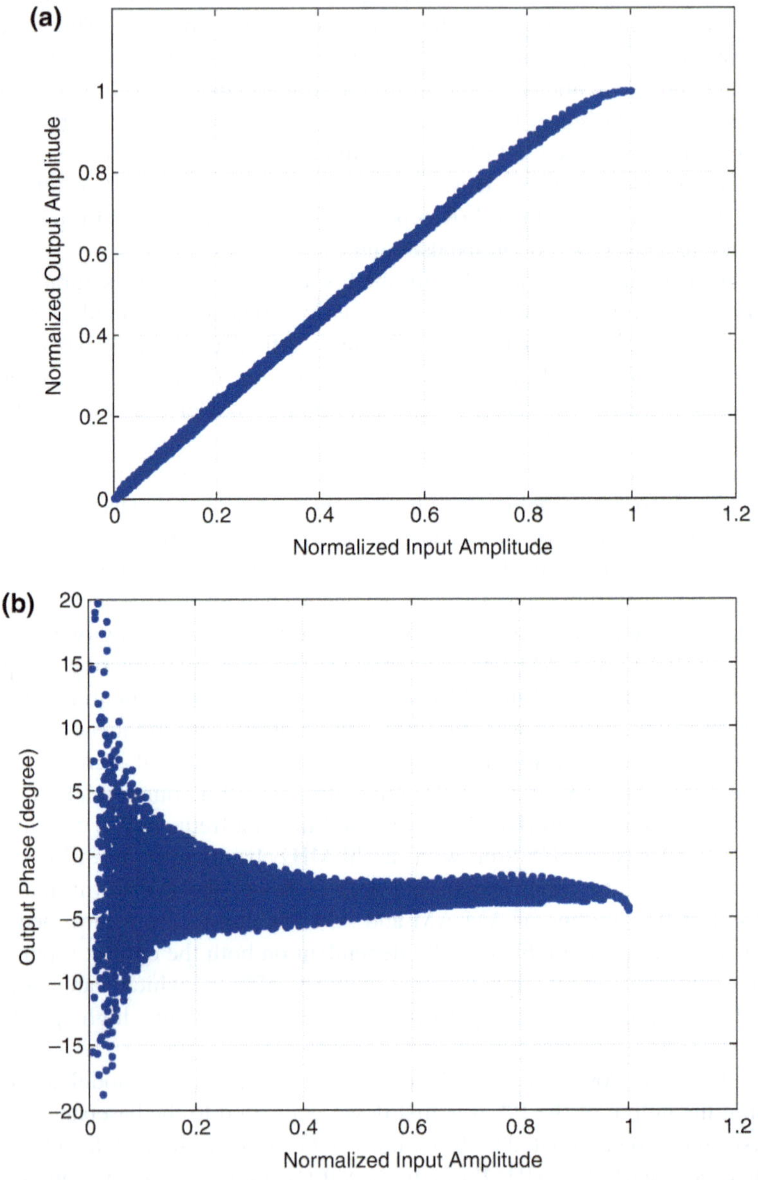

Fig. 6.16 Normalized characteristics of a RFIC commercial power amplifier used for Wi-Fi transceivers: (**a**) AM-AM, (**b**) AM-PM, and (**c**) amplitude voltage gain, where a WLAN OFDM signal at 5-GHz frequency with a 20-MHz bandwidth is used as input signal to PA

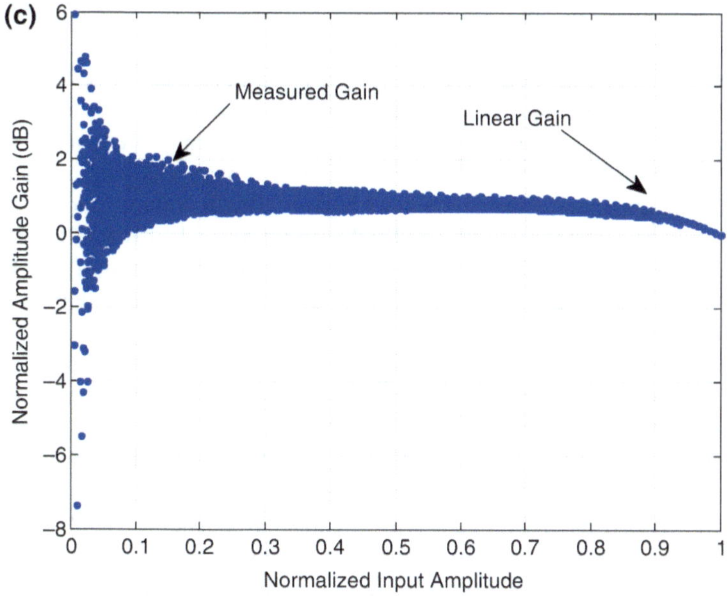

Fig. 6.16 (continued)

$$\mathrm{AM-AM}:\ \left(20\,\mathbf{log}_{10}(|u(n)|), 20\,\mathbf{log}_{10}\frac{|y(n)|}{|u(n)|}\right) \tag{6.33}$$

$$\mathrm{AM-PM}:\ (20\,\mathbf{log}_{10}(|u(n)|), \angle y(n) - \angle u(n)) \tag{6.34}$$

It can be clearly seen that the widths of the AM-AM and AM-PM curves are due to memory effects of the PA, and the width even spreads in a low input signal amplitude range for the AM-PM characteristic. Frequency-dependent memory effects may result in asymmetric power spectral density (PSD) of the RF transmitted signal, and also cause difficulties in the PA pre-distortion. Generally, the memory effects of the PA can arise from multiple sources, including bias circuit effects, self-heating, and trapping effects. Meanwhile the memory effects in RF power amplifiers are variations of the nonlinear gain due to the frequency of the signal, the frequency of the envelope of the signal, or temperature.

As the input signal amplitude increases, the PA gain goes into compression and drops by 1 dB from its linear gain at the normalized input signal amplitude of 1, as shown in Fig. 6.16c. Corresponding to such a 1-dB gain compression point, the input power of the PA is referred to as the input P1dB point, while the output power of the PA is called the output P1dB point. When the mean power that the PA delivers to its load is close to the output P1dB point, the PA can cause the nonlinear distortions of the amplified signal. Major effects of the nonlinearity of the PA on the system performance are spectral regrowth and EVM degradation of the RF-modulated signal.

To avoid such nonlinear distortion, it is preferable that a PA operates back from P1dB point by a PAPR value, which is dependent on the modulation format. Otherwise, the envelope amplitude of the amplified signal will be clipped or compressed by the PA. Therefore, it is very challenging for the PA to amplify high-speed data signals with large PAPR values, such as WLAN OFDM signals. For example, if an OFDM signal has a 10-dB PAPR value, as a rule of thumb the PA should operate from its P1dB point back-off by 10 dB in order to avoid either spectral regrowth or EVM degradation of the RF-modulated signal.

To see how nonlinearity affects the system performance, we tested PSDs and EVMs of a 802.11n WLAN OFDM-64QAM signal with a PAPR value of 10 dB versus different back-off values from a P1dB compression point that corresponds to an actual output P1dB power of 28 dBm, as illustrated in Figs. 6.17 and 6.18. First, we look at how PA back off affects PSD. Figure 6.17a, b illustrates the PSD of the RF OFDM-64QAM signal with a carrier frequency of 5500 MHz at the output of the PA when the power amplifier operates at a 12-dB back-off and a 4-dB back-off from the P1-dB compression point, respectively. The minimum margin of PSD from the transmitted spectrum mask is about 15 dB at the back-off of 12 dB, as shown in the bottom part of Fig. 6.17a and then is reduced to less than 4 dB at the back-off of 4 dB, as shown in the bottom part of Fig. 6.17b. Now let's see how EVM performance behaves for the same back-off values as those used in Fig. 6.17. Figure 6.18a illustrates that the RMS EVM is about −38.63 dB and meets the specification of -27.00 dB when the back-off is set to a 12 dB from P1dB compression point. In this test case, a 10-dB margin for EVM is achieved compared to the specification of −27.00 dB. Figure 6.18b shows that the RMS EVM is about −27.24 dB, which is almost equal to the specification of −27.00 dB, when the back-off is reduced to 4 dB. Table 6.1 summarizes the minimum margins of the tested PSD and EVM versus PA back-off (BO) from the P1dB compression point.

It can be seen from Table 6.1 that PSD has more margin from specification at certain PA BO values compared to RMS EVM. Hence, PSD has looser requirement when the PA operates close to saturation region or the PA has less back-off from the P1dB point compared to RMS EVM. Therefore, the nonlinearity of the PA has more severe effect on EVM than on PSD.

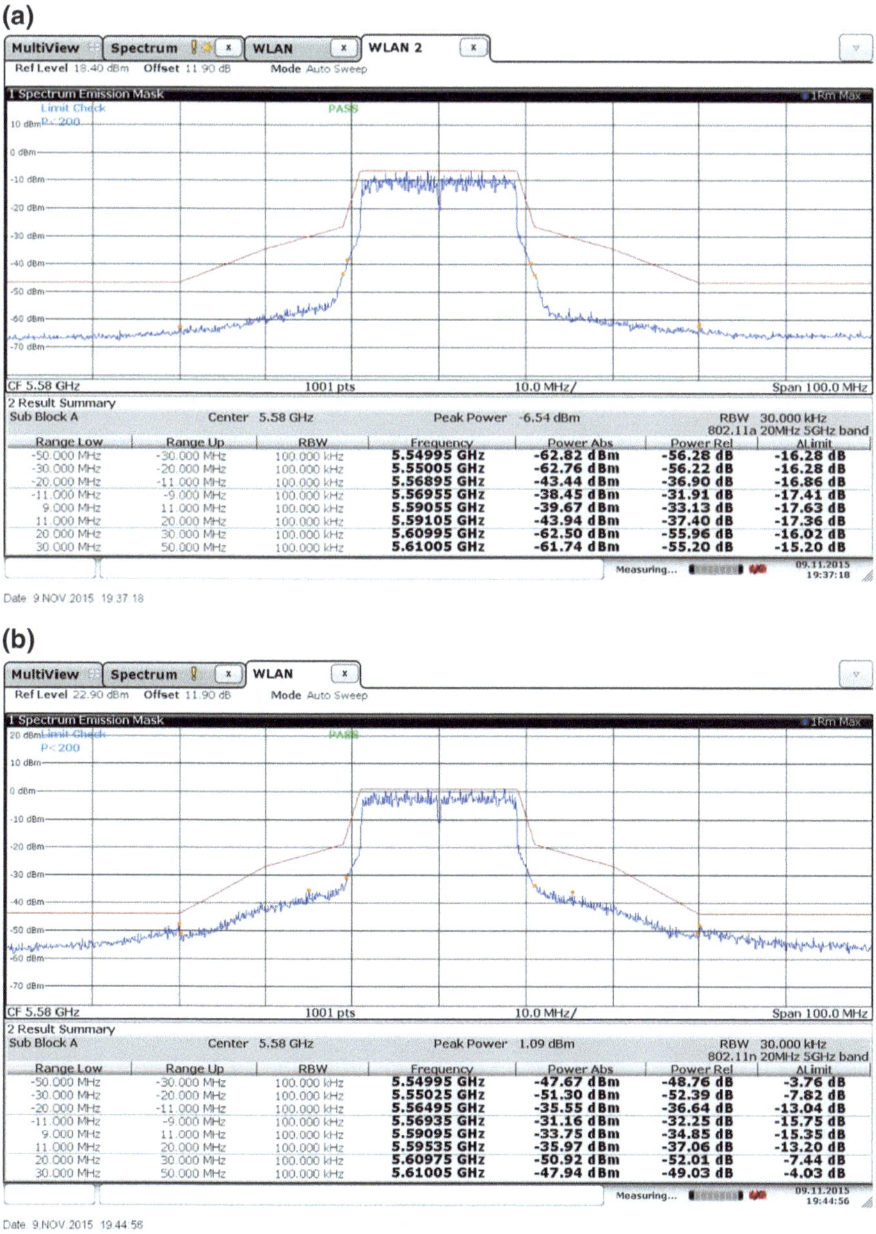

Fig. 6.17 PSD of a WLAN 802.11n OFDM-64QAM-modulated signal at 5580 MHz: (**a**) 12-dB back-off from P1dB of 28 dBm and (**b**) 4 dB back-off from P1 dB

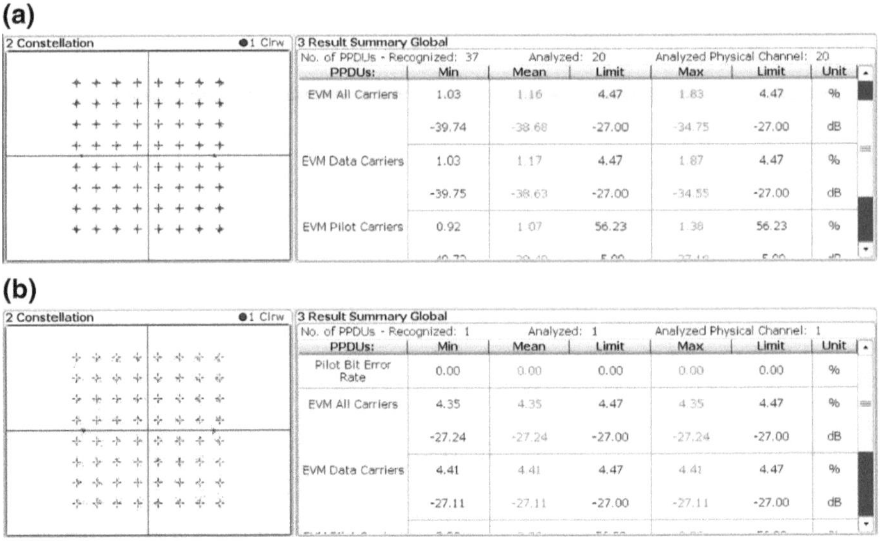

Fig. 6.18 EVM of an OFDM-64 QAM-modulated signal versus PA output BO from the P1dB compression point at 5500 GHz: (**a**) RMS data EVM of −38.63 dB at a 12-dB back-off from P1dB of 28 dBm and (**b**) RMS Data EVM of −27.11 dB at a 4-dB back-off from P1dB

Table 6.1 Minimum margins of the tested PSD and EVM of the 802.11n WLAN OFDM-64QAM signal versus PA back-off (BO)

Test item	PA back-off (BO) from P1dB @ 12 dB BO	PA back-off (BO) from P1dB @ 4 dB BO	Specification
Minimum margin[a] for PSD (dB)	15	4	−40.00 dBr $(f \geq \lvert 30MHz \rvert)^{b}$
Minimum margin for RMS EVM (dB)	11	0	−27.00 dB

[a]Minimum margin is the difference between specification value and tested value. A minimum margin of 11 means the test RMS EVM is −38.00 dB

[b]The most crucial range to meet the transmit PSD mask is in the range $f \geq \lvert 30\,\mathrm{MHz} \rvert$

References

1. Chen, S., & Zhao, J. (2014, May). The requirements, challenges, and technologies for 5G of terrestrial mobile telecommunication. *IEEE Communications Magazine 53*, 36–43.
2. Osseiran, A., Boccardi, F., Braun, V., Kusume, K., Marsch, P., & Maternia, M. (2014). Scenarios for 5G mobile and wireless communications: The vision of the METIS project. *IEEE Communications Magazine, 52*(5), 26–35.
3. Chih-Lin, I., Rowell, C., Han, S., Xu, Z., Li, G., & Pan, Z. (2014). Toward green and soft: A 5G perspective. *IEEE Communication Magazine, 52*(2), 66–73.
4. Muhammad, K., Staszewski, R. B., & Leipold, D. (2005, August). Digital RF processing: Toward low-cost reconfigurable radios. *IEEE Communications Magazine, 43*(8), 105–113.

5. Razavi, B. (1997, June). Design consideration for direct-conversion receiver. *IEEE Trans. Circuits and Systems-II: Analog and Digital Signal Processing, 44*(6), 428–435.
6. Cavers, J. K., & Liao, M. W. (1993, November). Adaptive compensation for imbalance and offset losses in direct conversion transceivers. *IEEE Transactions on Vehicular Technology, 42*(4), 581–588.
7. Gao, W. (2008, March 13). Compensation for gain imbalance, phase imbalance and DC offsets in a transmitter. US Patent, Pub. No.: US 20080063113 A1.
8. Mass, S. A. (1993). *Microwave mixers* (2nd ed.). Norwood, MA: Artech House.
9. Lanschutzer, C., Springer, A., Maure, L., Boos, Z., & Weigel, R. (2003). Integrated adaptive LO leakage cancellation for WCDMA direct upconversion transmitters. *2003 IEEE Radio Frequency Integrated Circuits Symposium* (pp. 19–22).
10. Asam, M. (1999). *Mismatch of current mirrors.* Internal technical report, Infineon Technology AG.
11. Universal Mobile Telecommunications System (UMTS); UE Radio Transmission and Reception (FDD), (3GPP TS 25.101 version 5.2.0 Release 5), ETSI TS 125 101, version 5.2.0, 2002-2003.
12. Mohindra, R., & Stroet, P. (2001, January 2). Quadrature modulator with set-and -forget carrier leakage compensation. United States Patent. Patent No.: US 6,169,463 B1.
13. Olgaad, C. (2004, October). Using advanced signal analysis to identify source of WLAN transmitter degradations. *RF Design* (pp. 28–36).
14. Zhang, P., Der, L., Guo, D., Sever, I., Bowdi, T., & Lam, C. (2003, February). A direct conversion CMOS transceiver for IEEE 802.11a WLANs. *ISSCC Digest of Technical Papers* (pp. 354–355).
15. Darabi, H., Khorram, S., Chien, H.-M., Pan, M.-A., Wu, S., & Moloudi, S. (2001). A 2.4 GHz CMOS transceiver for bluetooth. *IEEE Journal of Solid-State Circuits, 36*(12), 2016–2024.
16. Razavi, B. (2003). *RF microelectronics*. Taiwan: Pearson Education.
17. Galal, S. H., Galal, S. H., Ragaie, H. F., & Tawfik, M. S. (2000). RC sequence asymmetric polyphase networks for RF integrated transceivers. *IEEE Transaction on Circuits and Systems-II: Analog and Digital Signal Processing, 47*(1), 18–27.
18. Feher, K. (1995). *Wireless digital communications: Modulation & spread spectrum applications.* Upper Saddle River, NJ: Prentice Hall.
19. Kenney, J. S., & Leke, A. (1995, October). Power amplifier spectral regrowth for digital cellular and PCS applications. *Microwave Journal 38*, 74–92.

Chapter 7
Transceiver II: Receiver Architectures

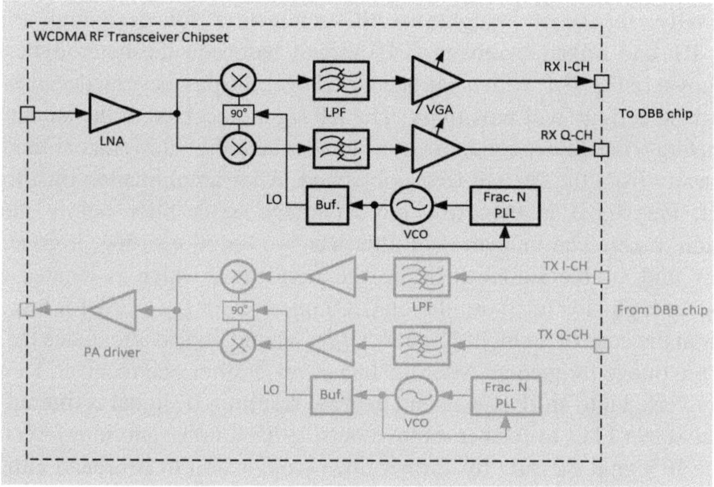

7.1 Introduction

Similar to transmitter architectures, receiver architectures are mainly classified into three types: a heterodyne (or high intermediate frequency [IF]) receiver, a low-IF receiver, and a direct down-conversion (or zero-IF) receiver. In a low-IF receiver, the RF signal is mixed down to a low, but non-zero, IF signal that is usually set to one or two times the channel-spacing frequency and is compatible with the bandwidth of the desired signal. To further suppress the adjacent channel

© Springer International Publishing Switzerland 2017

W. Gao, *Energy and Bandwidth-Efficient Wireless Transmission*, Signals and Communication Technology, DOI 10.1007/978-3-319-44222-8_7

interferers, the low-IF may be set to half the channel-spacing frequency such that the image signal is located at the adjacent channel. Because of some advantages over the zero-IF receiver, such as being insensitive to direct current (DC) offset and lessening the impact of the flicker noise, the low-IF receiver has been adopted by many integrated circuit (IC) design companies in the design of the radio frequency (RF) transceivers.

In the following sections, we present some general receiver architectures that are widely used in wireless communication standards. In addition, we focus on some common issues or challenges that RFIC designers face and introduce some effective techniques to minimize these effects on system performance.

7.2 Heterodyne Receiver

Traditionally, the most straightforward architecture for designing a wireless receiver RF and mixed baseband (BB) circuit has been the heterodyne receiver [1], as shown in Fig. 7.1, which has been used over the past several decades because of its high selectivity and sensitivity. The RF signal received at the antenna is first filtered using a pre-select bandpass filter to attenuate the interferers at the frequencies far away from the desired frequency band. After amplification by a low-noise amplifier, the signal is then fed into an image-reject filter before the down-conversion mixer. The image-reject filter has two functions: first, it greatly attenuates any undesirable signal at the image frequency, which is located at either $f_{RF} + 2f_{IF}$ for high-side local oscillation (LO) injection or $f_{RF} - 2f_{IF}$ for low-side LO injection in the case of an input frequency f_{RF}; second, it also attenuates the thermal noise at the image frequency. After the image-reject filter, the received RF signal is down-converted into an IF signal, and then the resulting IF signal is filtered through a channel-select filter to further remove outside band noise and interferers. At this point, the IF signal is ready for further down-conversion to baseband either in the analog domain or digital domain.

The important design choice in a heterodyne is the frequency of the IF signal. The separation between the RF signal and its image is $2f_{IF}$. When this separation is large it is referred to as high IF, whereas when it is small it is referred to as low-IF. In the former, the image is greatly attenuated with the image-reject filter, but close-in interference after down-conversion is not significantly suppressed using another bandpass filter unless a high-Q bandpass filter is used, such as an off-chip passive surface-acoustic wave or an LC filter that is comprised of the inductor (L) and capacitor (C), which increases the cost and size. In the latter, the attenuation of the image is poor because the separation is small, but the close-in interferer after down-conversion is greatly attenuated using a bandpass filter with sharp cut-off characteristics, where the bandpass filter with high-Q is easily implemented due to a low IF.

Traditionally, many down-mixing stages can relax the Q value required by each channel selection filter and linearity required by each amplification stage due to

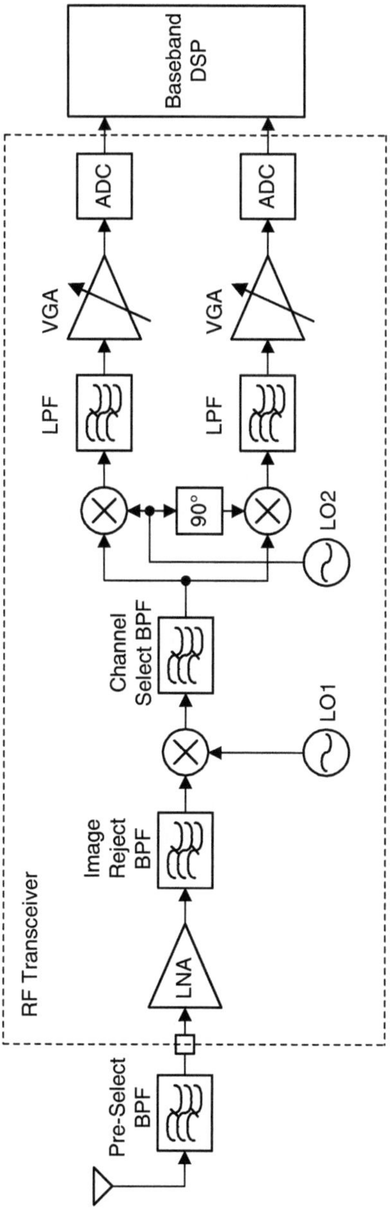

Fig. 7.1 A heterodyne (high-IF) receiver

large interferers at the input of the receiver antenna. However, the heterodyne receiver requires expensive and bulky external filters and extra internal local oscillators. Therefore, from a low-cost, high-integration, low-power-consumption point of view, the heterodyne receiver architecture is not an optimal solution for achieving such characteristics, especially for RF IC transceivers. Heterodyne receivers are, however, widely used in some special applications, such as satellite communication and microwave communication systems, where the implementations are mainly designed with discrete component units.

7.2.1 Image Rejection

Figure 7.2 shows a block diagram of the RF front-end circuits for a heterodyne receiver. The frequency translation from a RF to an IF is first carried out by the means of an RF mixer, in which the received RF signal at a frequency of ω_{RF} is mixed with a real LO signal cos($\omega_{LO}t$) with $\omega_{LO} = \omega_{RF} + \omega_{IF}$ to generate two major signal components, or one at a frequency of $\pm\omega_{IF} = \pm(\omega_{LO} - \omega_{RF})$ and another at a frequency of $\pm(\omega_{LO} + \omega_{RF})$. The latter can be removed by a lowpass filter. In the frequency domain, the frequency components of the real LO signal cos ($\omega_{LO}t$) are located at $-\omega_{LO}$ and ω_{LO}, respectively. The symmetrical property from zero frequency is due to the real signal property of the LO signal. In detail, the desired and image signals at the negative frequency are down-converted to the IF with the LO signal at the positive frequency, whereas the desired and image signals at the positive frequency are down-converted to the IF with the LO at the negative frequency, as indicated by arrows in Fig. 7.2.

In frequency translation, any frequency components located around the frequency of $\pm\omega_{im} = \pm(\omega_{LO} + \omega_{IF})$, which is called the "image frequency," is also translated to the same IF around $\pm\omega_{IF}$. The frequency component around the image frequency is called the "image signal." Within the desired signal bandwidth, the image signal distorts the desired signal.

A simple method to remove the image signal is to add an RF image reject filter before the mixer, as shown in Fig. 7.2. If the chosen IF is high enough, the image signal can be significantly attenuated, and the image rejection filter can be implemented relatively simply. But the drawback for a high IF receiver is that it is challenging for an RF designer to design an IF channel select bandpass filter (BPF) with sharp attenuation or a higher Q after the mixer because of the limitation of the B_{IF}-to-f_{IF} ratio, especially for narrow-band signal reception. On the contrary, a certain amount of residual image signal may be down-converted to the band around the IF because of a small frequency spacing $2\omega_{IF}$ between the wanted signal and image signal if the choice of IF is too low, as shown by the image rejection filter (dashed line) in Fig. 7.2b.

The pre-select filter before the low-noise amplifier (LNA) typically acts to select the desired signal and suppress the out-of-band interferers and blockers in order to prevent the LNA from being saturated in the presence of larger interferers and

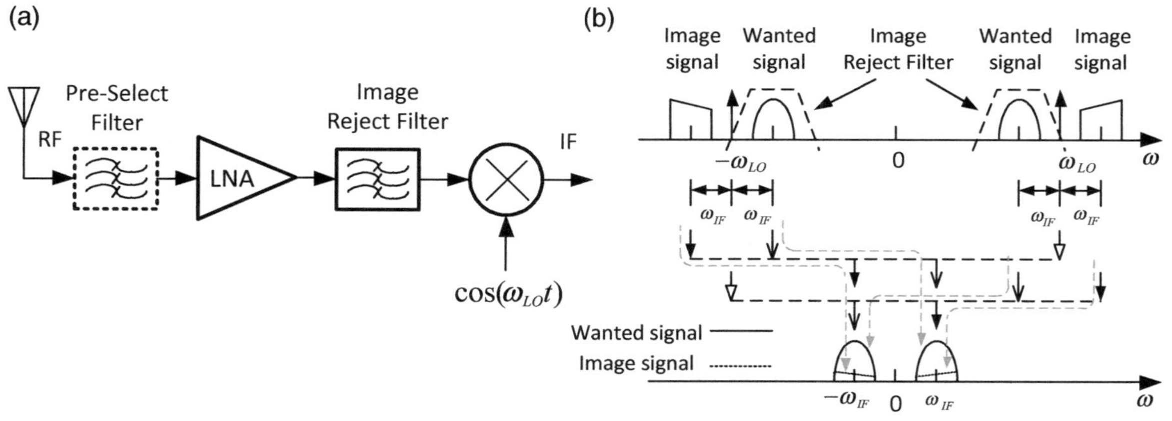

Fig. 7.2 Front-end of a super-heterodyne receiver: (**a**) first stage of RF to IF down-conversion and (**b**) concept of image rejection with a bandpass filter

blockers. However, the in-band loss of the passive pre-select filter can contribute to the noise figure degradation of the system. Therefore, the use of the pre-select filter is an optional choice, as shown with a dashed line.

The choice of the IF depends on an actual application associated with the signal bandwidth and the amount of acceptable image noise. The trade-off between image rejection and channel selection, however, can be ignored by using multiple down-conversions, which are widely used in the digital microwave and satellite communication systems that are mostly implemented with the discrete components. Figure 7.3 illustrates a dual-IF superheterodyne receiver, where the last down-conversion from the second IF_2 to the BB signal is performed in a quadrature demodulator with a pair of orthogonal carrier signals at the frequency of ω_{IF_2}. Hence, this superheterodyne receiver performs frequency down-conversion from RF to BB in three stages, or two stages from the RF to the second IF_2 and one stage from the second IF_2 to BB.

One of these applications is a satellite communication system using a single channel per carrier (SCPC)/frequency-division multiple access (FDMA) technology. The primary advantage of the SCPC system is that the architecture allows full connectivity between any sites in the network and also allows quick set-up to a satellite link as needed. This system is especially useful for remote area communications with a relatively small capacity needed for each user. The frequency of the up-link from the earth stations to the satellite is 6 GHz, while the frequency of the down-link is 4 GHz. A total of 800 data/voice channels occupy a 36-MHz transponder bandwidth of the satellite. Each user can transmit and receive either data at the rate of 64 kbps with a QPSK modulation or voice at the rate of 32 kbps with a BPSK modulation in a channel with an equivalent noise bandwidth of 38.4 kHz. The channel spacing is 45 kHz (or 36 MHz/800).

The received RF signal with a frequency of 4 GHz at the down-link is first down-converted to the IF_1 signal of 70 MHz with the first LO_1 after LNA amplification and image rejection. After passing through a wide channel select filter, the IF_1 signal is further down-converted to an IF_2 signal of 512 kHz with the second LO_2. The narrow channel select filter with a noise bandwidth of 38.4 kHz centered at a frequency of 512 kHz significantly suppresses outside channel interferers and noise that are close to the desired signal before the quadrature demodulation. This narrow channel select filter determines the noise bandwidth of the receiver, and therefore it plays an important role in determining the sensitivity of the receiver. Further down-conversion from IF_2 to the BB domain, carrier recovery, symbol timing recovery, and BB data detection all are performed after the narrow channel selection filter, and hence their performances greatly depend on the amplitude and group delay responses of the narrow channel select filter.

This narrow channel selection BPF was implemented in the late 1980s with a passive BPF filter with sharp attenuation or a high quality factor Q. A group delay equalizer was built with the BPF to minimize intersymbol interference (ISI) because of the non-constant group delay property of the analog BPF. From the SCPC system described above, it is clear that this dual-IF heterodyne receiver has

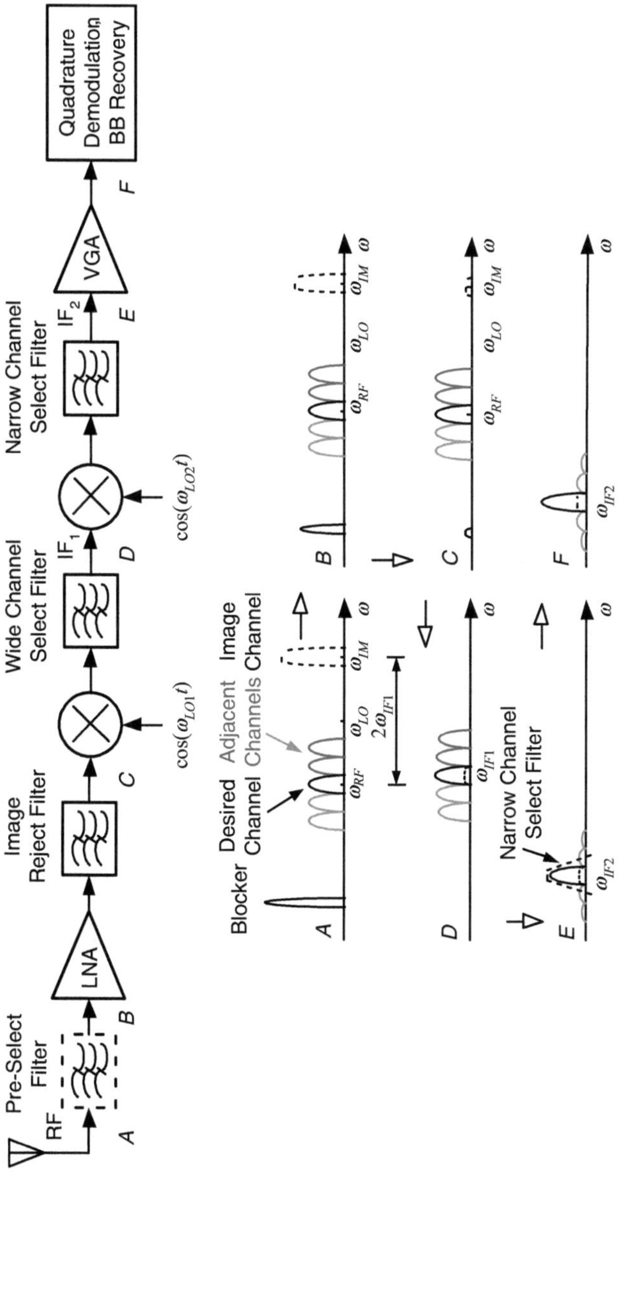

Fig. 7.3 A dual-IF superheterodyne receiver

advantages of significant image rejection and great channel selection. It should be noted that the heterodyne receiver usually has such a narrow channel select filter in the receive chain to determine a noise bandwidth of the receiver before the demodulation.

7.3 Low-IF Receiver and Zero-IF Receiver

Low cost and low power consumption are key factors for wireless IC vendors in choosing low-IF and zero-IF (or direct-conversion) receiver architectures, which have shown good performance with regard to low power consumption and small chip area as well as system bit error rate. A number of articles [2–6] describe and report their applications in wireless communications systems, such as IEEE 802.11a/b/g/n/ac WLAN, IEEE 802.15.4 Zigbee, GSM/EDGE, and 3GPP WCDMA.

Figure 7.4 shows a block diagram of a typical receiver, using either the zero-IF or the low-IF architecture. The difference between the zero-IF and the low-IF is that the RF signal in the former is down-converted directly to baseband, while the RF signal in the latter is mixed down to non-zero IF or a low-IF. In practice, depending on the bandwidth of the transmitted signal, the low-IF frequency can be in a range from a half to two times (or even more) the bandwidth of the transmitted signal. After down-conversion, the baseband/low-IF I–Q signals are passed through the analog selection lowpass filters to attenuate out-of-band noise and adjacent channel interferers or blockers so that the linearity requirements of the variable gain amplifier (VGA) and analog-to-digital converter (ADC) are relaxed. For a low-IF receiver, the digitized low-IF I–Q signals are fed to a set of complex digital down- converters with a digital frequency that is the same as the low-IF frequency, where the actual image rejection and the frequency translation from a low-IF frequency to DC are performed. For a zero-IF receiver, the digital frequency is set to 0 Hz.

Starting from this point or at the output of the digital down-converter, the digital back-end is the same for both low-IF and zero-IF receivers. Subsequently, the BB signals are passed though high-order digital lowpass filters that actually select the channels and also determine an equivalent noise bandwidth of the receiver. The equivalent noise bandwidth is explained in detail later in the section on receiver sensitivity. After being passed through the digital channel selection filters, the I and Q baseband signals are processed in a digital demodulator, where down-sampling, symbol timing and carrier synchronizations, and channel compensation are performed if coherent detection is used, and finally the data bits are recovered.

Compared with the heterodyne receiver in Fig. 7.1, the low-IF and zero-IF receivers eliminate the need for the image rejection BPF and channel selection BPF before the quadrature down-converter. In the low-IF receiver, the image rejection is achieved by using either a complex digital down-converter in the digital domain, as shown in Fig. 7.4, or a complex bandpass channel select or polyphase

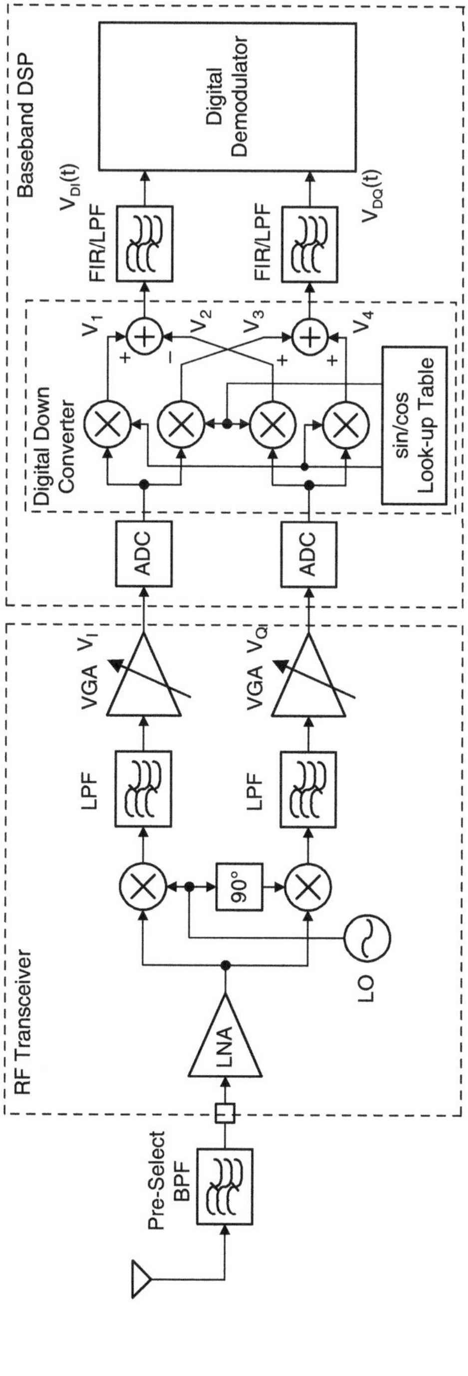

Fig. 7.4 Block diagram of a zero-IF (without digital down-converter) and low-IF (with digital down converter) receiver

filter at the outputs of the mixers in the analog domain, as illustrated in Fig. 7.5. In the zero-IF receiver, there is no explicit image problem, since the image to the desired channel is the desired signal itself. Hence, the zero-IF receiver has no specific need for the image rejection. As a result, the zero-IF receiver is very suitable for integration as well as multi-band and multi-standard operation.

Now, the question is whether a low-IF receiver or a zero-IF receiver is the best architecture to choose as a receiver. Choosing either a low-IF or a zero-IF receiver structure usually depends on the modulation format and the transmitted signal bandwidth. In general, the low-IF receiver is suitable for a wireless system when the modulation signal has a relatively low data rate or narrow bandwidth and has plenty of low-frequency energy around DC in its power spectral density (PSD). Thus, the low-IF architecture can be chosen such that the low-IF receiver can avoid distorting the desired signal around DC frequency by eliminating DC removal with highpass filtering. For example, in the GSM system, the received RF GMSK signal with a data rate of 270.833 kbps is down-converted either at the low-IF of 100 kHz [7] or at the low-IF of 160 kHz [8] in channel spacing of 200 kHz. The PSD of the GMSK modulation signal has plenty of low-frequency energy around DC. In the Bluetooth system, the received RF GFSK signal with a data rate of 1 Mbits/s is down-converted to a low-IF of 1 MHz in a bandwidth of 1 MHz. Actually, both GMSK and GFSK signals have similar properties or plenty of energy in their PSDs concentrated around DC.

Otherwise, the zero-IF receiver can be chosen if the modulation signal has a relatively flat PSD shape and a wide bandwidth or a high data rate. Thus, DC removal using highpass filtering does not damage signal components around the DC frequency too much, and therefore the performance degradation due to DC cancellation can be minimized. Meanwhile, choosing a zero-IF receiver can avoid the need to use a relatively high IF frequency due to the larger bandwidth associated with high data rates. For example, for the 3GPP WCDMA system with channel spacing of 5 MHz and the 802.11ac WLAN system with channel spacing of 20/40/80/160 MHz, the zero-IF receiver is used for these systems due to the characteristics of their flat PSD shapes and larger bandwidths.

The zero-IF receiver, however, has some disadvantages. Two serious weaknesses are DC offset and I–Q gain and phase imbalances even though $1/f$ noise is also troublesome. The DC offset and the I–Q gain and phase imbalances related to circuit mismatch and path mismatch become more serious due to large gain values of the variable gain amplifier (VGA). This is because the amplification gain of the receive chain is mainly performed by the baseband VGAs compared with the heterodyne architecture. The gain from the front-end LNA to mixers is typically around 30 dB, while the gain of the baseband VGAs, including the gain of the LPF can be up to about 65 dB. Thus, small DC offset, and the I–Q gain and phase imbalances at the output of the mixers can become extremely large at the input of the ADC after amplification. These problems, however, can be overcome or minimized by using DC offset cancellation and I–Q imbalance calibration techniques. Detailed problem-solving methods shall be described in the following sections.

On the contrary, the low-IF receiver topologies have some advantages over the zero-IF receiver architectures. Two desirable properties that the low-IF receiver has are that it is much less vulnerable to the DC offset and the 1/*f* noise because they are located away from the low-IF signal. Any such DC offsets will be frequency-converted to low-IF after the complex digital down-converter and easily removed in the digital domain. The low-IF receiver, however, introduces an image problem at the quadrature output of the first stage down-converter. Any real filter cannot remove the image signal at this point. Fortunately, the image signal can be suppressed in the complex digital down-converter in the baseband DSP, as shown in Fig. 7.4. Of course, there is another way to suppress the image signal at the low-IF frequency: using an asymmetric *polyphase* filter as shown in Figs. 7.5 and 7.6 [2]. The image rejection methods for the low-IF receiver will be discussed in more detail later in this section.

To minimize the performance degradation of the receiver due to the image signal, the image rejection has to be performed with different implementation strategies, and meantime the I–Q gain and phase imbalances should be minimized by means of the calibration. The effects of the image signals on the performance degradations of the receivers are primarily dependent on the chosen receiver architecture. Therefore the methods and strategies used for the image rejections are quite different and closely associated with the actual receiver architecture.

7.3.1 Image Rejection in the Low-IF Receiver

The biggest challenge in a low-IF receiver is image rejection. In a heterodyne receiver, the image rejection can be performed by using a bandpass filter before the down-converter. In a low-IF receiver, the image rejection, however, can be achieved through either a complex digital down-converter in the digital domain or a complex bandpass filter in the analog domain [2]. The difference between them mainly lies in where the image signal-suppression processing takes place, either before the ADC or after the ADC. The basic concept of the low-IF receiver is that an RF signal can be down-converted with a single positive frequency component (or a complex LO signal) to a low-IF signal without causing image distortion. Figure 7.5 illustrates the down-conversion operation in the frequency domain for a complex LO signal at a positive frequency of ω_{LO}. Two major receiver structures regarding different image rejection topologies are introduced below.

7.3.1.1 Complex Polyphase Filtering

As shown in Fig. 7.5, the desired signal and the image signal at the negative frequencies of $-\omega_{\mathrm{LO}}+\omega_{\mathrm{IF}}$ and $-\omega_{\mathrm{LO}}-\omega_{\mathrm{IF}}$ are respectively down-converted to the low-IF signals on either side of DC frequency by multiplying a complex LO

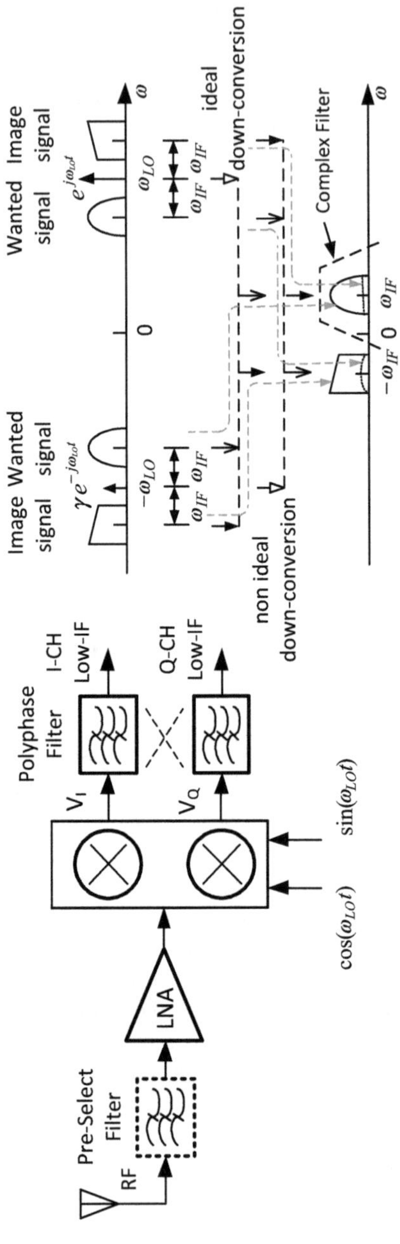

Fig. 7.5 Frequency down-conversion in a low-IF receiver

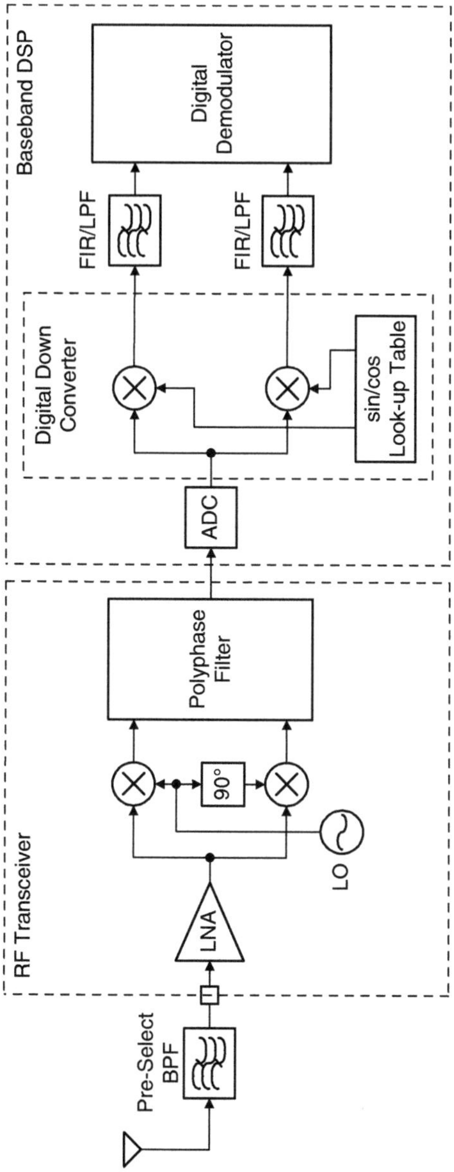

Fig. 7.6 Block diagram of a low-IF receiver with polyphase filter

signal located at a positive frequency of ω_{LO}, which is represented by an ideal complex local carrier of $e^{j\omega_{\mathrm{LO}}t} = \cos(\omega_{\mathrm{LO}}t) + j\sin(\omega_{\mathrm{LO}}t)$. After down-conversion, the desired low-IF signal and image low-IF signal are located at the positive and negative frequencies, respectively. Before they are further down-converted to the baseband signals, the image signal needs to be suppressed first to avoid the superimposition of the desired signal and image signal at the baseband. A complex asymmetric polyphase filter [2] implemented with active components is capable of suppressing the image signal and passing the desired signal. Its transfer function is asymmetric from the zero frequency, as shown in Fig. 7.5. This complex BPF suppresses not only the image signal, but also adjacent channel interferers and other larger blockers. As a result, the complex BPF reduces the dynamic range requirement for the ADC. Figure 7.6 shows a block diagram of a low-IF receiver with a complex BPF for a Bluetooth standard application. The inputs of the complex BPF are the low-IF signals on the I and Q channels, respectively, and the output is a real IF signal. After the image signal suppression, the desired low-IF signal is digitized through the ADC and then is digitally down-converted to the I and Q baseband signals.

To be understood the procedure of the frequency down-conversion in Fig. 7.5, image rejection based on complex polyphase filtering can be analyzed graphically in Fig. 7.7. A pair of desired and image signals are located at $\pm(\omega_c - 2\omega_{\mathrm{IF}})$ and $\pm\omega_c$, respectively. A LO signal is located at $\pm\omega_{\mathrm{LO}} = \pm(\omega_c - \omega_{\mathrm{IF}})$, where ω_{IF} represents the low-IF. Figure 7.7a illustrates the down-conversion procedures by convolving the input spectrum with $\cos(\omega_{\mathrm{LO}}t)$ on the I channel, while Fig. 7.7b displays the down-conversion steps by convolving the input spectrum with $\sin(\omega_{\mathrm{LO}}t)$ on the Q channel. Then, after their combinations as a complex low-IF signal, the desired and image signals at the input of the complex Polyphase BPF are separated without distortion at both sides' zero frequency, as shown in Fig. 7.7c. Hence, the image signal is removed if the center frequency of the complex BPF is located at the positive low-IF frequency of ω_{IF}.

Due to the I and Q gain and phase imbalances, a very small LO signal with an amplitude of γ is also located at a negative frequency of $-\omega_{\mathrm{LO}}$, as shown in Fig. 7.5. The non-ideal complex LO is now represented by $e^{j\omega_{\mathrm{LO}}t} + \gamma \times e^{-j\omega_{\mathrm{LO}}t}$. Hence, the desired signal and the image signal at the positive frequencies of $\omega_{\mathrm{LO}} - \omega_{\mathrm{IF}}$ and $\omega_{\mathrm{LO}} + \omega_{\mathrm{IF}}$ are down-converted to the low-IF on either side of DC frequency with $\gamma e^{-j\omega_{\mathrm{LO}}t}$, as shown with the dotted curves in Fig. 7.5, and are superimposed with those down-converted from the negative frequencies of $-\omega_{\mathrm{LO}} + \omega_{\mathrm{IF}}$ and $-\omega_{\mathrm{LO}} - \omega_{\mathrm{IF}}$. Therefore, the calibrations of the I–Q gain and phase imbalances are needed in order to minimize the image distortion signal represented by the dotted curve within complex filter.

Considering that a pair of non-ideal LO signals have a gain mismatch of α and a phase mismatch of θ between the quadrature LO paths, the non-ideal LO signals can be redefined as

$$\mathrm{LO}_{\mathrm{NIDE\_I}} = \left(1 + \frac{\alpha}{2}\right)\cos\left(\omega_{\mathrm{LO}}t + \frac{\theta}{2}\right) = \left(1 + \frac{\alpha}{2}\right)\left(\frac{e^{j\left(\omega_{\mathrm{LO}}t + \frac{\theta}{2}\right)} + e^{-j\left(\omega_{\mathrm{LO}}t + \frac{\theta}{2}\right)}}{2}\right) \tag{7.1}$$

$$\mathrm{LO}_{\mathrm{NIDE\_Q}} = \left(1 - \frac{\alpha}{2}\right) \sin\left(\omega_{\mathrm{LO}} t - \frac{\theta}{2}\right) = \left(1 - \frac{\alpha}{2}\right) \left(\frac{\mathrm{e}^{j\left(\omega_{\mathrm{LO}} t - \frac{\theta}{2}\right)} - \mathrm{e}^{-j\left(\omega_{\mathrm{LO}} t - \frac{\theta}{2}\right)}}{2j}\right) \tag{7.2}$$

Assuming α and θ are small, as they usually are, then the non-ideal complex LO signal $\mathrm{LO}_{\mathrm{NIDE\_I}} + j\mathrm{LO}_{\mathrm{NIDE\_Q}}$ can be approximated using Taylor series as

$$\begin{aligned} \mathrm{LO}_{\mathrm{NIDE}} &= \mathrm{LO}_{\mathrm{NIDE\_I}} + j\mathrm{LO}_{\mathrm{NIDE\_Q}} \\ &\approx \mathrm{e}^{j\omega_{\mathrm{LO}} t} + \left(\frac{\alpha - j\theta}{2}\right) \mathrm{e}^{-j\omega_{\mathrm{LO}} t} \\ &= \mathrm{LO}_{\mathrm{IDE}} + \gamma \times \mathrm{conj}[\mathrm{LO}_{\mathrm{IDE}}] \end{aligned} \tag{7.3}$$

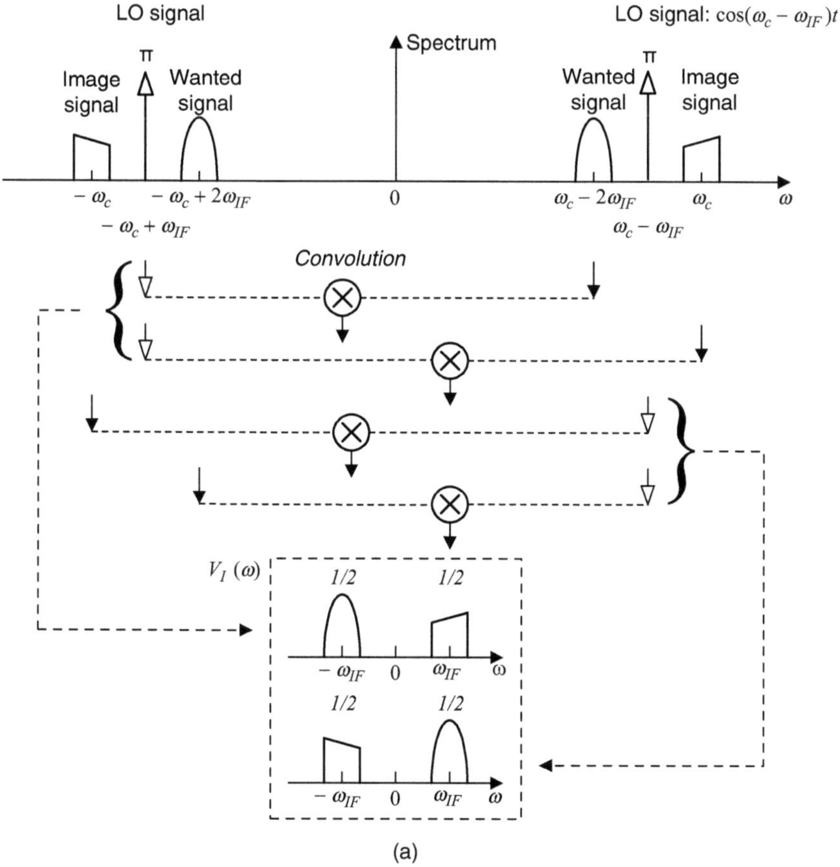

Fig. 7.7 Spectrum analysis of a low-IF receiver in Fig. 7.5: (**a**) down-conversion scheme with a $\cos(\omega_c - \omega_{IF})t$ LO signal on the I channel in a low-IF receiver, (**b**) down-conversion scheme with a $\sin(\omega_c - \omega_{IF})t$ LO signal on the Q channel in a low-IF receiver, and (**c**) spectrum of a complex low-IF signal constructed with the I and Q channel signals at the input of the complex bandpass filter

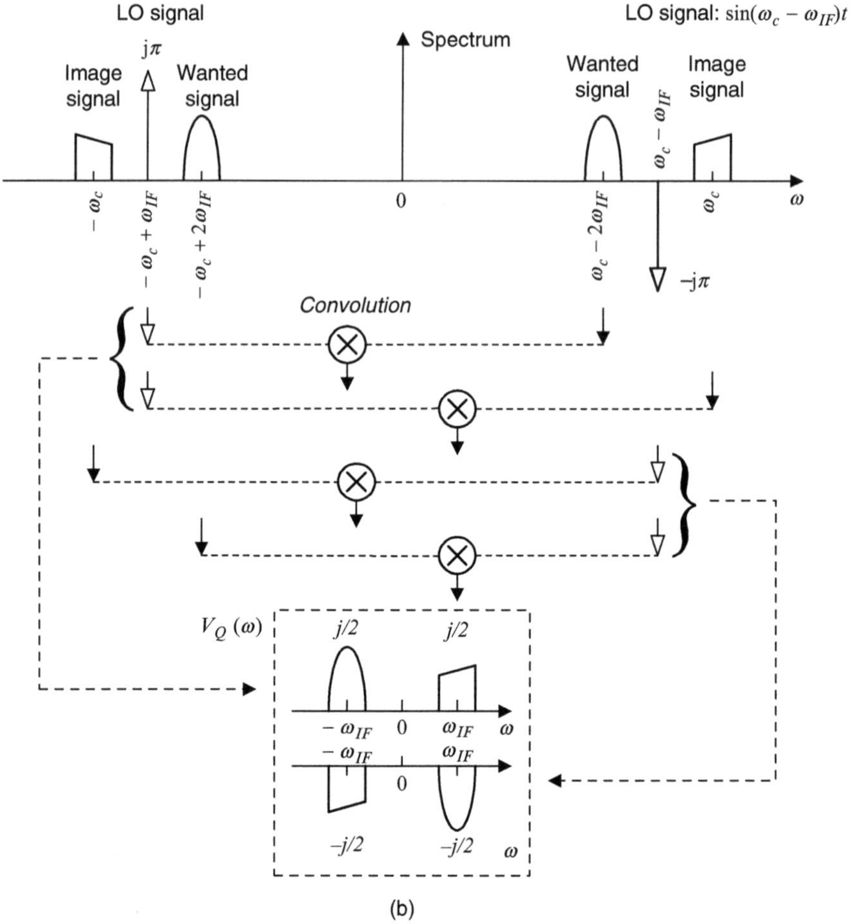

Fig. 7.7 (continued)

where the ideal complex LO_{IDE} signal and a complex leakage coefficient of γ are given as [9]

$$LO_{IDE} = e^{j\omega_{LO}t} \tag{7.4}$$

$$\gamma = \left(\frac{\alpha - j\theta}{2}\right) \tag{7.5}$$

Thus, the non-ideal complex signal LO_{NIDE} is approximated by the ideal complex signal LO_{IDE} plus its conjugate multiplied by γ.

One advantage that this low-IF receiver with a complex BPF has is to relax the dynamic range requirement for the ADC because the image signal and adjacent channel interferers are suppressed together before the ADC. The reduced dynamic range avoids the need to use a high-resolution ADC. One other advantage is that

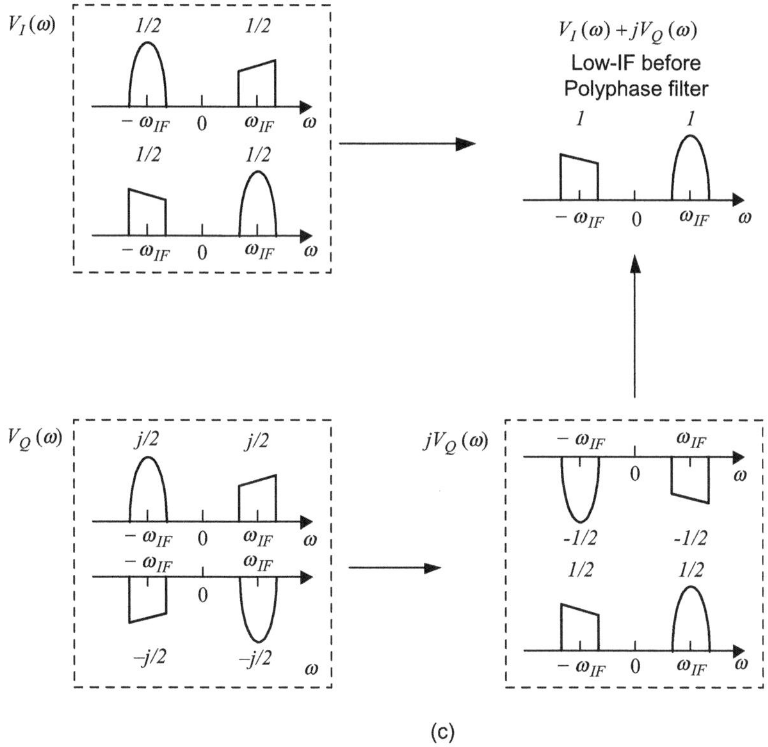

Fig. 7.7 (continued)

only one ADC is required in the receiver, as shown in Fig. 7.6. As a result, the current consumption is improved. However, the biggest challenge in designing a low-IF receiver with a complex BPF is that higher-order BPF is required to sufficiently suppress the image signal when the low-IF is relatively low. The complex analog BPF with a higher order has worse group delay variation within the transmission bandwidth, which in turn results in severe ISI, such that the receiver sensitivity could degrade. To compensate for the group delay variation, an analog group delay equalizer may be needed after the complex BPF. Furthermore, in order to reduce the effect of the analog component tolerance on the accuracy of the transfer function, the calibration of the complex BPF is also needed. All of these factors mentioned above cause an increase in die size in the IC design.

7.3.1.2 Complex Digital Down-Conversion

Compared with the image rejection through a complex analog BPF, the image signal suppression performed through a digital complex down-converter in the digital domain in Fig. 7.4 is more accurate than that in the analog domain. Furthermore, channel selection filtering can be easily and accurately implemented with a digital lowpass filter that has sharp attenuation and constant group delay, while analog filtering only performs a coarse and anti-aliasing filtering. All these merits make the low-IF topology shown in Fig. 7.4 more preferable and popular than the one shown in Fig. 7.6 in the hardware implementation of a fully integrated circuit. For example, the low-intermediate frequencies of 100 kHz and 2 MHz are chosen for the 2G GSM system with channel spacing of 200 kHz [5] and the Zigbee system with channel spacing of 5 MHz [10], respectively. The low-IF receiver with this topology, however, requires that the ADCs have a higher dynamic range and higher bandwidth than that with a complex analog BPF structure due to the fact that the analog filtering performs only a coarse and anti-aliasing filtering. To achieve the required high dynamic range for handling larger interferers and blockers, bandpass $\Sigma\Delta$ converters are usually used in this type of receiver. Pushing the ADC closer to the front-end circuits conforms to a current development tendency in today's CMOS technology.

The principle of image rejection in a low-IF receiver with complex digital down-conversion is easily understood via a graphical analysis. First, we illustrate a pair of orthogonal carrier signals at the frequency of ω_c to be modulated by the baseband I–Q signals at the frequency of ω_s, as shown in Fig. 7.8. In this illustration, the cosine and sinusoidal signals are used as the baseband signals because of their simple Fourier transforms for a graphical explanation. Then, the RF-modulated signals at the outputs of the mixers are summarized to form the RF-modulated signal $\cos(\omega_c - \omega_s)t$, whose frequency components are located at $\omega_c - \omega_s$ and $-\omega_c + \omega_s$, respectively. It is supposed that this RF-modulated signal is received in

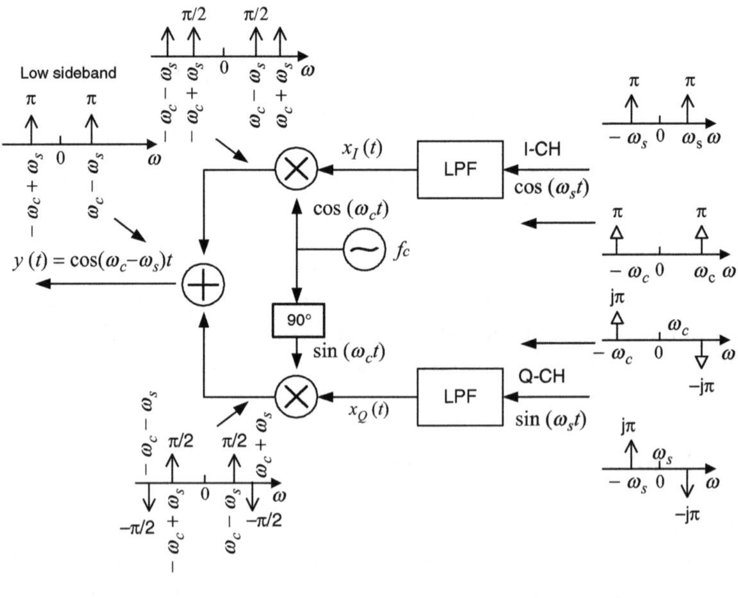

Fig. 7.8 An I–Q modulator with cosine and sine baseband input signals

the receiver as shown in Fig. 7.4 without any corruptions of Gaussian noise and other interferers. After passing the pre-select BPF, image rejection of the received RF-modulated signal is graphically analyzed in Fig. 7.9.

In Fig. 7.9, it is assumed that the LO frequency is $\omega_c - \omega_{IF}$, where $\omega_{IF} > \omega_s$ is the low-IF frequency. Because the frequencies of the received RF signal are $\omega_c - \omega_s$ and $-\omega_c + \omega_s$, the image signals are located at $\omega_c - 2\omega_{IF} + \omega_s$ and $-\omega_c + 2\omega_{IF} - \omega_s$,

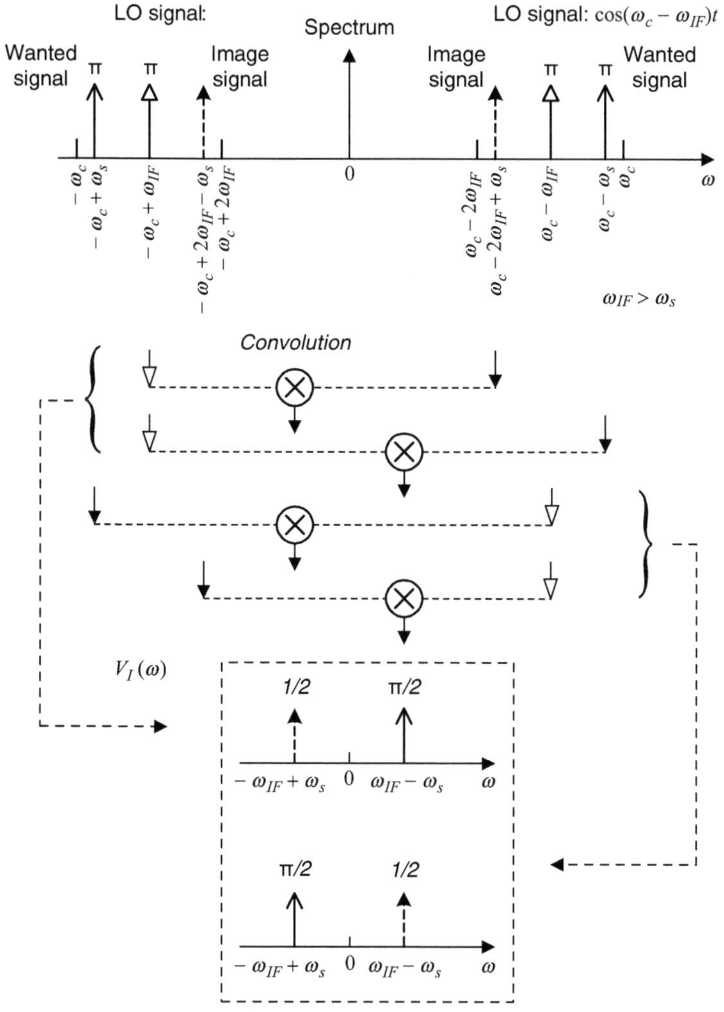

(a)

Fig. 7.9 Spectrum analysis of a low-IF receiver: (**a**) down-conversion scheme with a $\cos(\omega_c - \omega_{IF})t$ LO signal on the I channel in a low-IF receiver, (**b**) down-conversion scheme with a $\sin(\omega_c - \omega_{IF})t$ LO signal on the Q-channel in a low-IF receiver, and (**c**) spectrum recombination of the I–Q channels after a digital complex down-conversion

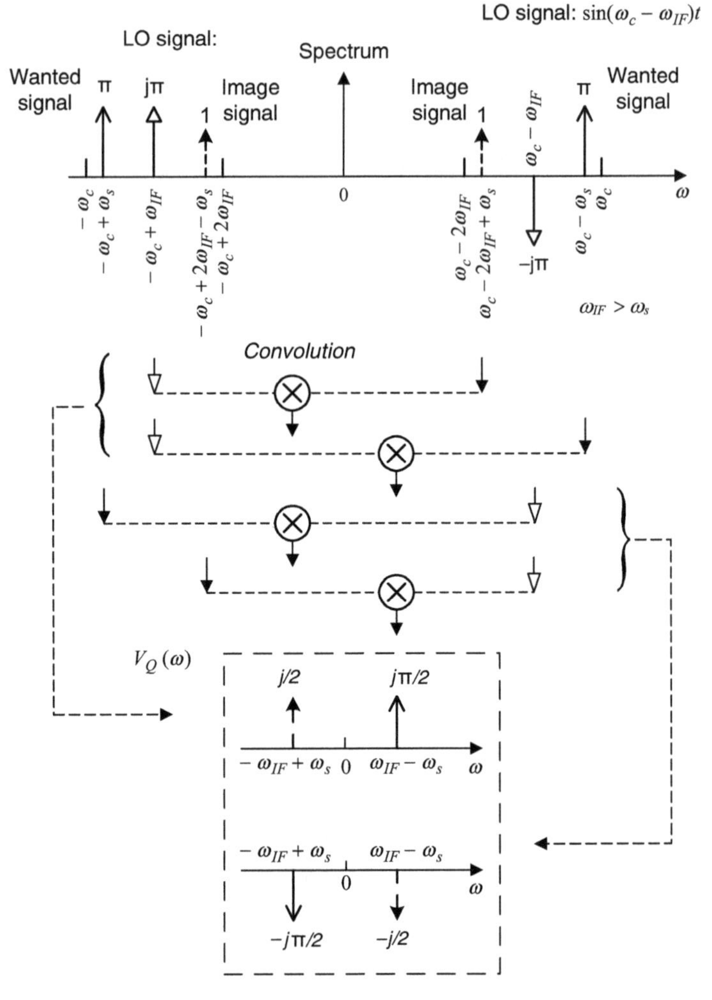

Fig. 7.9 (continued)

and they are mirrored in the frequency domain related to the desired signals from the LO signal, as shown in Fig. 7.9a. The image signal could be adjacent channel signals or any other interferers at the mirror frequencies. In Fig. 7.4, the low-IF I–Q signals of V_I and V_Q in the frequency domain are graphically analyzed in Fig. 7.9a, b, respectively.

It can be clearly seen that the image signals are also down-converted into the low-IF domain together with the desired signals at the outputs of the first down-converter, represented by the signals of V_I and V_Q. The low-IF analog I–Q signals are then converted to the low-IF digital I–Q signals through the ADCs.

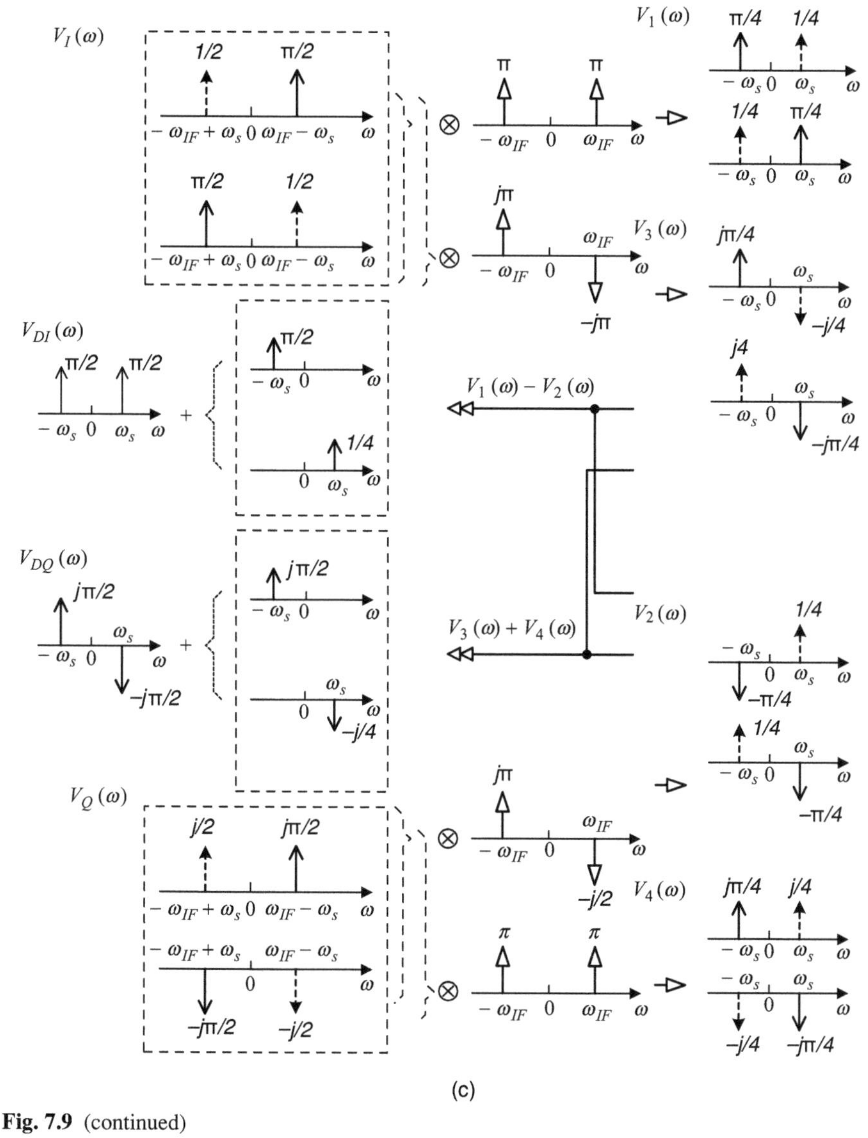

Fig. 7.9 (continued)

The digitized low-IF signals are further down-converted to the baseband I–Q signals with a complex low-IF signal through a complex digital down-converter. Finally, the baseband signals $V_{DI}(t)$ and $V_{DQ}(t)$ are recovered as the cosine and sine signals, as shown in Fig. 7.9c after passing through digital lowpass filters. The recovered $V_{DI}(t)$ and $V_{DQ}(t)$ are the same as the transmitted signals as shown in Fig. 7.8 except that their amplitudes are reduced by half, which can be avoided by increasing the amplitudes at the outputs of the sin and cos Look-up Table by 2.

Meanwhile, the image signals are suppressed by performing the combinations of $V_1(\omega) - V_2(\omega)$ and $V_3(\omega) + V_4(\omega)$ on the I–Q channels, respectively, as shown in Fig. 7.9c. Thus, in the ideal case the summed outputs are the desired signals with image-free. In this example, even though the cosine and sine signals are used as baseband signals here, the graphical analysis for the image suppression is also validly applied to random baseband signals.

7.3.1.3 Hilbert Transform Architecture

In fact, the Hilbert transformation-based architecture is the same as the Hartley architecture [11], where the RF input signal is down-converted to a pair of low-IF I-Q signals with the quadrature local oscillation signals of $\cos(\omega_{LO}t)$ and $\sin(\omega_{LO}t)$. The low-IF I–Q signals are passed through the lowpass filters, and then one of the low-IF I–Q signals is phase-shifted by 90° before being added together to form a real low-IF signal. Like the Hartley architecture, the Hilbert transformer performs a phase shift by 90° on one branch before adding two branches together to convert the low-IF I–Q complex signals into to a real IF signal. The Hilbert transformer has the advantage of reaching the precise phase shift of 90° within wide bandwidth around the low-IF over Hartley architecture, which is greatly beneficial to the received signal with wide bandwidth, such as digital TV (DTV) and digital video broadcasting (DVB) tuners. Another advantage is that it can be implemented digitally.

As mentioned previously, the image-reject ratio (IRR) is defined as the ratio of the desired signal gain to the image signal gain and is infinite ideally. IRR is limited in practice by the gain and phase mismatches between the I–Q branches. Therefore, the I–Q mismatch should be compensated prior to Hilbert transformer in the digital domain. Figure 7.10 shows a block diagram of a low-IF DTV tuner with Hilbert transformer.

The frequency response of the Hilbert transformer $H(\omega)$ shown in Fig. 7.11 has unity magnitude, a phase angle of $-\pi/2$ for $0 < \omega < \pi$, and a phase angle of $\pi/2$ for $-\pi < \omega < 0$. Usually, such a circuit is called an ideal 90° phase shifter. Alternatively, it is also called a *Hilbert transformer* if this characteristic is applied to a digital sequence. The Hilbert transformer $H(\omega)$ is expressed as

$$H(\omega) = \begin{cases} -j, & 0 < \omega < \pi \\ +j, & -\pi < \omega < 0 \end{cases} \tag{7.6}$$

The impulse response $h[n]$ of a 90° phase shifter corresponding to (7.6) is

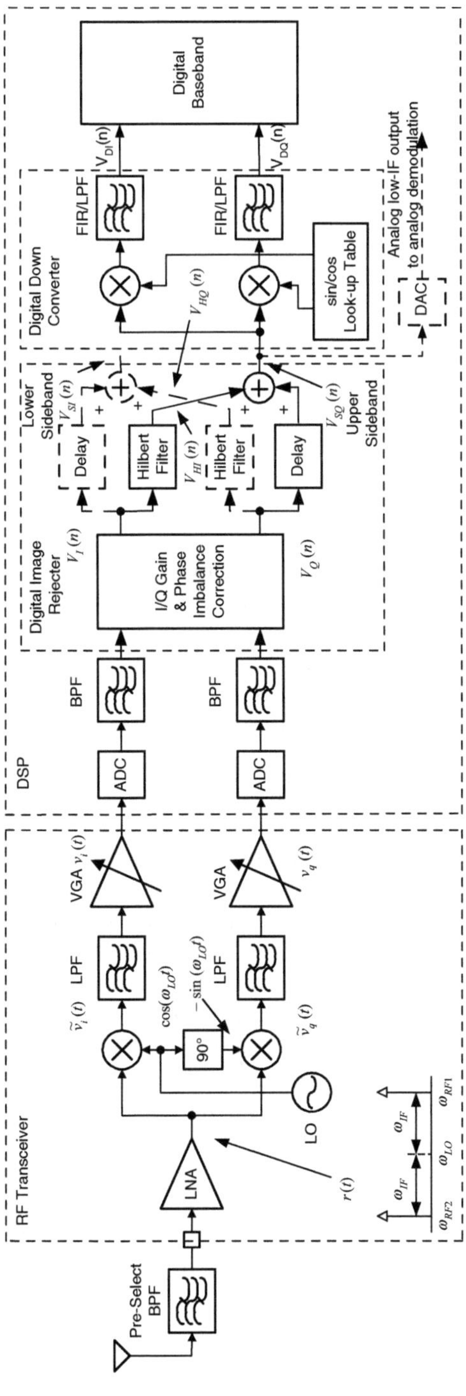

Fig. 7.10 Block diagram of a digital low-IF receiver with an image rejection filter

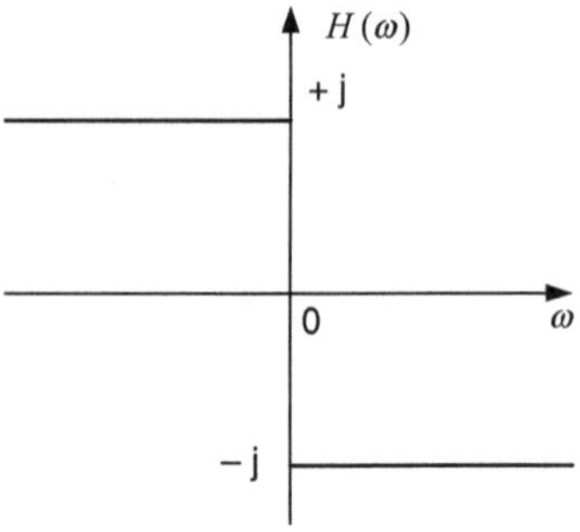

Fig. 7.11 Frequency response of the Hilbert transformer

$$h[n] = \frac{1}{2\pi}\int_{-\pi}^{\pi} H(\omega)e^{j\omega n}d\omega \\ = \begin{cases} \dfrac{2}{\pi}\dfrac{\sin^2(\pi n/2)}{n}, & n \neq 0 \\ 0, & n = 0 \end{cases} \tag{7.7}$$

Figure 7.12a illustrates the impulse response of an ideal Hilbert transformer and amplitude response of an FIR Hilbert transformer truncated by the Kaiser window with a length of 18. In practice, approximations to the ideal Hilbert transformer can be obtained by an FIR filter using the window method. From Fig. 7.12b, it can be seen that its amplitude response is nearly constant with small ripples within much wider frequency range, where the ripple can be reduced by increasing the window length.

In Fig. 7.10, the RF signal is down-converted to the low-IF I–Q signals after mixing with the quadrature phases of the local oscillation signals of $\sin(\omega_{LO}t)$ and $\cos(\omega_{LO}t)$. The I and Q low-IF signals are then filtered through lowpass filters and amplified by variable gain amplifiers (VGAs). Digital low-IF I–Q signals from ADCs are fed into the channel select bandpass filters to remove outside noise and adjacent channel interferers. The channel selection FIR filter is digitally programmable to have different bandwidths for multiple data rates. The filtered I–Q signals then pass through the image rejection filter that consists of the I–Q gain and phase imbalance compensator and Hilbert transformer. The I–Q gain and phase imbalances can be compensated using either the calibration during a power-on period or the adaptive compensation algorithm during the data transmission period [12]. The delay in the image rejecter is used to compensate for the delay caused by the Hilbert transformer.

Compared with the analog Hartley image-reject filter [11], the digital Hilbert transformer based image reject filter can achieve very high IRR and high adjacent channel interferer attenuation. To understand its principle of the image rejection, we assume that the RF input signal that consists of the desired signal with the amplitude of A_1 at the frequency of ω_{RF1} and image signal with the amplitude of

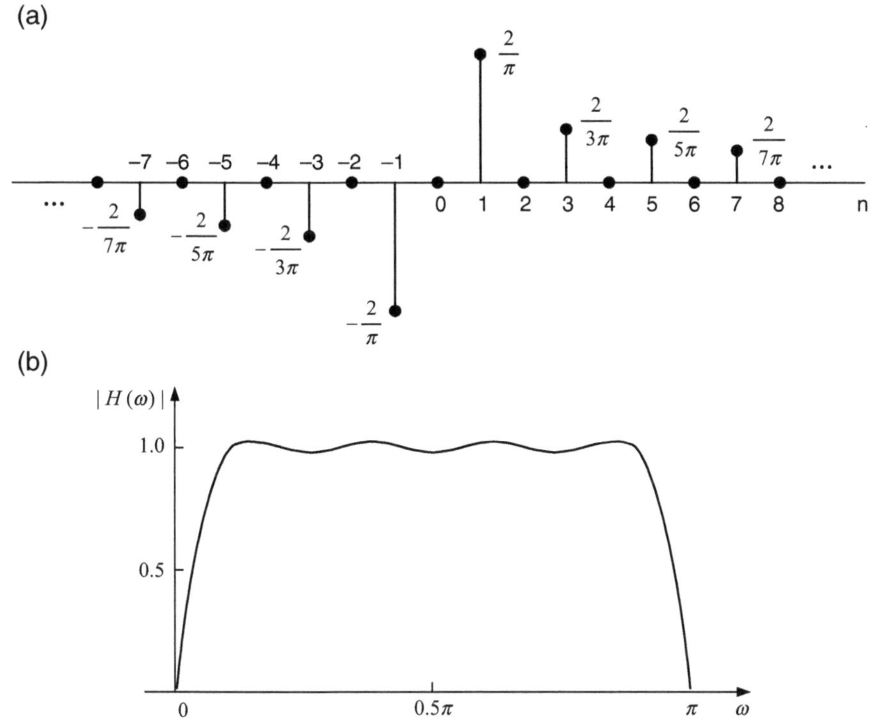

Fig. 7.12 Hilbert transformer: (**a**) impulse response of an ideal Hilbert transformer and (**b**) magnitude response of an FIR Hilbert transformer using a Kaiser window

A_2 at the frequency of ω_{RF2} is expressed as at the LNA output after ignoring Gaussian noise

$$r(t) = A_1 \cos(\omega_{\mathrm{RF1}} t) + A_2 \cos(\omega_{\mathrm{RF2}} t) \tag{7.8}$$

After down-conversion, the low-IF I–Q signals at the outputs of the mixers are, respectively,

$$\begin{aligned} \widetilde{v}_{\mathrm{i}}(t) &= r(t) \times \cos(\omega_{\mathrm{LO}} t) \\ &= \frac{A_1}{2}[\cos(\omega_{\mathrm{LO}} + \omega_{\mathrm{RF1}})t + \cos(\omega_{\mathrm{LO}} - \omega_{\mathrm{RF1}})t] \\ &\quad + \frac{A_2}{2}[\cos(\omega_{\mathrm{LO}} + \omega_{\mathrm{RF2}})t + \cos(\omega_{\mathrm{LO}} - \omega_{\mathrm{RF2}})t] \end{aligned} \tag{7.9}$$

$$\begin{aligned} \widetilde{v}_{\mathrm{q}}(t) &= -r(t) \times \sin(\omega_{\mathrm{LO}} t) \\ &= -\frac{A_1}{2}[\sin(\omega_{\mathrm{LO}} + \omega_{\mathrm{RF1}})t + \sin(\omega_{\mathrm{LO}} - \omega_{\mathrm{RF1}})] \\ &\quad - \frac{A_2}{2}[\sin(\omega_{\mathrm{LO}} + \omega_{\mathrm{RF2}})t + \sin(\omega_{\mathrm{LO}} - \omega_{\mathrm{RF2}})t] \end{aligned} \tag{7.10}$$

Through the lowpass filters, the second-order harmonics are removed and the signals at the output of VGAs are expressed:

$$v_i(t) = \left(\frac{A_1}{2} + \frac{A_2}{2}\right) \cos(\omega_{IF} t) \tag{7.11}$$

$$v_q(t) = \left(\frac{A_1}{2} - \frac{A_2}{2}\right) \sin(\omega_{IF} t) \tag{7.12}$$

where $\omega_{IF} = \omega_{RF1} - \omega_{LO}$ or $\omega_{IF} = \omega_{LO} - \omega_{RF2}$.

Without losing their general property and assuming no I–Q gain and phase imbalance errors, the low-IF I–Q signals $V_I(n)$ and $V_Q(n)$ at the inputs of Hilbert transformers in the digital domain are identical to those in (7.11) and (7.12) except for the amplitude scale factor of G.

$$V_I(n) = G\left(\frac{A_1}{2} + \frac{A_2}{2}\right) \cos(\omega_{IF} n) \tag{7.13}$$

$$V_Q(n) = G\left(\frac{A_1}{2} - \frac{A_2}{2}\right) \sin(\omega_{IF} n) \tag{7.14}$$

The low-IF I–Q signals at the outputs of Hilbert transformers are given by

$$V_{HI}(n) = G\left(\frac{A_1}{2} + \frac{A_2}{2}\right) \sin(\omega_{IF} n) \tag{7.15}$$

$$V_{HQ}(n) = -G\left(\frac{A_1}{2} - \frac{A_2}{2}\right) \cos(\omega_{IF} n) \tag{7.16}$$

The lower and upper sideband signals are, respectively,

$$V_{SI}(n) = V_I(n) + V_{HQ}(n) = GA_2 \cos(\omega_{IF} n) \tag{7.17}$$

$$V_{SQ}(n) = V_Q(n) + V_{HI}(n) = GA_1 \sin(\omega_{IF} n) \tag{7.18}$$

It can be seen from (7.17) and (7.18) that the desired signal A_1 and image signal A_2 are separated at the output of Hilbert transformer. In this case, the desired signal is the upper sideband real signal, while the image signal is the lower sideband signal. As shown in Fig. 7.10, the desired signal $V_{SQ}(n)$ is only sent to the following stage of the digital down-converter, where the digital down-conversion is performed by transferring the low-IF I–Q signals to the baseband I–Q signal.

7.3.2 Image Rejection in Zero-IF Receiver

In a direct down-conversion receiver, as shown in Fig. 7.13, the received RF signal is amplified by a LNA and then is down-converted to the baseband I–Q signals with quadrature local carriers of $\cos(\omega_{\mathrm{LO}}t)$ and $\sin(\omega_{\mathrm{LO}}t)$ without image filtering. The baseband I–Q signals are passed through anti-aliasing lowpass filters such that high-order harmonics and larger interferers are significantly attenuated to reduce the dynamic requirement of the ADCs. As seen in Fig. 7.13b, the image signal is the desired signal itself due to the band overlapping of the desired signal and image signal. To make down-conversion procedure more visible, an asymmetrical waveform from the local oscillator frequency of ω_{LO} for both the desired signal and image signal caused by multipath fading is used. After the signals are down-converted to the baseband domain, such an asymmetrical property can be only observed on a spectrum analyzer with an option for complex signal analysis.

In an ideal case when the I–Q gain and phase imbalance errors are zero, the desired signal located at the negative frequency of $-\omega_{\mathrm{LO}}$ is down-converted to DC with a complex positive frequency local carrier of $\omega^{j\omega_{\mathrm{LO}}t} = \cos(\omega_{\mathrm{LO}}t) + j\sin(\omega_{\mathrm{LO}}t)$. The down-conversion mixing process presented by a complex local carrier is equivalent to individually mixing the real RF input signal with two real quadrature carriers, as shown in Fig. 7.13a. In this case, only the desired signal is down-converted to DC. In practice, however, gain and phase imbalances in the quadrature LO paths cause the image signal to leak into the desired baseband. The image signal leakage is better understood by the fact that imbalances cause a complex negative frequency local carrier of $\gamma\omega^{-j\omega_{\mathrm{LO}}t} = \gamma[\cos(\omega_{\mathrm{LO}}t) - j\sin(\omega_{\mathrm{LO}}t)]$ that is γ times smaller than the desired positive frequency carrier, and its amplitude determines the image rejection ratio (IRR). As a result, the non-ideal complex carrier can be expressed as $e^{j\omega_{\mathrm{LO}}t} + \gamma e^{-j\omega_{\mathrm{LO}}t}$, in which the small LO component of the negative frequency of $\gamma e^{-j\omega_{\mathrm{LO}}t}$ results in the complex image.

Without the I–Q gain and phase imbalance calibration, direct down-conversion receivers have an IRR of about below −30 dB due to a smaller leakage coefficient γ. There are no explicit image rejection requirements for a direct down-conversion receiver due to the fact that the image signal has the same power as the desired signal, and hence the performance of the receiver does not degrade too much without the calibration. However, with the I–Q gain and phase calibration in the receiver, the IRR can be further reduced to below −50 dB, so that SNR or EVM can be improved by more than 1 dB. Similar to a receiver, a transmitter also suffers EVM degradation from a poor IRR due to the I–Q gain and phase imbalances. Therefore, the calibration of the I–Q gain and phase imbalances is necessary for the transmitters to improve EVM and sideband suppression as well.

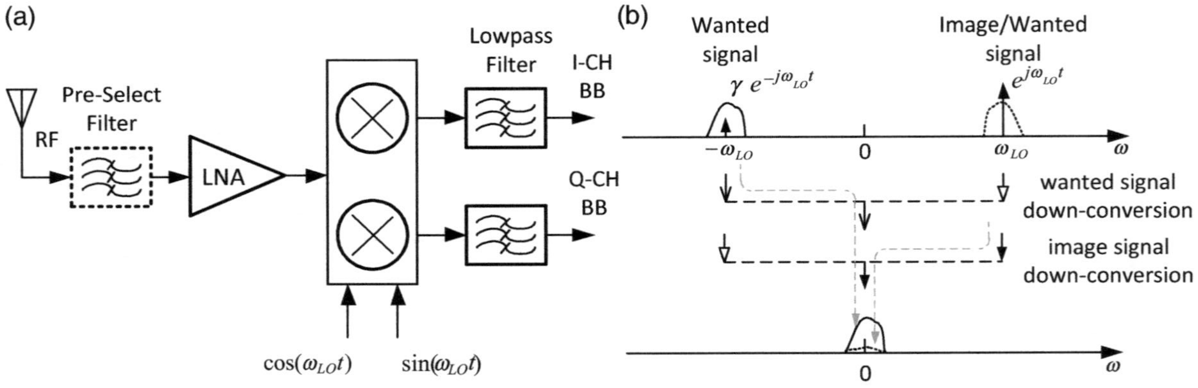

Fig. 7.13 A direct-conversion receiver with quadrature down-conversion: (**a**) block diagram of the front-end and quadrature down-conversion, and (**b**) waveforms of the down-conversion procedure in a frequency domain

7.4 Receiver Impairments

In this section, the effects of real-world signal impairments on the performance of the receiver will be discussed. There are various RF impairments in real radio receivers. Some of impairments can be significantly reduced by using calibration methods, while some of them can be minimized through either careful circuit design or proper operation conditions. Common impairments that will be introduced in this section are I–Q imbalance, DC offset, and nonlinearity. We will discuss each of them in turn and introduce some strategies to reduce their effects on the performance of the receiver.

7.4.1 I–Q Imbalance Compensation

Similar to I–Q imbalance generated in the transmitter, I–Q imbalance is one of the major RF impairments encountered in the design of direct conversion receivers. Generally, the I–Q gain imbalance occurs due to a gain difference between either the baseband I–Q signal paths or the in-phase LO signal and quadrature LO signal paths, while the I–Q phase imbalance happens because of non-ideal phase quadrature between the in-phase LO signal and quadrature LO signal or a non-exact 90° phase difference between them. I–Q phase imbalance generated in the RF domain can be equivalently transferred into the baseband I–Q signals in the baseband domain. If any another distortions and additive Gaussian noise are ignored in the received RF signal for the simplicity, the received modulation signal can be expressed as

$$r(t) = I(t)\cos\left[\omega_c t + \varphi(t)\right] - Q(t)\sin\left[\omega_c t + \varphi(t)\right] \tag{7.19}$$

where $I(t)$ and $Q(t)$ are the transmitted baseband signals and $\varphi(t)$ is the carrier phase due to the propagation delay through the channel. If this signal is multiplied by the two local quadrature carriers $c_I(t)$ and $c_Q(t)$ with the carrier phase $\hat{\varphi}(t)$ that is ideally synchronized to the received signal carrier phase or $\hat{\varphi}(t) = \varphi(t)$ and with the carrier phase imbalance ϕ at the receiver

$$c_{\mathrm{I}}(t) = 2\cos\left[\omega_c t + \hat{\varphi}(t) + \phi/2\right] \tag{7.20}$$

$$c_{\mathrm{Q}}(t) = -2\sin\left[\omega_c t + \hat{\varphi}(t) - \phi/2\right] \tag{7.21}$$

we obtain the following baseband I–Q signals after harmonic components are removed through LPFs, as shown in Fig. 7.14

$$v_{\mathrm{I}}(t) = G_{\mathrm{I}}\cos(\phi/2)I(t) + G_{\mathrm{I}}\sin(\phi/2)Q(t) \tag{7.22}$$

$$v_{\mathrm{Q}}(t) = G_{\mathrm{Q}}\sin(\phi/2)I(t) + G_{\mathrm{Q}}\cos(\phi/2)Q(t) \tag{7.23}$$

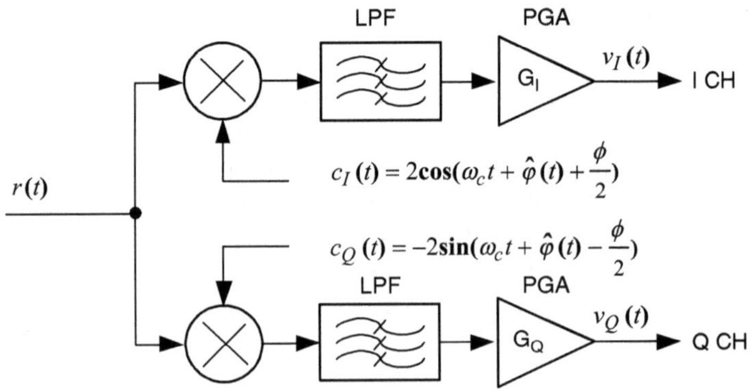

Fig. 7.14 I–Q gain and phase imbalances caused by possible sources in a receiver

where gain imbalance occurs if $G_\mathrm{I} \neq G_\mathrm{Q}$. We can see from the equations above that the carrier phase imbalance ϕ from the RF domain is transferred to the baseband domain as the scaling factors of $\cos(\phi/2)$ and $\sin(\phi/2)$ for the baseband I–Q signals and results in cross-talk between the I–Q channels due to $\sin(\phi/2)$ appearance. If the imbalance ϕ is zero, the received baseband I–Q signals in (7.22) and (7.23) are equal to the transmitted baseband signals multiplied by the I–Q gains: G_I and G_Q:

$$v_\mathrm{I}(t) = G_\mathrm{I} I(t) \tag{7.24}$$

$$v_\mathrm{Q}(t) = G_\mathrm{Q} Q(t) \tag{7.25}$$

To simplify the expression, we can define ε as a gain error related to G_I and G_Q such that they have the following relationship, or

$$\varepsilon = \frac{G_\mathrm{I}}{G_\mathrm{Q}} - 1 \tag{7.26}$$

$$G_\mathrm{I}^2 + G_\mathrm{Q}^2 = 2 \tag{7.27}$$

From these relationships above, we have the normalized gain as

$$G_\mathrm{I} = (1+\varepsilon)\sqrt{\frac{2}{2+2\varepsilon+\varepsilon^2}} \tag{7.28}$$

$$G_\mathrm{Q} = \sqrt{\frac{2}{2+2\varepsilon+\varepsilon^2}} \tag{7.29}$$

If ε is very small quantity, as it usually is, using the Taylor series approximations to G_I and G_Q with the first-order term above gives the following expressions:

$$G_I \approx 1 + \frac{\varepsilon}{2} \tag{7.30}$$

$$G_Q \approx 1 - \frac{\varepsilon}{2} \tag{7.31}$$

The baseband signal equations in (7.22) and (7.23) can be also expressed as a matrix format as

$$\begin{bmatrix} v_I(t) \\ v_Q(t) \end{bmatrix} = \begin{bmatrix} G_I & 0 \\ 0 & G_Q \end{bmatrix} \begin{bmatrix} \cos(\phi/2) & \sin(\phi/2) \\ \sin(\phi/2) & \cos(\phi/2) \end{bmatrix} \begin{bmatrix} I(t) \\ Q(t) \end{bmatrix} \tag{7.32}$$

To compensate the I–Q gain and phase imbalances, we can add a post-compensator or an equalizer at the outputs of $v_I(t)$ and $v_Q(t)$ and thus the outputs of the equalizer are

$$\begin{bmatrix} u_I(t) \\ u_Q(t) \end{bmatrix} = \begin{bmatrix} G_I & 0 \\ 0 & G_Q \end{bmatrix} \begin{bmatrix} \cos(\phi/2) & \sin(\phi/2) \\ \sin(\phi/2) & \cos(\phi/2) \end{bmatrix} \begin{bmatrix} \cos(\phi_c/2) & \sin(\phi_c/2) \\ \sin(\phi_c/2) & \cos(\phi_c/2) \end{bmatrix} \times \begin{bmatrix} G_{CI} & 0 \\ 0 & G_{CQ} \end{bmatrix} \begin{bmatrix} I(t) \\ Q(t) \end{bmatrix} \tag{7.33}$$

where ϕ_c and G_C is the compensator phase and gain used to compensate for the I–Q phase and gain imbalances. Equation (7.33) can be further expressed as

$$\begin{bmatrix} u_I(t) \\ u_Q(t) \end{bmatrix} = \begin{bmatrix} G_I & 0 \\ 0 & G_Q \end{bmatrix} \begin{bmatrix} \cos(\phi/2 - \phi_c/2) & \sin(\phi/2 + \phi_c/2) \\ \sin(\phi/2 + \phi_c/2) & \cos(\phi/2 - \phi_c/2) \end{bmatrix} \begin{bmatrix} G_{CI} & 0 \\ 0 & G_{CQ} \end{bmatrix} \begin{bmatrix} I(t) \\ Q(t) \end{bmatrix} \tag{7.34}$$

If $\phi_c = -\phi$ is met, the expression above becomes

$$\begin{aligned} \begin{bmatrix} u_I(t) \\ u_Q(t) \end{bmatrix} &= \begin{bmatrix} G_I & 0 \\ 0 & G_Q \end{bmatrix} \begin{bmatrix} \cos(\phi) & 0 \\ 0 & \cos(\phi) \end{bmatrix} \begin{bmatrix} G_{CI} & 0 \\ 0 & G_{CQ} \end{bmatrix} \begin{bmatrix} I(t) \\ Q(t) \end{bmatrix} \\ &= \begin{bmatrix} G_I & 0 \\ 0 & G_Q \end{bmatrix} \begin{bmatrix} G_{CI}\cos(\phi) \\ G_{CQ}\cos(\phi) \end{bmatrix} \begin{bmatrix} I(t) \\ Q(t) \end{bmatrix} \\ &= \begin{bmatrix} G_I G_{CI}\cos(\phi) \\ G_Q G_{CQ}\cos(\phi) \end{bmatrix} \begin{bmatrix} I(t) \\ Q(t) \end{bmatrix} \end{aligned} \tag{7.35}$$

It is seen that the cross-talk terms between the I–Q channels are cancelled. If $G_{\mathrm{CI}} = 1/G_{\mathrm{I}}$ and $G_{\mathrm{CQ}} = 1/G_{\mathrm{Q}}$ are met, the expression above becomes

$$\begin{bmatrix} u_{\mathrm{I}}(t) \\ u_{\mathrm{Q}}(t) \end{bmatrix} = \begin{bmatrix} \cos(\phi) I(t) \\ \cos(\phi) Q(t) \end{bmatrix} \tag{7.36}$$

Now the recovered baseband I–Q signals are equal to the transmitted baseband I-Q signals multiplied by a scaling factor of $\cos(\phi)$, and therefore the I–Q gain and phase imbalances are completely corrected.

Similar to the normalized gain expressions in (7.28) and (7.29) for the I–Q gain imbalance of the receiver, the I–Q compensation gain corresponding to the ideal compensation condition can be normalized as follows:

$$G_{\mathrm{IC}} = \frac{1}{G_{\mathrm{I}}} = \frac{\sqrt{1+\varepsilon_{\mathrm{c}}+\varepsilon_{\mathrm{c}}^2/2}}{(1+\varepsilon_{\mathrm{c}})} \tag{7.37}$$

$$G_{\mathrm{QC}} = \frac{1}{G_{\mathrm{Q}}} = \sqrt{1+\varepsilon_{\mathrm{c}}+\varepsilon_{\mathrm{c}}^2/2} \tag{7.38}$$

where $\varepsilon_{\mathrm{c}} = \varepsilon$ is assumed at the ideal compensation condition. After the Taylor series approximations to G_{IC} and G_{QC}, they can be expressed as

$$G_{\mathrm{IC}} \approx 1 - \frac{1}{2}\varepsilon_{\mathrm{c}} \tag{7.39}$$

$$G_{\mathrm{QC}} \approx 1 + \frac{1}{2}\varepsilon_{\mathrm{c}} \tag{7.40}$$

Equation (7.33) can be rewritten by substituting (7.39) and (7.40) for G_{IC} and G_{QC} as

$$\begin{bmatrix} u_{\mathrm{I}}(t) \\ u_{\mathrm{Q}}(t) \end{bmatrix} \approx \underbrace{\begin{bmatrix} G_{\mathrm{I}}\cos(\phi/2) & G_{\mathrm{I}}\sin(\phi/2) \\ G_{\mathrm{Q}}\sin(\phi/2) & G_{\mathrm{Q}}\cos(\phi/2) \end{bmatrix}}_{\text{I–Q imbalance}} \times \underbrace{\begin{bmatrix} \left(1-\frac{1}{2}\varepsilon_{\mathrm{c}}\right)\cos(\phi_{\mathrm{c}}/2) & \left(1+\frac{1}{2}\varepsilon_{\mathrm{c}}\right)\sin(\phi_{\mathrm{c}}/2) \\ \left(1-\frac{1}{2}\varepsilon_{\mathrm{c}}\right)\sin(\phi_{\mathrm{c}}/2) & \left(1+\frac{1}{2}\varepsilon_{\mathrm{c}}\right)\cos(\phi_{\mathrm{c}}/2) \end{bmatrix}}_{\text{I–Q compensator}} \begin{bmatrix} I(t) \\ Q(t) \end{bmatrix} \tag{7.41}$$

If the gain and phase imbalances of ε and ϕ are both very small, the gain and phase values of ε_{c} and ϕ_{c} at the compensator are also very small after the compensation. Thus, the matrix of the compensator in (7.41) can be further approximated as follows:

$$\begin{bmatrix} u_{\mathrm{I}}(t) \\ u_{\mathrm{Q}}(t) \end{bmatrix} \approx \underbrace{\begin{bmatrix} G_{\mathrm{I}} \cos(\phi/2) & G_{\mathrm{I}} \sin(\phi/2) \\ G_{\mathrm{Q}} \sin(\phi/2) & G_{\mathrm{Q}} \cos(\phi/2) \end{bmatrix}}_{\text{I–Q imbalance}} \underbrace{\begin{bmatrix} 1 - \frac{1}{2}\varepsilon_{\mathrm{c}} & \frac{\phi_{\mathrm{c}}}{2} \\ \frac{\phi_{\mathrm{c}}}{2} & 1 + \frac{1}{2}\varepsilon_{\mathrm{c}} \end{bmatrix}}_{\text{I–Q compensator}} \begin{bmatrix} I(t) \\ Q(t) \end{bmatrix} \tag{7.42}$$

It is clear from the equation above that one element $\phi_c/2$ at the first row and second column and another element $\phi_c/2$ at the second row and the second column are used to compensate for the phase imbalance ϕ generated from the down-conversion processing due to non ideal orthogonal local oscillator signals at the receiver. Another two elements of $1 - \varepsilon_c/2$ and $1 + \varepsilon_c/2$ are employed to compensate for the gain imbalance ε produced on the baseband I–Q channels due to gain difference between G_{I} and G_{Q} as shown in Fig. 7.14.

Usually, the I–Q gain and phase imbalances of ε and ϕ generated at the receiver are unknown. In order to compensate for these two unknown gain and phase imbalances, an adaptive equalizer with four real coefficients can be used as a compensator. The coefficients can be adaptively updated by adding a calibration signal to the receiver. For example, a QPSK modulation signal can be used as the calibration signal for the I–Q imbalance compensation at the receiver due to its simple decision rule. During the calibration period, the QPSK modulation signal can be externally added at the input port of the LNA or can be internally connected to the input of the quadrature down-conversion mixers through a loop back from the transmitter path after the I–Q imbalance compensation at the transmitter.

Due to the digital implementation of the equalizer, the matrix of the compensator in (7.42) can be expressed with four real coefficients C_i, $i = 0, 1, 2, 3$ of the equalizer in the digital domain

$$\begin{bmatrix} u_{\mathrm{I},k} \\ u_{\mathrm{Q},k} \end{bmatrix} \approx \underbrace{\begin{bmatrix} G_{\mathrm{I}} \cos(\phi/2) & G_{\mathrm{I}} \sin(\phi/2) \\ G_{\mathrm{Q}} \sin(\phi/2) & G_{\mathrm{Q}} \cos(\phi/2) \end{bmatrix}}_{\text{I–Q imbalance}} \underbrace{\begin{bmatrix} C_0 & C_1 \\ C_2 & C_3 \end{bmatrix}}_{\text{I–Q compensator}} \begin{bmatrix} I_k \\ Q_k \end{bmatrix} \tag{7.43}$$

The structure of the adaptive equalizer implemented in the digital domain is shown in Fig. 7.15. The cross coefficients of C_1 and C_2 mainly correct the quadrature phase imbalance ϕ while the I–Q branch coefficients of C_0 and C_3 primarily compensate for the gain imbalance ε.

The coefficients of an adaptive equalizer with four independent real coefficients can be updated by using the ZF algorithm in (4.56)

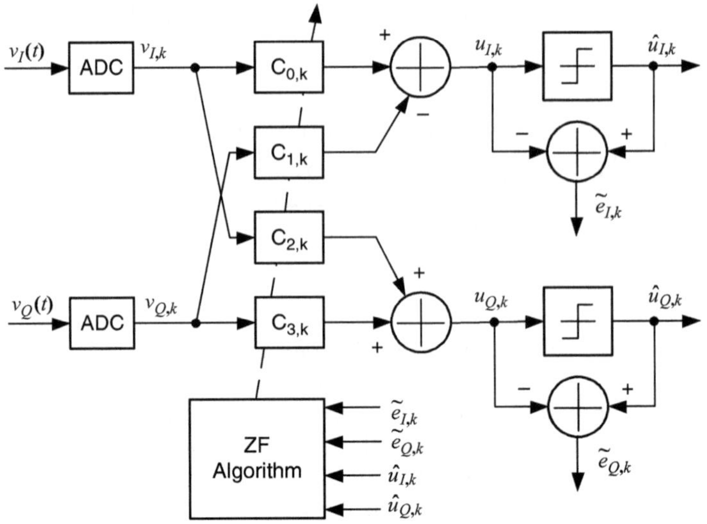

Fig. 7.15 A block diagram of an adaptive equalizer with four real taps for I–Q imbalance compensation

$$\begin{aligned} c_{0,k+1} &= c_{0,k} + \lambda \tilde{e}_{\mathrm{I},k} \hat{u}_{\mathrm{I},k} \\ c_{1,k+1} &= c_{1,k} + \lambda \tilde{e}_{\mathrm{I},k} \hat{u}_{\mathrm{Q},k} \\ c_{2,k+1} &= c_{2,k} + \lambda \tilde{e}_{\mathrm{Q},k} \hat{u}_{\mathrm{Q},k} \\ c_{3,k+1} &= c_{3,k} + \lambda \tilde{e}_{\mathrm{Q},k} \hat{u}_{\mathrm{I},k} \end{aligned} \tag{7.44}$$

where the error signal is $\tilde{e}_k = \tilde{e}_{\mathrm{I},k} + j\tilde{e}_{\mathrm{Q},k} = \hat{u}_{\mathrm{I},k} - u_{\mathrm{I},k} + j(\hat{u}_{\mathrm{Q},k} - u_{\mathrm{Q},k})$. It can be seen from Fig. 7.15 that the decision signals of $\hat{u}_{\mathrm{I},k}$ and $\hat{u}_{\mathrm{Q},k}$ become the sign signals of either +1 or -1 to avoid four times multiplication operations between $\tilde{e}_k$ and $\hat{u}_k$ in (7.44) if a type of QPSK signal is used as the calibration signal at the receiver. After the I–Q imbalance compensation, the coefficients of the equalizer are fixed during the receiver operation until the next calibration procedure.

Figure 7.16 illustrates the WCDMA QPSK constellation plots before and after compensations. One case corresponds to the phase imbalance of $-5°$ only, as shown in Fig. 7.16a, b while another case is for the gain and phase imbalances of 3 dB and $-5°$, as shown in Fig. 7.16c, d. It clearly shows that the I–Q imbalance is completely corrected in two different cases. The convergence of the equalizer is about ten iterations or ten chip intervals when a WCDMA QPSK signal with one sample per symbol duration at the chip rate of 3.84 Mchips/s is used for the calibration. In the simulation, a normalized QPSK signal with values of ± 0.707 is used as the calibration signal and decisions in the equalizer are made at the normalized values.

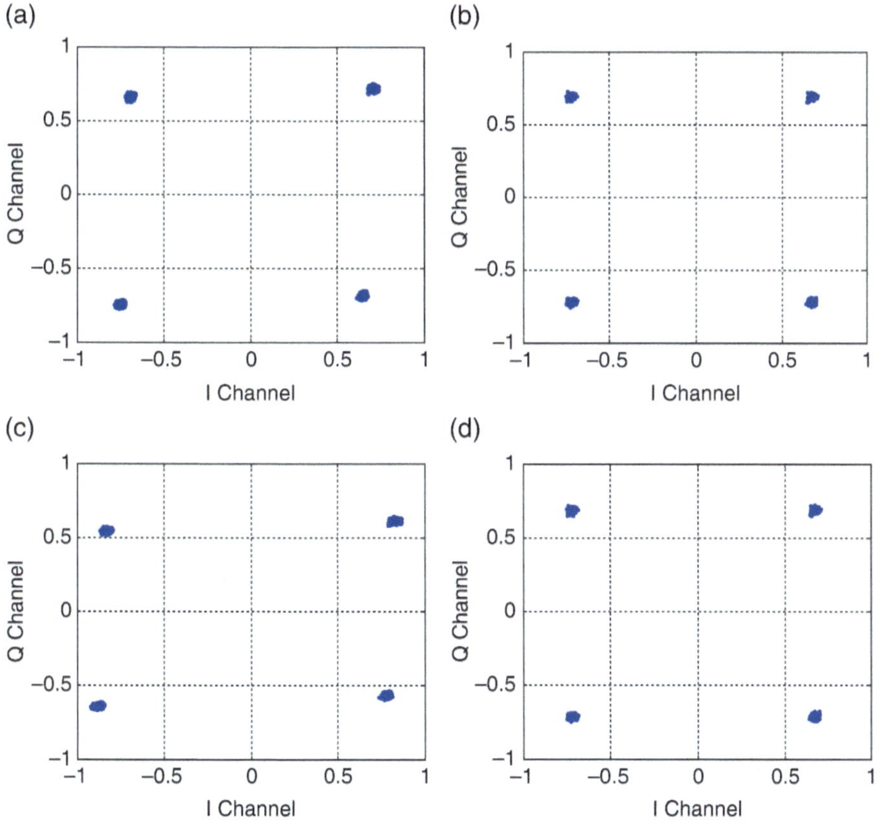

Fig. 7.16 Constellation of SRRC filtered QPSK with $\alpha = 0.22$ at the receiver: (**a**) I–Q phase imbalance of $-5°$ before compensation, (**b**) I–Q phase imbalance of $-5°$ after compensation, (**c**) I–Q gain and phase imbalances of 3 dB and $-5°$ before compensation, and (**d**) I–Q gain and phase imbalances of 3 dB and $-5°$ after compensation

7.4.2 DC Offset Cancellation

Besides I–Q imbalance, DC offset is another major drawback in a direct down-conversion receiver. In direct down-conversion, as the desired RF signal is down-converted to the baseband I–Q signals in the receive chain, any small DC offsets at the output of the mixer may result in very large DC offsets at the inputs of the ADCs because the baseband amplification stages may offer a gain value up to more than 60 dB. Such large DC offsets at the input of the ADC can either saturate the ADC or reduce the dynamic range of the ADC. Furthermore, DC offsets also degrade the signal-to-noise ratio. Various phenomena contribute to the creation of DC offsets. In most integrated wireless transceiver designs, two major sources are involved: LO Self-Mixing and Blocker Self-Mixing.

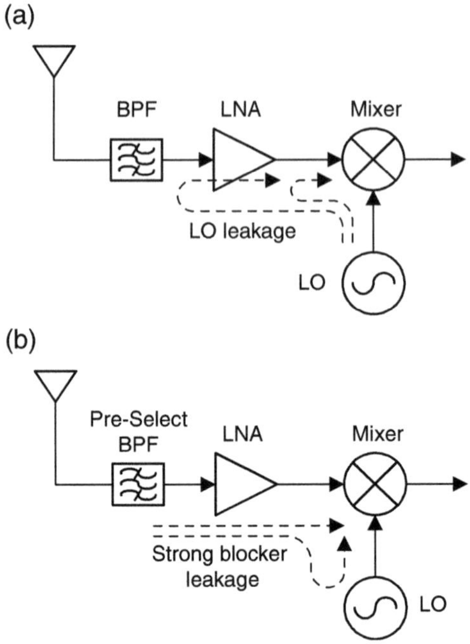

Fig. 7.17 Self-mixing block diagram: (**a**) DC offset produced by LO leakage and (**b**) DC offset produced by blocker or interferer leakage. Referenced from [13]

- **LO Self-Mixing**: The LO signal may be conducted or radiated through unintended paths to another input port of the mixer. As a result, the LO signal effectively mixes with itself, producing an undesirable DC component at the mixer output, as shown in Fig. 7.17a. In the worst case, the LO leakage may reach the LNA input, producing an even stronger DC offset at the mixer output due to a larger LNA gain. This LO-leakage–based self-mixing problem is mainly caused by poor silicon isolation. These include substrate coupling, ground bounce, bond wire radiation, and capacitive and magnetic coupling [13].
- **Blocker Self-Mixing**: As with LO self-mixing, a DC offset may be produced when a stronger in-band interferer, once amplified by the LNA, leaks into the LO input port of the mixer, and then mixes with itself, as shown in Fig. 7.17b.

It should be mentioned that LO or RF signal leakage to the opposite mixer port is not the only way in which undesirable DC can be generated. In the RF transceiver chip, any stage that exhibits even-order nonlinearity or differential circuit imbalance also produces a DC offset. The earlier the DC offset is generated in the receive chain, the more critical the DC offset to system performance due to larger gain value.

In a direct-conversion receiver, DC offset should be removed by using DC offset cancellation (DCOC) circuits to avoid saturating subsequent stages., Two topological structures performing DCOC are generally defined according to their circuit structures as follows:

1. Alternative current (AC) coupling: The AC coupling method means that DCOC is performed by using circuits that are capable of highpass filtering functions from the output of the mixer to the input of the ADC.
2. Direct current (DC) coupling: The DC coupling method means that the DC offset is removed by subtracting the estimated DC offset from the output of the mixer.

7.4.2.1 DCOC with Highpass Filtering

AC coupling can be implemented by using either a highpass filter with a feed-forward structure as shown in Fig. 7.18a or one with a lowpass filter on the feedback path to perform the subtraction from the signal on the feed-forward path, as shown in Fig. 7.18b, which is called *servo-loop*–based highpass filtering (HPF).

Usually it is preferable for an HPF to have a large corner frequency or a short time constant so that it can reach its steady state quickly. An HPF with a large corner frequency, however, removes too many low frequency components of the received signal, so that BER of the received signal degrades severely. On the other hand, it would take a long time for an HPF with a small corner frequency to reach its steady state, so some data may be lost due to the incorrect decision level. To avoid such problems, analog IC design engineers prefer HPF with a switched-resistance structure to accelerate the HPF response time without severely distorting the received signal, as shown in Fig. 7.18c, where switches $S1$, $S2$, and $S3$ control which resistor is connected to form different time constants.

The transfer function of the HPF in Fig. 7.18b is derived in the following. The DC feedback loop forms a negative feedback at low frequency close to DC, thus filtering out the DC offsets. *Both the DC offset of the input signal and the DC offsets of the amplifier* $K1$ *are canceled at the output.* The circuit $K1$ can be either a filter or an amplifier, but an amplifier is usually used here to simplify the derivation. The circuit $K2$ usually is a buffer with the unit gain, or $K2=1$. The transfer function of the block diagram is

$$A(s)=\frac{K1}{1+K1K3H(s)} \tag{7.45}$$

where $H(s)$ can be an integrator or lowpass pole. $H(s)$, however, has typically a first-order transfer function to ensure stability. In the case of an ideal integrator, i.e., $H(s)=-\omega_o/s$, where $\omega_o=1/RC$, the transfer function in (7.45) can be written as

$$A(s)=\frac{K1}{1-\dfrac{K1K3\omega_o}{s}} \tag{7.46}$$

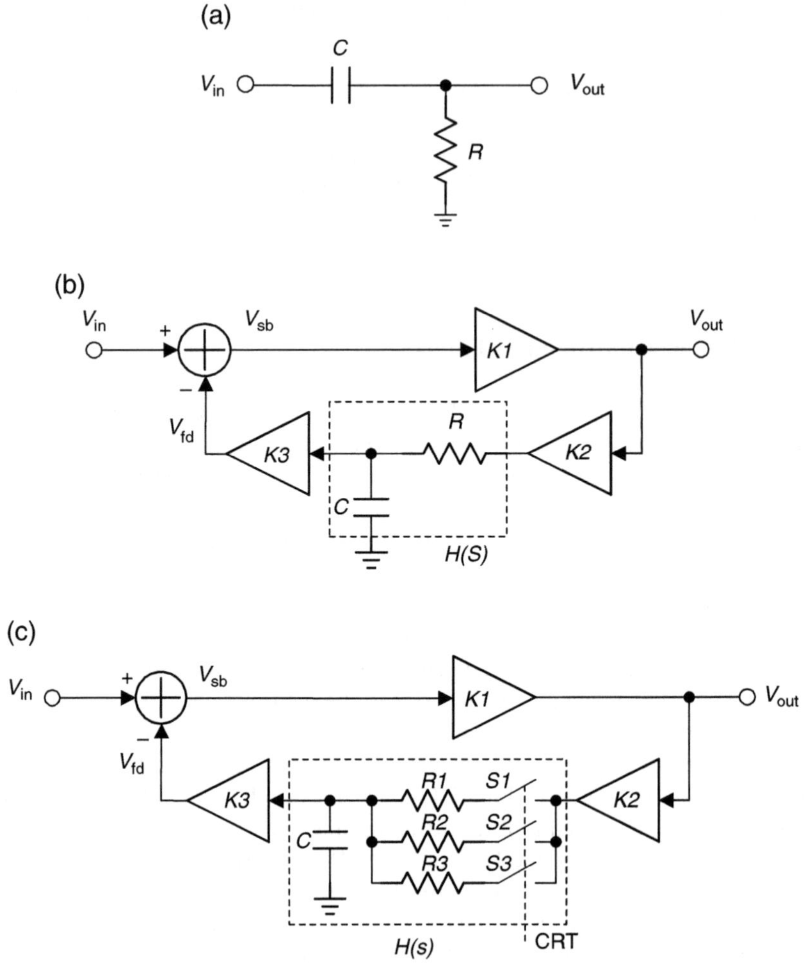

Fig. 7.18 AC coupling block diagram: (**a**) AC coupling with a highpass filter, (**b**) AC coupling with fixed servo loop parameter (or feedback loop), and (**c**) AC coupling with variable servo loop parameters

The amplitude response of $A(s)$ is expressed as

$$|A(\omega)| = \frac{K1}{\sqrt{1 + \left(\frac{K1K3\omega_o}{\omega}\right)^2}} \tag{7.47}$$

The DC gain $|A(0)| = 0$ and the corner frequency of the resulting first-order HPF is given by

$$\omega_{3\,\mathrm{dB}} = K1K3\omega_o \tag{7.48}$$

If $K1 = 1/K3$ is met, then $\omega_{3dB} = \omega_o = 1/RC$ is obtained. The relationship between $K1$ and $K3$ is useful when the gain $K1$ is programmable and the corner frequency must remain unchanged. It is clear that the corner frequency of the HPF-based servo loop is dependent on RC production on the feedback loop. Therefore, the adjustable corner frequency of the servo-loop–based HPF can be implemented with the switched-resistance structure, as shown in Fig. 7.18c, where three different corner frequencies can be realized. In most applications, three different corner frequencies selected by a state machine are usually accepted.

For example, three corner frequencies of around 8 MHz, 500 kHz, and 3 kHz are suitable to DCOC in the 802.11b WLAN application. Upon entering the RX mode, the corner frequency of 8 MHz is momentarily chosen for a very short programmable duration. The HPF quickly removes all the DC offsets from the down-converted baseband I–Q signals. After this initial duration, the corner frequency is automatically reduced to 500 kHz, and remains at it until the digital baseband processor finishes AGC testing and adjusting on the receiver chain. Then, a lowest corner frequency of 3 kHz is switched to avoid filtering out more signal components around DC before the end of the 10 short training symbols. The procedure for switching three corner frequencies should be settled down less than 5 μs within the ten short training symbols after each frame is detected at the receiver.

7.4.2.2 DCOC with DC Coupling

In DCOC with DC coupling, DC offsets are removed without using any highpass filtering circuitry during the information reception operation. The basic idea of removing DC offsets is based on following two steps:

- Before receiving the information sequence, any DC offsets generated in the receive circuitries are estimated, and then stored in a capacitor or a look-up table (LUT).
- During receiving information period, the received signal is subtracted by the estimate DC offset. The subtraction may be performed at one place or several places along the receive path.

As mentioned previously, DC offsets are mainly generated in the front-end circuit of the mixer, and appear at the output of the mixer. Since the DC offset changes with the LNA gain setting, the DC offset can be measured at different LNA gain settings during the calibration period, and then be stored in the LUT. During the receive operation, the received signal is subtracted with the output of the LUT addressed by the LNA gain settings. The calibration procedure is controlled by the digital baseband processor. The calibration can be automatically performed whenever the receiver is in an idle mode and no signal is detected in order to track any change with temperature. After DCOC is performed in the analog domain, any residual DC offset can be further removed in the digital domain. This LUT-based DCOC has been reported in the WLAN system [14] and the Ultrawide Band (UWB) system [15, 16].

Figure 7.19 shows a block diagram of DCOC implementation based on a DC coupling method for a UWB receiver. DACs are placed along the receive chain to

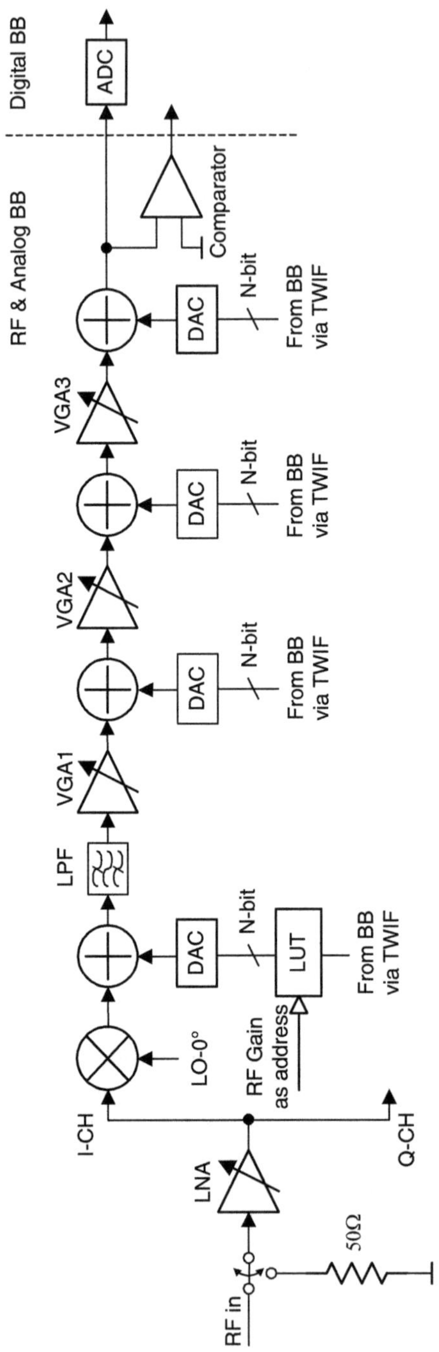

Fig. 7.19 DC coupling block diagram by means of LUT in a direct down-conversion receiver. Redrawn from [15]

remove DC offsets. The number of DACs usually ranges from one to three or even more depending on the application. One placed at the output of the mixer plays the most important role compared with any other DACs. A very small DC offset at the output of the mixer may become large at the input of the ADC because of a large BB gain along the receive chain. Without DCOC, the amplified DC offset can result in saturating subsequent stages.

Before the calibration, the input of the LNA is connected to a dummy load 50 ohm to avoid any input signals or interferers. The calibration is performed by means of the binary search algorithm that is updated by measuring the comparator output shown in Fig. 7.19. Binary N-bits at the DAC inputs are updated through the three-wire interface (TWIF). For an N-bit DAC, $N-1$ steps are required to finish each calibration. The DC offset calibration is performed from the back to the front [15]. For example, the calibration processing begins with the rightmost-side DAC when turning on VGA3 and turning off the LPF, VGA1, and VGA2. Then, any DC offset measured in the digital domain is generated by VGA3. After the binary search, the estimated DC offset value for VGA3 is stored in register 3. Next, the estimate DC offset values for VGA2 and VGA1, respectively, are stored in the register 2 and 1 when turning on VGA2 first, and then VGA1 next. Finally, the estimated DC offsets generated in the LNA and mixer associated with the High Gain (HG) Mode, Middle Gain (MG) Mode, and Low Gain (LG) Mode are calibrated and stored in the registers respectively.

During the reception operation, the estimate DC values stored in the LUT are read out corresponding to the LNA HG, MG, and LG modes, and then are passed through the DAC to subtract DC offset from the down-converted signal at the output of the mixer. In the following stages, VGA1, VGA2, and VGA3 can be implemented in such a way that DC offsets at their outputs are independent of gain values [15]. In more detail, these VGAs are implemented with a current steering DAC, which is used to inject a corrective current to the resistive loads so that offset is minimized. Hence, each DAC associated with VGA1, VGA2, or VGA3 needs only one group estimate DC value because of independence of gain setting.

In the DC coupling method, a DAC can be replaced with a capacitor in TDD systems in order to reduce the die area. For example, in the GSM system, periodic DC offset cancellation can be performed during idle times where the estimate DC offset is stored on a capacitor and then subtracted from the received signal during actual reception. Figure 7.20 shows two kinds of basic block diagrams of DC offset cancellation circuits [14]. In Fig. 7.20a, during the idle time the output of the lowpass filter is connected to a pair of capacitors C such that the output DC offset voltage V_{DCOFF} in the output signal V_{OUT} via a transconductance $Gm1$ is sampled and stored onto capacitors C when a switch S is closed. The sampled and stored voltage is then fed back to the output of the lowpass filter to subtract DC offset voltage V_{DCOFF} from V_{OUT} via a transconductance amplifier $Gm2$ when the switch S is opened during the signal reception time. In order to cancel different DC offsets corresponding to different LNA gain settings, the amplifier $Gm2$ is designed such that its gain can also be programmed by an LNA gain control signal CTR. In Fig. 7.20b, the procedures of sampling and subtracting are the same as that shown in

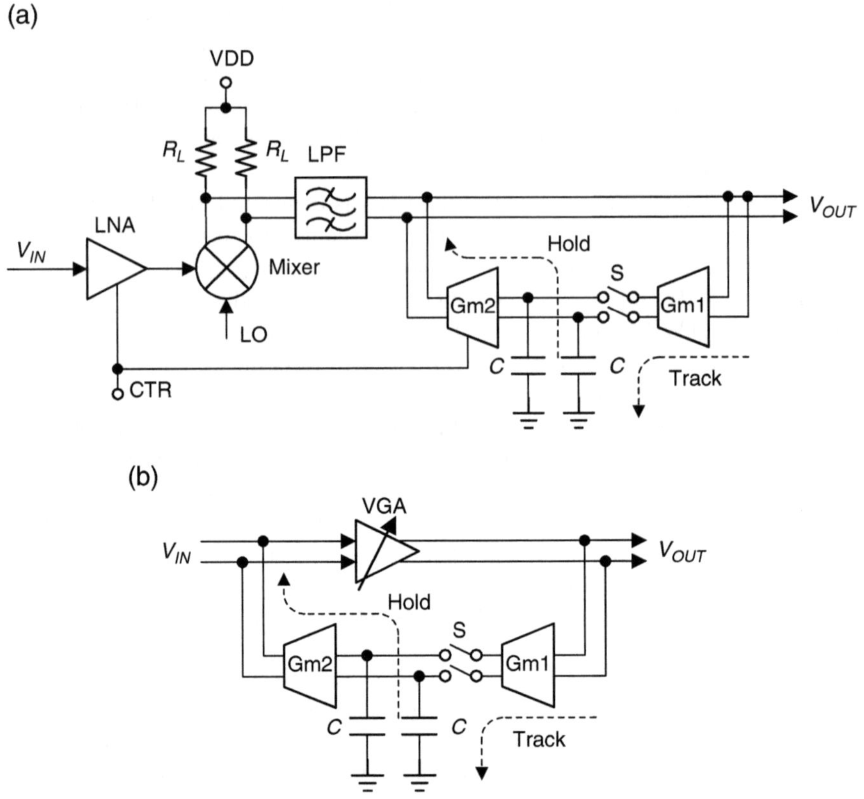

Fig. 7.20 DC-coupling–based DC offset cancellation circuits: (**a**) one used at mixer output [14], and (**b**) one associated with VGA. Redrawn from [14]

Fig. 7.20a, except that the sampling and subtracting points are at different places or at the input and output of the VGA.

In the GSM system, for example, each frame contains eight bursts, and each burst has 148 bits. Each user can receive only one burst message in one frame, and in the meantime can transmit information at different burst times within the same frame. Thus, it is completely possible for the receiver to have time to sample or track DC offset before receiving the desired time slot. Furthermore, there is the guard time period of 30.46 μs (or 8.25 bits) between two bursts, which can also be used for the receiver to sample DC offsets. This type of DC offset cancellation has been used in the Skyworks' CX7444 transceiver chip [13].

7.4.3 Nonlinear Distortion

At a receiver, nonlinear distortions are mainly generated through the front-end blocks, such as a LNA and mixers, due to strong adjacent channel interferers and larger out-of-band blockers. In an RF transceiver, there is no front-end passband filter used for attenuating these interferers and blockers prior to the LNA and mixers in order to reduce size and cost. If nonlinear distortions occur, nonlinear components or products that fall in the desired band degrade the received SNR and result in the degradation of BER or PER in the receiver. In the following section, nonlinear distortions related to the second-order intermodulation (IM2) and third-order intermodulation (IM3) distortions, which are the most severe nonlinear distortions, are discussed in detail.

7.4.3.1 Second-Order Distortion

In a direct down-conversion receiver, low even-order distortion, especially second-order distortion or second-order intermodulation (IM2) products, can severely degrade the receiver's performance due to the presence of strong blockers or adjacent modulated signals, as well as the leakage of the transmitted signal in an SoC transceiver. As a result, a spurious baseband signal that is directly proportional to the squared version of the blocker envelope is generated at baseband due to the second-order distortion, especially for direct down-converter receivers. The bandwidth of these second-order spectral productions at baseband can be up to twice the bandwidth of the blocker's amplitude envelope because of the squared property and the modulation index doubled [17]. Sometimes these blockers have the same modulation format as the desired signal.

The IM2 distortion products that occur at baseband in a zero-IF receiver may degrade the receiver performance. Even though there are many mechanisms that may generate the IM2 products in the zero-IF receiver, three main mechanisms are as follows [17]:

- **Blocker Self-Mixing:** This happens when a large RF blocker enters the front end of a receiver, and a small portion of it could leak into the LO port due to parasitic coupling, as illustrated in Fig. 7.17b. Once the RF blocker leakage component appears at the LO port, such a self-mixing product is directly proportional to the mixing of the RF blocker input and the RF-to-LO leakage component. Hence, low-frequency IM2 products, including a DC component, are generated at the output of the mixer.
- **Down-Converter-Stage Second-Order Nonlinearity**: A DC component can be also generated at baseband because of the second-order nonlinearity in the active devices of the mixer or RF stage when a strong signal or a modulated blocker occurs at the input of the receiver. This phenomenon is easily explained by the fact that even-order nonlinearities cause asymmetrical distortion when a strong continuous wave (CW) is applied to the device, as illustrated in Fig. 7.21, where a DC component is produced by such an asymmetrical output waveform. A small DC signal may be amplified e enough to saturate the ADC after the amplification of the VGA with a gain value of up to 60 dB.
- **Unbalanced Mixer:** In a perfectly balanced mixer, the equivalent differential *IM2* products are translated to high frequencies and the equivalent common-

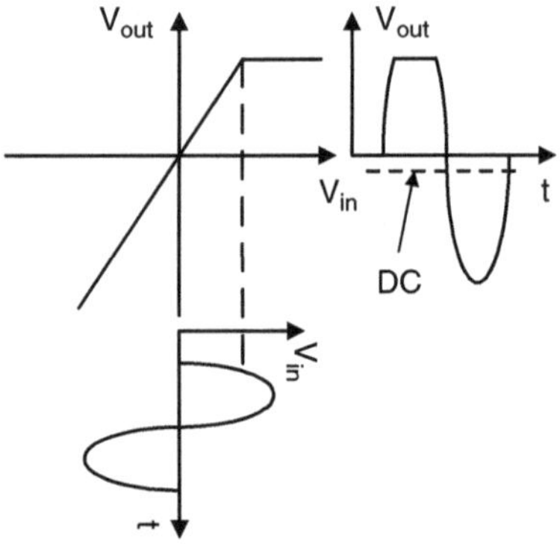

Fig. 7.21 DC component produced by even-order nonlinearity. Redrawn from [18]

mode *IM2* products are canceled out at the mixer differential output [17]. However, in actual designs, the mismatches usually exist in the LO stage devices and result in the equivalent common-mode *IM2* products at the outputs of the I-Q mixers. For example, the deviation of the LO duty cycle from 50% produces the low frequency *IM2* products at the outputs of the I–Q mixers.

The second-order distortion is associated with the second-order intercept point (IP2) requirement that is a key specification for the direct conversion receivers, especially for wireless cellular phone systems, like GSM and WCDMA systems. This is because a large interferer or "blocker" with a power level 76 dB greater than the desired signal would be received with the desired signal at the receiver input. For example, in the blocking test of the GSM900, the worst case occurs for blockers with power levels up to –23 dBm, which are inside the GSM receive band, but greater than 3 MHz away from the desired signal with a power level of –99 dBm. Such a large blocker demands the receiver with a high IP2 specification to meet the performance requirements.

Due to the nonlinearities of the front-end units in a direct conversion receiver, including the LNA and mixer, a second-order intermodulation (IM2) product of a large blocker may drop into the desired signal band at the output of the mixer. The second-order nonlinearity of the receiver is usually evaluated with IP2. IP2 can be tested by using two CW tone signals that have the same power levels at different frequencies of f_1 and f_2; but the frequency difference $\Delta f = |f_1 - f_2|$ should be less than the bandwidth of the baseband signal in order to test the IM2 component within the bandwidth of the baseband signal.

The input IP2 (IIP2) referred to the input of the LNA is expressed as (see the Appendix for more detail)

$$\mathrm{IIP2(dBm)} = 2P_{1\mathrm{T}}(\mathrm{dBm}) - P_{\mathrm{IM2,AC}}(\mathrm{dBm}) \tag{7.49}$$

where $P_{1\mathrm{T}}$ is the power level of one of two CW test tones with equal power at the input of the LNA, and $P_{\mathrm{IM2,AC}}$ is the power level of the IM2 component at the frequency of either $f_2 - f_1$ or $f_2 + f_1$ referred to the input of the LNA. In the IIP2 test with a two-tone, there are a total of three IM2 products, located at the frequencies of DC, f_2-f_1, and f_2+f_1, respectively. The total power of the IM2 products is composed of 50% (−3 dB) IM2 product at DC, 25% (−6 dB) IM2 product at (f_2-f_1), and 25% (−6 dB) IM2 product at (f_2+f_1). Thus, IIP2 calculation related to IM2 product at DC is given by

$$\mathrm{IIP2(dBm)} = 2P_{1\mathrm{T}}(\mathrm{dBm}) - P_{\mathrm{IM2,DC}}(\mathrm{dBm}) + 3\,\mathrm{dB} \tag{7.50}$$

The IM2 product at DC can be cancelled by using a DCOC method, while IM2 product at f_2+f_1 can be removed due to outside band. The IM2 at f_2-f_1 falls into the band if the frequency difference Δf is less than the bandwidth of the baseband signal of $\mathrm{BW_{BB}}$ and then distorts the desired signal.

In practice, the worst-case IM2 interferers at the input of the receiver are not the two-tone type, but modulated-type blockers with non-constant envelopes or AM envelope characterization. This type of modulated signals as blockers may come from either adjacent channels or transmitted signal leakage and may have much larger power than the received weak signal. Therefore, it is a very large challenge in designing the front end circuitry of the receiver to minimize the impact of the modulated blockers on receiver performance.

An amplitude and phase modulation format like QPSK can be realized by using the I and Q baseband signals of $I(t)$ and $Q(t)$ to modulate quadrature carrier signals of $\cos(\omega_c t)$ and $\sin(\omega_c t)$: i.e.,

$$\begin{aligned} x(t) &= I(t)\cos(\omega_c t) - Q(t)\sin(\omega_c t) \\ &= a(t)\cos[\omega_c t + \varphi(t)] \end{aligned} \tag{7.51}$$

where the signal envelope $a(t)$ and phase $\varphi(t)$ are given as

$$a(t) = \sqrt{I^2(t) + Q^2(t)} \tag{7.52}$$

$$\varphi(t) = \arctan\left(\frac{Q(t)}{I(t)}\right) \tag{7.53}$$

When this input signal is passed through a front-end device like an LNA or a mixer that is characterized as a transfer function given in (A.1), the second-order nonlinear item at the output of the device is generated as

$$\begin{aligned} g_2 x^2(t) &= g_2\{a(t)\cos[\omega_c t + \varphi(t)]\}^2 \\ &= \frac{g_2 a^2}{2}\{1 + \cos[2\omega_c t + 2\varphi(t)]\} \end{aligned} \tag{7.54}$$

After lowpass filtering, high-frequency components are removed and only low-frequency components (LFC) that are DC and low-frequency AC components at the output of the lowpass filter remain:

$$\mathrm{IM2_{LFC}} = \frac{g_2 a^2(t)}{2} \tag{7.55}$$

From (7.55) above, it can be seen that the low-frequency component of the $IM2_{LFC}$ distortion is a scaled version of the *squared envelope* of the baseband signal. Thus, the $IM2_{LFC}$ distortion type is highly dependent on the envelope property of the baseband signal. In the following example, three types of different modulation signals are used to generate the low-frequency product $IM2_{LFC}$.

Constant Envelope Modulation Signal: In the case of a constant envelope modulation format, such as the GMSK baseband I–Q signals of $u_i(t)$ and $u_q(t)$ in (4.33) and (4.34), the envelope signal $a(t)$ expressed in (7.52) is a constant of A if $I(t)$ and $Q(t)$ in (7.52) are replaced by $u_i(t)$ and $u_q(t)$ in (4.33) and (4.34), respectively, and any distortions and Gaussian noise are ignored for simplicity's sake. From (7.55), the IM2 product is

$$\mathrm{IM2_{LFC}} = \frac{g_2 A^2}{2} = \text{constant} \tag{7.56}$$

The constant IM2 product corresponds to a DC offset, which can be removed by using a DCOC circuit like any other sources of DC offsets.

Two-Tone Modulation Signal: In an I–Q quadrature modulation structure with ideal I–Q gain and phase imbalances, two-tone baseband signals at the frequencies of ω_{b1} and ω_{b2}, both having the same amplitude A on the I and Q channels, can be expressed as

$$I(t) = A\cos(\omega_{b1}t) + A\cos(\omega_{b2}t), \quad Q(t) = A\sin(\omega_{b1}t) + A\sin(\omega_{b2}t) \tag{7.57}$$

The envelope $a(t)$ of the complex composite baseband signal is

$$a(t) = \sqrt{I^2(t) + Q^2(t)} \tag{7.58}$$

and

$$\begin{aligned} a^2(t) &= I^2(t) + Q^2(t) \\ &= 2A^2[1 + \cos(\omega_{b2} - \omega_{b1})t] \end{aligned} \tag{7.59}$$

The IM2 product given in (7.55) is

$$\mathrm{IM2_{LFC}} = \frac{g_2 a^2(t)}{2} = g_2 A^2[1 + \cos(\omega_{b2} - \omega_{b1})t] \tag{7.60}$$

It is clear from the equation above that $IM2_{LFC}$ product at the low frequency contains a DC component and an AC component at the frequency difference of $\omega_{b2} - \omega_{b1}$, which is expected from the classical two-tone characterization calculation in (A.4) and (A.7).

Non-Constant Envelope Modulation Signal: For a non-constant envelope modulation format, such as the filtered QPSK baseband I–Q signals of $x_i(t)$ and $x_q(t)$ in (2.38), the envelope signal $a(t)$ in (7.52) can be expressed by substituting $I(t)$ and $Q(t)$ for $x_i(t)$ and $x_q(t)$, respectively, and any distortions and Gaussian noise are ignored for simplicity's sake, or

$$a(t) = \sqrt{x_i^2(t) + x_q^2(t)} \tag{7.61}$$

and

$$a^2(t) = x_i^2(t) + x_q^2(t) \tag{7.62}$$

The IM2 product is

$$\mathrm{IM2_{LFC}} = \frac{g_2 a^2(t)}{2} = \frac{g_2\left[x_i^2(t) + u_q^2(t)\right]}{2} \tag{7.63}$$

From the equation above, since the mean value of $\mathrm{IM2_{LFC}}$ is not equal to zero, the $\mathrm{IM2_{LFC}}$ product has a DC component. An exact spectrum of $\mathrm{IM2_{LFC}}$ depends on the nature of either $x_i^2(t)$ or $x_q^2(t)$ due to their similar characteristic of random stationary process. The spectrum of $\mathrm{IM2_{LFC}}$ also contains a sum of self-convolution of the spectrum of $u_i(t)$ and self-convolution of the spectrum of $u_q(t)$ due to the fact that the multiplication of two signals in the time domain results in the convolution of their spectrums in the frequency domain. Thus, the spectrum of either $x_i^2(t)$ or $x_q^2(t)$ is twice as wide as the spectrum of either $u_i(t)$ or $u_q(t)$. Hence, a bandwidth of power spectral density (PSD) of $\mathrm{IM2_{LFC}}$ is twice as wide as that of either $u_i(t)$ or $u_q(t)$ because the PSD of a signal is directly proportional to its squared Fourier transform.

Among the three different types of baseband I–Q signals described above, the worst $\mathrm{IM2_{LFC}}$ distortion is non-constant envelope modulation, which can appear at the input of the receive front-end devices from either a strong adjacent channel interferer or a large leakage of the RF transmitted modulation signal. A two-tone modulation signal (or I–Q signal) described above is usually used for valuating IP2 performance at a frequency difference of $f_2 - f_1$ in the receiver to avoid testing a DC due to many source contributions to DC. In the following section, we give an example of how to decide the IIP2 requirement for a receiver.

Design Example 7.1 The IIP2 requirement for GSM receivers is derived from the AM suppression test case. The GSM standard requires that under a test case, where a useful signal for the DCS 1800 band is set up to 3 dB above the reference sensitivity level of −102 or −99 dBm, and a modulated interfering signal is set up to −29 dBm at the frequency offset of greater than 6 MHz, a receiver must be able to achieve its minimum performance criteria (2% residual bit error rate [RBER]) in the presence of the modulated interferer.

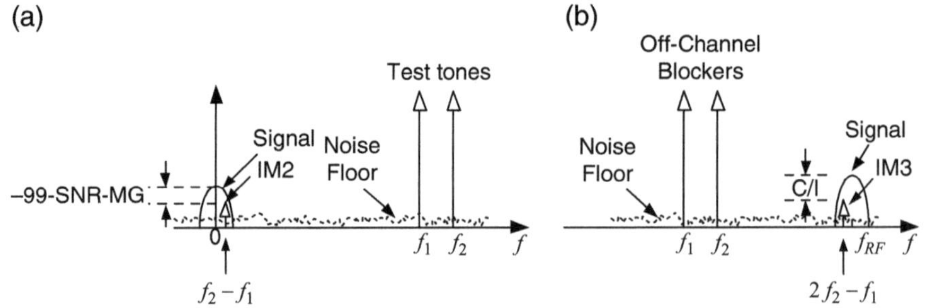

Fig. 7.22 Intermodulation products: (**a**) IM2 product referred to the input of a nonlinear device and (**b**) IM3 product referred to the input of a nonlinear device as an in-band interferer due to two off-channel blockers

Solution To evaluate the IIP2 requirement for a receiver, we use two tone signals with a power level of −32 dBm/tone at the frequencies of $f_1 = f_c + 6$ MHz and $f_2 = f_c + 6\ \text{MHz} + \Delta f$, respectively. The frequency difference of Δf should be less than one-half of the bandwidth (double-side) of 200 kHz, or less than 100 kHz. As shown in Fig. 7.22a, the IM2_{LFC} product referred to the input of the receiver is should be equal and less than the desired signal level of −99 dBm by the amount of SNR = 7 dB required for 2% BER [1], plus a 3-dB margin. Thus, $\text{IM2}_{\text{LFC}} = -99 - 7 - 3 = -109$ dBm is required at least. This 3-dB margin is considered for one other interferer with the same power level as the IM2 product, such as an IM3 product.

The required IIP2 for a GSM receiver can be calculated by using an equation in (7.49) as

$$\begin{aligned} \text{IIP2} &= 2P_{1\text{T}} - P_{\text{IM2},(\omega_2-\omega_1)} \\ &= 2 \times (-32) + 109 \\ &= 45\,\text{dBm} \end{aligned} \tag{7.64}$$

Actually, the GSM standard does not specify an IIP2 requirement, but an AM suppression requirement. The IIP2 characteristic through testing IM2 helps RF IC designers in the troubleshooting of the second-order nonlinearity of the front-end blocks before the AM suppression test. However, it is more challenging for RF IC designs to pass the AM suppression test because a modulated interfering signal, whose IM2 product is given in (7.63), has a significant effect on SNR compared to using a two-tone test signal.

7.4.3.2 Third-Order Distortion

The third-order nonlinear item among all odd-order nonlinear items in (A.1) dominates the degradation of the receiver's performance. With two strong interferers or blockers at the frequencies of f_1 and f_2, which are separated from the assigned input signal frequency, the third-order intermodulation products (IM3) at $2f_1 - f_2$ and $2f_2 - f_1$ at the LNA output can fall in the band of the desired signal, as shown in Fig. 7.22b, depending on their frequency difference. This in-band IM3 product reduces the carrier-to-interference ratio (C/I) at the input of the receiver's demodulator, and therefore they degrade the performance of the receiver.

This IM3 product referred to the input of the front-end block can be determined by the IIP3 requirement from the equation below (see Appendix A.2 for detail), in which two-tone signals have the same power level of $P_{\mathrm{INT,IN}}$.

$$\mathrm{IIP3(dBm)} = P_{\mathrm{INT,IN}}(\mathrm{dBm}) + \frac{1}{2}\left(P_{\mathrm{INT,IN}}(\mathrm{dBm}) - P_{\mathrm{IM3,IN}}(\mathrm{dBm})\right) \qquad (7.65)$$

and the output IP3(OIP3) (dBm) is equal to IIP3(dBm) plus power gain in dB.

It should be noted that it is more convenient to calculate the IIP3 from all input parameters. Hence, the IIP3 is equal to the input power level of each of the two tones plus half the difference between the input power level of each of the two tones and the input power level of the IM3 product $P_{\mathrm{IM3,IN}}$. In practice, the IIP3 is calculated from (7.65) by measuring the IM3 product $P_{\mathrm{IM3,OUT}}$ at the output of the device under test (DUT) and then dividing it by its power gain G to be referred to the input of the DUT, or $P_{\mathrm{IM3,IN}} = P_{\mathrm{IM3,OUT}} - G$.

IM3 Distortion Due to TX Leakage: Figure 7.23 illustrates two possible cases when IM3 products can fall in the band of the received signal in a 3G WCDMA receiver. The IM3 products can be created by each of two blockers at $f_{\mathrm{BL1}} = 2f_{\mathrm{TX}} - f_{\mathrm{RX}}$ (also called RX image frequency) and $f_{\mathrm{BL2}} = (f_{\mathrm{TX}} + f_{\mathrm{RX}})/2$ (also called half-duplex frequency), where f_{TX} is the TX up-link frequency and f_{RX} is the RX down-link frequency, with a TX leaked signal at the frequency of f_{TX} at the LNA output. Here, assume that a WCDMA transceiver operates in the band 1 and f_{TX} is set to 1950 MHz in the up-link range from 1920 to 1980 MHz and f_{RX} is equal to 2140 MHz in the down-link range from 2110 to 2170 MHz.

Consider both blocker1 and TX leakage signal appearing at the input of the LNA and let the combination of these two input signals at the LNA input be $x(t) = A\cos(\omega_{\mathrm{BL1}}t) + B(t)\cos(\omega_{\mathrm{TX}}t)$, where the first term is an unmodulated blocker signal and the second term is a TX-leakage-modulated signal with the envelope $B(t)$. The received signal is ignored in the input signal expression because it is assumed to be too weak compared with the blocker and TX leakage signal. The output of the LNA can be obtained by substituting $x(t)$ into (A.1):

$$\begin{aligned} y(t) = {} & g_1[A\cos(\omega_{\mathrm{BL1}}t) + B(t)\cos(\omega_{\mathrm{TX}}t)] + g_2[A\cos(\omega_{\mathrm{BL1}}t) + B(t)\cos(\omega_{\mathrm{TX}}t)]^2 \\ & + g_3[A\cos(\omega_{\mathrm{BL1}}t) + B(t)\cos(\omega_{\mathrm{TX}}t)]^3 + \cdots \end{aligned} \qquad (7.66)$$

The third-order distortion term of interest here is one with the frequency $2\omega_{\mathrm{TX}} - \omega_{\mathrm{BL1}}$, which falls in the band of the desired signal at the output of the LNA.

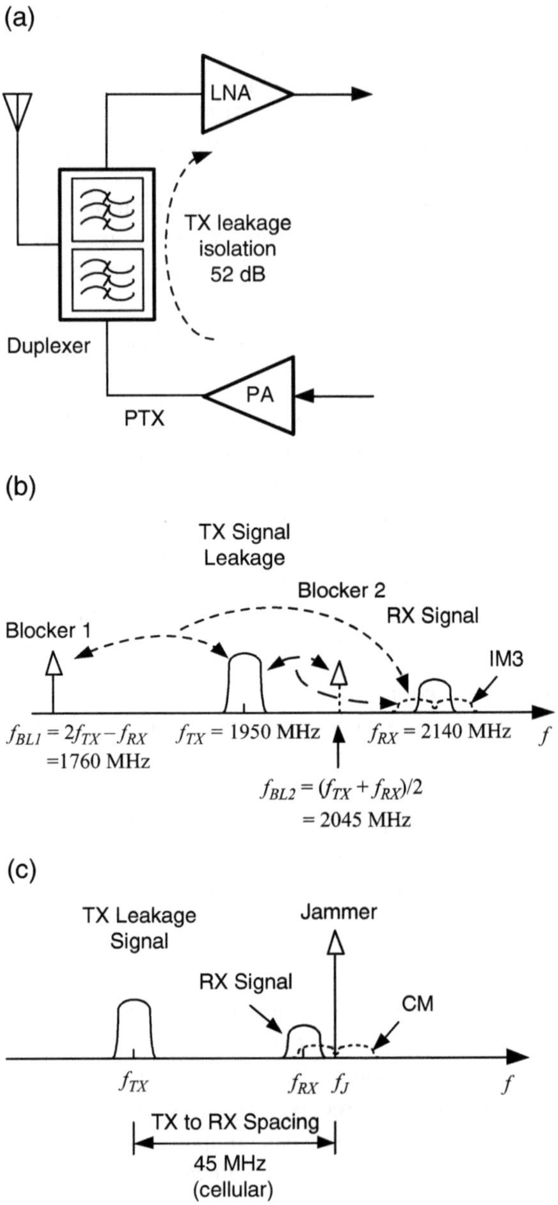

Fig. 7.23 IM3 products in the front-end devices: (**a**) TX leakage and blocker in the front-end devices, (**b**) two possible IM3 products created by a blocker and TX leakage at the LNA output in a WCDMA system, and (**c**) cross-modulation product created by a jammer and TX leakage at the LNA output due to IIP3 in a CDMA (IS-95) system

Leaving such a third-order distortion term and ignoring others in the output expression $y(t)$, we have the following expression at the output of the LNA:

$$y(t) = \cdots + \frac{3}{4} g_3 A B^2(t) \cos\left(2\omega_{\mathrm{TX}} - \omega_{\mathrm{BL1}}\right) + \cdots \tag{7.67}$$

Here, the third-order term at $2\omega_{TX} - \omega_{BL1}$ can be expressed by using IM3 notation:

$$\text{IM3} = \frac{3}{4} g_3 A B^2(t) \cos(2\omega_{TX} - \omega_{BL1}) \tag{7.68}$$

It can be clearly seen that the IM3 product has the same frequency as the received signal in this example, or $2f_{TX} - f_{BL1} = f_{RX}$, and its amplitude shape is dependent on both the amplitude A of the blocker and the squared amplitude $B^2(t)$ of the TX leakage. Similar to the IM2 case where the blocker is a TX leakage signal at the input of the LNA, the bandwidth of this IM3 product is twice that of the TX leakage signal due to the squared amplitude $B^2(t)$ of the TX leakage, as shown in Fig. 7.23b, thereby directly overlapping the spectrum of the desired channel.

There is another case of the third-order nonlinear product. If a blocker2 is located between the TX leakage signal and RX received signal and its frequency is equal to $\omega_{BLK2} = (\omega_{TX} + \omega_{RX})/2$, the IM3 product dropped inside band of the desired signal is generated between the blocker2 and TX leakage signal as shown in Fig. 7.23b. The bandwidth of this IM3 product is also twice that of the TX leakage signal.

Design Example 7.2 In the 3GPP WCDMA standard, third-order intermodulation response rejection is a measure of the capability of the receiver to receive a desired signal on its assigned channel frequency in the presence of two interfering signals. In this measurement, one blocker (CW) and one modulated WCDMA signal both have power levels of −46 dBm at frequency offsets of 10 and 20 MHz, respectively, from the desired channel frequency of 1950 MHz. Thus, the frequency of the IM3 product that falls in the band of the desired channel is $2f_1 - f_2 = 2 \times 1960 - 1970 = 1950$ MHz, which is the same as the desired channel frequency. What is the required IIP3 for the receiver?

Solution In the 3GPP WCDMA specification, the minimum sensitivity level DPCH_E_C <Ref. sensitivity> in the band I is −117 dBm at the UE antenna port. A typical SNR requirement DPCH_E_C/I_{OC} of −19.6 dB at the input of the demodulator is required to achieve a 0.1% BER. The allowable noise power (I_{OC}) regarding such a SNR is $I_{OC} = -117 + 19.6 = -97.4$ dBm, as shown in Fig. 7.24.

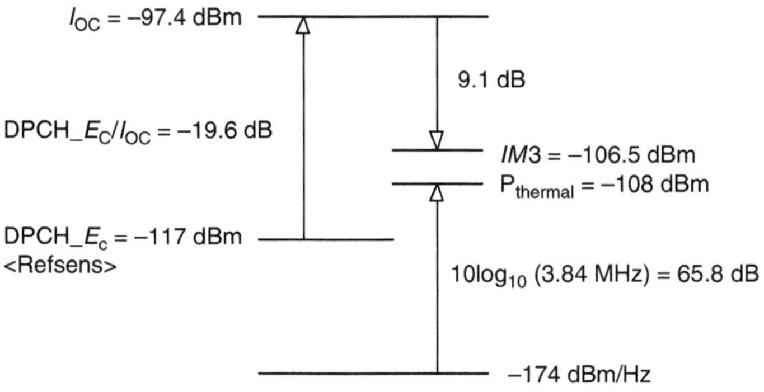

Fig. 7.24 Power level diagram for WCDMA band I reference sensitivity. Referenced from [19]

If it is assumed that IM3 product causes 0.5-dB receiver desensitization, which requires the power level of IM3 product dropping to the desired channel to be at least 9.1 dB lower than I_{OC}; the maximum allowable IM3 referred to the input of the LNA is $\mathrm{IM3} = I_{OC} - 9.1 = -97.4 - 9.1 = -106.5$ dBm. The receiver IIP3 is calculated using (7.65) as

$$\begin{aligned}\mathrm{IIP3} &= P_{\mathrm{INT,IN}} + \frac{1}{2}(P_{\mathrm{INT,IN}} - P_{\mathrm{IM3,IN}}) \\ &= -46 + 0.5 \times (-46 - (-106.5)) = -15.75\,\mathrm{dBm}\end{aligned} \tag{7.69}$$

Cross Modulation: In a WCDMA system or a CDMA system, the transmitter and receiver operate simultaneously, and are connected to the antenna through a duplexer. Usually duplexers used in the cellular mobile applications can provide about 45–55-dB isolation. When a transmitter transmits the signal with a power level close to its maximum level, for example, 25 dBm, the TX leakage signal that appears at the input of the LNA would be −25 dBm at the duplexer isolation of 50 dB. Such a TX leakage signal alone does not cause a problem for the received signal, but when combined with a strong adjacent single-tone jammer it poses a big design challenge for the linearity requirement of the LNA. The TX leakage signal can cause the strong single-tone jammer to be cross-modulated due to the third-order nonlinearity of the LNA; or the single-tone jammer is modulated by the squared envelope of the TX leakage signal. As a result, the bandwidth of the cross-modulated jammer is twice of the bandwidth of the TX-transmitted signal due to the squared-envelope modulation. Because the jammer is close to the desired channel, a large part of the cross-modulation signal power falls in the band of the desired signal at the output of the LNA as shown in Fig. 7.23c.

Consider a TX-modulated signal and an unmodulated blocker or a jammer that are both present at the input of a LNA; the output $y(t)$ of the LNA can be calculated by using (7.66). The third-order distortion term of interest here is one with the frequency of ω_j. Leaving such a third-order distortion term out of the expression and ignoring others in the output expression $y(t)$ after mathematical expansion and combination, we have the following expression at the output of the LNA:

$$y(t) = \left[\frac{3g_3}{4}A^3 + \frac{3g_3}{2}AB^2(t) + g_1 A\right]\cos\left(\omega_j t\right) + \cdots \tag{7.70}$$

From (7.70), the squared envelope $B^2(t)$ of the TX leakage signal now modulates the blocker signal at ω_j. Here the cross-modulation term can be expressed as an amplitude modulation format:

$$\mathrm{XMOD} = \frac{3g_3}{2}AB^2(t)\cos\left(\omega_j t\right) \tag{7.71}$$

Compared with (7.68), it can be seen that the IM3 low-frequency signal at $2\omega_{\mathrm{TX}} - \omega_{BL1}$ and XMOD signal at ω_j are both modulated by the squared envelope

$B^2(t)$ of the TX-modulated signal due to the third-order nonlinearity of a device. Similar to the IM3 product, the cross-modulation in (7.71) is an AM modulation with twice the bandwidth of the original TX-modulated signal due to the squared envelope $B^2(t)$.

The cross-modulation interferers can occur in the CDMA cellular radio system, as shown in Fig. 7.23c, which is designed to operate within the same radio-frequency spectrum as the older Advanced Mobile Phone System (AMPS). The AMPS RF scheme employs many closely spaced and relatively narrow band FM channels, while the CDMA RF scheme uses fewer but wider-band RF channels. AMPS channels are spaced 30 kHz apart, and each occupies roughly 24 kHz at peak deviation. CDMA service occupies the same US cellular band as AMPS, and each CDMA channel occupies a bandwidth of 1.23 MHz. The nearest AMPS channel is set 285 kHz away from the edge of the nearest CDMA channel, or 1.23/2 + 285 = 900 kHz away from the center of the desired CDMA channel. The power level of this interferer in the worst case is established in the 3GPP2 Air-Interface Standard as a test tone at −30 dBm, while the desired CDMA signal is −101 dBm, or 3 dB higher than its sensitivity level of −104 dBm.

The effect of the cross-modulation on the performance of the receiver can be evaluated by measuring receiver desensitization under a test case where a TX leakage signal and a jammer are added to the input of the receiver. The receiver desensitization test is described in the following example.

Design Example 7.3 The received CDMA signal is an OQPSK signal with a bandwidth of 1.23 MHz, and is set to −101 dBm at the input of the LNA. A TX CDMA reverse-channel–modulated signal at its maximum level of 23 dBm that is referred to the antenna port leaks to the input of the LNA as

$$P_{\mathrm{TX\_LK}} = +23\,\mathrm{dBm} + 2\,\mathrm{dB} - 52\,\mathrm{dB} = -27\,\mathrm{dBm} \tag{7.72}$$

Here a TX to RX rejection of 52 dB at the RX port of the duplexer and a 2-dB loss from the TX port to antenna are assumed. Thus, the transmitted signal at the TX port is 23 + 2 = 25 dBm. A CW jammer tone at the frequency offset of 900 kHz from the center of the desired signal band is −30 dBm at the input of the LNA.

Without turning on either the jammer tone or the TX leakage signal, the receiver should meet the frame error rate (FER) less than 0.5% for 95% of the time at the level of −101 dBm due to a 3-dB higher than required sensitivity level of −104 dBm. When both of them are turned on together, the baseband signal envelope of the TX leakage modulated signal is transferred to the jammer tone such that a part of the cross-modulation noise power falls into the band of the CDMA desired signal, and hence the increased noise power degrades the sensitivity.

Then, the CW jammer level is adjusted for a noise-floor rise or fall until FER $\leq$ 0.5% for 95% of the time is met. Record this CW jammer power level as P_{j}. The difference between P_{j} and −30 dBm is called the single-tone desensitization margin. If the margin is a positive number or zero, it means that the applied jammer level is greater than or equal to the required −30 dBm and meets the desensitization requirement. Otherwise, it does not meet the desensitization requirement.

P1dB Compression Point: A closely related quantity to IIP3 used in a receiver for measuring the nonlinearity of a circuit is the 1-dB compression point on its gain curve, or P1dB. Similar to the P1dB description for a transmitter, the P1dB is the point at which the output power of a circuit is 1 dB lower than the ideal output power of a linear circuit when the input power increases. It has been analyzed in [11] that the P1dB of an amplification block is about 9.6 dB lower than its IP3 point for a third-order nonlinearity case using single-tone as the input for the P1dB test. Unlike IIP3, the P1dB can be measured in a real circuit or a block by using a single CW tone or a real modulated signal. But the measured P1dB could be slightly different in these two kinds of tests, with one using a single tone and the other using a modulated signal. Definitely, P1dB with the real modulated signal is more accurate.

Furthermore, a RMS power operation point of a circuit for an in-band modulated signal with a PAPR value is highly related to its P1dB and signal PAPR value. In general, the RMS power operation point of the circuit should be back off from its P1dB point by a PAPR value of its input-modulated signal in order to avoid nonlinear distortion. For a large PAPR value of the received signal, the circuit needs more back-off from its P1dB point. The requirement of the back-off is actually required for each stage along the receive chain, especially for the front-end circuits, such as an LNA and mixer. This is because large interferers and blockers can appear at the input of the front-end circuit without enough attenuation by using appropriate bandpass filters. Considering multipath fading circumstances, the circuit needs an extra back-off from the P1dB compression point to reduce any distortion due to multipath fading. Hence, the total back-off for each stage along the receive chain is the sum of the PAPR and the extra back-off due to multipath fading.

The operation point of the circuit from the P1dB point can be represented by the ratio of P1dB to signal power S, or P1dB/S. Like SNR versus the RF input power level, the ratio of P1dB/S that is measured at the input of the ADC versus the RF input signal is also one of the most important evaluation parameters in the receiver. P1dB/S versus the RF input power level will be discussed in more detail later in this chapter.

7.5 Channel Selection Filtering

After down-conversion, the baseband I–Q signals or low-IF I–Q signals are passed through the lowpass filters to remove high-order harmonics and out-of-channel noise and then are amplified by the VGAs to the desired level at the inputs of the subsequent ADCs. Either a lowpass filter or a bandpass filter can function as channel selection filtering, depending on the actual receiver architecture. The main purpose of the channel select filtering is to select the desired channel signal and meanwhile to attenuate the adjacent channel interferers and blockers as well as

Gaussian noise in order to maximize the signal-to-interference-pulse-noise ratio (SINR) at the input of the demodulation. The combination of channel select filtering and VGA amplification reduces the dynamic range requirements of the ADCs.

In practice, channel selection can be performed in either the analog domain only or both the analog and digital domains with a partition approach. In the analog domain, the analog filter must have enough of a dynamic range and linearity to select the desired channel signal in the presence of strong adjacent channel interferers and out-of-band blockers. The benefit of analog channel selection with large attenuation for out-of-band interference is to reduce the dynamic range requirement of the ADCs and then to require lower-resolution ADCs. This analog channel selection filter usually determines an equivalent noise bandwidth in the receive chain.

In the case of being partitioned between the analog and digital domains, firstly, channel selection filtering is performed by a low-order analog filter as an anti-alias filter to partially attenuate the interfering signals to avoid aliasing distortion and also to reduce the dynamic range requirement of the ADCs. But higher-resolution ADCs are usually required in order to handle large out-of-band interference. Secondly, a digital channel selection filter implemented with sharp attenuation and a short transition region is able to further attenuate most out-of-band interferers and noise to achieve an optimal SNR before the digital demodulation. This digital channel selection filter also determines an equivalent noise bandwidth in the receiver chain. This equivalent noise bandwidth is used as the overall bandwidth of the receiver to calculate the sensitivity of the receiver.

Now, the question is why channel selection filtering cannot be completely performed in the digital domain. If the ADC were able to handle every interferer and blocker appearing in the front-end circuits without needing significant attenuation for interferers and blockers from analog filtering, channel selection filtering could be completely implemented in the digital domain by moving the ADC close to the front-end circuits as much as possible. This is a desirable case since digital filters are very accurate without tuning circuitry, and are readily integrated together with the front-end circuits of the receiver. Without any anti-alias analog filtering prior to the ADC, however, it is very difficult to design such an ADC that has extremely high dynamic range and linearity with low power consumption. Therefore, this approach is not suitable for wireless portable devices with a low power-consumption requirement. In the following section, as mentioned above, two different approaches to performing channel select filtering in wireless receiver systems are introduced in detail.

7.5.1 Channel Selection Filtering With Partition

In the partition-based channel selection filtering approach, the third and fourth-order analog lowpass filter used as an anti-alias filter as is able to achieve some rejection for the adjacent channel interferers and out-of-band blockers in the analogy domain. The filter type can be either Butterworth or Chebyshev. The ADC with a higher dynamic range of more than 70 dB and high sampling rate

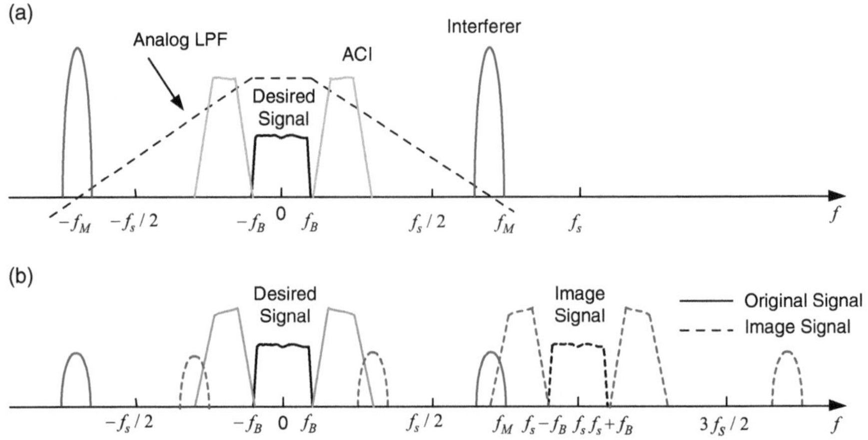

Fig. 7.25 Frequency responses of a desired signal and some interferers: (**a**) before a lowpass filter and (**b**) after an ADC

allows a large portion of the channel selection filtering and amplification to be performed in the digital domain. The ADC with a 70-dB dynamic range that provides a 12-bit resolution or effective number of bits (ENOB) can be implemented with oversampling frequency in a sigma-delta structure.

The attenuation of the analog filter mainly depends on the strength of the adjacent channel interferers, adjacent channel spacing, and the sampling frequency of the ADC. A higher sampling frequency on the ADC, such as oversampling, can not only relax the attenuation requirement of the analog filter, but also improve the SNR. As shown in Fig. 7.25a, there is an adjacent channel interfering (ACI) signal, which is close to the desired signal, and an interferer, which is far away from the desired signal and has a maximum frequency of f_M. The desired signal has a limited bandwidth of f_B. If a higher sampling frequency is used, the attenuation of the analog filter can be relaxed, as shown in Fig. 7.25b, where the ACI signal is slightly attenuated and the strong interferer is greatly attenuated. But the attenuated interferers are not small enough compared with the desired signal, and its maximum frequency of f_M is greater than a half-sampling frequency of $f_s/2$. After ADC, the frequency response of the original signal, however, is replicated at multiples of the sampling rate called "images" of the original signal.

It is clearly shown in Fig. 7.25b that there is no alias distortion falling to the band of the desired signal as along as the maximum frequency of f_M at the input of the ADC is less than the sampling frequency f_s minus the maximum frequency f_B of the original signal, or $f_M < f_s - f_B$. Even in the case that f_M is greater than $f_S - f_B$, as long as the interference attenuation requirements of the anti-alias filter from the frequency of $f_M < f_S - f_B$, upwards, is achieved with a certain amount of rejection [20], the distortion caused by aliasing can be neglected. After ADC, ACI can be significantly attenuated by the digital FIR filter with sharper attenuation and an accurate cut-off frequency in the digital domain without causing extra group delay variation.

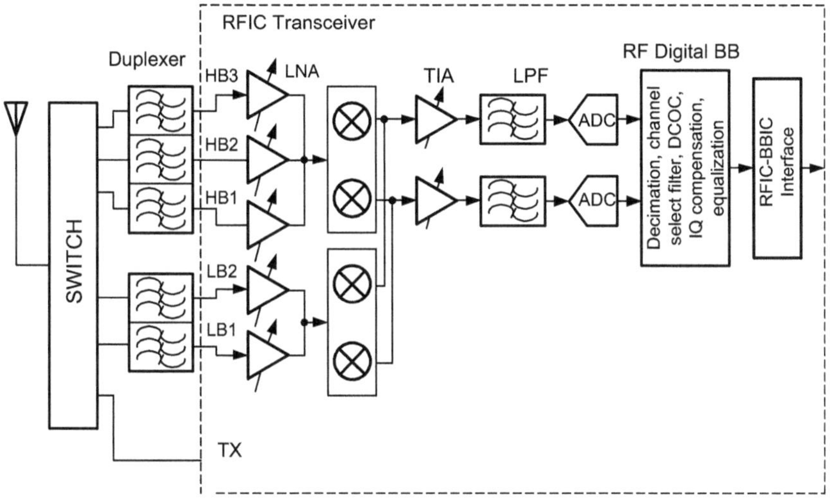

Fig. 7.26 Receiver block diagram of the RFIC transceiver

As one application example, Fig. 7.26 shows the overall receiver block integrated as a part of an RFIC transceiver, which supports four GSM/GPRS/EDGE bands and six WCDMA bands (I–V and VIII) [21]. The receiver uses a third-order Chebyshev lowpass filter as an anti-alias filter in the analog domain, and a sigma-delta ADC with a dynamic range of 71 dB. This high dynamic range ADC with oversampling rate allows a large portion of the channel select filtering to be performed by a channel selection digital filter (CSDF) in the RF digital BB block in Fig. 7.26. The CSDF implemented with the FIR filter performs large attenuation on the in-band interferers compared with the analog select filter. After the CSDF, a complex-tap compensator performs overall magnitude and group delay compensation for impairments, mainly caused by the analog channel selection filter due to component tolerance variation. The goal of the compensation is to minimize EVM in a static channel in order to enhance SNR, especially for supporting HSDPA+ 64QAM reception while providing excellent adjacent channel interferer rejection.

To have accurate compensation, the receiver performs both coarse and fine calibration *offline*. The goal of the coarse calibration is to minimize the in-band amplitude ripple and the group delay variations of the analog filter. In the design, the overall transfer response of the receive channel selection filter, including both analog and digital filters, should be close to the SRRC filter as much as possible, or

$$H_{\mathrm{SRRC}}(f) = H_{\mathrm{Cas}}(f) \times H_{\mathrm{Err}}(f) \tag{7.73}$$

where $H_{\mathrm{Err}}(f) = |H_{\mathrm{Err}}(f)|e^{j\phi_{\mathrm{Err}}(f)}$ is the error transfer response depending on the design accuracy, and the cascaded transfer response is given by

$$\begin{aligned} H_{\mathrm{Cas}}(f) &= H_{\mathrm{Sim}}(f) \times H_{\mathrm{Dig}}(f) \times H_{\mathrm{Comp}}(f) \\ &\approx H_{\mathrm{Adju}}(f) \times H_{\mathrm{Dig}}(f) \times H_{\mathrm{Comp}}(f) \end{aligned} \tag{7.74}$$

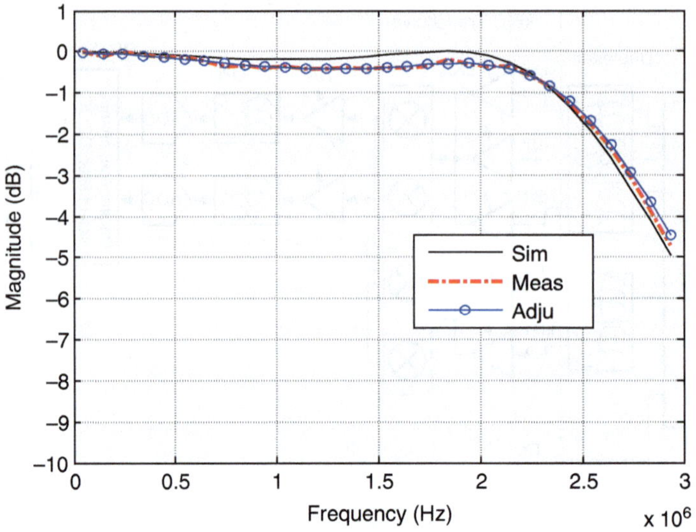

Fig. 7.27 Simulated, measured, and adjusted RX analog baseband lowpass filter characteristics after and before coarse calibration [22]

In (7.74), $H_{\text{Adju}}(f)$, representing the actually measured analog filter $H_{\text{Meas}}(f)$ due to RC component tolerance variation, is obtained by adjusting the poles of the simulation transfer response $H_{\text{Sim}}(f)$ to approximate to $H_{\text{Meas}}(f)$ in the MATLAB simulation. $H_{\text{Dig}}(f)$ is the digital channel selection FIR filter, and $H_{\text{Comp}}(f)$ is the digital compensation filter. Figure 7.27 shows the passband response of the analog lowpass filter with a 3-dB corner frequency of 2.7 MHz after and before the coarse calibration. It can be seen that the amplitude response of $H_{\text{Adju}}(f)$ precisely matches the amplitude response of $H_{\text{Meas}}(f)$ such that actual pole locations of the analog channel selection filter through $H_{\text{Sim}}(f)$ are obtained. Thus, the group delay characteristic of $H_{\text{Adju}}(f)$ is known, and can be used for the compensator to perform the group delay compensation.

The amplitude response $H_{\text{Comp}}(f)$ of the compensator can be obtained from (7.73) and (7.74) through MATLAB optimization:

$$|H_{\text{Comp}}(f)| \approx \frac{|H_{\text{SRRC}}(f)|}{|H_{\text{Adju}}(f)| \times |H_{\text{Dig}}(f)| \times |H_{\text{Err}}(f)|} \tag{7.75}$$

The group delay of the compensator implemented with the fifth-order allpass filter, mainly targeting the group delay of the adjusted transfer function of $H_{\text{Adju}}(f)$, can be calculated to achieve the smallest variation of the overall group delay within a bandwidth of 1.92 MHz (or Nyquist frequency), as shown in Fig. 7.28. Here the absolute group delay is not important because it does not affect the system performance. Finally, these coefficients of the amplitude and group delay of the compensator are stored in the registers as the coarse compensation values.

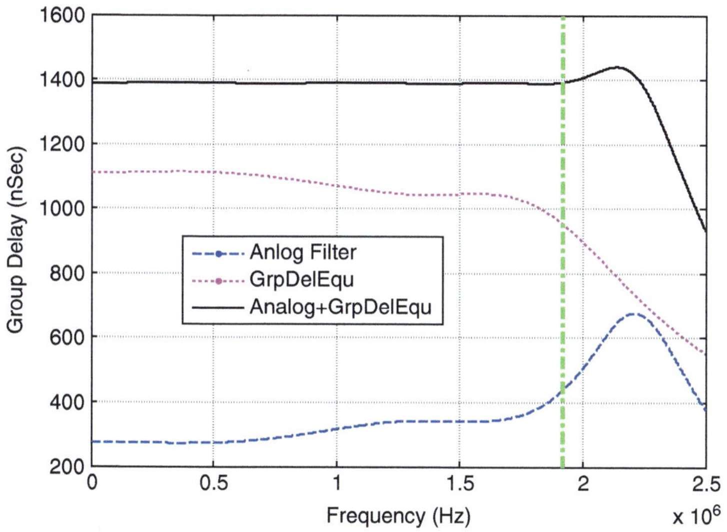

Fig. 7.28 Group delay characteristics of channel selection filtering after and before digital group delay equalization [22]

The objective of fine calibration is to minimize EVM for a WCDMA-modulated signal by automatically fine-tuning the real pole of the third-order Chebyshev lowpass filter within a certain range after the coarse calibration. Fine-tuning the real pole slightly changes the group delay response of the overall channel selection filtering such that minimal EVM can be achieved. RX EVM versus the real pole code on a RFIC chip is illustrated in Fig. 7.29. It can be seen that the minimal EVM around 2% can be achieved within a wide code range from 60 to 70 with a step size of 1, where the optimal code is 65. Each code corresponds to a certain capacitance. Once data are collected based on a few RFIC chips during the fine calibration, the optimal code value can be statistically calculated and stored in the register.

Figure 7.30 illustrates the measured EVM versus the RF QPSK input signal in bands II and V, where SNR is calculated using $\mathrm{SNR} = -20 \times \log(\mathrm{EVM})\mathrm{dB}$. It can be seen that the measured EVM is less than 3% or SNR is greater than 30 dB when the RF input is greater than −75 dBm. Such very small RX EVM values are achieved by optimal digital compensation realized with fine-tune calibration. These optimal compensation parameters had been verified to achieve similar EVM on other RF transceiver IC chipsets before mass production.

This partition approach is also applied to the low-IF receiver architecture [5], where the low-IF signals on both I–Q channels are centered at 100 kHz with a bandwidth of 200 kHz and are passed through lowpass filters on the I–Q branches to attenuate out-of-band interferers, harmonics, and noise after down-conversion. Then, a 12-bit sigma-delta ADC with a large dynamic range can handle large interferers and blockers. Thus, these interferers and blockers can be significantly

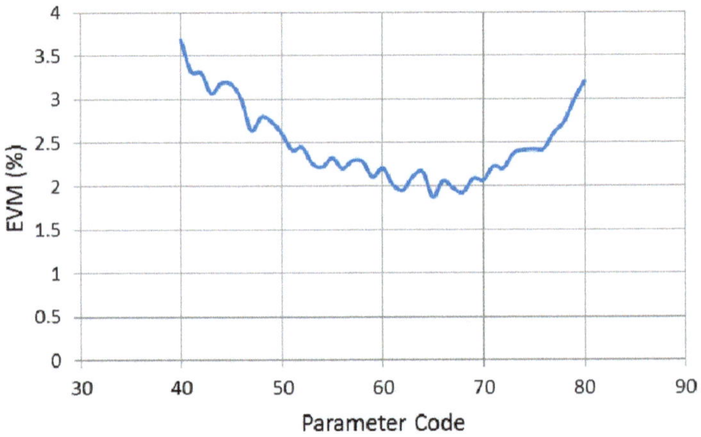

Fig. 7.29 RX EVM versus a fine-tune parameter code in band II

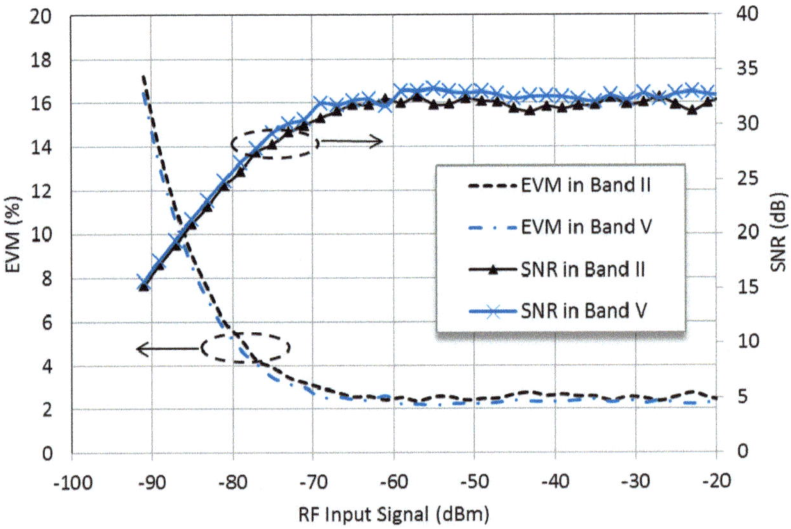

Fig. 7.30 Measured EVM and SNR for a WCDMA QPSK signal versus RF input signal in bands II (1960 MHz) and V (884 MHz) [22]

attenuated by the digital filter in the digital domain. It is obvious from the discussions above that the benefit of digital filtering comes at the expense of the need for high-dynamic-range ADCs, which can be achieved with an oversampling process. As a general rule, every doubling of the sampling frequency yields approximately a 3-dB improvement in noise performance.

7.5.2 Channel Selection Filtering in the Analog Domain

Besides channel selection filtering with the partition approach, channel selection filtering can be mainly performed by analog filters before the ADCs. As the attenuation of the analog channel select filter increases, nonlinear distortions of the following PGAs are minimized and the dynamic range requirements of the ADCs are relaxed. Attenuation of the filter is primarily determined by three important parameters: filter prototype, 3-dB corner frequency, and filter order. Four widely used filter prototypes are Butterworth, Chebyshev, Inverse Chebyshev, and Elliptic filters. Each filter type has its advantages and disadvantages, regarding the characteristics of attenuation and group delay variation. In terms of attenuation, the Elliptic filter provides the largest attenuation and a sharper transition region among the four types of filters, but has the worst group delay variation within a 3-dB corner frequency. The Butterworth filter shows the smallest group delay variation, but it requires higher orders than others to achieve the same attenuation. Because of the importance of attenuation for adjacent channel interferers, the last three types of filter are popular in the realization of the channel selection filter. In general, due to its characteristic nature, a filter with large attenuation and a sharp transition region for a fixed order usually has large group delay variation within a 3-dB corner frequency. The higher the order, the larger the group delay variation.

The 3-dB corner frequency and filter order are generally determined by signal bandwidth, channel spacing, and blocking profiles in actual applications. Several design examples of the analog filters in realization of overall channel selection filtering have been reported in the literature for the 3GPP WCDMA receivers. A seventh-order analog filter as a channel selection filter was reported in [22]. This analog filter consists of two untuned single-pole RC sections followed by the fifth-order Chebyshev prototype with a 3-dB cutoff at 1.92 MHz and a low-ripple of 0.001 dB. The single-pole RC sections have a cascaded cut-off frequency of about 2.8 MHz, relaxing the linearity and dynamic range requirements of the following Gm-C filter due to attenuating large interferers. The analog channel-select filter provides a minimal selectivity of 40 dB at $\pm$5-MHz adjacent channels by using two transmission zeros located slightly below 4 MHz. With tuning circuitry, the filter cut-off frequency can be tuned to within $\pm 5\,\%$ of 1.92 MHz, including temperature drift.

Another reported analog channel select filter is a fifth-order Chebyshev lowpass with a 0.01-dB passband ripple and a 3-dB frequency of 1.92 MHz [23, 24], where a real pole is implemented at the output of a mixer to attenuate in-band blockers. Minimum power consumption is achieved with this analog channel selection filter and a 7–8-bit resolution ADC with a sampling frequency at a four times chip rate of 15.36 MHz. To compensate the group delay variation of the filter, a first-order allpass filter with a pole and a zero located at 1.4 MHz as a group delay equalizer is cascaded with the lowpass filter so that the receive EVM can be significantly reduced.

For a 3G WCDMA receiver, the analog filter can realize the functions of both channel selection filtering and matched filtering at the cost of slightly increasing the order of the lowpass filter and cascading it with the group delay compensator. Figure 7.31a illustrates amplitude responses of the SRRC filter implemented with a 41-tap FIR filter with $\alpha = 0.22$ and a fifth-order Chebyshev1 filter with a 0.5-dB ripple within the passband and the Nyquist frequency of $f_N = (1/2)T_{chip}$ or equivalent to a 3-dB frequency of 1.92 MHz for the chip rate of 3.84 Mchips/s. This SRRC filter is designed with two samples per chip. The fifth-order Chebyshev1 filter approximates to the SRRC filter pretty well up to –10 dB attenuation. In an adjacent channel of the frequency offset of 5 MHz, the attenuation is about 55 dB. The analog filter, however, creates large group delay variation within a 3-dB bandwidth, especially for higher-order type. Therefore, a group delay equalizer (or an allpass filter) needs to be cascaded with the analog filter. Figure 7.31b shows group delays of the fifth-order Chebyshev1 lowpass filter with a 0.5 dB ripple and a 3-dB frequency of 1.92 MHz without and with three different allpass filters. If we minimize the maximum group delay variation from DC to 1.8 MHz, the maximum group delay variations for four different curves of A to D are 425, 315, 252, and 122 ns, respectively. It should be pointed out that an absolute constant group delay does not degrade the performance of the digital communication system. As a rule of thumb, the range that needs to be compensated is about 75% of its –3 dB frequency, or 1.5 MHz for the –3-dB frequency of 2.0 MHz. It should be noted, however, that the smallest group delay within the bandwidth of 1.92 MHz does not necessarily give the minimum EVM. Therefore, the optimal group delay variation should be obtained by minimizing EVM through slightly adjusting the locations of the poles and the zeros of the allpass filter. Since the first-order allpass filter has been verified to improve EVM by a significant amount, it is usually used to compensate for the group delay variation of the analog channel selection filter due to its low complexity. Figure 7.32a shows the simulated constellation in a WCDMA receiver by using a FIR filter with 81-tap to implement a SRRC filter, where EVM = 3.1% is obtained. With the fifth-order analog Chebyshev1 filter approximation to a SRRC filter and the first-order allpass filter, EVM = 5.5% can be achieved, as shown in Fig. 7.32b. The EVM difference above shows that a close approximation to the SRRC FIR filter with 81-tap is obtained by using the analog Chebyshev1 filter cascaded with the first-order allpass filter. The EVM difference also indicates that the analog channel selection filtering relaxes the requirement of the ADCs at the expense of slightly sacrificing EVM.

Most wireless standards specify the requirements of the adjacent channel interference rejection to ensure that the adjacent channel interfering signals can be attenuated enough through the channel selection filter to prevent the performance of the receiver from suffering significant degradation. For example, the 3GPP WCDMA standard defines the minimum adjacent channel selectivity (ACS) requirement of 33 dB, which is the ratio of the receiver filter attenuation on the desired channel frequency to the receiver filter attenuation on the adjacent channel. In the ACS test, the desired signal power in Band 1 is set −103 dBm or 14 dB above

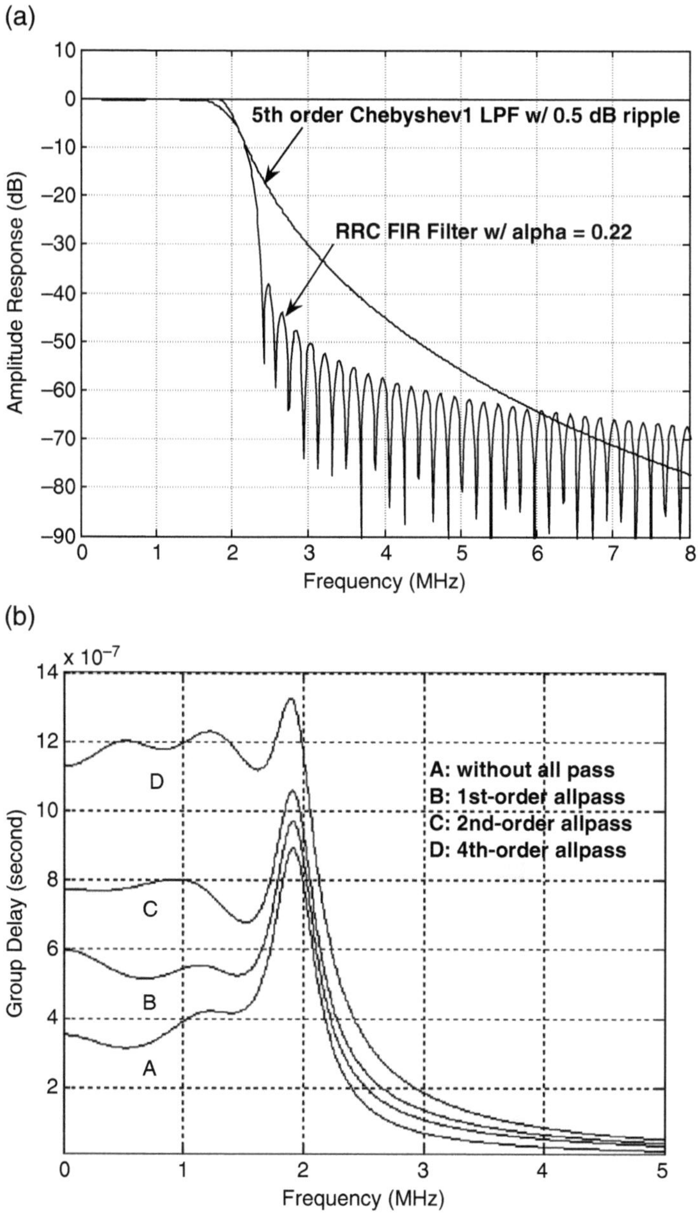

Fig. 7.31 Frequency response of a digital SRRC filter and an analog Chebyshev1 filter: (**a**) amplitude responses of a FIR filter implemented by 81-tap and a fifth-order Chebyshev 1 with 0.5 dB ripple within passband and passband frequency of 1.92 MHz; (**b**) group delay responses of a fifth-order Chebyshev1 filter with different order allpass filter

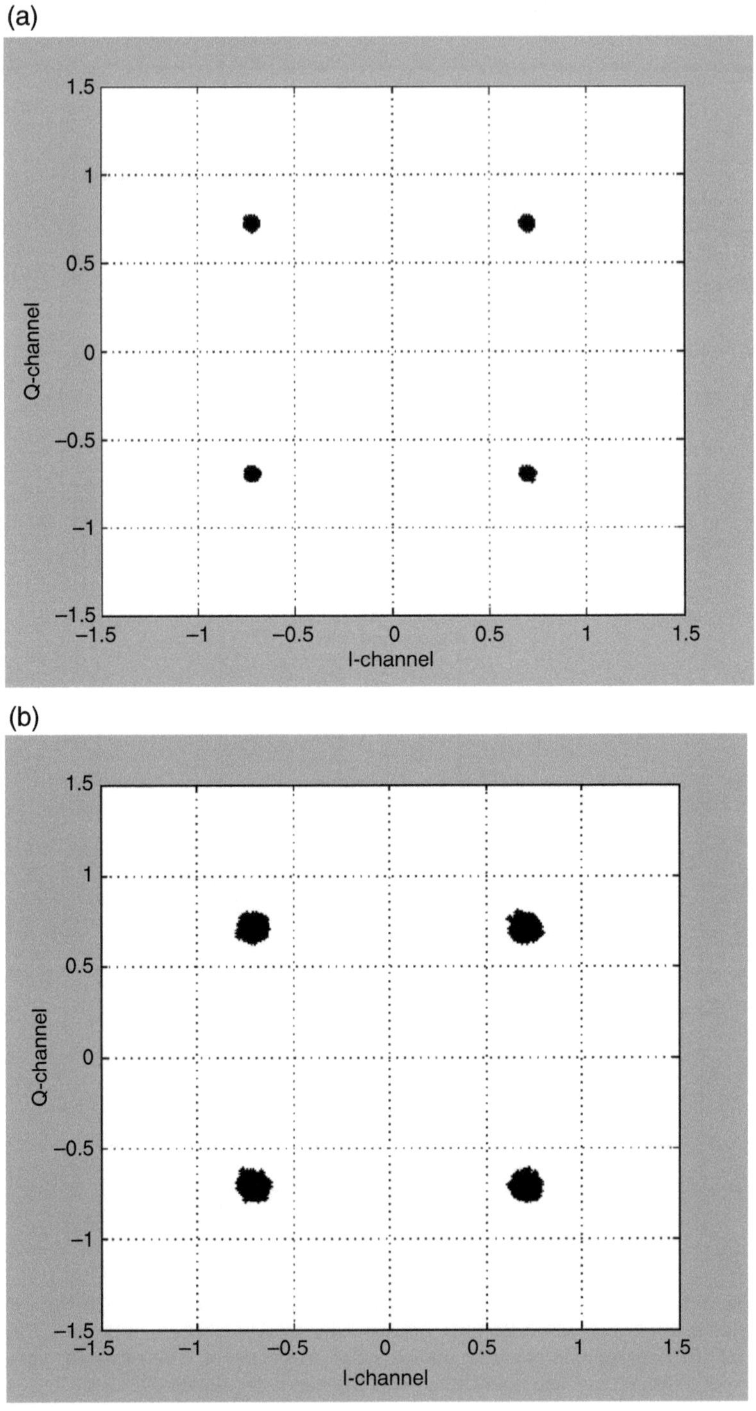

Fig. 7.32 Receive constellation of QPSK signal: (**a**) EVM = 3.1%, an 81-tap FIR filter based SRRC, alpha = 0.22. (**b**) EVM = 5.5%, a fifth-order Chebyshev1 filter approximation to SRRC with a passband frequency of 1.92 MHz and a first-order allpass filter

the sensitivity level, and the adjacent channel interfering signal is set to −52 dBm; the BER will not exceed 0.01%.

It can be seen from Fig. 7.31a that average attenuation of the analog Chebyshev1 filter is about 50 dB at a frequency of 5 MHz. Hence, the power of the adjacent channel interfering signal at the frequency offset of 5 MHz would be attenuated more than the requirement of 33 dB down if the Chebyshev1 prototype were to be used as the channel selection filter.

From the discussion above, the channel selection filtering that is primarily performed in the analog domain can reduce the dynamic range requirement of the ADC. Of course, some amount of channel selection filtering can be further realized in the digital domain if the adjacent channel interfering signals are not sufficiently attenuated by the analog channel selection filter. One of the main issues to address when using analog channel selection filtering is how to accurately approximate to the frequency response of a desired filter such as a SRRC filter in the WCDMA receiver and how to accurately calibrate a 3-dB corner frequency due to component tolerance up to 25%. This issue becomes more complicated in multi-mode and multi-band transceivers, where different bandwidths of the channel selection filters are needed to support different data rates. On the contrary, with a filtering partition between the analog and digital domains, digital FIR filters with large taps can be easily realized without problems of device tolerance and stability. Additionally, the FIR filters do not introduce ISI due to the property of constant group delay. Furthermore, it is very convenient for the FIR filters to adjust the cut-off frequency and to change the tap length when facing the large in-band interferers. Therefore, it is preferable to realize as much filtering as possible in the digital domain so that the complexity of the analog design and calibration can be reduced.

7.6 Automatic Gain Control

Automatic gain control (AGC) is a critical building block in modern wireless transceivers, especially in the receiver. In the receiver, the received signal can reach the input of the antenna with different power levels due to a distance between the transmitter and the receiver, and can vary drastically in a mobile environment. AGC circuitry in the receiver is used to adjust the received signal power level to a desired and fixed level at the inputs of the ADCs. A combination of channel selection filtering and AGC setting can reduce the dynamic range requirement of the ADC.

The overall RF analog gain is distributed from the LNA to the last stage driver in the receive chain before the ADC. The ADC is usually built in a baseband (BB) chip rather than a RF transceiver chip to minimize the pin count and substrate noise that might be introduced by relatively large switching power in the ADCs/DACs. Besides the analog gain in the RF transceiver chip, the digital BB chip also provides some amount of the digital gain. The purpose of the AGC is to properly amplify the voltage amplitude of the received signal at the LNA input within the desired input voltage amplitude at the input of the ADC. Generally, the overall gain of the direct-down conversion receiver consists of a LNA gain, a mixer gain, a programmable lowpass filter (LPF) gain, and a variable gain amplifier (VGA) gain. The combination of these block gains insures that the input levels of the ADCs stay within the desired range. In most direct-conversion receivers, the total gain value of the LPF and VGA blocks can be up to more than 60 dB in addition to the LNA gain and mixer gain. The maximum gain of the receiver is mainly dependent on the sensitivity specification, while the minimum gain is largely determined by the maximum input signal level specification. Both maximum and minimum gain values, on the other hand, are closely related to the sensitivity and the dynamic range of the receiver. Hence, it is necessary to first introduce the sensitivity and dynamic range before going further.

7.6.1 Receiver Sensitivity

Receiver sensitivity defines the ability of a receiver to reliably demodulate the received signal with the minimum level at the input of the receiver in the absence of interfering signals and to recover the transmitted data with a BER less than and equal to the required specification. The sensitivity can be derived from the noise figure, which is in turn obtained from the noise factor expressed in decibels. The noise factor F of a network or circuitry is defined to be the ratio of the input signal-to-noise power ratio $\mathrm{SNR_{in}}$ to the output signal-to-noise power ratio: $\mathrm{SNR_{out}}$:

$$F = \frac{\mathrm{SNR_{in}}}{\mathrm{SNR_{out}}} \tag{7.76}$$

For a receiver, $\mathrm{SNR_{in}}$ is the signal-to-noise ratio measured at the input of the receiver, and $\mathrm{SNR_{out}}$ is the signal-to-noise ratio measured at the output of the channel selection filter with *a double-sideband* equivalent noise bandwidth of B_{enb} before the demodulator.

If the network were perfect, then no noise created inside the system would be added to the signal when it passed through the network and the signal-to-noise ratio at the output would be the same as that at the input. As we know, this is not the case and noise is always added to the signal after passing through the network. This means that the signal-to-noise ratio at the output is worse than the signal-to-noise ratio at the input, or F is always greater than 1. The noise factor is rarely defined in standard specifications. Instead the noise figure is always simply expressed in decibels.

$$\mathrm{NF} = 10\log F = 10\log\frac{S_\mathrm{i}/N_\mathrm{i}}{S_\mathrm{o}/N_\mathrm{o}} \tag{7.77}$$

The noise factor in (7.76) can be rewritten as

$$F = S_i/N_i \\ = \frac{S_i/(kT \times B_{enb})}{SNR_{out}} \tag{7.78}$$

where the noise power $N_i = kT \times B_{enb}$ is measured within the double-sideband equivalent noise bandwidth B_{enb} at the input of the receiver, k is the Boltzmann constant (1.38×10^{-23} J/K), and T is the absolute temperature in degrees Kelvin under test. Hence, the signal power S_i at the input of the receiver is expressed as

$$S_i = kT \times B_{enb} \times F \times SNR_{out} \tag{7.79}$$

The equivalent noise bandwidth of a filter is defined as the bandwidth of a brick wall or rectangular filter that passes the same amount of power as the cumulative bandwidth of the channel selection filter in the receiver. Usually, an equivalent noise bandwidth B_{enb} is slightly greater than a 3-dB bandwidth of the filter. For a lowpass filter with an order equal to or greater than 3, the equivalent noise bandwidth B_{enb} of the lowpass filter is approximately equal to $1.1 \times f_{3dB}$. The equivalent noise bandwidth B_{enb}, however, is the composite bandwidth of the overall channel filters in a receive chain back up to the input of the demodulator and is most likely dominated by a filter with the narrowest bandwidth. Usually, the equivalent noise bandwidth can be approximated by 1.1 times a −3-dB bandwidth of the dominated lowpass filter. In channel selection filtering with partition, the equivalent noise bandwidth B_{enb} is usually determined by the digital channel selection filter. In practice, a channel selection filter can be realized with either an IF bandpass filter or a BB lowpass filter, depending on an actual receiver architecture. In the BB lowpass filter case, a double-sided equivalent noise bandwidth should be used in the sensitivity calculation.

Noise factor F expressed in (7.79) is the minimally achievable value in a receiver with a larger gain along a receive chain. Therefore, the signal power S_i represents the minimum signal power, or the sensitivity. Thus, (7.79) can be rewritten using a sensitivity notation, and meanwhile can be expressed in logarithmic units as

$$S_{sen}(\text{dBm}) = -174\,\text{dBm/Hz} + 10\log B_{enb} + \text{NF(dB)} + \text{SNR}_{out}(\text{dB}) \tag{7.80}$$

where $kT = -174$ dBm/Hz is used in (7.80). The first two terms represent the integrated thermal noise power in the equivalent noise bandwidth and are referred to as "thermal noise" because of the dependency on temperature. The third term of the noise figure generated by the network in the receiver, such as amplifiers and mixers ,adds additional noise to the thermal noise within the receive channel to give the total noise power at the input of the demodulator. To understand the total noise power, we can write (7.76) as

$$F = \frac{S_i/N_i}{S_o/N_o} = \frac{N_o}{GkTB_{enb}} \tag{7.81}$$

In (7.81), the expression form is obtained by replacing S_o with $G \times S_i$ and N_i with kTB_{enb}, respectively, where G is the power gain of the network. From (7.81), the total noise power at the output of the network can be rewritten as

$$N_o = GFkTB_{enb} \tag{7.82}$$

Thus, the total noise power calculated at the output of the network is referred to the input noise power N_{ref_in} at the input of the network as

$$N_{ref\_in} = N_o/G = FkTB_{enb} \tag{7.83}$$

By moving the first three terms on the right side of (7.80) to the left side, we have the following expression:

$$S_{sen}(\text{dBm}) - [-174\,\text{dBm/Hz} + 10\log B_{enb} + \text{NF(dB)}] = \text{SNR}_{out}(\text{dBm}) \tag{7.84}$$

Comparing (7.83) with (7.84), we can see that in (7.84) the three terms in square brackets represent the total input referred noise power that is expressed in (7.83). Thus, the difference between the first term and the rest of the terms on the left side in (7.84) is the signal-to-noise ratio (SNR) at the input of the receiver, called SNR_{ref_in}, and the term on the right side is the SNR at the output of the receiver or the input of the demodulator renamed SNR_{dem}. Hence, (7.84) is rewritten as

$$\text{SNR}_{ref\_in} = \text{SNR}_{dem} \tag{7.85}$$

Therefore, SNR_{ref_in} at the input of the receiver is equal to SNR_{dem} at the input of the demodulator after the noise figure is referred to the input of the receiver. In other words, the SNR at the input of the RF receiver is equal to the SNR at the input of the demodulator if the noise factor is referred to the input of the RF receiver. Hence, the expression given in (7.85) has a more specific system meaning compared to the sensitivity expression in (7.80).

Design Example 7.4 The 3GPP WCDMA standard specifies the sensitivity level for the three different operation bands listed in Table 7.1, where each test is carried out with 12.2-kbps test reference data and a BER that should not exceed 0.1%. What is the noise figure required by the operation band I receiver?

Table 7.1 3GPP receiver sensitivity level requirements

Operating band	Unit	DPCH_Ec	$\hat{I}_{or}$
I	dBm/3.84 MHz	−117	−106.7
II	dBm/3.84 MHz	−115	−104.7
III	dBm/3.84 MHz	−114	−103.7

Solution In a case of the band I receiver, the sensitivity level of DPCH_E_c is −117 dBm, and the required SNR to meet a BER requirement of 0.1% can be obtained from the simulation, which is −19.6 dB, from Fig. 2.11 [20]. Using a band of 3.84 MHz as an approximate noise bandwidth, the noise figure referred to the antenna port is calculated from (7.80) by replacing S_{sen} with DPCH_E_c as

$$\mathrm{NF} = -117 + 174 - 10\log\left(3.84 \times 10^6\right) + 19.6 = 10.8\,\mathrm{dB} \tag{7.86}$$

This noise figure is the maximum value. With a 3-dB insertion loss caused by a RF switch and a duplexer between the antenna and LNA, the noise figure referred to the LNA input is 7.8 dB, which is easily achievable. Actually, most RF IC vendors can provide the RF WCDMA transceivers with a 3-dB margin at least regarding the NF of 7.8 dB for the receivers.

The required SNR at BER = 0.1% can be also calculated using a data-processing gain G_P that can be calculated from the ratio of a spreading chip rate R_C to a data rate R_D.

$$\begin{aligned} G_P &= 10\log\left(\frac{R_C}{R_D}\right) \\ &= 10\log\left(\frac{3.84\,\mathrm{Mcps}}{12.2\,\mathrm{kbps}}\right) = 25\,\mathrm{dB} \end{aligned} \tag{7.87}$$

The required SNR at BER = 0.1% without spreading code is SNR = 5.2 dB [25]. Thus, SNR with spreading code denoted by $\mathrm{SNR}_{\mathrm{Spread}}$ is given by

$$\begin{aligned} \mathrm{SNR}_{\mathrm{Spread}} &= \mathrm{SNR} - G_P \\ &= 5.2 - 25 = -19.8\,\mathrm{dB} \end{aligned} \tag{7.88}$$

The $\mathrm{SNR}_{\mathrm{Spread}}$ value of −19.8 dB in (7.88) is very close to the SNR value of −19.6 dB used in (7.86) from two different reference sources. The advantage of a spreading system is to significantly reduce the SNR requirement for achieving a specified BER. But such a benefit comes at the cost of extending a transmission bandwidth.

7.6.2 *Receiver Dynamic Range and Total Analog Gain*

In today's radio communications environment where there are many transmitters either near by or further away, a receiver with the ability to handle strong adjacent channel signals and blockers, both in- and out-bands, is needed to achieve good performance. The dynamic range of the receiver is very important because it is able to not only handle strong signals but also pick up weak ones. In general, receiver dynamic range is defined as the difference in dB between the strongest input signals both on- and off-channel that the receiver can reliably handle and the weakest input signal on-channel that the receiver can reliably detect.

The question is to how to define the reliability. In a reliable detection for the weakest input signal, we can use the sensitivity as the low bound of the dynamic range. In a reliable detection for the strongest input signal, it is more complicated to define the high bound of the dynamic range. One common definition regarding the high bound of the dynamic range is called the "spurious-free-dynamic range" (SFDR).

In such an SFDR definition, the high bound of the dynamic range is limited by the third-order IM products of a two-tone test added at the input of the receiver. In detail, the high bound of the dynamic range is equal to the strongest input level of a two-tone test signal with equal power when their IM3 products falling into the desired channel are the same as the noise floor power, which is equal to $-174\,\text{dBm} + \text{NF} + 10\log B_{\text{enb}}$.

The high bound of SFDR is mainly limited by the IM3 products highly related to the IIP3 of the receiver and in-band noise floor power, while the low bound of SFDR is dependent upon the noise figure of the receiver and the signal-to-noise ratio required for the modulation format of the received signal. In addition to dynamic range, the receiver should have enough gain to cover the sensitivity requirements of the receiver. The total analog gain is calculated by bring the received RF signal from the sensitivity level at the antenna port to the desired I–Q baseband signal level at the input of the ADC. Because either the output impedance of the last VGA stage before the ADC or the input impedance of the ADC is not 50 Ω, the rms voltage gain is preferred along the receive chain.

Design Example 7.5 Determine the maximum gain for the 802.11b/g WALN receiver. The IEEE 802.11b/g WLAN standard specifies a sensitivity of −80 dBm at the antenna port for the DSSS/CCK signal with data rates of 1 and 2 Mbps, which is the minimum sensitivity level for multiple data rates of the IEEE 802.11 b/g WALN systems.

Solution In today's markets, most RF IC venders can provide 802.11b/g WLAN transceivers with sensitivity levels much lower than −80 dBm for rates of 1 and 2 Mbps. Considering a design margin of 15 dB, we can target the minimum sensitivity of −95 dBm for the 802.11 b/g WLAN receiver. With a 3-dB front-end loss from the antenna port to the LNA input, including a diplexer and matching circuit, the minimum signal level at the input of the LNA is −98 dBm, which is equal to $2.8\mu V_{\text{rms}}$ or $-111\ \text{dB}V_{\text{rms}}$ referred to a 50 Ω load. If the desired baseband signal on either the I channel or Q channel is $110mV_{\text{rms}}$ at the input of the ADC in a typical CMOS process, therefore, the maximum analog gain that the receiver needs to provide is $20 \times \log(110mV_{\text{rms}}/2.8\mu V_{\text{rms}}) = 92$ dB.

7.6.3 AGC Setting Strategy

A RF front-end of the receiver usually consists of the low noise amplifier (LNA) and mixer, and its gain mode is slightly dependent from different wireless systems. In general, in the 802.11 WLAN systems the LNA has three gain-setting modes to cover the required dynamic range with optimized current consumption [26], while the front-end gain modes of the 2G and 3G mobile systems have more than three gain-setting modes. In the 802.11 systems, the high-gain and middle-gain modes of the front end are usually implemented with active gain formats, while the low gain mode is designed as a passive attenuator. Over the entire receiver band, the LNA with a low noise figure (NF) plays an important role in achieving low sensitivity. Therefore, the LNA is required to have very low NF with reasonable current consumption. Meanwhile, the linearity of the receive chain is also crucial in order to have a high signal-to-noise ratio (SNR) when either the received input signal is close to its maximum input signal specification or there exist stronger adjacent channel blockers and interferers. The linearity requirement of the receive chain leads to high IIP3 or P1dB and IIP2 specifications over the dynamic range. In addition, other useful specifications in the receiver are SNR and EVM over the dynamic range.

To set up correct AGC settings along the receive chain, most RF transceivers have a received signal strength indictor (RSSI) function to estimate the power level of the received signal in dBm at the input of the LNA. The RSSI circuits are usually implemented in either analog baseband before the BB amplifiers for a zero-IF receiver or intermediate frequency (IF) before the IF amplifiers for a low-IF receiver. The RSSI has a large signal dynamic range to cover the received signal strength at the input of the LNA to assist AGC units. The RSSI output reflects on a logarithmic scale the amplitude of the instantaneous modulated RF signal (envelope) or provides a logarithmic to linear conversion, corresponding to the input power level in dBm on the x-axis to the RSSI output voltage on the y-axis. The RSSI output voltage can be sampled through an ADC and then be encoded for the gain control via a 3-wire interface or internal processor bus.

Figure 7.33 illustrates a direct conversion receiver with a wideband RSSI (WRSSI) and a narrowband RSSI (NRSSI) block along the receive chain in the analog domain used for the 802.11n WLAN transceiver [27]. There might be one more a RSSI block in the digital domain, termed a digital RSSI (DRSSI). The

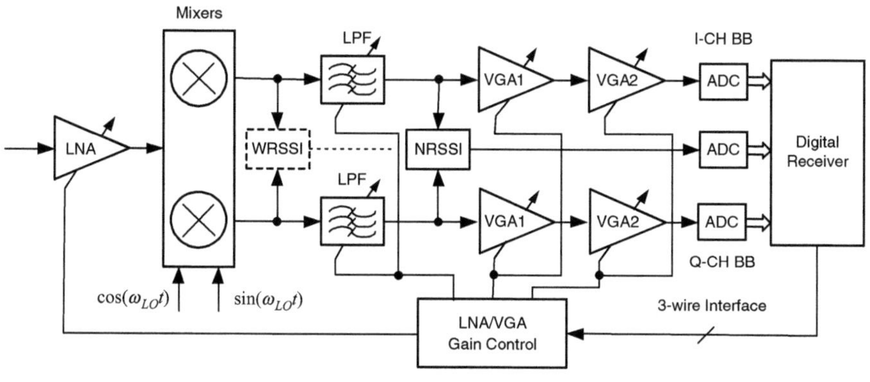

Fig. 7.33 Block diagram of a receiver with analog channel selection filters, VGAs, and RSSI circuits in an 802.11n WLAN transceiver. Referenced from [27]

WRSSI block mainly detects the voltage envelope level of the adjacent channel interferers and in-band interferers and blockers, while the NRSSI block measures the voltage envelope level of the desired signal. If the received signal contains large interferers or blockers, the WRSSI value should be either much greater than the NRSSI value or slightly greater than the NRSSI value, depending on how strong the interferers are. If the received signal does not have interferers or blockers, their values should be roughly be the same. Hence, based on the different strengths between their values, a gain control block in the digital domain sends the control command signal to a LNA/VGA gain control block through a three-wire interface to adjust the analog gain values. For example, in the case that the WRSSI value is much larger than NRSSI value due to larger interferers, the receiver can decrease the LNA gain to prevent the following stages from severe saturating. In addition to WRSSI and NRSSI blocks, a DRSSI block after the digital channel selection filter is capable of distinguishing in-band signal from adjacent channel interfering signals in a case of using a channel selection filtering partition such that the AGC loop can operate on the desired in-band signal only.

The priority order of the AGC setting in the receive chain is to set the LAN gain first, the LPF gain and VGA gain next. The gain of the LNA may first need to be adjusted based on the measured RSSI value corresponding to the received signal level at the LNA input. Then, the gain values of the LPF and VGA are adjusted to ensure the signal level at the input of the ADC around the desired level as closely as possible. The minimum gain step is 1 dB over the input signal range. The gain values can be optimized for either SNR or EVM and P1dB/S at each input signal level. With the optimized AGC-setting table, the best receiver's EVM or SNR can be maintained and P1dB/S can be retained to meet the required minimal value of PAPR + MFM over the entire dynamic range. Here, PAPR is the peak-to-average power ratio of the modulated signal, and MFM is the multipath fading margin that depends on an actual type of multipath fading channel. A 10-dB fading margin would be reasonable for most applications.

The basic idea for the AGC setting is to keep the gain of the front-end (LNA, Mixer) in the receive chain as high as possible in order to achieve a low noise figure until the received input signal becomes large. Usually, the front-end gain has three gain modes: high gain (HG), middle gain (MG) and low gain (LG), which cover the entire input signal range. There are three optimal switching points corresponding to three front-end gain modes. The optimal switching points are crucial for achieving good SNR performance. If the LNA gain is switched from HG to MG too early, then SNR or RX EVM would degrade due to an increased noise figure, while the received signal level is still not large enough to produce any significant noise compression. If the LAN gain, on the other hand, is switched from HG to MG too late, then the RX EVM would also degrade due to slight saturation or nonlinear distortion of the receiver while the received signal level is larger. Hence, the optimal gain switching points of the front-end gain are very important to ensure the smooth transitions of the RX EVM between these switching points. The final gain table would achieve the optimal EVM and reasonable P1dB/S ratio both inside and outside band.

Assume the front-end adjustable gain that consists of a variable-gain LNA and a fixed-gain mixer has three combination gain modes of HG, MG and LG. The entire dynamic range of the input signal can be mainly divided into three ranges, which correspond to three the front-end gain modes. Gain switching points between these ranges are determined by spreadsheet calculation and are finalized by actual testing verification. Staring with the achievable minimum received signal level or achievable sensitivity level and continually increasing the received signal from the sensitivity level, the adjustment procedure of the AGC gain table staring with its maximum gain value is as follows:

1. *Maximum voltage gain value along the receive chain.*
   In order to detect the received RF signal arrival around the desired sensitivity level at the antenna port, the receiver's AGC along the receive chain is initially set by having the front-end block at a HG mode and the following stages of LPF and VGA blocks at an appropriate gain value. The total gain setting value should be able to bring the RF signal with the minimum sensitivity level at the antenna port to the IF I–Q or baseband I–Q signals with the desired level at the inputs of the ADCs. The required maximum gain calculation, plus design margin, was introduced in Design Example 7.5 of the previous section. Whenever the received signal is reliably detected, the received RF signal is further processed to make sure whether it is the desired signal or not. If it is not, the receiver goes back to a detection state to continually monitor the desired signal arrival.
2. *Gain distribution along the receive chain.*
   Usually, the gain values along the receive chain are distributed on the LNA, the mixer, the LPF, multi-stage VGAs, and possibly the output buffer. The LNA and mixer gains are called the front-end gain, while the LPF, VGA, and buffer gains are called the back-end gain. The LNA gain needs to be large enough to bring the smallest desired signal sufficiently above the noise floor of the mixers [1]. In designing gain distribution along the receive chain due to linearity consideration, the system is designed to allow back-off from a P1dB compression point at each stage output. The back-off is usually greater than the PAPR value of the modulation signal, possibly plus an additional fade margin.
3. *Adjustment procedure of AGC gain table.*
   According to the received signal strength at the antenna input port, which is detected by the RSSI circuit, the front-end gain is set first. Whenever the front-end gain is switched among HG, MG, and LG modes due to the received signal level changes, the gain of the LPF following the mixer should be primarily adjusted in order to reduce the effect of the large gain transition on the front-end gain either decrease or increase. Then, other back-end gains can be adjusted to keep the I and Q baseband signals at the inputs of the ADCs close to the desired level.
4. *Verification of AGC gain table.*
   First, a spreadsheet is used to verify that the curves of SNR, and P1dB/S versus the RF input power level are close to the design expectations. In the spreadsheet, each circuit block along the receive chain has its own NF, IIP3, and gain values. Then, the AGC gain table is finalized by actually measuring SNR and P1dB/S curves versus the RF input power level at the I and Q baseband

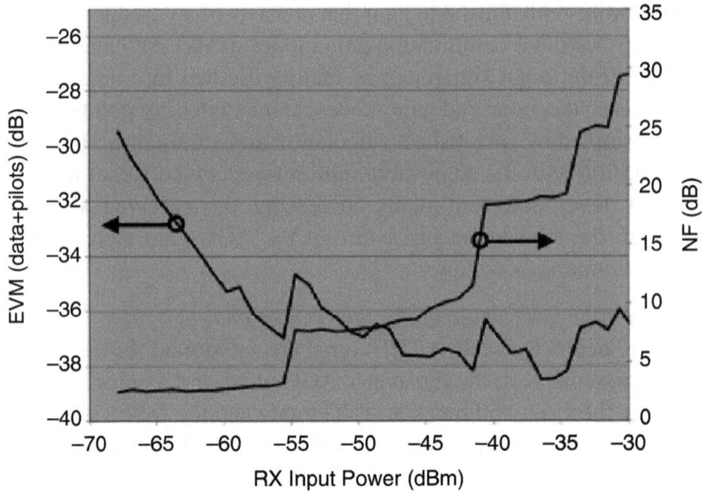

Fig. 7.34 Measured NF and RX EVM versus RX input power at LNA input in an IEEE802.16e WiMAX receiver. Referenced from [26]

outputs without adjacent interference when either a single tone or a real modulated signal is added at the input of the LNA. In measurements of SNR and P1dB/S at the in-channel, the signal power S and compression point P1dB can be measured by adding a single tone with the frequency located at the center frequency of the measured channel at the input of the LNA. The noise power is measured by integrating noise power density within the channel bandwidth, excluding the signal power if the signal is still added during noise power measurement. After the initial SNR measurement by using the single tone, accurate SNR measurement can be obtained by using the real modulated signal to directly measure RX EVM and then to convert it to SNR [21]. Measuring EVM can take all impairments into account, such as LO phase noise, I–Q gain and phase imbalances, amplitude and phase variations of the channel selection filter, and nonlinearity of the front-end circuits.

Figure 7.34 illustrates the measured RX EVM and noise figure (NF) versus the RX input power with a gain table used in a RF transceiver for the IEEE802.16e WiMAX standard in the frequency range from 2.3 to 2.7 GHz [26]. From the curves, we can see that NF has three jumping-up points around the input power levels of −55, −42, and −34 dBm, corresponding to the front-end gain switching points. As described in [26], the LNA has three gain modes of high gain (HG), middle gain (MG), and low gain (LG) plus two gain modes in the LPF stage. Either NF instantly degrades about 5 dB or EVM sharply degrades about 2 dB when the LNA gain is reduced from HG to MG modes at a switching point of about −55 dBm due to a 1-dB input power increase or from −55 to −54 dBm. Meanwhile, the VGA or output buffer gain should be increased by the same amount of dB minus 1 dB in order to keep a 1-dB reduction in total gain. As a result, such an instant degradation of the NF is due to a cliff reduction of the front-end gain. Similarly, the cliff

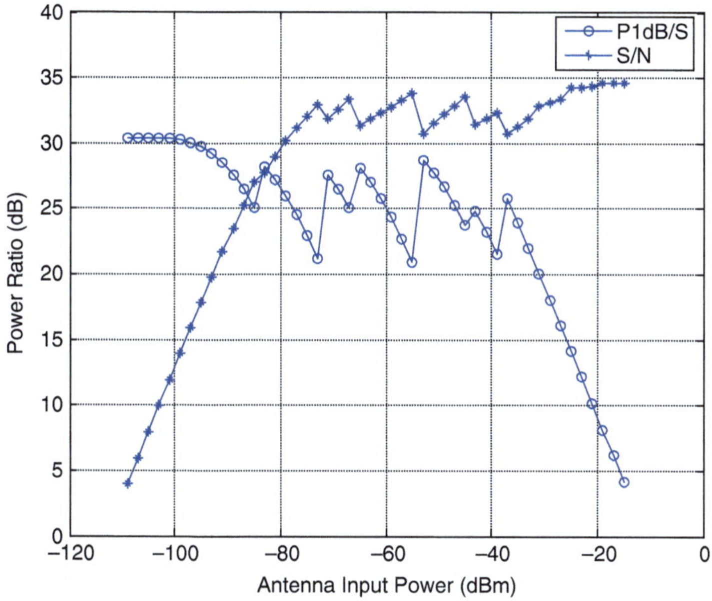

Fig. 7.35 Typical SNR and P1dB/S versus antenna input power in the 2G GSM850 band

degradations of the NF at the RX input power values of around −42, −34, and −32 dBm are also related to either LNA gain reductions or LPF gain reductions.

It should be mentioned that instantaneous NF degradations at the LNA gain switching points are caused by the LNA gain instantly decreasing from a high-gain mode to a low-gain mode, while the received signal level is still not large enough to produce any significant noise compression. But the LNA gain reduction is used to ensure that the P1dB/S ratio would not be reduced too much in the receiver chain, which will be introduced in the following section.

As mentioned above, the AGC gain table should be optimized for both SNR and P1dB/S over the RF input dynamic range. Figure 7.35 illustrates the typical SNR and P1dB/S in the 2G GSM850 band over the entire RF input signal range on a commercially available 2G GSM transceiver chip. It can be seen that the P1dB/S ratio stays greater than 20 dB up to the input power of about −30 dBm at the antenna input port. At the input power of −30 dBm, P1dB/S still has a 5-dB margin if the requirement of 15 dB for the P1dB/S ratio is assumed after considering that the received GMSK signal has less than a 5-dB envelope fluctuation and the transmission channel is supposed to have less than a 10-dB multipath fading variation. The SNR is greater than 30 dB after the input signal is larger than −80 dBm at the antenna input. The total voltage gain, which consists of the LNA gain, the mixer gain, the LPF gain, and three-stage VGA gain, is more than 100 dB and is capable of detecting a weak signal down to about −110 dBm at the input of the antenna.

The optimal AGC gain table should be experimentally verified by applying a single test tone to the input of the LAN to meet both SNR and P1dB/S requirements over the entire input signal range in the absence of interferers and blockers. In order to take the I–Q imbalance error and any other errors into account such as VCO phase noise and nonlinearity in the receiver, we prefer using an actual modulation signal to verify SNR by measuring RX EVM and then converting it to RX SNR, as shown in Fig. 7.30.

References

1. Janssens, J., & Steyaet, M. (2001). *CMOS cellular receiver front-ends*. Boston: Kluwer Academic.
2. Crols, J., & Steyaert, M. (1995). A single-chip 900 MHz CMOS receiver front-end with a high performance low-IF topology. *IEEE Journal of Solid-State Circuits, 30*(12), 1483–1492.
3. Crols, J., & Steyaert, M. (1998). Low-IF topologies for high performance analog front ends of fully integrated receivers. *IEEE Transactions on Circuits and Systems-II: Analog and Digital Signal Processing, 45*(3), 269–282.
4. Le, K. T. (2005). Transceiver design for IEEE 802.15.4 and Zigbee-Compliant Systems. *Microwave Journal, 48*, 160–171.
5. Levy, G. (2002). From performance to manufacturing implementation and yield: A receiver comparison for GSM/GPRS applications. Application note, Silicon Laboratories.
6. Silicon Laboratories. (2002). *Aero+GSM/GPRS transceiver, chipsets application*. Austin, TX: Silicon Laboratories.
7. Janssens, J., & Steyaert, M. (2002). *CMOS cellular receiver front-ends from specification to realization* (p. 25). New York: Kluwer Academic.
8. Darabi, H., Chang, P., Jensen, H., Zolfaghari, A., Lettieri, P., Leete, J. C., et al. (2011). A quad-band GSM/GPRS/EDGE SoC in 65 nm CMOS. *IEEE Journal of Solid-State Circuits, 46*(4), 870–882.
9. Lerstaveesin, S., & Song, B. (2006). A complex image rejection circuit with sign detection only. *IEEE Journal of Solid-State Circuits, 41*(12), 2693–2702.
10. ZigBee Alliance. (2005, April). *ZigBee specifications, version 1.0*.
11. Razavi, B. (2003). RF microelectronics. Taiwan: Pearson Education.
12. Lerstaveesin, S., Gupta, M., Kang, D., & Song, B.-S. (2008). A 48-460 MHz CMOS low-IF direct-conversion DTV tuner. *IEEE Journal of Solid-State Circuits, 43*(9), 2013–2024.
13. Conexant Systems, Inc. (2001, July 20). Application note 101799.
14. Razavi, B., & Zhang, P. (2003, January 21). Method and apparatus for reducing DC Offset. *Unite States Patent* (Patent No.: US 6,509,777 B2).
15. Aytur, T., Kang, H. C., Mahadevappa, R., Altintas, M., ten Brink, S., Diep, T., et al. (2006, February). A fully integrated UWB PHY in 0.13μm CMOS. *ISSCC Digest of Technical Papers* (pp. 124–125).
16. Lo, S., Lo, S., Sever, I., Ma, S.-P., Jang, P., Zou, A., et al. (2006). (2006, February). A dual-antenna phased-array UWB transceiver in 0.18μm CMOS. *ISSCC Digest of Technical Papers* (pp. 118–119).
17. Ali-Ahmad, W. Y. (2004, April). Effective IM2 estimation for two-tone and WCDMA modulated blockers in zero-IF. *RF Design* (pp. 32–40)
18. Conexant Systems, Inc. (2001, September). Application note 101771B.
19. Holma, H., & Toskala, A. (2010). *WCDMA for UMTS: HSPA evolution and LTE* (5th ed.). New York: Wiley.
20. Tanner, R., & Woodard, J. (2004). *WCDMA requirements and practical design* (p. 32). England: Wiley.

21. Gao, W. (2013). Performance enhancement of a WCDMA/HSDPA+ receiver via minimizing error vector magnitude. *IEEE International Test Conference (ITC'13) 2013, Anaheim, California, September 8–13, 2013*.
22. Reynolds, S. K., Floyd, B. A., Beukema, T. J., Zwick, T., & Pfeiffer, U. R. (2005). Design and compliance testing of a SiGe WCDMA receiver IC with integrated analog baseband. *Proceedings of the IEEE, 93*(9), 1624–1636.
23. Jussila, J., & Halonen, K. (2002). WCDMA channel selection filter with high IIP2. 2000 *IEEE International Symposium on Circuit and Systems (*2000 *ISCAS)* (Vol. 1, pp. 533–536).
24. Jussila, J. (2003, June). Analog baseband circuits for WCDMA direct conversion receivers. *Ph. D. Dissertation*, Helsinki University of Technology, Finland.
25. Jensen, O. K., Kolding, T. E., Iversen, C. R., Laursen, S., Reynisson, R. V. Mikkelsen, J. H. (2000). RF receiver requirements for 3G WCDMA mobile equipment. *Microwave Journal*.
26. Locher, M., Kuenen, J., Daanen, A., Visser, H., Essink, B. H., Vervoort, P. P., et al. (2008). A versatile, low power, high performance BiCMOS MIMO/diversity direct conversion transceiver IC for WiBro/WiMAX (802.16e). *IEEE Journal of Solid-State Circuits, 43*(8), 1731–1740.
27. Behzad, A., Carter, K. A., Chen, H. M., Wu, S., Pan, M. A., & Lee, C. P. (2007). A fully integrated MIMO multiband direct conversion CMOS transceiver for WLAN applications (802.11n). *IEEE Journal of Solid-State Circuits, 42*(12), 2795–2808.

Chapter 8
Applications for RF Transceiver ICs

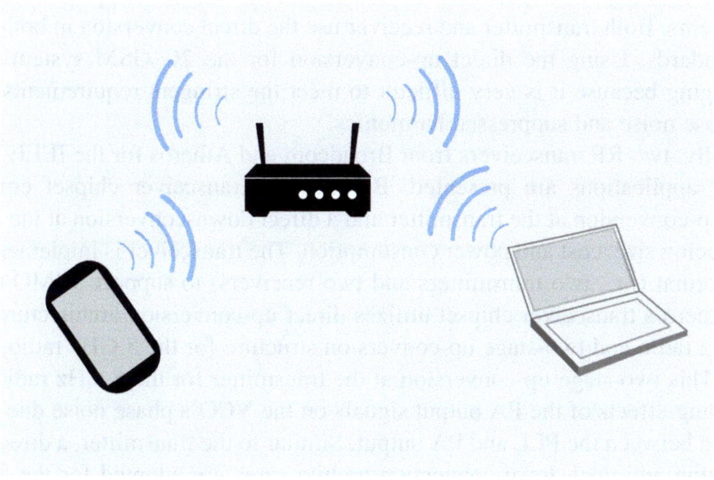

8.1 Introduction

So far, we have described some main system design ideas and technologies used in wireless cellular and 802.11 RF transceivers. We have also discussed radio frequency (RF) analog impairments that can affect system performance, and introduced various calibration techniques that can minimize analog impairments. In this chapter, we introduce several practical transceiver products that have been or currently are available in cellular and Wi-Fi markets. The purpose of this chapter is to provide readers with an overview about how the actual integrated circuit (IC) products perform and how the described technologies' advantages are applied to these products. The products chosen are among the most popular and advanced technologies used in wireless RF transceivers.

© Springer International Publishing Switzerland 2017

W. Gao, *Energy and Bandwidth-Efficient Wireless Transmission*, Signals and Communication Technology, DOI 10.1007/978-3-319-44222-8_8

Firstly, two 2G GSM RF transceivers from Silicon Lab and Conexant for 2G GSM system are introduced. In the Silicon Lab's transceiver, the transmitter uses an OPLL in the transmit path as a frequency up-converter for achieving both frequency up-conversion and narrow bandpass filtering, and the receiver employs a low-IF architecture because of being less vulnerable to the DC offset. The image rejection is carried out in the digital domain, accompanied with digital down-conversion processing. In the Conexant's 2G GSM transceiver, the transmitter also uses the OPLL in the transmit path as a frequency up-converter for achieving both frequency up-conversion and narrow bandpass filtering, but the receiver adopts a direct down-conversion architecture for eliminating the image rejection problem that the low-IF structure must face.

Secondly, another two RF transceivers from MediaTek and Skyworks for the 3G cellular standard are discussed. MediaTek's transceiver adopts a direct up-conversion and direct down-conversion architectures at the transmitter and receiver, respectively. Skyworks provides the dual-mode solution to support both 2G and 3G systems. Both transmitter and receiver use the direct conversion in both 2G and 3G standards. Using the direct up-conversion for the 2G GSM system is more challenging because it is very difficult to meet the stringent requirements of both low phase noise and suppressed harmonics.

Finally, two RF transceivers from Broadcom and Atheros for the IEEE 802.11n WLAN applications are presented. Broadcom's transceiver chipset employs a direct up-conversion at the transmitter and a direct down-conversion at the receiver for reducing size, cost and power consumption. The transceiver is implemented in a 2×2 format (i.e., two transmitters and two receivers) to support MIMO technology. Athero's transceiver chipset utilizes direct up-conversion architecture for the 2.4 GHz radio and two-stage up-conversion structure for the 5 GHz radio, respectively. This two-stage up-conversion at the transmitter for the 5 GHz radio avoids the pulling effects of the PA output signals on the VCO's phase noise due to finite isolation between the PLL and PA output. Similar to the transmitter, a direct down-conversion and dual down-conversion architectures are adopted for the 2.4 GHz radio and the 5 GHz radio, respectively.

8.2 Cellular Communication Transceivers

Wireless cellular communication systems were developed in the early 1980s with the first-generation (1G) standard of Advanced Mobile Phone System (AMPS), which is an analog mobile cell phone system standard developed by Bell Labs for speech services. Second-generation (2G) mobile systems launched in the early 1990s included the first digital cellular systems and networks with time division multiplexing (TDM), frequency division multiplexing (FDM), or code division multiple access (CDMA). Two important systems in the 2G standard are the global system for mobile (GSM) in Europe and an early version of Qualcomm's CDMA, known as IS-95 in the USA. Compared with the 1G systems, 2G systems provide better sound quality and security, higher spectrum efficiency, and larger total capacity. Then the third-generation (3G) standard wireless systems were introduced during the late 1990s and the early 2000s, including the WCDMA system in Europe, the TD-SCDMA system in China, and the CDMA2000 system in USA, to provide interactive multimedia, including teleconferencing and internet access with higher data transmission rates. In the early 2010s, the real fourth-generation (4G) standard represented by Long Term Evolution (LTE) Advanced was released to offer peak download rates up to 1 Gbit/s fixed speed and 100 Mbits/s to mobile users. Since early 2013 [1, 2], with the maturing of the 4G standard and its worldwide development and commercial operation, research activities on fifth-generation (5G) advanced technologies have continued worldwide in both academic and commercial communities. From the data-rate perspective, it is expected that 5G systems will be able to offer a minimum of a 1-Gbits/s data transmission rate anywhere to all users and up to 5- and 50-Gbits/s data transmission rates for high-mobility and pedestrian users, respectively [2]. The 5G era is expected to emerge in 2020.

8.2.1 2G GSM Transceivers

GSM (Global System for Mobile Communications) is a standard developed by the European Telecommunication Standards Institute (ETSI) for 2G digital cellular networks used by mobile phones in July 1991. It became the global standard for mobile communications worldwide. The GSM frequency bands are the cellular frequencies designed by the International Telecommunication Union (ITU) for the operation of GSM mobile phones. The four major bands that operate globally are listed in Table 8.1. Bands 2 and 5 have been deployed in North American Region (Canada and the USA), Caribbean, and Latin America. Bands 3 and 8 have been deployed in Europe, the Middle East and Africa, and the Asia-Pacific region.

The GSM standard adopts the GMSK modulation technique as its format because of its constant envelope feature, which enables the power amplifier to operate in a saturated or near-saturated region in order to achieve energy efficiency. In addition to its constant envelope property, the GMSK modulation signal adopted by the GSM system standard also has some other specific characteristics: it has a relatively low data rate, narrow bandwidth, and plenty of low-frequency energy

Table 8.1 GSM frequency bands

GSM band	Frequency (MHz)	Up-link frequency (MHz)	Down-link frequency (MHz)	Equivalent LTE band
GSM-850	850	824.2–849.2	869.2–893.8	5
E-GSM-900	900	880.0–915.0	925.0–960.0	8
DCS-1800	1800	1710.2–1784.8	1805.2–1879.8	3
PCS-1900	1900	1850.2–1909.8	1930.2–1989.8	2

around DC in its power spectral density (PSD). In addition to these features, the GSM standard stringently specifies the output RF modulation spectrum. Unlike the case of conventional direct-up transmitter and direct-down receivers, these specific features and stringent spectral specification mean that the transmitter adopts the translation loop of the phase locked loop (PLL) to transfer the frequency of the GMSK-modulated signal from the IF to the RF, and the receiver utilizes a low-IF architecture to perform the frequency down-conversion from the RF to the low-IF.

The main advantages of the PLL translation loop approach, in contrast to a conventional up-conversion mixer, are that the translation loop does not produce an image signal and other harmonic components because it performs both frequency transfer processing and narrow bandpass filtering. On the receiver side, the low-IF receiver is suitable for GSM systems because it avoids distorting the desired signal around DC frequency by eliminating DC removal with highpass filtering; this is possible because the GMSK-modulation signal has plenty of low-frequency energy around DC in its PSD.

In this section, we will introduce two different GSM transceiver architectures. The first one, designed by Silicon Labs, has the frequency translation loop at the transmitter and the low-IF structure at the receiver. The second one, implemented by CONEXANT Inc., also has the frequency translation loop at the transmitter and the direct-down conversion structure with the charge-and-hold DC-based DCOC method at the receiver.

8.2.1.1 Silicon Labs' GSM Transceiver

Silicon Labs provided a single-chip transceiver with a true quad-band design for 850-, 900-, 1800- and 1900-MHz frequency bands, as illustrated in Fig. 8.1, in the early 2000s [3]. LO generation for both transmit and receive bands is built with a fast-locking fractional-N PLL synthesizer, which has integrated loop filters, Tx and Rx VCOs, and tank circuits. The transmitter provides two steps to transfer the baseband I–Q signals into the RF signal. First, the baseband I–Q signals modulate a pair of orthogonal carrier signals with an intermediate frequency (IF) around 100 MHz. The baseband I–Q GMSK signal generation is described in the Chap. 4. In order to avoid an external surface acoustic wave (SAW) filter between the transmitter and power amplifier (PA), a translation-loop (or an offset phased-locked loop)–based architecture is used to further transfer the IF-modulated signal into the RF signal at the frequencies of 800/900 MHz or 1800/1900 MHz by means of the PLL, which performs both frequency up-conversion and narrow bandpass filtering after the I–Q modulation. The detailed operation of the translation loop can be reviewed in Chap. 4.

The RF signal at the output of the VCO is divided by two for the high band signals at the frequency of either 1800 or 1900 MHz, while it is divided by four for the low band signals at a frequency of either 850 or 900 MHz. The PA can operate in a saturation mode to achieve energy efficiency and extend the battery duration

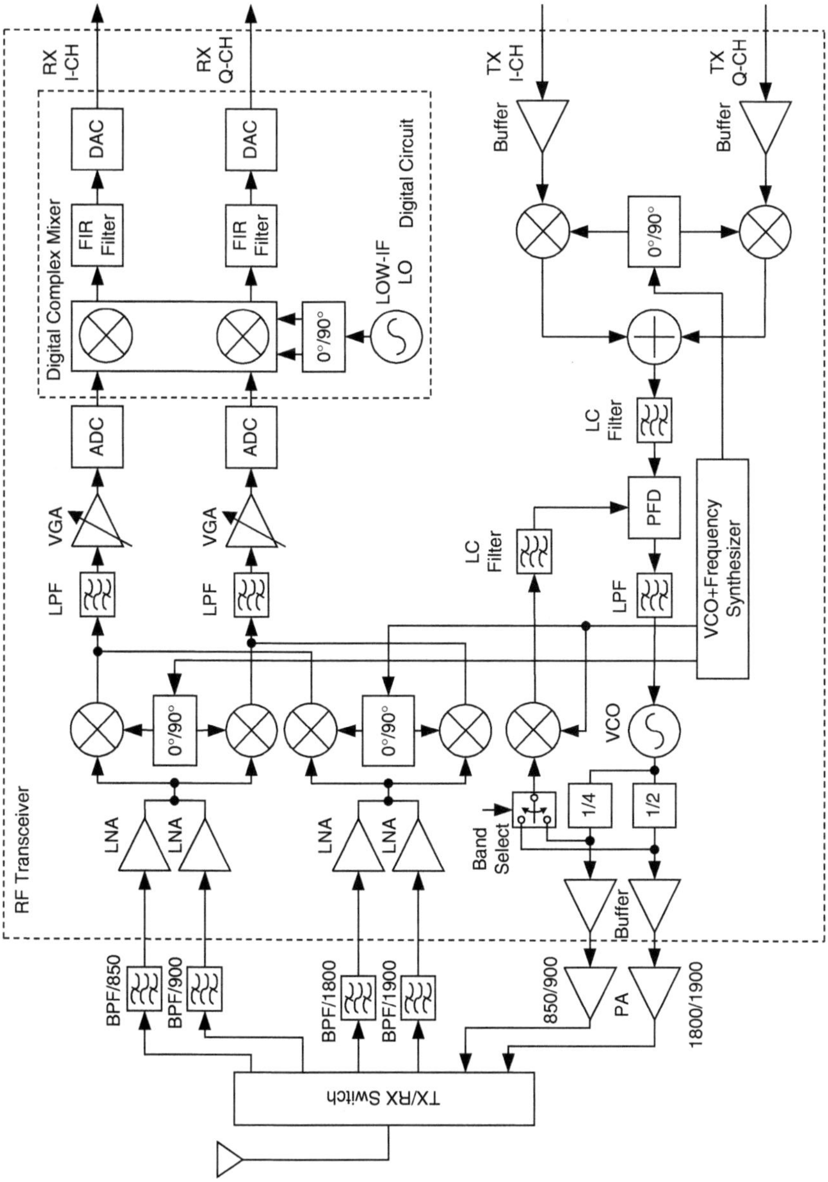

Fig. 8.1 Silicon Labs' GSM quad-band transceiver. Referenced from [3]

because of the constant envelope property of the GMSK signal. The RF-modulated signal is also down-converted back to the IF signal with the VCO signal, and then phase comparison is performed with the IF-modulated signal in the phase and frequency detector after the signal passes through a loop filter.

On the receiver side, the received signal is selected by a corresponding external SAW filter and an on-chip LNA. The SAW filter used in the frond-end attenuates strong block signals. Two I–Q down-converters are used for both bands—900 MHz and 1900 MHz. After passing through the quadrature down-converter, the RF signal is down-converted into low-IF I–Q signals at the center frequency of 100 kHz to eliminate the need for the DC offset cancellation. After passing through the lowpass filters and variable gain amplifiers, the low-IF signals are sampled, and then are fed to a set of complex digital demodulators. At this point, the image signals are also translated to the low-IF domain. The quadrature demodulation process implements the actual image rejection and translates the low-IF signals on the I–Q channels from 100 kHz to the baseband I–Q signals at DC. The, the baseband signals are passed through digital lowpass filters that actually function the channel selection filtering. The lowpass filters also remove the DC offset, which is located at 100 kHz. After digital process and amplification, the digital IF I–Q signals are converted back to the differential I–Q analog signals through ADCs and are sent to the BB chip for further digital carrier and timing symbol recovery circuits, where the frequency offset and phase errors are compensated and final data decision is made. The reason for converting the I–Q digital signals back to the I–Q analog signals is to reduce the number of the interface pins between the RFIC transceiver chip and BBIC chip as few as possible. With the analog interface between them, there are a total of four analog signal interface pins, or one couple of the differential I signals and one couple of the differential Q signals.

In this receiver, channel selection filtering is partitioned into both analog and digital domains. One advantage of the partition is that the digital filter performs primary filtering such that the receiver can achieve significant adjacent channel interference rejections in the digital domain. On the other hand, a high dynamic range requires a high equivalent-resolution ADC, at least 12-bit ADC, to handle large adjacent channel interferers and blockers due to mild attenuation of the analog channel selection filter.

8.2.1.2 Conexant's GSM Transceiver

The CX74017 manufactured by CONEXANT Inc., now SKYWORKS Inc., is a highly integrated transceiver chip for multi-band GSM or General Packet Radio Service (GPRS) applications [4]. As shown in Fig. 8.2, the receiver path has a direct down-conversion architecture that eliminates the need for Intermediate Frequency (IF) components. In order to meet the requirements quad-band applications, the receiver consists of three integrated Low Noise Amplifiers (LNAs), two quadrature demodulators, tunable receiver baseband filters, and DC-offset cancellation circuits. Three separate LNAs are integrated for different bands of operation. At

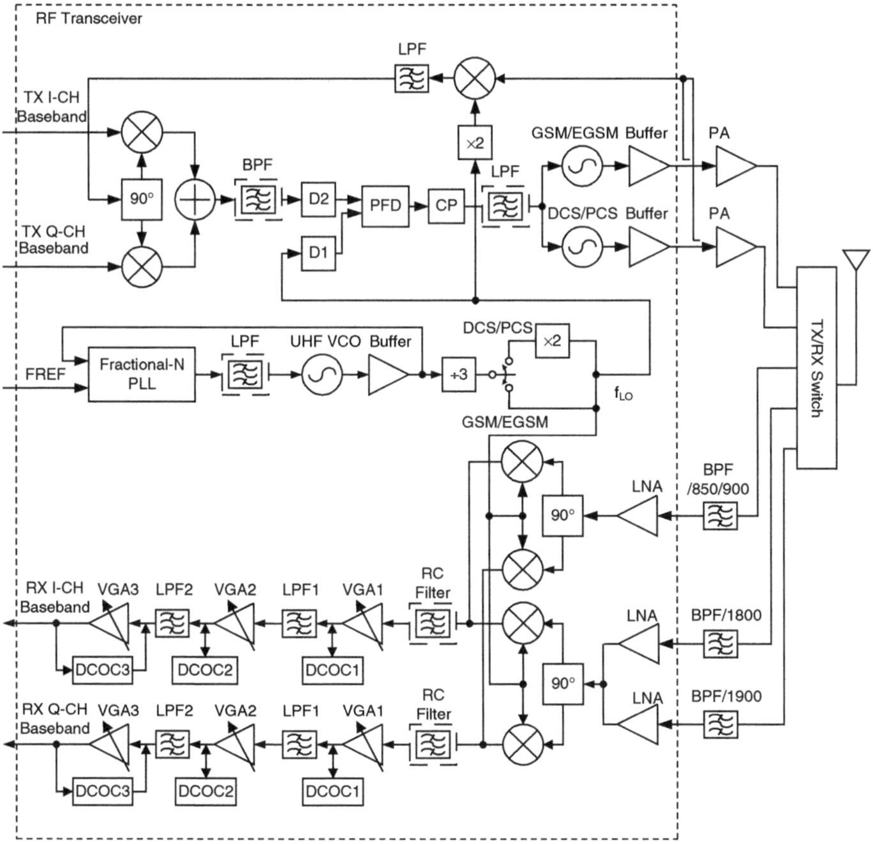

Fig. 8.2 Conexant's GSM transceiver. Redrawn from [4, 5]

the outputs of two quadrature demodulators, two RC lowpass filters on the I–Q channels are used to suppress the out-of-band blockers so that the following amplifiers can avoid operating in the nonlinear range. These filters, marked with dash lines, mean that some external components such as capacitor must be used. In addition, LPF1 and LPF2 together perform the sixth-order lowpass filtering to reject the in-band and out-of-band stronger interferers and blockers. In order to achieve minimum ISI, the cascaded lowpass filter should have the lowest small group delay variation possible within the desired bandwidth. The channel selection filtering is obviously performed by the analog selection filters with a total of eight poles, including an RC filter.

The receiver frequency plan makes it difficult to generate quadrature through the LO path; therefore the quadrature phase is performed by means of a two-stage polyphase filter in the RF path. After passing through lowpass filters in different stages, the baseband I–Q signals are then amplified by different variable gain amplifiers (VGAs), which have programmable gains from 9 to 125 dB, with a gain step of 2 dB [5]. Three-stage DC offset loops that are connected with each

VGA are implemented with the intent of charging before and holding during the receive slot.

The transmitter consists of a frequency translation loop or an offset PLL with in-loop in-phase and quadrature (I–Q) modulation that performs frequency up-conversion from the IF around 100 MHz to RF 850/900/1800/1900 MHz, with the lowest output spectral spurs and phase noise. The frequency translation loop eliminates the need for an external SAW filter. One LO provides both the down-converted IF signal to the I–Q modulator and reference IF signal to the phase and frequency detector. By using a divide-by-three circuit and a frequency doubler in the LO generation, the transmitter is able to prevent any pulling problems between the transmitter output and VCO. The VCO can be centered at any frequency in the range from 1.2 to 1.55 GHz. Selectable frequency dividers add flexibility for frequency planning. In the DCS/PCS mode, a frequency doubler is switched, while it is not in GSM/EGSM mode. Hence, the receive frequency is two-thirds of the VCO output frequency for GSM/EGSM and four-thirds for DCS/PCS [5]. In the transmitter, the LO frequency is one-third of the VCO output frequency for GSM/EGSM mode and two-thirds for DCS/PCS mode. The RF output frequency is equal to $f_{\mathrm{Lo}}(2D1 - D2)/D1$, where D1 and D2 are divider parameters. Two on-chip transmit VCOs correctly select parameters of the frequency doubler and divider, cover all four bands, and meet the stringent out-of-band noise specifications of the GSM standard.

With a 64-pin Land Grid Array (LGA) 9×9 mm-device package, the CX74017 transceiver draws a typical supply current of 41 mA in the receive GSM/EGSM mode and 118 mA in the transmit GSM/EGSM mode, including TX VCO, while operating from a 2.7-V supply voltage.

8.2.2 3G WCDMA Transceivers

Unlike the 2G GSM system, which makes use of both FDD and TDD techniques simultaneously, the transmitters and receivers in the 3G WCDMA and CDMA 2000 systems operate simultaneously in different frequency bands using an FDD technique to achieve higher data speeds and support more users. One of the bigger challenges for FDD systems when both transmitter and receiver operate simultaneously is that the leakage signal from the transmitter to the receiver input on a RF transceiver IC chip can desensitize the performance of the receiver due to possible cross-modulation distortion. To minimize such signal leakage, an external duplexer is usually used as a combiner and splitter to separate the transmission and reception signals.

In this section, two WCDMA transceivers made by two different vendors—MediaTek and Skyworks Solutions—are introduced. MediaTek's WCDMA transceiver adopts a direct-up conversion architecture to support three transmission bands at the transmitter, and performs the primary channel selection filtering by using a fifth-order analog filter to reduce the requirement of ADC dynamic range at the receiver. Skyworks offers a single-chip transceiver that operates as a multimode multiband radio to support both 2G and 3G systems. One of the biggest challenges for this transceiver is that the transmitter utilizes passive mixers and a summer as a quadrature modulator for the GSM signal in order to perform the frequency transfer from the baseband frequency to the RF because this kind of frequency transfer is very difficult to meet the stringent spectral specification of −60 dBc/Hz at 400-kHz frequency offset and phase-noise requirements of the GSM system. At the receiver, a low-IF structure is used for GSM due to easy DC offset removal, direct-down conversion architecture is chosen for WCDMA.

8.2.2.1 MediaTek's WCDMA Transceiver

MediaTek developed a tri-band WCDMA transceiver [6], where the three bands are 850/1900/2100 MHz (2100 MHz is primarily a European and Asian band) as shown in Fig. 8.3. In the receiver, three single-ended LNAs that cover three bands are used to interface with standard single-ended duplexers. The LNA has a notch filter to provide additional attenuation of the TX leakage to prevent nonlinear distortion from the cross-modulation at the receiver. The LNA is followed by a

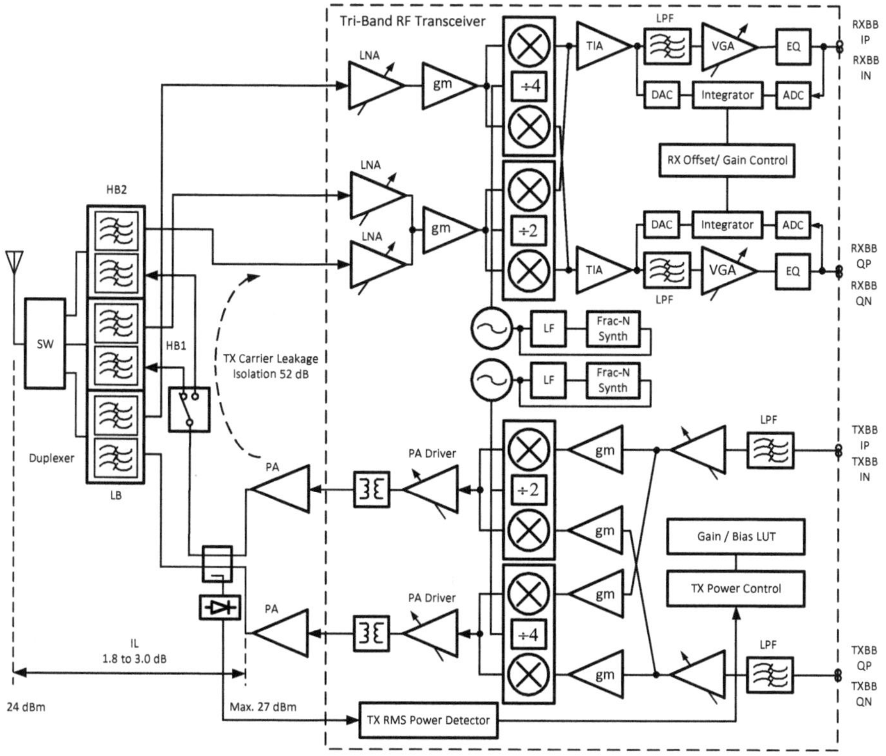

Fig. 8.3 MediaTek's tri-band WCDMA simplified transceiver. Referenced from [6]

high IIP3 transconductor and a passive mixer optimized for achieving high IIP2. Unlike a channel selection filtering partition strategy, a fifth-order analog filter is chosen as a channel selection filter in the analog domain. A single real pole is inserted between the passive mixer and the transimpedance amplifier (TIA) to attenuate large blockers in order to further improve out-of-band IIP3. The remaining four poles are implemented with two biquads. To reduce group delay variation caused by the fifth-order analog lowpass filter, a first-order equalizer or allpass filter is added after a variable gain amplifier (VGA). The overall baseband filter response is optimized for minimizing amplitude ripple and group delay variation within the signal band to achieve low EVM, especially for HSDPA applications, where a 64-QAM modulation format is used. DC-offset correction (DCOC) is performed by subtracting the estimated DC offset at the output of the DAC from the actual DC offset at the output of the TIA. The estimated DC offset is obtained from the output of the last stage of the group delay equalizer through an ADC and is processed inside a mixed-signal feedback loop that implements a highpass filter with fast automatic settling after gain and frequency changes. With analog channel selection filtering, the measured EVM of the receiver for the 3GPP Test-model 5 with 8 HSDPA channels enabled is less than 5% over a wide input-power range, from −70 to −20 dBm, which indicates that the receiver achieves better in-band linearity and SNR.

Since WCDMA is an FDD mode system, the transmitter and receiver operate simultaneously. To reduce the effect of the TX leakage on the receive band, a duplexer is used at the output of the transmitter to isolate the receiver from the transmit signal. The receiver noise figure is measured when the transmitter is turned on for transmitting a modulated signal at a maximum power level of +24 dBm at the antenna port in the UE power class 3. With consideration of 1.8–3.0 dB insertion loss between the PA and the antenna port, the PA should have a power output level of +27 dBm. With an isolation of 52 dB on the TX carrier and an isolation of 43 dB on the TX noise in the RX frequency band, the TX carrier leakage to the receive band, or frequency separation of 190 MHz for the band I, is −26 dBm, which is equal to +27 dBm (PA output power) −52 dB (duplexer IL) −1 dB (power detection circuit loss). In such a test case, the measured *NF* is less than 3 dB.

The VCO, running at around a frequency of 4 GHz, is divided by two and by four to provide the LO frequencies for the high band and the low band, respectively. The maximum TX power accuracy is +1/−3 dB with a 1-dB power step. The RF transceiver is manufactured in 0.18 μm CMOS.

Design Example 8.1 To analyze a realistic noise target due to TX leakage at the RX frequency of the WCDMA transceiver, the following assumptions are made: PA output power is +27 dBm, power detection circuit IL is 1 dB, the duplexer has a 52 dB isolation on the TX carrier and a 43-dB isolation on the TX noise at the RX frequency, and the worst-case RX noise figure is 3.3 dB referred to the LNA input with the transmitter off. What is the maximum TX leakage noise power spectral density allowed at the input of the duplexer for a 4 dB NF with a maximum desensitization of 0.7 dB from a 3.3-dB NF when the transmitter is turned on?

Solution Figure 8.4 shows the calculation steps for a maximum TX noise leakage at the RX frequency to cause a 0.7-dB desensitization of NF by using a method described in [7]. Noise power referred to the input of the LNA with the transmitter off is −104.7 dBm (step 1 of Fig. 8.4). With 0.7-dB desensitization, the TX noise leakage power P_{TX_n} to the RX band can be expressed as

$$10\,\mathbf{log}10\left(P_{TX\_n} + 10^{-104.7/10}\right) = -104\text{dBm} \tag{8.1}$$

Solving P_{TX_n} in (8.1), we have the TX leakage noise power expressed in logarithmic units as (step 2)

$$P_{TX\_n}(\text{dBm}) = 10\,\mathbf{log}10(P_{TX\_n}) = -112.3\,\text{dBm} \tag{8.2}$$

This value corresponds to 7.6-dB noise correction from $N_o = -104.7$ dBm. The noise power spectral density (PSD) within the receive bandwidth of 3.84 MHz is calculated in step 3. Worst-case 43-dB minimum duplexer noise isolation is assumed, so the maximum acceptable noise PSD of −134 dBm/Hz at

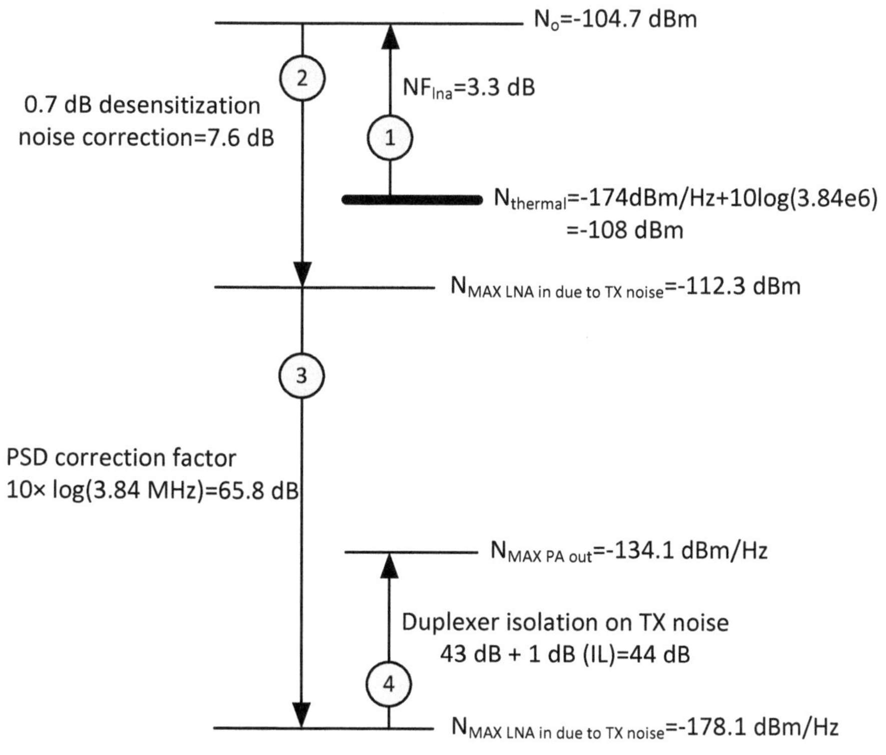

Fig. 8.4 Transmitter noise leakage and impact on LNA noise figure. Referenced from [7]

the output of the PA is obtained (step 4). This requirement can be achieved by most PA vendors. In addition, the noise floor at the LAN input can be also increased due to the IM2 effect of a strong blocker or interferer arising at the input of the LNA.

8.2.2.2 Skyworks Solutions' WCDMA Transceiver

SKYWORKS Solutions presented a single-chip transceiver that operates as a multimode multiband radio in the 2009 I.E. International Solid-State Circuits Conference (ISSCC) [8]. Figure 8.5 shows a block diagram of the RF transceiver that supports 7 primary and 4 diversity bands in WCDMA, including HSDPA/HSUPA and quad band in GSM.

At the transmitter, the baseband I–Q signals in WCDMA are generated by passing the original I–Q data through SRRC filters with a roll-off factor of $\alpha = 0.22$ in the digital domain, while GMSK and EDGE baseband signals in GSM are created with a digital signal processor in the digital domain as well. The digital baseband I–Q signals are then passed through oversampled current steering DACs with 10-bit accuracy. The outputs of the DACs are fed to a third-order Chebyshev filter of the I–Q channels to filter out the image signals. The filters in the I–Q channels can be configured with different corner frequencies to support multi-mode WCDMA/GSM/GPRS/EDGE. Passive mixers driven by either LO or $2 \times$ LO frequencies are needed to achieve the stringent linearity and noise requirements of GSM and WCDMA, because it is usually most difficult for the output RF spectrum (ORFS) of the GMSK signal in GSM to meet the −60 dBc/Hz specification at 400-kHz frequency offset, or for the ORFS of 8-PSK in EDGE to meet the −56 dBc/Hz specification at 400-kHz frequency offset. Actually, this kind of a quadrature modulator for GSM and EDGE baseband signals is more challenge design than an offset-PLL– based modulator (described in the section on 2G GSM transceivers), because the former needs to achieve more stringent noise requirements for the GMSK signal. The differential output of the I–Q modulator is converted to a single-ended output with an on-chip balun in order to drive an external signal-ended power amplifier. The transmitter provides a gain-control range of 80 dB in the WCDMA mode due to a near–far effect related to CDMA-type signaling and 40-dB in the EDGE mode. At high power ranges, from 0 to 24 dBm, a closed-loop power control scheme in the transmitter is adopted to achieve accurate power control, and a conventional power control method through the base station is used when the power range is below 0 dBm.

In the receiver, the passive mixer outputs are connected to the virtual grounds created by the transimpedance amplifiers (TIA), and the mixer inputs are driven by the output current for the LNA. Such a current-driven passive mixer topology leads to improved linearity of the front-end circuits. The noise contribution of the TIA can decrease by adjusting an LC tank resonant frequency at the LAN output according to the operation frequency. In GSM/EDGE mode, the RF signal is down-converted to a low-IF of 135 kHz while in WCDMA mode the RF signal is down-converted to DC. After down-conversion, a third-order Chebyshev filter is

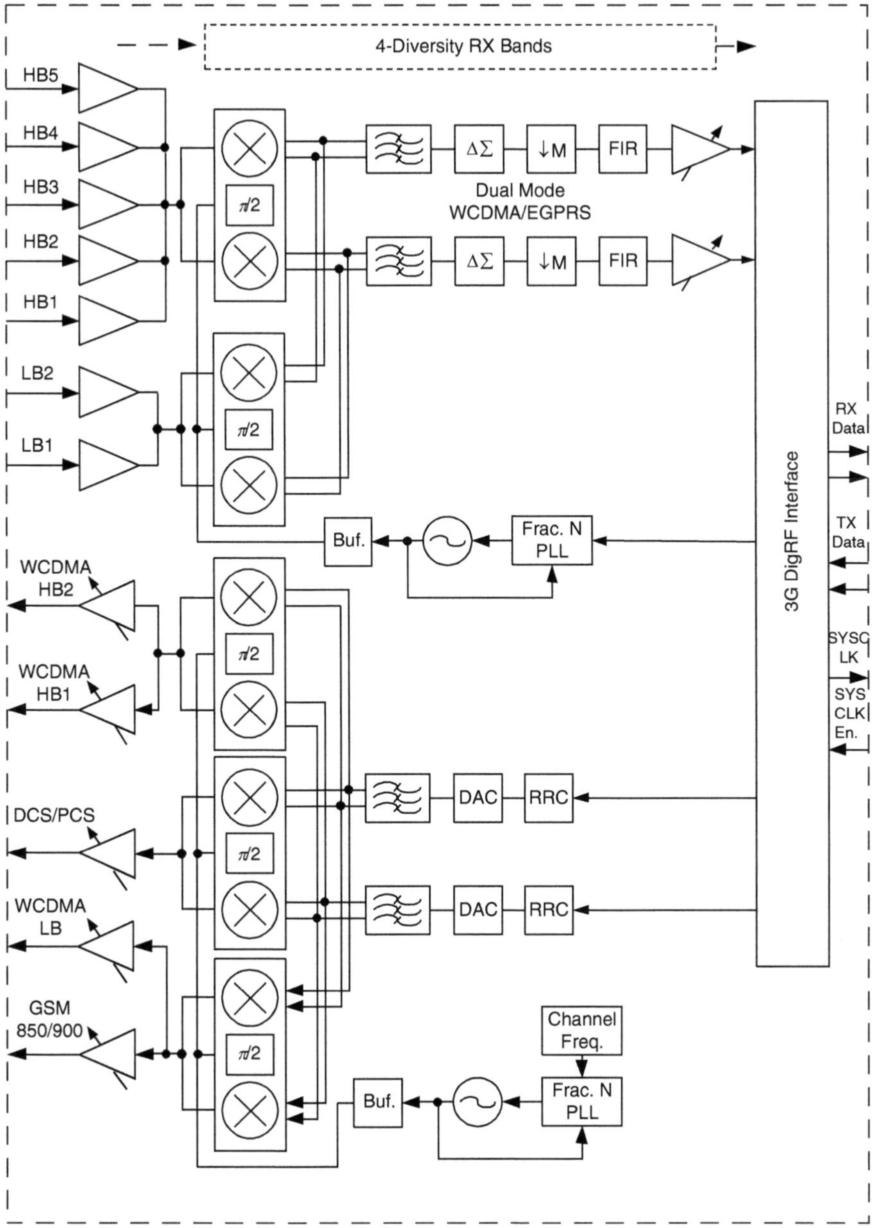

Fig. 8.5 Simplified Skyworks' multiband WCDMA/GSM transceiver. Redrawn from [8]

used to perform anti-alias filtering. To handle the large dynamic requirement due to large blockers, a third-order 4-bit ΔΣ continuous-time ADC with an SNDR of 85 dB for GSM and 71 dB for WCDMA is used. The digital signals are then down-sampled and primary channel-selection filtering is performed in the digital domain.

Finally, the I–Q signals are coherently demodulated and data sequences are covered in a 3G Dig RF interface unit.

A typical noise figure (NF) of about 2 dB in low bands and about 2.5 dB in high bands in the presence of the TX modulated signal with a power level of 24 dBm at the antenna port of the duplexer. NF degrades about 0.6 dB at an isolation of 52 dB compared to an isolation of 57 dB. Now duplexers with more than 52-dB isolation from TX to RX are commercially available. The transceiver die, fabricated in a 0.13 μm CMOS process, occupies an area of 25 mm^2.

8.3 WLAN Transceivers

Wireless Local Area Networks (WLAN) described in this section are standardized by the IEEE 802.11 as a set of media access control (MAC) and physical layer (PHY) specifications in the 2.4- and 5-GHz frequency bands. The first 802.11b standard in the IEEE 802.11 family was released in 1997 for data communications over short distances; subsequent standards were 802.11a, 802.11g, 802.11n, and 802.11ac. Since the IEEE 802.11a and 802.11g standards were launched in the early 2000s, WLAN products have been widely used to connect people to the world and have become a necessary part of people's lives.

IEEE 802.11b and 802.11g use the 2.4-GHz ISM band—as does as Bluetooth standard—where they adopt direct-sequence spread spectrum (DSSS) and orthogonal frequency division multiplexing (OFDM) signaling methods, respectively. The IEEE 802.11a uses the 5-GHz band and adopts the OFDM signaling method to effectively combat multipath fading distortion. The IEEE 802.11n standard, released in October 2009, is an amendment that improves upon the previous 802.11 standards by adding multiple-input multiple-output (MIMO) antennas to enhance the overall throughput. The IEEE 802.11n can operate on both the 2.4-GHz and 5-GHz bands. The IEEE 802.11ac is an amendment to the IEEE 802.11 family and was updated based on 802.11n, but it only operates in the 5-GHz band. Compared to the IEEE 802.11n, the updated standard includes wider transmission bands (up to 80 and 160 MHz [option] versus 40 MHz in the 5-GHz band), more spatial steams (up to eight versus four), higher-order modulation formats (up to 256-QAM versus 64-QAM), and the addition of multi-user MIMO (MU-MIMO) technology (versus single-user MIMO [SU-MIMO]). In 2015, IEEE 802.11ac products entered our lives and brought us higher data throughput for the internet and other communication devices as well as home entertainment systems.

The newest 802.11 standard, 802.11ax, is planning to deliver wireless connectivity faster over the 2.4 GHz and 5 GHz bands by utilizing OFDM up to higher order modulation of 1024-QAM and multi-user MIMO. The 802.11ax draft specification that has been available since January 2016 builds on 802.11ac by doubling the number of spatial streams and significantly improving the spectral efficiency of the existing IEEE 802.11ac. Although still in the early stages of development, the

IEEE 802.11ax will replace both IEEE 802.11n and IEEE 802.11ac as the next high throughput IEEE 802.11 standard in the year 2019 [9].

8.3.1 Broadcom's WLAN Transceiver

Broadcom announced the industry's first Wi-Fi solutions designed to comply with the 802.11n Multiple Input Multiple Output (MIMO) draft specification in 2006. Broadcom's AirForce^{TM} WLAN product family offers the broadest line of Wi-Fi integrated circuits in the industry for system designs ranging from PC and consumer devices to access points and routers. Its WLAN products included 802.11a, 802.11b, 802.11g, and 802.11n solutions up to the late 2000s. After 802.11n, Broadcom continued to develop the next generation of Wi-Fi 802.11ac MIMO at operating at the 5-GHz band. In April 2014, Broadcom announced the industry's first six-stream 802.11ac MIMO for home networks, called the 5G Wi-Fi XStream platform, to be available on the market. The highly integrated architecture enabled Broadcom to deliver the industry's first single-chip Wi-Fi solution for several WLAN based 802.11 standards.

A high-level block diagram of a single-chip fully integrated multiband direct-conversion 802.11n MIMO WLAN transceiver [10] is shown in Fig. 8.6. This transceiver is implemented in a 2×2 format (i.e., two transmitters and two receivers) with a low-cost 0.18 μm CMOS technology and is capable of operating at the 2.4–2.5 GHz (or 802.11g band) as well as 4.9–5.9 GHz (or 802.11a band) bands to support data rates up to 300 Mbps (utilizing short GI) with the bandwidths of 20 and 40 MHz, corresponding to the data rates.

The full 2×2 MIMO transceiver is composed of two multiband TX chains, two RX chains, a PLL and LO generation, digital control section, and various calibration blocks. At the receiver, the received 5-GHz or 2.4-GHz signal is first differentially amplified by the corresponding LNA, then directly down-converted to the baseband by the corresponding quadrature LO signals associated with that band. The down-converted signal is passed through the first programmable gain amplifier (PGA1) built with a highpass filter, where the baseband signal is amplified and the DC offsets are removed. The signal is then filtered by a forth-order Butterworth filter to attenuate the out-of-band interferers. After the signal is further amplified by the PGA2 and PGA3 to achieve the desired signal level at the input of ADCs, where each PGA has its own DC offset cancellation loop, the baseband I–Q signals are sent to the ADCs on the MAC digital baseband chip.

To save space, the 2.4- and 5-GHz RF paths in both transmitters and receivers share the same baseband circuitries, which can be configured to either transmit or receive. To support different bandwidths, the corner frequency of the LPF is adjustable and also calibrated to the desired bandwidths of 5, 10, and 20 MHz to ensure that the desired signal will pass through and the adjacent channel interferers will be rejected. In order to ensure the proper gain setting along the RX chain over a very wide dynamic range in the presence of large interferers, two wideband received signal strength indication (WRSSI) blocks are inserted before and after the PGA1, while one narrowband RSSI (NRSSI) block is inserted after the LPF. The WRSSI detects all signal strengths, including the outside band interferers and inside band signal, while the NRSSI detects only the inside band signal. Based on the WRSSI and NRSSI values, appropriate gain distributions along the RX chain are set. The total gain control range, including RF gain and baseband gain, is greater than 100 dB, and the noise figures of 4 and 4.5 dB are achieved at the maximum gain for the 2.4–2.5-GHzand 4.8–4.9-GHz bands, respectively.

In the transmitter, the digital baseband I–Q signals are converted to analog I–Q signals through the ADCs, and then amplified and lowpass filtered by the lowpass filters with programmable bandwidth to ensure proper data-rate operation and the image-signal rejections. The filtered baseband I–Q signals are then directly up-converted to the RF band by modulating a pair of quadrature LO signals. The RF signal is amplified through the PA driver to achieve an output P1dB power of +14 dBm in the 5-GHz band and an output P1dB power of +16 dBm in the 2.4-GHz band.

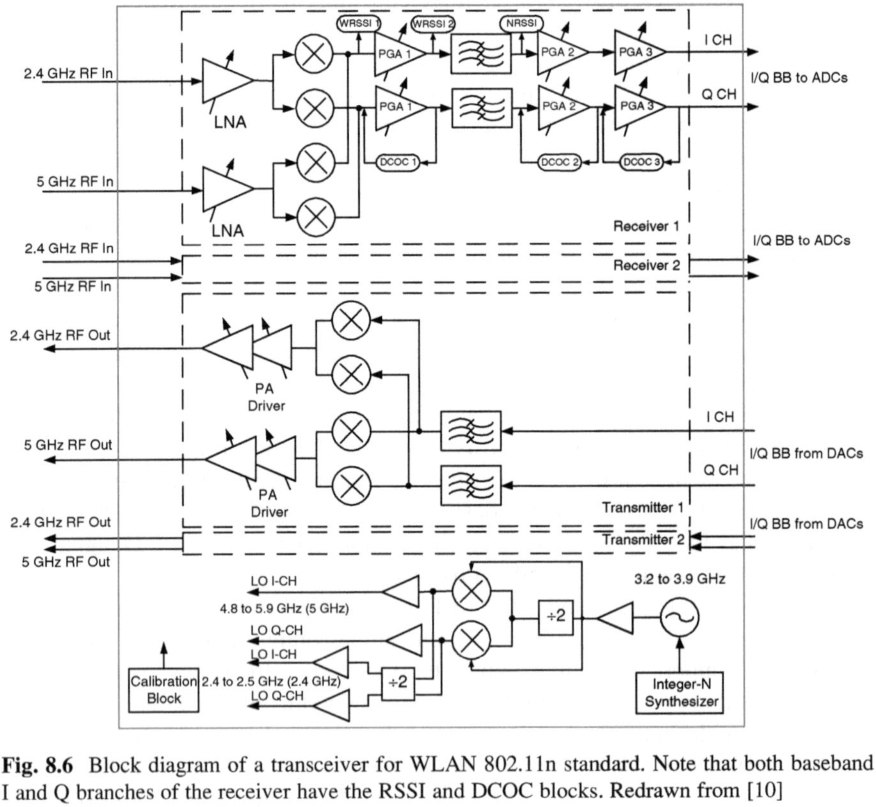

Fig. 8.6 Block diagram of a transceiver for WLAN 802.11n standard. Note that both baseband I and Q branches of the receiver have the RSSI and DCOC blocks. Redrawn from [10]

In order to avoid the pulling effects of the PA output signals on the VCO's phase noises, the VCO operates at a frequency range of 3.2–3.9 GHz, which is two-thirds of the channel frequency for the 802.11a band and four-thirds of the channel frequency for the 802.11b band. As illustrated in Fig. 8.6, the 3.2–3.9-GHz synthesizer output is divided by two and then mixed with the 3.2–3.9-GHz synthesizer output to generate a pair of the quadrature LO signals at 4.8–5.9 GHz for the 802.11a band. The 4.8–5.9-GHz output is then further divided by two to generate a pair of the quadrature LO signals at 2.4–2.9 GHz for the 802.11b/g bands. In order to minimize the impact of VCO noise, the comparison frequency of the phase-frequency detection (PFD) is chosen to be as high as possible to allow wide PLL loop bandwidth, functioning like a highpass filter. Meanwhile, in order to minimize the phase noise arising from the loop filter and charge pump, a low programmable K_{VCO} parameter (30 MHz/V at 3.5 GHz, typically) and high charge pump current are used. A single VCO with a wide tuning range to cover both 2.4- and 5-GHz bands is used in the PLL loop.

The transceiver is capable of supporting PHY rates of >270 Mbps and an effective throughput of >90 Mbps in real-world over-the-air testing. The RF transceiver IC occupies a total die area of 18 mm^2 in a digital 0.18-μm CMOS process and draws 275 mA in RX mode and 280 mA in TX mode [10]. The chip is designed in such a configuration to build a larger MIMO system (e.g., 4×4).

In Mobile World Congress 2015, March 2–5, Barcelona, Spain, Broadcom announced it's a high-end 2×2 MIMO combo chip of BCM4359 for high-performance smart phones. The BCM4359 has several innovation new features, the most significant one being the inclusion for the first time of Real Simultaneous Dual Band (RSDB). RSDB enables the chip to operate at both 2.4 GHz and 5 GHz bands simultaneously. This is achieved by doubling up on the baseband processors on the combo chip. With the RSDB mode, the BCM4359 potentially enhances the speed as well as the quality of the connection. The BCM4359 can support the dual-stream (2×2) setup of 802.11ac that achieves a speed up to 867 Mbps. It can quickly switch between a traditional single-band mode and RSDB mode depending on applications that require concurrent multi-band connections for higher quality of service. Mobile devices equipped with BCM4359 on the market are expected to be available in 2016.

8.3.2 Atheros' WLAN 802.11n Transceiver

Atheros introduced a three-stream, 3×3 MIMO 802.11n transceiver, which is backward-compatible with legacy 802.11a/b/g standards, to enhance throughput, covering range, and link robustness [11]. Figure 8.7 shows a block diagram of the dual-band 3×3 MIMO WLAN transceiver. The 2.4-GHz radio employs a direct-conversion architecture with an LO frequency at twice the channel frequency, while the 5-GHz radio utilizes a traditional superheterodyne architecture, or two

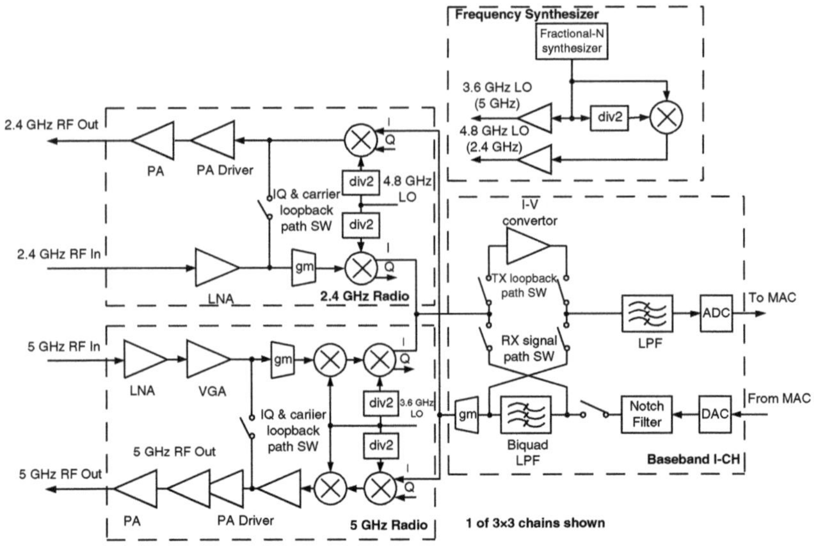

Fig. 8.7 Block diagram of a transceiver for WLAN 802.11n standard: (**a**) block diagram of one of 3×3 chains with I channel shown and (**b**) the detailed dual up-conversion for the 5-GHz band. Redrawn from [11]

stages of frequency conversion, by mixing an LO frequency at two-thirds the channel frequency with one-half the LO frequency. Similar to Broadcom's PLL strategy above, the VCO operates at a frequency range of 3.2–3.9 GHz, which is two-thirds of the channel frequency for the 802.11a band and four-thirds of the channel frequency for the 802.11b band, to avoid the pulling effects of the PA output signals on the VCO's phase noise. The 2.4-GHz quadrature carrier frequencies for 802.11b/g are the same as those described in the previous section. The up-conversion and down-conversion for 5-GHz radio are performed by mixing 1.8-GHz sliding IF, which is half of the LO RF frequency of 3.6 GHz, with LO RF as shown in Fig. 8.7.

In the transmitter, the baseband I–Q signals are first passed through the notch filters to attenuate the image signals located at the sampling frequency, and then passed through the second-order LPFs to further attenuate the image signals after DACs. The filtered I–Q signals are directly up-converted to the RF signal for the 2.4 GHz radio, and twice up-converted to the RF signal for the 5-GHz radio. The modulated RF signal is amplified by the power amplifier driver and then by the power amplifier (PA). The integrated PAs for both bands can achieve a saturated output power greater than 25 dBm.

In the receiver, the received RF signal for the 2.4 GHz radio is directly down-converted to the baseband I–Q signals, while for the 5 GHz radio it is dual-down-converted to the baseband I–Q signals. After DCOC processing at the output of the mixers, the baseband signals are fed to a Biquad LPF in the transmit path, and then feed back to the receive path, controlled by the switches. The receiver utilizes the Biquad LPF as a part of its receive LPF to save space.

EVM performance directly indicates the transmit modulation accuracy and indirectly denotes the receive signal-to-noise ratio as well even though there is no EVM requirement for the receiver in 802.11 specifications. The I–Q gain and phase imbalances are among the major contributing factors to EVM, which can be minimized by using the I–Q calibration. The I–Q calibration used here is accomplished by enabling a phase shift added in the RF loopback path to calibrate the I–Q mismatch in the transmitter first, independent of the I–Q mismatch in the receiver. The I–Q calibration is carried out by closing the I–Q loopback switches and adjusting the phase shift to minimize the I–Q mismatch in the transmitter. Then the transmit data can be digitally pre-distorted (or compensated) before the DAC. Finally, the I–Q calibration for the receiver is performed with the loopback and then the receive data is digitally corrected after the ADC. The calibration for the 2.4- and 5-GHz bands is performed separately due to its frequency dependency.

The transceiver achieves an over the-air TCP throughput rate greater than 310 Mbits/s and PHY rates of up to 450 Mbits/. The die area is 22 mm^2 in a digital 65-nm CMOS process. TX EVM for the 2.4-and 5-GHz bands, respectively, is −39 dB and −36 dB at a PA driver output of −5 dBm. RX NF for both the 2.4- and 5-GHz bands is 4 dB. The saturation power output of the PA for the 2.4 GHz and 5 GHz bands is 26 dBm and 25.5 dBm.

References

1. Chih-Lin, I., Rowell, C., Han, S., Xu, Z., Li, G., & Pan, Z. (2014). Toward green and soft: A 5G perspective. *IEEE Communication Magazine, 52*(2), 66–73.
2. Roh, W., Seol, J.-Y., Park, J. H., Lee, B., Lee, J., & Kim, Y. (2014). Millimeter-wave beamforming as an enabling technology for 5G cellular communications: Theoretical feasibility and prototype results. *IEEE Communications Magazine, 52*(2), 106–113.
3. Silicon Laboratories. (2002). *Aero + GSM/GPRS transceiver chipsets applications*. Austin, TX: Author.
4. Conexant Corporation. (2001). *CX74017 RF transceiver for GSM/GPRS/EDGE application data sheet*. Irvine, CA: Author.
5. Molnar, A., Magoon, R., Zachan, J., Hatcher, G., & Rhee, W. (2002). A single-chip quad-band (850/900/1800/1900 MHz) direct-conversion GSM/GPRS RF transceiver with integrated VCOs and fractional-N synthesizer. *IEEE International Solid-State Circuits Conference (ISSCC) 2002* (pp. 14.2).
6. Tenbroek, B., Strange, J., Nalbantis, D., Jones, C., & Fowers, P. (2008). Single-chip tri-band WCDMA/HSDPA transceiver without external SAW filters and with integrated TX power control. *ISSCC 2008* (pp. 202–204).
7. Holma, H., & Toskala, A. (2010). *WCDMA for UMTS: Evolution and LTE* (5th ed.). New York, NY: Wiley.
8. Sowlati, T., Agarwal, B., Cho, J., Obkircher, T., El Said, M., & Vasa, J. (2009, February). Single-chip multiband WCDMA/HSPA/EGPRS transceiver with diversity receiver and 3G

DigRF interface without SAW filters in transmitter and receiver paths. *I.E. International Solid-State Circuits Conference* (pp. 116–118).

9. Bellalta, B. (2016, February). IEEE 802.11ax: High-efficiency WLANs. *IEEE Wireless Communications* (pp. 38–46).
10. Behzad, A., Carter, K. A., Chien, H.-M., Wu, S., Pan, M.-A., & Paul Lee, C. (2007). A fully integrated MIMO multiband direct conversion CMOS transceiver for WALN applications (802.11n). *IEEE Journal of Solid-State circuits, 42*(12), 2795–2805.
11. Abdollahi-Alibeik, S., Webe, D., Dogan, H., Si, W. W., Baytekin, B., & Komijani, A. (2011, February). A 65 nm dual-band 3-steam 802.11n MIMO WLAN SoC. *ISSCC Digest of Technical Papers* (pp. 170–172).

Tutorial Appendices

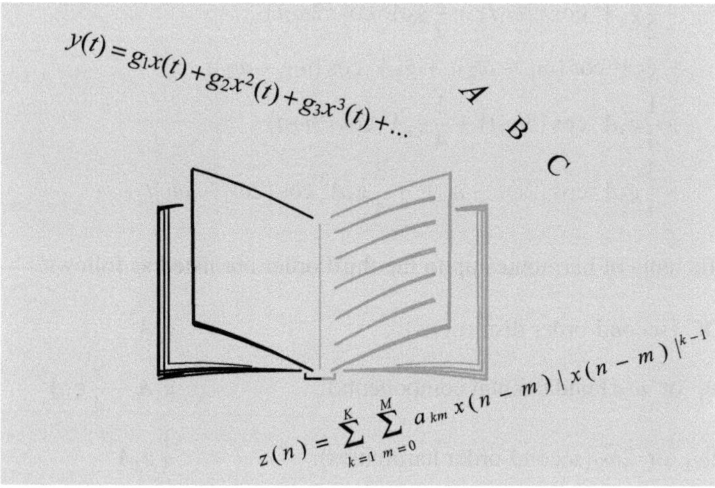

Appendix A: Nonlinear Distortion

When a device or circuitry without memory effect shows nonlinearity, its output is usually expressed as a series expansion of power terms:

$$y(t) = g_1 x(t) + g_2 x^2(t) + g_3 x^3(t) + \ldots \tag{A.1}$$

where g_i, $i = 1, 2, 3, \ldots$ is the coefficient for the fundamental and second and third harmonics, respectively. Even though the output contains an infinite number of terms, the first three terms have important effects on the device's performance and are accurate enough to characterize the device's nonlinearity.

© Springer International Publishing Switzerland 2017

W. Gao, *Energy and Bandwidth-Efficient Wireless Transmission*, Signals and Communication Technology, DOI 10.1007/978-3-319-44222-8

In the nonlinearity measurement of power amplifiers (PAs), the intermodulation (IM) distortions (IMDs) at the output of PAs are usually tested by adding two tones with equal amplitude A and different frequencies ω_1 and ω_2, which are closely spaced frequency carriers. This is called a two-tone test. In the two-tone test, the input is given by

$$x(t) = A\cos(\omega_1 t) + A\cos(\omega_2 t) \tag{A.2}$$

When this input signal is applied to the device with the transfer function in time domain given in (A.1), the output of the device is given by

$$\begin{aligned} y(t) = {} & g_2A^2 + \left(g_1A + \frac{9}{4}g_3A^3\right)\cos(\omega_1 t) + \left(g_1A + \frac{9}{4}g_3A^3\right)\cos(\omega_2 t) \\ & + \frac{1}{2}g_2A^2\cos(2\omega_1 t) + \frac{1}{2}g_2A^2\cos(2\omega_2 t) \\ & + g_2A^2\cos(\omega_1 + \omega_2)t + g_2A^2\cos(\omega_1 - \omega_2)t \\ & + \frac{1}{4}g_3A^3\cos(3\omega_1 t) + \frac{1}{4}g_3A^3\cos(3\omega_2 t) \\ & + \frac{3}{4}g_3A^3\cos(2\omega_1 - \omega_2)t + \frac{3}{4}g_3A^3\cos(2\omega_2 - \omega_1)t + \cdots \end{aligned} \tag{A.3}$$

Coefficients of harmonics up to the third order are listed as follows:

$$\text{DC (second-order distortions):} \quad g_2A^2 \tag{A.4}$$

$$\omega_1 \text{ or } \omega_2 \text{(Fundamental components):} \quad g_1A + \frac{9}{4}g_3A^3 \tag{A.5}$$

$$2\omega_1 \text{ or } 2\omega_2 \text{(second-order harmonics):} \quad \frac{1}{2}g_2A^2 \tag{A.6}$$

$$\omega_1 - \omega_2 \text{ or } \omega_1 + \omega_2 \text{ (second-order distortions):} \quad g_2A^2 \tag{A.7}$$

$$3\omega_1 \text{ or } 3\omega_2 \text{ (third-order harmonics):} \quad \frac{1}{4}g_3A^3 \tag{A.8}$$

$$2\omega_1 - \omega_2 \text{ or } 2\omega_2 - \omega_1 \text{(third-order distortions):} \quad \frac{3}{4}g_3A^3 \tag{A.9}$$

One DC product in (A.4) and two second-order intermodulation (IM2) products in (A.7) are at $(\omega_1 - \omega_2)$ and $(\omega_1 + \omega_2)$ because of the second-order distortion. The IM2 product at $\omega_2 - \omega_1$ is more critical to the desired signal in direct-conversion receivers when the frequencies of these two tones are close. Usually, the DC component can be removed with a DC cancellation circuit, whereas the $\omega_1 + \omega_2$ component is located outside the bandwidth.

The two third-order intermodulation (IM3) products are at $2\omega_1 - \omega_2$ and $2\omega_2 - \omega_1$ because of the third-order distortion. One of these two products may fall in the band of the desired output signal if ω_1 and ω_2 are close to each other, and also if they are close to the band of the desired input signal. Therefore, these IM3 products distort the desired signal because they are not easily filtered out.

The nonlinearity is mainly characterized by IM3, which is described by a third-order intercept point (IP3). However, the IM3 production power level at the PA output does not follow the 3:1 slope of the amplification when the output power level is close to the P1dB compression point [1]. Hence, the two-tone test is not an accurate method to characterize the nonlinearity of a PA by measuring IM3 products when it operates close to the saturation range. Alternatively, the measurement of the P1dB compression point is a good method to characterize the nonlinearity of a PA by using either a single tone or a modulated signal. With the knowledge of the P1dB and the peak-to-average power ratio (PAPR) value of the modulation signal, the PA can be set up to operate at a back-off from the P1dB point by a maximum value up to the PAPR value without significantly degrading the spectral regrowth and error vector magnitude (EVM) at the output of the PA. The operation back-off value is also dependent on tolerances of these performance degradations. The two-tone test is usually used to roughly evaluate the nonlinearity in the case where the modulated signal is not available, especially in the RF integrated circuit design phase.

A.1 Second-Order Distortion

The second-order distortion is generated by the second term in (A.1). The second-order distortion products at the output of a device are derived by substituting a two-tone signal in (A.2) into the second term in (A.1) as

$$\begin{aligned} g_2x^2(t) = g_2A^2\big[1 + \cos(\omega_2 - \omega_1)t + \cos(\omega_2 + \omega_1)t \\ + \frac{1}{2}\cos(2\omega_1 t) + \frac{1}{2}\cos(2\omega_2 t)\big] \end{aligned} \tag{A.10}$$

The resultant IM2 products include first three distortions at the frequencies of $(\omega_1 + \omega_2)$ and $(\omega_2 - \omega_1)$ and a DC in (A.10). After being referred to device impedance R, the total power in the IM2 products is calculated as

$$P_{\mathrm{IM2,OUT}} = |g_2|^2A^4\left(\frac{1}{R} + \frac{1}{2R} + \frac{1}{2R}\right) = \frac{2|g_2|^2A^4}{R} \tag{A.11}$$

From (A.2), the total two-tone power at the device input is equal to A^2/R. The output power in the fundamental components is equal to the total input two-tone power of A^2/R multiplied by the power gain of $|g_1|^2$, or

$$P_{\mathrm{FD,OUT}} = \frac{|g_1|^2A^2}{R} \tag{A.12}$$

Based on the definition of the output second-order intercept point OIP2, at the second-order intercept point the total output signal power in the fundamental components (A.12) is equal to the total power in the IM2 products in (A.11), or

$$\frac{|g_1|^2 A_{\mathrm{iip2}}^2}{R} = \frac{2|g_2|^2 A_{\mathrm{iip2}}^4}{R} \tag{A.13}$$

In (A.13), the amplitude A is replaced with A_{iip2} at the second-order intercept point. Thus, the input amplitude at the input IP2 is

$$A_{\mathrm{iip2}} = \frac{1}{\sqrt{2}} \bullet \left|\frac{g_1}{g_2}\right| \tag{A.14}$$

The IIP2 that presents the power of the second-order intercept point is obtained by letting $A_{\mathrm{iip2}} = A$ in the input two-tone power A^2/R and using (A.14), or

$$\mathrm{IIP2} = \frac{A_{\mathrm{iip2}}^2}{R} = \frac{1}{2R} \bullet \left|\frac{g_1}{g_2}\right|^2 \tag{A.15}$$

It can be seen from (A.15) that the IIP2 is independent of the amplitude of A, but dependent on the ratio of g_1 to g_2. The smaller the value of g_2, the larger the parameter of IIP2. The output power in the IM2 products (A.11) can be also written as the input power by dividing the power gain $|g_1|^2$:

$$P_{\mathrm{IM2,IN}} = \frac{P_{\mathrm{IM2,OUT}}}{|g_1|^2} = 2\left|\frac{g_2}{g_1}\right|^2 \bullet \frac{A^4}{R} \tag{A.16}$$

Using (A.15) and the total two-tone input power $P_{\mathrm{2T,IN}} = A^2/R$, we can rewrite (A.16) as [2]

$$P_{\mathrm{IM2,IN}} = \frac{P_{\mathrm{2T,IN}}^2}{\mathrm{IIP2}} \tag{A.17}$$

Thus, IIP2 is expressed in dBm as

$$\mathrm{IIP2(dBm)} = 2P_{\mathrm{2T,IN}}(\mathrm{dBm}) - P_{\mathrm{IM2,IN}}(\mathrm{dBm}) \tag{A.18}$$

and the output IP2 (OIP2) is given by

$$\mathrm{OIP2(dBm)} = \mathrm{IIP2(dBm)} + |g_1|^2(\mathrm{dB}) \tag{A.19}$$

In (A.18), the second-order intermodulation product power $P_{\mathrm{IM2,IN}}$ represents the total input power at DC, $\omega_2 - \omega_1$, and $\omega_1 + \omega_2$, and $P_{\mathrm{2T,IN}}$ stands for the total input two-tone power or $P_{\mathrm{2T,IN}}(\mathrm{dBm}) = P_{\mathrm{1T}}(\mathrm{dBm}) + 3(\mathrm{dB})$.

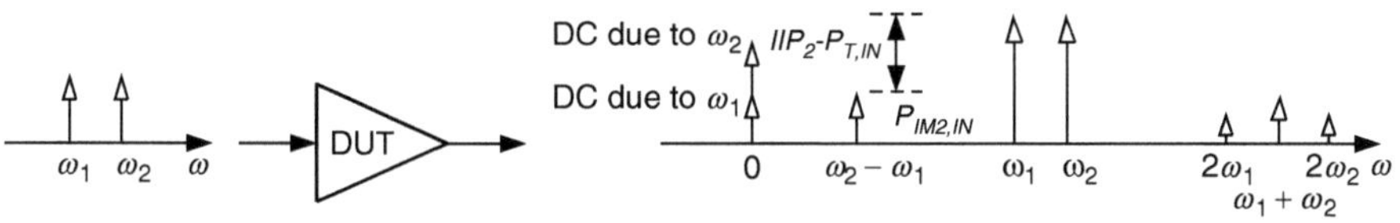

Fig. A.1 Second-order intermodulation products generated from two input tones

It can be noted from (A.11) that the total power of the IM2 products is distributed with 50% (−3 dB) power of the IM2 product at DC, 25% (−6 dB) power of the IM2 product at $\omega_2 - \omega_1$, and 25% (−6 dB) power of the IM2 product power at $\omega_1 + \omega_2$, respectively. The IM2 product at $\omega_2 - \omega_1$ can fall into the signal band to distort the desired signal if the frequency difference $|\Delta\omega| = |\omega_2 - \omega_1|$ is less than the bandwidth of the baseband signal. Thus, IIP2 in (A.18) can be expressed with the IM2 product at $\omega_2 - \omega_1$ by

$$\mathrm{IIP2(dBm)} = 2P_{2\mathrm{T}}(\mathrm{dBm}) - \left[P_{\mathrm{IM2},(\omega_2-\omega_1)}(\mathrm{dBm}) + 6(\mathrm{dB})\right] \tag{A.20}$$

When one-tone power replaces two-tone power ($P_{2\mathrm{T,IN}} = 2P_{1\mathrm{T}} + 3$ dB), IIP2 above can be rewritten as

$$\mathrm{IIP2(dBm)} = 2P_{1\mathrm{T}}(\mathrm{dBm}) - P_{\mathrm{IM2},(\omega_2-\omega_1)}(\mathrm{dBm}) \tag{A.21}$$

Compared with (A.18), the IIP2 calculation in (A.21) has a similar format, except for using one-tone power instead of two-tone power and using the power of the IM2 product at $\omega_2 - \omega_1$ instead of the total power of the IM2 products. The relationship among these parameters in (A.21) is illustrated in Fig. A.1.

Similar to the IIP2 calculation related to the IM2 product at $\omega_2 - \omega_1$, the IIP2 calculation associated with the IM2 product at DC is given by

$$\mathrm{IIP2(dBm)} = 2P_{2\mathrm{T}}(\mathrm{dBm}) - \left[P_{\mathrm{IM2,DC}}(\mathrm{dBm}) + 3(\mathrm{dB})\right] \tag{A.22}$$

With one-tone power expression rather than two-tone power, IIP2 above is rewritten as

$$\mathrm{IIP2(dBm)} = 2P_{1\mathrm{T}}(\mathrm{dBm}) - P_{\mathrm{IM2,DC}}(\mathrm{dBm}) + 3(\mathrm{dB}) \tag{A.23}$$

A.2 *Third-Order Distortion*

A third-order intercept point is a measure for the third-order distortion of weakly nonlinear systems and devices and is mainly used to characterize the third-order distortion. The third-order intercept point is measured by applying two tones or sinusoidal signals with equally small amplitudes and close frequency offsets, say,

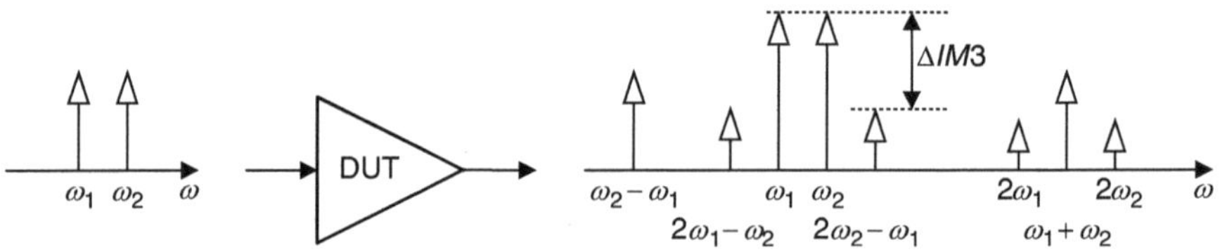

Fig. A.2 Output spectrum at the DUT output for two-tone inputs

ω_1 and ω_2, to a PA or device under test (DUT), and testing the fundamental signal output and third-order intermodulation (IM3) product outputs as a function of the input power as a logarithmic scale plot, as shown in Fig. A.2, where power is expressed in dBm. The fundamental component rises with a slope of gain G_1 in dB (20 dB/decade in Fig. A.2), and the IM3 products at the frequencies of $2\omega_1 - \omega_2$ and $2\omega_2 - \omega_1$ rise with a slope of gain $3G_1$ in dB (60 dB/decade). In other words, the output power of the IM3 products grows at a rate of three times that at which the fundamental increases. Theoretically, as the input signal power increases, these two lines would intersect. The intercept point is called *the third-order intercept point* (IP3). The corresponding input power at this point is called *the input third-order intercept point* (IIP3), and the corresponding output power is called *the output third-order intercept point* (OIP3). The larger the OIP3, the better the large signal capability of the PA.

To make the measurement correct, the input signal with the amplitude A must be small enough, or $A \ll 1$ so that DUT operates in the linear range. As the input signal level or the amplitude A increases, the output amplitudes of the fundamental signal and the IM3 products also increase. By definition, at the IIP3 power level these two output power levels referred to the DUT impedance R are equal to each other from (A.5) and (A.9) if $g_1 \gg 9g_3A^3/4$ due to the small input signal [3]:

$$|g_1|^2 \frac{A_{\mathrm{IIP3}}^2}{2R} = \left(\frac{3}{4}\right)^2 |g_3|^2 \frac{A_{\mathrm{IIP3}}^6}{2R} \tag{A.24}$$

or

$$|g_1|^2 \mathrm{IIP3} = \left(\frac{3}{4}\right)^2 |g_3|^2 4R^2 \mathrm{IIP3}^2 \tag{A.25}$$

where the input IP3 power is $\mathrm{IIP3} = A_{\mathrm{IIP3}}^2/2R$. Hence, the IIP3 can be simplified as

$$\mathrm{IIP3} = \frac{2}{3R} \left|\frac{g_1}{g_3}\right| \tag{A.26}$$

The OIP3 can be obtained by $\mathrm{OIP3} = g_1^2 \mathrm{IIP3}$. The input amplitude can be derived from the equation $\mathrm{IIP3} = A_{\mathrm{IIP3}}^2/2R$, and is given as

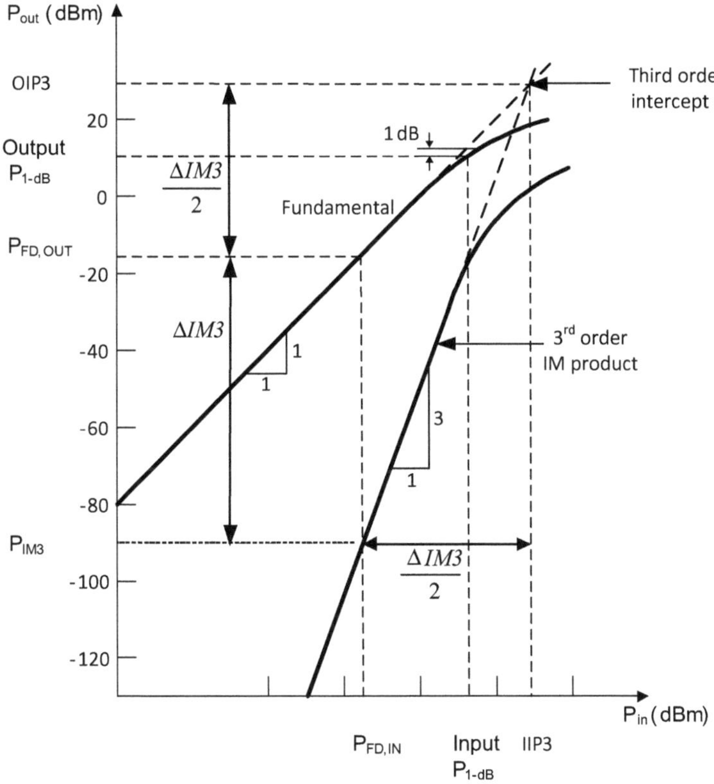

Fig. A.3 Output power of fundamental and IM3 versus input power

$$A_{\mathrm{IIP3}} = \sqrt{\frac{4}{3}\left|\frac{g_1}{g_3}\right|} \tag{A.27}$$

It is important to understand that in practice the IP3 values cannot be measured since the DUT will saturate before it reaches the IP3 (see Fig. A.3). Hence, it is common practice to extrapolate or calculate the IP3 values from a few data measured for small input-signal power levels, at least 10 dBm below P1dB. Such small input-signal levels are used to make sure that the DUT operates completely in a linear region.

Denote the output power of the fundamental signal at frequencies of ω_1 and ω_2 by $P_{\mathrm{FD,OUT}}$, the output power of the IM3 products at frequencies of $2\omega_1 - \omega_2$ and $2\omega_2 - \omega_1$ by $P_{\mathrm{IM3,OUT}}$, and the input power of the fundamental signal by $P_{\mathrm{FD,IN}}$. Then from (A.24), the ratio of the $P_{\mathrm{FD,OUT}}$ to the $P_{\mathrm{IM3,OUT}}$ is calculated as

$$\frac{P_{\text{FD,OUT}}}{P_{\text{IM3}}} = \frac{\dfrac{|g_1|^2 A_{\text{IIP3}}^2}{2R}}{\left(\frac{3}{4}\right)^2 |g_3|^2 \dfrac{A_{\text{IIP3}}^6}{2R}} = \left(\frac{2}{3R}\right)^2 \left|\frac{g_1}{g_3}\right|^2 \frac{1}{P_{\text{FD,IN}}^2} \tag{A.28}$$

where $P_{\text{FD,IN}} = A_{\text{IIP3}}^2/2R$. By substituting (A.26) into (A.28), we have

$$\frac{P_{\text{FD,OUT}}}{P_{\text{IM3}}} = \frac{\text{IIP3}^2}{P_{\text{FD,IN}}^2} \tag{A.29}$$

Equation (A.29) can be also expressed as

$$10\log P_{\text{FD,OUT}} - 10\log P_{\text{IM3}} = 20\log \text{IIP3} - 20\log P_{\text{FD,IN}} \tag{A.30}$$

Then, the IIP3 is calculated by

$$10\log \text{IIP3} = 10\log P_{\text{FD,IN}} + \frac{1}{2}(10\log P_{\text{FD,OUT}} - 10\log P_{\text{IM3}}) \tag{A.31}$$

or

$$\begin{aligned} \text{IIP3(dBm)} &= P_{\text{FD,IN}}(\text{dBm}) + \frac{1}{2}(P_{\text{FD,OUT}}(\text{dBm}) - P_{\text{IM3}}(\text{dBm})) \\ &= P_{\text{FD,IN}}(\text{dBm}) + \frac{1}{2}\Delta\text{IM3(dB)} \end{aligned} \tag{A.32}$$

and the OIP3(dBm) is equal to $10\log|g_1|^2 + \text{IIP3(dBm)}$.

Hence, the IIP3 is equal to the input power level of one of the two tones plus half the difference ΔIM3 between the output power level of the fundamental and the output power level of the IM3 products as illustrated in Fig. A.3.

In practice, it is more convenient to calculate the IIP3 from all input parameters, which are obtained by dividing $P_{\text{FD,OUT}}$ and P_{IM3} with power gain $|g_1|^2$. Thus, (A.32) can be also expressed as

$$\text{IIP3(dBm)} = P_{\text{FD,IN}}(\text{dBm}) + \frac{1}{2}(P_{\text{FD,IN}}(\text{dBm}) - P_{\text{IM3,IN}}(\text{dBm})) \tag{A.33}$$

From (A.32), we can plot lines of the fundamentals and the IM3 products from the following derivation:

$$\begin{aligned}\mathrm{OIP3(dBm)} &= G_1(\mathrm{dB}) + \mathrm{IIP3(dBm)} \\ &= G_1(\mathrm{dB}) + P_{\mathrm{FD,IN}}(\mathrm{dBm}) + \frac{\Delta\mathrm{IM3(dB)}}{2} \\ &= P_{\mathrm{FD,OUT}}(\mathrm{dBm}) + \frac{\Delta\mathrm{IM3(dB)}}{2}\end{aligned} \tag{A.34}$$

where $G_1 = 20\log10(g_1)$ is the power gain in dB for the fundamental. We can also express (A.34) as

$$\begin{aligned}\mathrm{OIP3(dBm)} &= P_{\mathrm{IM3}}(\mathrm{dBm}) + \Delta\mathrm{IM3(dB)} + \frac{\Delta\mathrm{IM3(dB)}}{2} \\ &= P_{\mathrm{IM3}}(\mathrm{dBm}) + \frac{3}{2}\Delta\mathrm{IM3(dB)}\end{aligned} \tag{A.35}$$

From (A.33) and (A.35), it is clear that the line of the fundamental has a slope of 1 because the IIP3 value is reached by increasing ΔIM3/2 from the input power point of $P_{\mathrm{FD,IN}}$ while the OIP3 value is reached by increasing the same amount of ΔIM3/2 from $P_{\mathrm{FD,OUT}}$. Compared (A.32) with (A.35), it is evident that the line of the IM3 has a slope of 3 because the IIP3 value is reached by increasing ΔIM3/2 from the input power point of $P_{\mathrm{FD,IN}}$ while the OIP3 value is reached by increasing three times amount of ΔIM3/2 (or 3ΔIM3/2) from P_{IM3}. Figure A.3 shows a geometric extrapolation of the fundamentals and the IM3 products.

In the measurements of the IP3 values, it is common practice to calculate them from a few data taken at least 10 dB below P1dB in order to ensure that a PA or a device operates in the linear range. One should check the slopes of the fundamentals and the IM3 products to verify that the data obey the expected slope of 1 for the former and the slope of 3 for the latter. Then, use (A.32) and (A.34) to calculate the IIP3 and OIP3, respectively.

A.3 P1dB Compression Point

In addition to the nonlinearity characteristic of the IP3 of a DUT, the P1dB compression point is another parameter to describe the nonlinearity property. The P1dB compression point is the output power level of the DUT, which is 1 dB less than an ideal linear value, as shown in Fig. A.3. It is called that the output power level is compressed by 1 dB at the P1dB compression point. The compression point corresponding to the input power axis is the input P1dB compression point, while the corresponding output power axis is the output P1dB compression point. Both of them are shown in Fig. A.3.

This parameter is very important for a power amplifier when it is used to amplify a modulated RF signal. This is because the average output power is usually determined by the P1dB compression point of the amplifier and the PAPR value of the modulated signal, especially for non-constant envelope RF-modulated signals. Unlike the indirect measurement of the IP3 through measuring the IM3 products, the measurement of the P1dB compression point requires only one tone rather than two tones, such as a real modulated signal.

For a one-tone test signal, the ratio of the actual output power to the ideal output power at the frequency of ω_1, which corresponds to a P1dB compression point, is

$$10\log\frac{P_{\mathrm{ACT}}}{P_{\mathrm{IDL}}} = 20\log\frac{v_{\mathrm{ACT-RMS}}}{v_{\mathrm{IDL-RMS}}} = -1\ \mathrm{dB} \tag{A.36}$$

or

$$\frac{v_{\mathrm{ACT-RMS}}}{v_{\mathrm{IDL-RMS}}} = 0.89125 \tag{A.37}$$

where $v_{\mathrm{ACT-RMS}}$ and $v_{\mathrm{IDL-RMS}}$ stands for the RMS values of the actual output voltage v_{ACT} and the ideal output voltage v_{IDL} at the frequency of ω_1, respectively, which are expressed as

$$v_{\mathrm{ACT}} = \left(g_1 A_{1\mathrm{dB}} + \frac{3}{4} g_3 A_{1\mathrm{dB}}^3\right)\cos(\omega_1 t) \tag{A.38}$$

and

$$v_{\mathrm{IDL}} = (g_1 A_{1\mathrm{dB}})\cos(\omega_1 t) \tag{A.39}$$

The RMS values of these two sinusoidal signals are $v_{\mathrm{ACT-RMS}} = |v_{\mathrm{ACT}}|/\sqrt{2}$, and $v_{\mathrm{IDL-RMS}} = |v_{\mathrm{IDL}}|/\sqrt{2}$. So, (A.37) can be rewritten as

$$\frac{g_1 + \frac{3}{4} g_3 A_{1\mathrm{dB}}^2}{g_1} = 0.89125 \tag{A.40}$$

Note that g_3 must be negative so that the numerator can be less than the denominator. Then, the amplitude of the input single tone is given by

$$A_{1\mathrm{dB}} = \sqrt{0.145\left|\frac{g_1}{g_3}\right|} \tag{A.41}$$

From (A.41), the input $P1$dB referred to the input impedance R is calculated by

$$P1\mathrm{dB} = \frac{0.145}{2R}\left|\frac{g_1}{g_3}\right| \tag{A.42}$$

Relationship between *P*1dB and IP3: Even though the $P1$dB is measured with a single tone test while the IP3 is measured with the two-tone test, their relationship is derived by dividing (A.26) by (A.42)

$$\frac{\mathrm{IIP3}}{P1\mathrm{dB}} = \frac{\frac{2}{3R}\left|\frac{g_1}{g_3}\right|}{\frac{0.145}{2R}\left|\frac{g_1}{g_3}\right|} = 9.2 \tag{A.43}$$

or

$$\mathrm{IIP3(dBm)} \approx P1\mathrm{dB(dBm)} + 9.6\mathrm{(dB)} \tag{A.44}$$

Equation (A.44) shows IIP3 is larger than the input $P1$dB by 9.6 dB for a single-tone signal. This relationship is also applied to OIP3 and the output $P1\mathrm{dB}_{\mathrm{out}}$, or

$$\mathrm{OIP3(dBm)} \approx P1\mathrm{dB_{out}(dBm)} + 9.6\mathrm{(dB)} \tag{A.45}$$

It has also been reported that the IP3 power is about 14.4 dB above the $P1$dB compression point in the case where the two tones are applied. In reality, the IP3 power is within a range from 10 to 15 dB higher than the $P1$dB compression power.

Appendix B: Transmit Modulation Accuracy

The quality of the RF modulated signal can be measured by the error vector magnitude (EVM). Using phasors in the I-Q plane, EVM is calculated in the digital baseband domain by comparing the vector difference between the actual signal vector and the reference signal vector. The concept of the EVM calculation for a 16-QAM signal is graphically illustrated in Fig. B.1.

In the 802.11a system, a root-mean-square (RMS) EVM is defined as [23]:

$$\mathrm{EVM_{RMS}}(\%) = 100 \times \frac{\sum_{i=1}^{N_{\mathrm{F}}} \sqrt{\frac{\sum_{k=1}^{N_{\mathrm{P}}}\left[\sum_{l=1}^{52}\left\{[\hat{x}_{\mathrm{i}}(i,k,l) - x_{\mathrm{i}}(i,k,l)]^2 + \left[\hat{x}_{\mathrm{q}}(i,k,l)\right] - x_{\mathrm{q}}(i,k,l)\right]^2\right\}\right]}{52 \times N_{\mathrm{P}} \times P_{\mathrm{O}}}}}{N_{\mathrm{F}}} \tag{B.1}$$

where the definition of each parameter is as follow:

The number 52 represents the total 52 subcarriers;
N_{P} is the length of the packet;
N_{F} is the number of frames used for the measurement;
$\hat{x}_{\mathrm{i}}(i,k,l)$ and $\hat{x}_{\mathrm{q}}(i,k,l)$ denote the observed I-Q points of the i-th frame, k-th OFDM symbol of the i-th frame, l-th subcarrier of the OFDM symbol in the frequency domain, respectively;

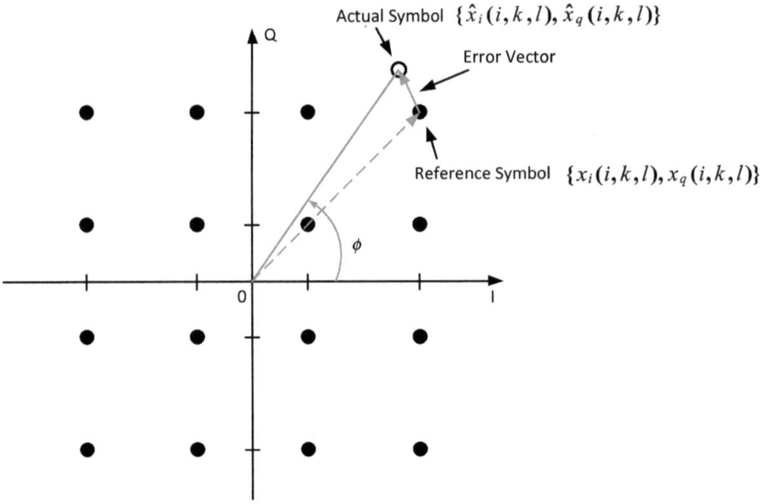

Fig. B.1 Error vector definition for the 16-QAM constellation

$x_{\mathrm{i}}(i,k,l)$ and $x_{\mathrm{q}}(i,k,l)$ represent the reference I-Q points of the i-th frame, k-th OFDM symbol of the i-th frame, l-th subcarrier of the OFDM symbol in the frequency domain, respectively;

P_0 is the average power of the constellation.

It is suggested that the EVM test be measured over at least 20 frames, the RMS average be taken, and the packets under measurement be at least 16 OFDM symbol long.

In some other applications such as WCDMA and TD-SCDMA systems, EVM calculations are simply defined as a RMS average normalized to the reference signal, or

$$\mathrm{EVM_{RMS}}(\%) = 100 \times \sqrt{\frac{\sum_{k=1}^{L} \left\{ [\hat{x}_{\mathrm{i}}(kT_{\mathrm{s}}) - x_{\mathrm{i}}(kT_{\mathrm{s}})]^2 + \left[\hat{x}_{\mathrm{q}}(kT_{\mathrm{s}})\right] - x_{\mathrm{q}}(kT_{\mathrm{s}})\right]^2 \right\}}{\sum_{k=1}^{L} \left[x_{\mathrm{i}}^2(kT_{\mathrm{s}}) + x_{\mathrm{q}}^2(kT_{\mathrm{s}}) \right]}} \quad \text{(B.2)}$$

where L is the length of symbols for the measurement and the actual and reference signals are sampled once per symbol at the maximum eye opening instant.

Because EVM is calculated at the baseband domain, the transmitted RF signal should be down-converted to the baseband signal before EVM calculation. Therefore, the transmitted RF signal is first down-converted to the baseband signal vector (or the I-Q signals) in the baseband domain, and then the error vector magnitude

(EVM) is measured by comparing the vector difference between the actual signal vector and the reference signal vector. To make sure that the EVM measurement value accurately embodies all impairments at the transmitter, a receiver with high performance is required to perform the frequency down-conversion, analog-to-digital conversion, and digital demodulation including carrier phase and data symbol clock synchronizations.

Most Vector Signal Analyzers can perform the EVM measurement. The basic process after the analog-to-digital conversion (ADC) of the frequency down-converted baseband signal is illustrated in Fig. B.2. In practice, EVM measurement includes three major steps in the digital domain as follows:

1. **Coherent demodulation**
   After ADCs on the I-Q channels, the carrier phase and symbol timing synchronizations are performed from the digitally sampled complex signal and then data symbol sequences are coherently recovered to the original symbol sequences at a corresponding symbol rate. The recovered symbol sequences go to the I-Q waveform regeneration block to create the reference signal vector while the synchronized I-Q signals pass through a delay block as the actual signal vector.
2. **Reference signal vector regeneration**
   One of the outputs of the demodulator is the recovered symbol sequences that are used to generate the original baseband waveforms in the I-Q waveform regeneration block as the reference signal vector. Pulse-shaping process may be included in the I-Q waveform regeneration block, depending on the original modulation signal. The accuracy of the reference signal vector can be achieved by digital signal processors with high resolution and high speed. Another one of the outputs of the demodulator is the synchronized I-Q waveforms that are used as the actual signal vector. The actual signal vector is delayed to compensate for the latency caused by the I-Q waveform regeneration block. The reference signal vector needs to be aligned with the actual signal vector by performing the cross-correlation between them and then setting an appropriate delay value in the delay block.
3. **Error vector calculation**
   The error vector can be calculated by comparing the reference signal vector and the actual signal vector. For the amplitude and phase modulation formats, such as M-QAM and QPSK signals, the error vector is calculated by comparing the sampled values of the reference signal vector and the sampled values of the actual signal vector, where both vectors are sampled once per symbol at the maximum eye opening instants.

It should be noted that the phase error rather than EVM is used to evaluate the quality of the transmitted GMSK signal. Phase error is the instantaneous angle difference between the actual signal and the reference signal and is calculated at every sample per symbol rather than one sample per symbol. The instantaneous angle is calculated through the relationship between the complex number written in Cartesian coordinates and polar coordinates.

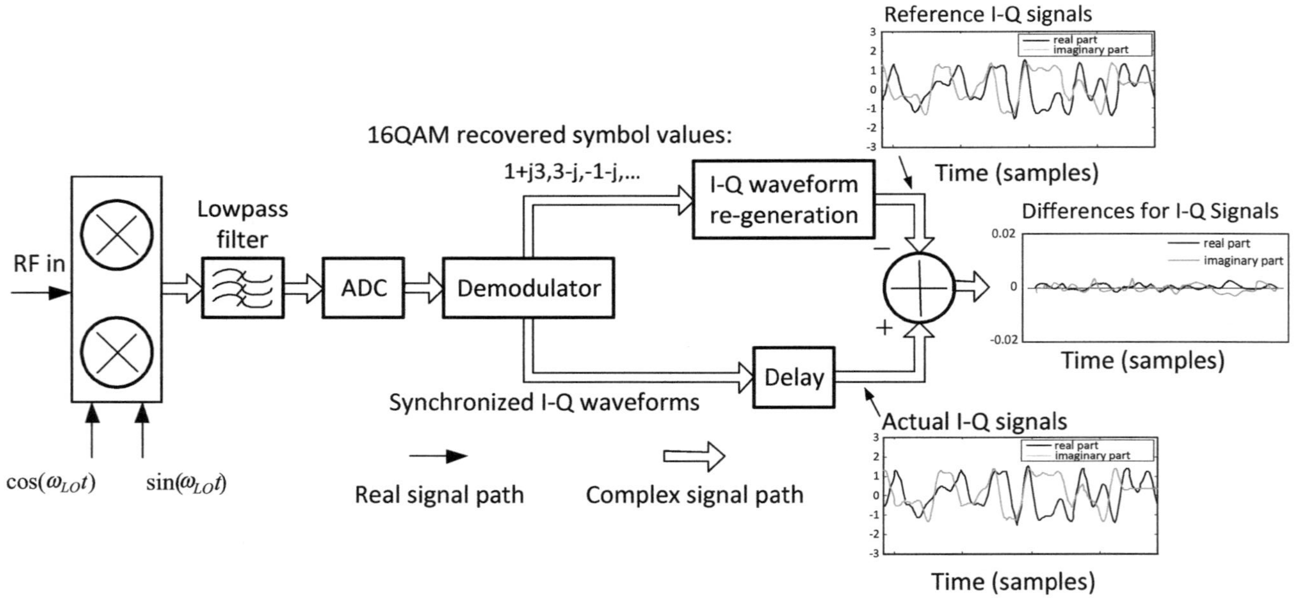

Fig. B.2 A general block diagram of EVM measurement process

Considering that eye diagrams of the actual signal vector are usually quite open due to a high SNR except full saturation of the power amplifier at the transmitter, the reference signal vector can be replaced with the decision signal vector that is generated from the actual signal vector through a decision algorithm. Thus, the I-Q waveform regeneration block can be omitted. But this replacement can be only applied to the EVM measurement where only one sampled value per symbol is used.

Impairment Diagnosis via EVM or Signal Constellation: An EVM value or an actual signal constellation indicates not only the quality of the transmitted signal, but also indirectly diagnoses what type of impairment sources may cause the performance degradation of the transmitted signal. As previously described in Sect. 3.4.1.2, there are major four types of impairments that degrade the EVM at the transmitter; I-Q imbalance, LO phase noise, nonlinear distortion, and crest factor reduction. The first type of impairment can be minimized through the calibration. The second type of impairment, or LO phase noise, cannot be minimized by using the calibration, but can be diagnosed from banana shapes of the actual signal constellation diagram when LO phase noise is relatively poor. Similar to thermal noise, the third type of impairment, or nonlinear distortion, results in random scatter of the constellation dots and may be identified by reducing the input signal power to a concern amplification block such as a power amplifier on the transmit path such that it completely operates in a linear region. If random scatter is reduced after the input signal power to this amplification block decreases, it can be concluded that such impairment that degrades EVM is related to nonlinear distortion. Otherwise, this type of impairment could be related to something else rather than nonlinear distortion. The fourth type of impairment hardly occurs because of rarely being adopted, especially for high-order QAM signals.

Receiver Optimization via Minimizing EVM: Even though EVM specification is not required for the receiver by most wireless communication standards, the EVM measurement at the receiver, however, provides a low cost and an effective approach to evaluating the performance of the receiver and also possibly diagnosing what types of signal impairments may degrade the performance of the receiver. Receiver optimization via minimizing EVM avoids a high cost and complicated bit error rate (BER) or packet error rate (PER) test. Most importantly, with a capability of possibly diagnosing the type of signal impairment, EVM measurement can help RFIC designers in the troubleshooting of signal impairments along a receiver chain. Similar to major four types of impairments at the transmitter, there are also major four types of impairments at the receiver. These impairments are the I-Q imbalance, DC offset, LO phase noise, and variations of the amplitude and group delay of the analog lowpass filter. The first two impairments can be minimized by the calibration while the third impairment can be diagnosed by means of the same approach as one used at the transmitter. The last impairment can be diagnosed by turning on or turning off an equalization function either in a measurement instrument if it has such an option or in a self-developed test program. For the latter case, the interested reader is referred to [24] in detail.

Appendix C: FQPSK Modulation Family

C.1 History of FQPSK Development

Feher-patented Quadrature Phase Shit Keying (FQPSK) family has been developed for more than 30 years. Its applications have ranged from satellite earth station digital communication systems to the latest telemetry systems due to its properties of energy- and bandwidth-efficient transmission. The FQPSK family has experienced four important influences or developments in the past, and they can be distinguished as different "periods". The first period of FQPSK, originally called intersymbol interference- and jitter-free OQPSK (IJF-OQPSK) [4, 5], was began in 1982 and was proposed to replace QPSK/OQPSK and MSK modulations for low-cost power and bandwidth-efficient satellite earth stations, where the transmission channels exhibited nonlinear characteristics or fully saturated amplifications, due to its small envelope fluctuation of 3 dB when compared with other modulation formats. In order to further reduce the 3-dB envelope fluctuation of IJF-OQPSK, a superposed QAM (SQAM) modulation technique was introduced in 1983, which can be considered the next period of FQPSK [6, 7]. The maximum envelope fluctuation of the SQAM modulated signal was reduced from 3 ($A = 1$ for IJF-OQPSK) to 0.7 dB ($A = 0.7$). Hence, the SQAM-modulated signal shows further improvements over the IJF-OQPSK signal in energy and spectral efficiency and BER performance in a nonlinear channel. Due to such improvements, SQAM gained potential perspective in the applications of satellite earth stations. At almost the same time, a cross-correlated PSK modulation technique, called XPSK (the third period), was proposed in 1983 [8] by adding cross-correlation between the I–Q channels to obtain a nearly constant envelope. The spectral efficiency and BER performance of the XPSK signal in a nonlinear channel was almost the same as that in a linear channel due to its nearly constant envelope. The contribution of XPSK to FQPSK family was to lay a solid foundation for the birth of the fourth period of FQPSK. In 1996, it was discovered that the side-lobes of XPSK's PSD could have a fast roll-off with frequency even through a nonlinear channel after passing through Butterworth lowpass filters (with a proprietary parameters) in the I and Q branches. After that, the filtered XPSK modulation, invented by Dr. Kamilo Feher, was known as FQPSK-B.

FQPSK mainly embodies pulse-shaping to achieve compact spectrum and cross-correlation between the I and Q channels to significantly reduce envelope fluctuation of the modulated signal and to achieve high energy and spectral efficiency through nonlinear power amplifiers. FQPSK has been demonstrated and confirmed through extensive studies done by the US Department of Defense (DoD), National Aeronautics and Space Administration (NASA), and the International Consultative Committee for Space Data Systems (CCSDS) to be the most power- and spectral-efficient systems with robust BER performance when nonlinearly amplified. In 2000, FQPSK was adopted as a standard in the Aeronautical Telemetry Standard IRIG 106 [9].

Since a nonlinear amplifier is more RF-energy efficient and has a longer battery duration, a lower cost, and smaller form factor, it is highly desirable for applications that require high transmit-energy efficiency, such as satellite and cellular systems. In the following sections, relatively detailed descriptions of IJF-OQPSK, SQAM, XPSK and FQPSK-B modulations in the FQPSK family are presented.

C.2 IJF-OQPSK Modulation

A block diagram of a IJF-OQPSK modulator is shown in Fig. 4.15, excluding the blocks of a cross-correlator and LPFs. After a serial-to-parallel (S/P) converter, the input bit non-return-to-zero (NRZ) data with the bit interval T_b are converted into the I and Q NRZ symbol data $x_I(t)$ and $x_Q(t)$ with the symbol interval of $T_s = 2T_b$, which are expressed as

$$x_I(t) = \sum_{n=-\infty}^{+\infty} d_{In} g(t - nT_s) \tag{C.1}$$

$$x_Q(t) = \sum_{n=-\infty}^{+\infty} d_{Qn} g(t - nT_s) \tag{C.2}$$

where the pulse shaping is rectangular, or

$$g(t - nT_s) = \begin{cases} 1, & |t - nT_s| \le T_s/2 \\ 0, & |t - nT_s| > T_s/2 \end{cases} \tag{C.3}$$

and

$$d_{In} = \pm 1, \quad \text{with probability of } 1/2 \text{ for each}$$
$$d_{Qn} = \pm 1, \quad \text{with probability of } 1/2 \text{ for each}$$

The I-channel data $x_I(t)$ and the half-symbol interval delayed Q-channel data $x_Q(t - T_s/2)$ are then encoded into IJF baseband signals $b_I(t)$ and $b_Q(t)$, respectively,

$$b_I(t) = \sum_{n=-\infty}^{+\infty} b_{In}(t) \tag{C.4}$$

where

$$b_{\mathrm{I}n}(t) = \begin{cases} s_1(t - nT_s) = s_e(t - nT_s), & \text{if } d_{\mathrm{I},n-1} = d_{\mathrm{I},n} = 1 \\ s_2(t - nT_s) = -s_e(t - nT_s), & \text{if } d_{\mathrm{I},n-1} = d_{I,n} = -1 \\ s_3(t - nT_s) = s_o(t - nT_s), & \text{if } d_{\mathrm{I},n-1} = -1, d_{\mathrm{I},n} = 1 \\ s_4(t - nT_s) = -s_o(t - nT_s), & \text{if } d_{\mathrm{I},n-1} = 1, d_{\mathrm{I},n} = -1 \end{cases} \quad \text{(C.5)}$$

and the odd and even waveforms, $s_o(t)$ and $s_e(t)$, meet

$$\begin{aligned} s_o(t - nT_s) &= -s_o(-t + nT_s), && \text{for } |t - nT_s| < T_s/2 \\ s_e(t - nT_s) &= s_e(-t + nT_s), && \text{for } |t - nT_s| < T_s/2 \\ s_o(t - nT_s) &= s_e(t - nT_s), && \text{for } |t - nT_s| \geq T_s/2 \end{aligned} \quad \text{(C.6)}$$

and are defined by

$$\begin{aligned} s_o(t - nT_s) &= \sin\frac{\pi t}{T_s}, && \text{for } |t - nT_s| < T_s/2 \\ s_e(t - nT_s) &= 1, && \text{for } |t - nT_s| < T_s/2 \end{aligned} \quad \text{(C.7)}$$

These two fundamental waveforms are shown in Fig. C.1. The Q-channel waveform segment $b_{Qn}(t)$ can be generated by the same mapping as $b_{\mathrm{I}n}$(t) in (C.5), which is delayed by a half-symbol relative to $b_{\mathrm{I}n}(t)$. For random symbol sequences at the input of the encoder, the baseband waveforms at the output of the encoder are illustrated in Fig. 4.17.

The eye diagram of the IJF-OQPSK is the same as that for the SQORC [10], as shown in Fig. 2.13b. From Fig. 2.13b, it can be seen that there is no Intersymbol Interference (ISI) at the decision instants and no jitter at the jitter instants (or cross-zero points), also known as *Jitter-Free (JF)*. ISI causes system performance degradation, while jitter causes symbol timing jittering, and both of them can result in the system performance degradation. The constellation of IJF-OQPSK is the same as the one for SQORC, as shown in Fig. 2.13c, where the maximum envelope fluctuation is $20 \times \log10(\sqrt{2}/1) = 3\,\mathrm{dB}$.

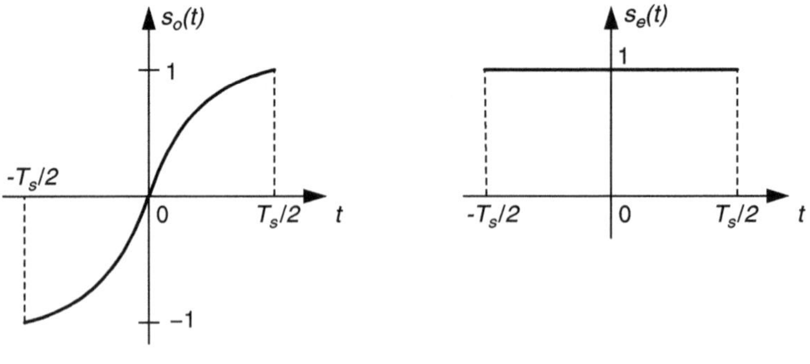

Fig. C.1 Odd and even waveforms of $s_o(t)$ and $s_e(t)$

A simply implementation of the IJF-OQPSK baseband signals based on switch-selecting scheme [11] is shown in Fig. 2.16. Four baseband waveforms expressed in (C.5) are generated from two basic waveforms in (C.7) and their inverse waveforms are then individually switched on as the output signals by the input logic combinations as defined in (C.5). Even though IJF-OQPSK has the same baseband signal shape as SQORC, the main difference between them is their different implementation. In Fig. 2.12, the baseband signal of SQORC within one symbol duration is generated by overlapping two raised cosine pulse waveforms with one symbol interval difference in time and with the same polarity as two consecutive NRZ input symbol bits, each with two symbol intervals of $2T_s$, while that of IJF-OQPSK is created based on switch-selecting scheme, as shown in Fig. 2.16.

C.3 SQAM Modulation

SQAM modulation was developed based on IJF-OQPSK modulation for the purpose of further reducing the maximum envelope fluctuation by 3 dB of the IJF-OQPSK. The maximum envelope amplitude happens when consecutive symbols with the same polarity in either the I channel or Q channel occur. Therefore, the key point is to reduce the overlapped amplitude of the baseband signals at the center of two consecutive symbols on the I channel or Q channel when two consecutive symbols have the same polarity.

To form a SQAM pulse waveform, two raised-cosine pulses, each having a symbol duration of T_s and adjustable amplitude parameter of A, are superposed to the original raised-cosine pulse with double symbol interval of $2T_s$. The quadrature modulation based on this superposed pulse is called the superposed quadrature modulation (SQAM) [7] and its pulse waveform is given by

$$s(t) = g(t) + d(t) \tag{C.8}$$

where

$$g(t) = \frac{1}{2}\left(1 + \cos\frac{\pi}{T_s}(t - T_s)\right) \tag{C.9}$$

$$d(t) = -\frac{1-A}{2}\left(1 - \cos\frac{2\pi t}{T_s}\right), \quad 0.5 \le A \le 1.0, \ \ 0 \le t \le 2T_s \tag{C.10}$$

In (C.10) A is an adjustable amplitude parameter. Note that the parameter A for SQAM signal has different meaning from the parameter A for XPSK signal. Figure C.2 illustrates the SQAM pulse-shaping process by adding two raised-cosine pulses $d(t)$ each with the period of T_s to one raised-cosine pulse $g(t)$with the period of $2T_s$.

Like the odd and even waveforms of IJF-OQPSK/SQORC, as shown in Fig. C.1, both odd and even waveforms of SQAM can be obtained by overlapping the

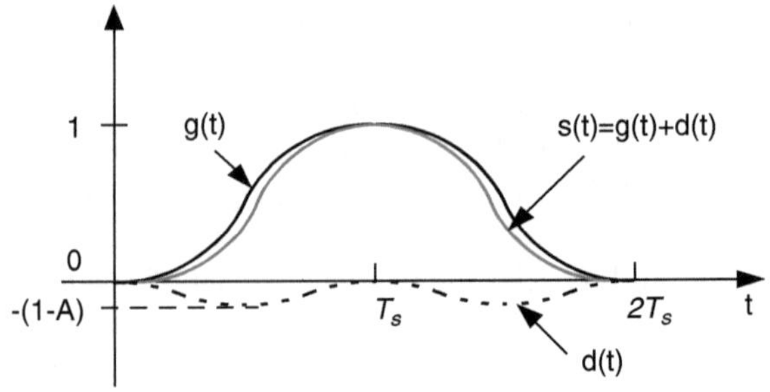

Fig. C.2 SQAM pulse shaping by superposing two raised-cosine pulses with symbol interval of T_s to one with $2T_s$

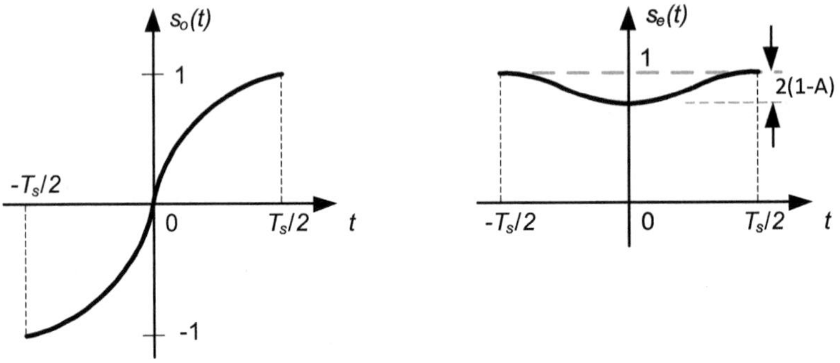

Fig. C.3 Odd and even waveforms of $s_o(t)$ and $s_e(t)$

double-interval pulses of $s(t - nT_s)$ and $s(t - (n + 1)T_s)$, as shown in Fig. C.3. Comparing Fig. C.1 with Fig. C.3, we can only see that the even segment is different between IJF-OQPSK and SQAM. The even segment with the valley at the center plays an important role in reducing the envelope fluctuation of the SQAM signal. The envelope fluctuations of the SQAM signal are controlled by the parameter A.

Figure C.4 shows computer-simulated constellations of the baseband SQAM signals at the transmitter for $A = 0.7$, 0.8, respectively. The envelope fluctuation is dependent on the parameter A. When A changes from 1 to 0.7, the envelope fluctuation is reduced from 3 to 0.7 dB [7]. Note that the SQAM at $A = 1$ becomes IJF-OQPSK. Therefore, IJF-OQPSK is a special case of the SQAM signal when A is equal to 1. Eye diagrams of the SQAM signal are shown in Fig. C.5. The PSD of the SQAM signal can be derived from (2.43) through the Fourier transform of the shaping pulse $s(t) = g(t) + d(t)$. The Fourier transform of $g(t)$ is given in (2.58), or

$$G(f) = \frac{\sin(2\pi f T_s)}{2\pi f\left(1 - 4f^2T_s^2\right)} e^{-j2\pi f T_s} \tag{C.11}$$

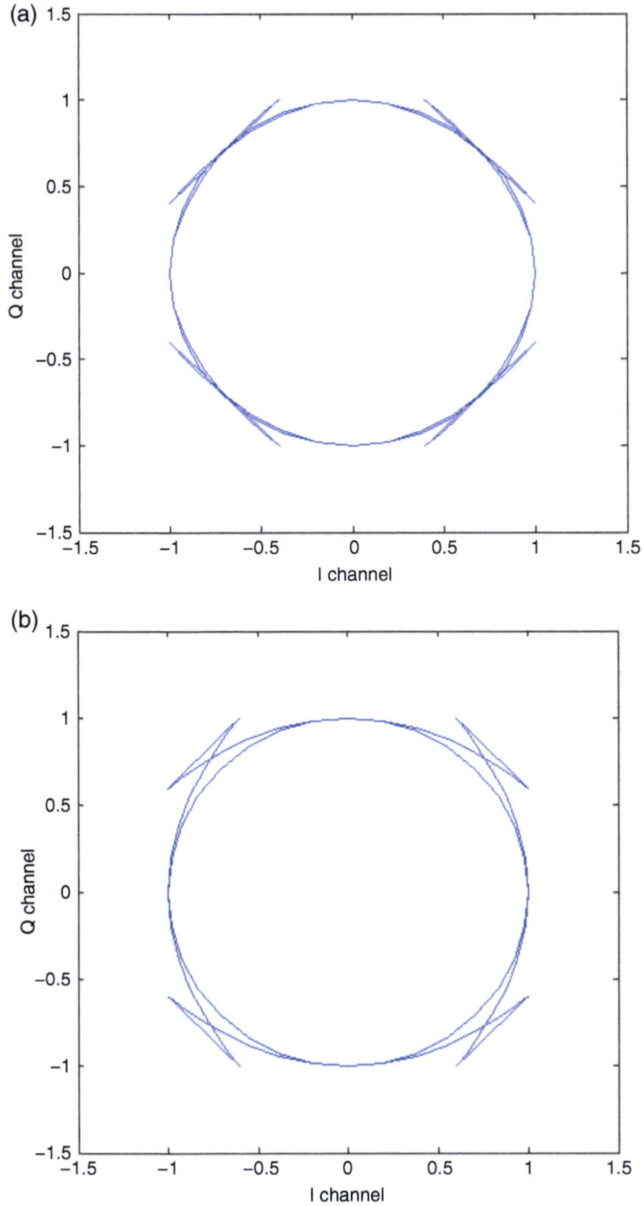

Fig. C.4 Constellations of the transmitted SQAM signal: (**a**) $A = 0.7$ and (**b**) $A = 0.8$

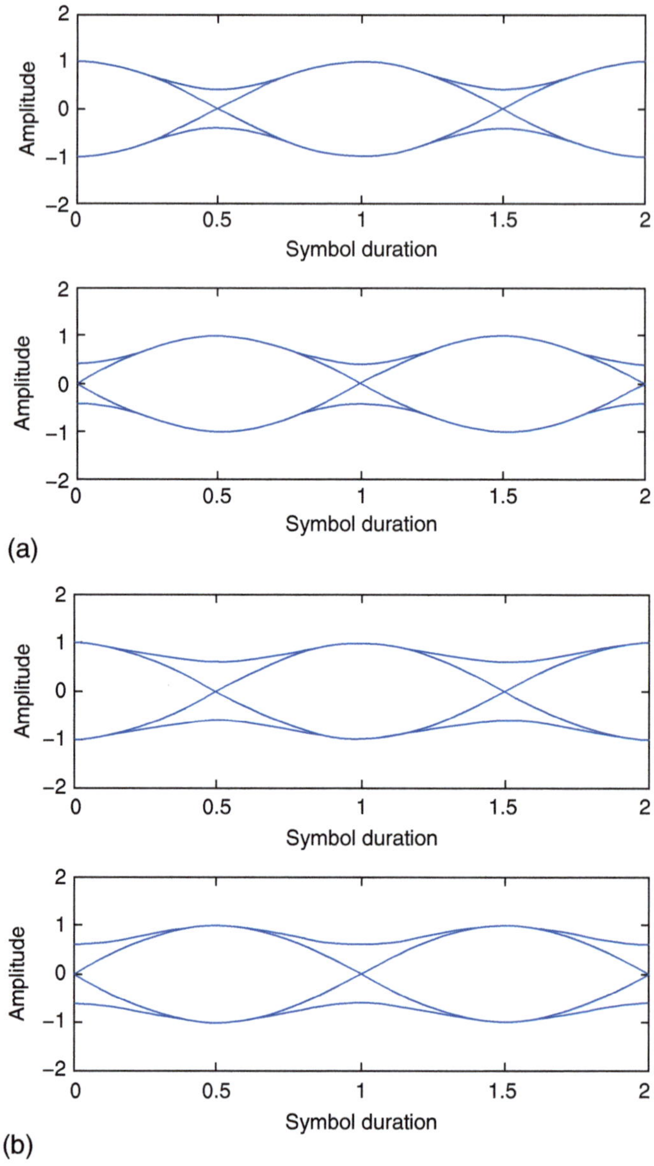

Fig. C.5 Eye diagrams of the transmitted SQAM signal: (**a**) $A = 0.7$ and (**b**) $A = 0.8$

Similar to the derivation of $G(f)$, the Fourier transform of $d(t)$ can be derived as

$$D(f) = \frac{(A-1)\sin\left(2\pi f T_{\mathrm{s}}\right)}{2\pi f\left(1 - f^2 T_{\mathrm{s}}^2\right)} \mathrm{e}^{-j2\pi f T_{\mathrm{s}}} \tag{C.12}$$

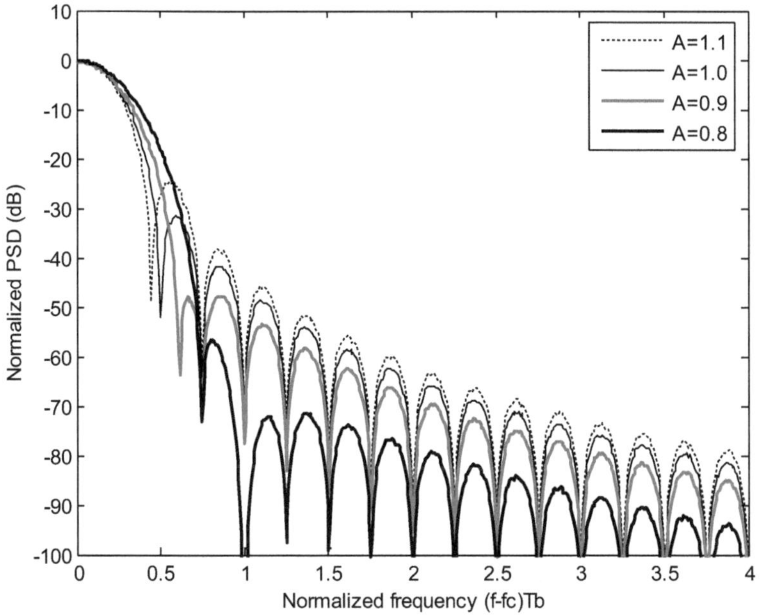

Fig. C.6 Power spectral density of SQAM with different parameters A in a linear channel. Here $T_b = T_s/2$ is the bit duration

From (C.11) and (C.12), the Fourier transform of $s(t)$ is expressed in the form

$$S(f) = T_s\left(\frac{1}{1-4f^2T_s^2} + \frac{A-1}{1-f^2T_s^2}\right)\frac{\sin(2\pi f T_s)}{2\pi f T_s}\,\mathrm{e}^{-j2\pi f T_s} \qquad \text{(C.13)}$$

By substituting $G(f)$ for $S(f)$ in (2.43), the normalized PSD of the SQAM is given as

$$\frac{\Psi_{\mathrm{SQAM}}(f)}{\Psi_{\mathrm{SQAM}}(0)} = \frac{1}{A^2}\left(\frac{1}{1-4f^2T_s^2} + \frac{A-1}{1-f^2T_s^2}\right)^2\left(\frac{\sin(2\pi f T_s)}{2\pi f T_s}\right)^2 \qquad \text{(C.14)}$$

When the parameter A is equal to 1, (C.14) is identical to (2.61), or the SQAM signal becomes SQORC/IJF-OQPSK signals. Figure C.6 shows the PSD curves of the SQAM signal with different parameters. Note that a decrease in the parameter A leads to faster side-lobe roll-off at the expense of a slightly wider main lobe. The side-lobes of the SQAM fall off at the rate of f^{-6}, which is the same as the SQORC/IJF-OQPSK signals. Figure C.7 shows the PSD of the SQAM with different A value in a nonlinear channel. One great advantage that the SQAM signal has is that its fast side-lobes roll off in a saturation (nonlinear) channel due to its small envelope fluctuation.

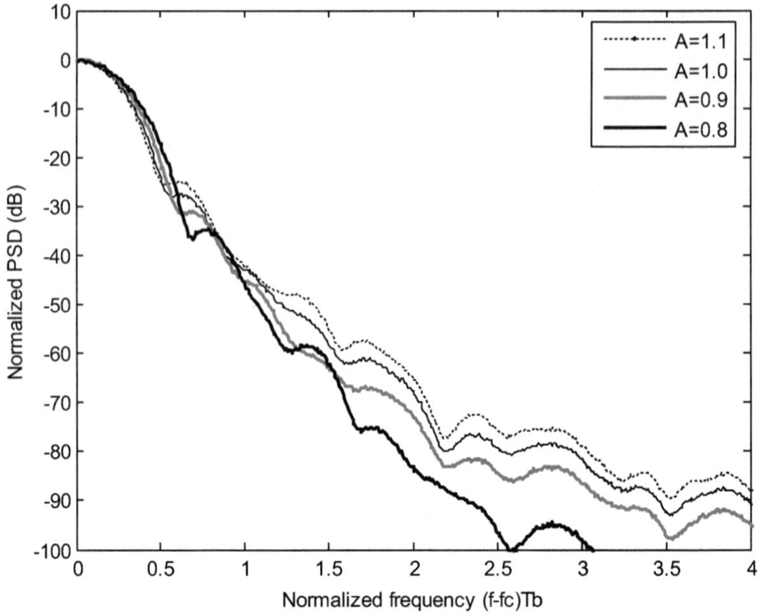

Fig. C.7 Power spectral density of SQAM signal in a nonlinear channel. Here $T_b = T_s/2$ is the bit duration

C.4 XPSK Modulation

From the previous section, we have seen that SQAM is still a non-constant modulation scheme even though SQAM baseband encoder has reduced the envelope fluctuation of the modulated SQAM signal compared with IJF-OQPSK. A nearly constant modulation XPSK [8] was proposed by introducing a cross-correlation operation performed on the pair of IJF encoder outputs at every half-symbol interval in order to achieve a constant envelope. Key points lie in that using different waveforms form constant envelopes as much as possible, except for alternative polarities on both I channel and Q channel at the same time in the half-symbol interval, in which the *Cosine* and *Sine* functions result in a constant envelope. The waveform of the I channel (or Q channel) in the half-interval is dependent on the Q channel (or I channel) in order to reduce the envelope fluctuation except for the case of alternative polarities in both the I channel and Q channel. As a result, a cross-correlator is proposed at the output of the IJF encoders, as shown in Fig. 4.15.

The basic idea of XPSK modulation is to reduce the envelope fluctuation by changing the peak amplitude of the baseband signal from 1 to A ($1/2 \leq A \leq 1$, note that A for XPSK has different meaning from A for SQAM) at the output of the IJF encoder except for the case of alternative polarities of the symbols on both the I channel and Q channel during any half-symbol interval. *Note that the parameter of A for XPSK signal is different from one for SQAM signal.* Besides the above waveforms with a peak amplitude of A, excluding the case of alternative polarities on both the I channel and Q channel, another four transition functions $f_1(t)$,$f_2(t)$, $f_3(t)$,$f_4(t)$ are defined in the interval $0 \leq t \leq T_s/2$ as [8]

$$\begin{aligned} f_1(t) &= 1 - (1 - A)\cos^2\left(\frac{\pi t}{T_s}\right) \\ f_2(t) &= 1 - (1 - A)\sin^2\left(\frac{\pi t}{T_s}\right) \\ f_3(t) &= -1 + (1 - A)\cos^2\left(\frac{\pi t}{T_s}\right) \\ f_4(t) &= -1 + (1 - A)\sin^2\left(\frac{\pi t}{T_s}\right) \end{aligned} \tag{C.15}$$

These transition functions are selected to have less envelope fluctuation than that of the IJF-OQPSK signal when they occur before $\cos(\pi t/T_s)$ or $-\cos(\pi t/T_s)$ and after $\cos(\pi t/T_s)$ or $-\cos(\pi t/T_s)$, as plotted in Fig. C.8. Note that the plots for $f_1(t)$ and $f_3(t)$ in Fig. C.8 are obtained from (C.15) by shifting $f_1(t)$ and $f_3(t)$ in the time domain by $-T_s/2$.

In the XPSK encoding scheme, the current output waveforms of the I–Q channels are determined by the present and immediately preceding symbols of

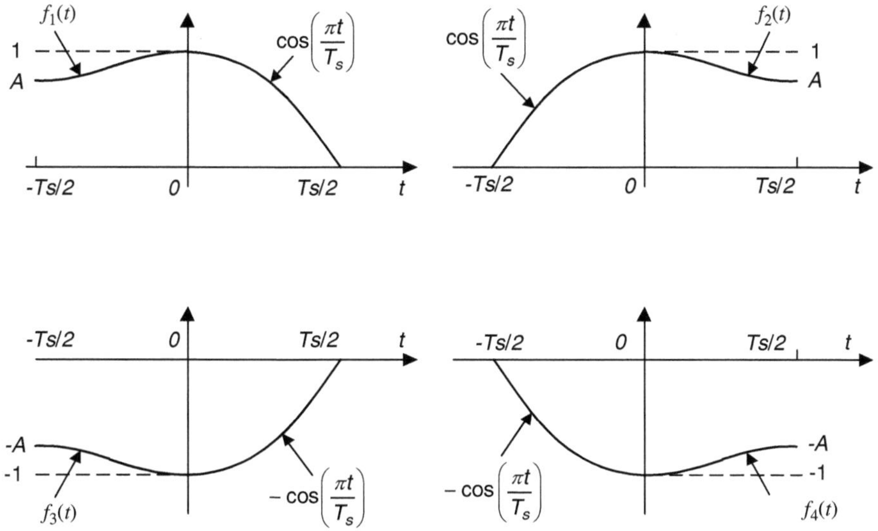

Fig. C.8 Examples of four transition waveforms $f_1(t)$,$f_2(t)$,$f_3(t)$,$f_4(t)$

Table C.1 Sixteen-pair waveforms of I- and Q-channel outputs

$c_{In}(t)$(or $c_{Qn}(t)$)	$c_{Qn}(t)$(or $c_{In}(t)$)	Number of combination
$\pm\cos\left(\frac{\pi t}{T_s}\right)$	$\pm\sin\left(\frac{\pi t}{T_s}\right)$	4
$\pm A\cos\left(\frac{\pi t}{T_s}\right)$	f_1 or f_3	4
$\pm A\sin\left(\frac{\pi t}{T_s}\right)$	f_2 or f_4	4
$\pm A$	$\pm A$	4

the respective I–Q channels. These two I-channel and two Q-channel symbols are serial-parallel converted from four input bit patterns. Therefore, the total 16 pairs of waveforms over the interval $0 \le t \le T_s/2$ are determined by four input bits of the serial-to-parallel input. These 16 different pair functions are listed in Table C.1, and 16 combinations of waveform are plotted in Fig. C.9.

In Fig. C.9, in a half-symbol interval $T_s/2$, all dashed-line curves represent the baseband waveform segments of the IJF-OQPSK, while all solid-line curves stand for the baseband waveform segments of the XPSK. Note that the first four combinations of 1, 2, 3, and 4 only have the solid-line waveforms because the two types of waveform segments overlap. It can be seen from the differences among each symbol pair, $C_{In}(t), C_{Qn}(t)$, that we know how to choose one pair of waveform segments from Table C.1 to generate the nearly constant envelope for the XPSK signal. Thus, every group of four input NRZ data at the input of the serial-to-parallel determines one corresponding pair of waveforms in a half-symbol interval. The envelope fluctuation of the XPSK signal is reduced to approximately 0 dB (nearly constant envelope) at $A = 1/\sqrt{2}$. Actually the envelope fluctuation is 0.18 dB at $A = 1/\sqrt{2}$ [8].

Instead of performing the XPSK waveform outputs at every half-symbol interval, a mapping performed directly on the input I- and Q-symbol sequences at every full symbol interval was proposed by Simon [12]. The sixteen waveforms, or $w_i(t); i = 0, 1, 2, \ldots, 15$, are defined over the interval $-T_s/2 \le t \le T_s/2$, which collectively forms the transmitted baseband signals on the I–Q channels. These waveforms are plotted in Fig. C.10, and their functions are given as follows:

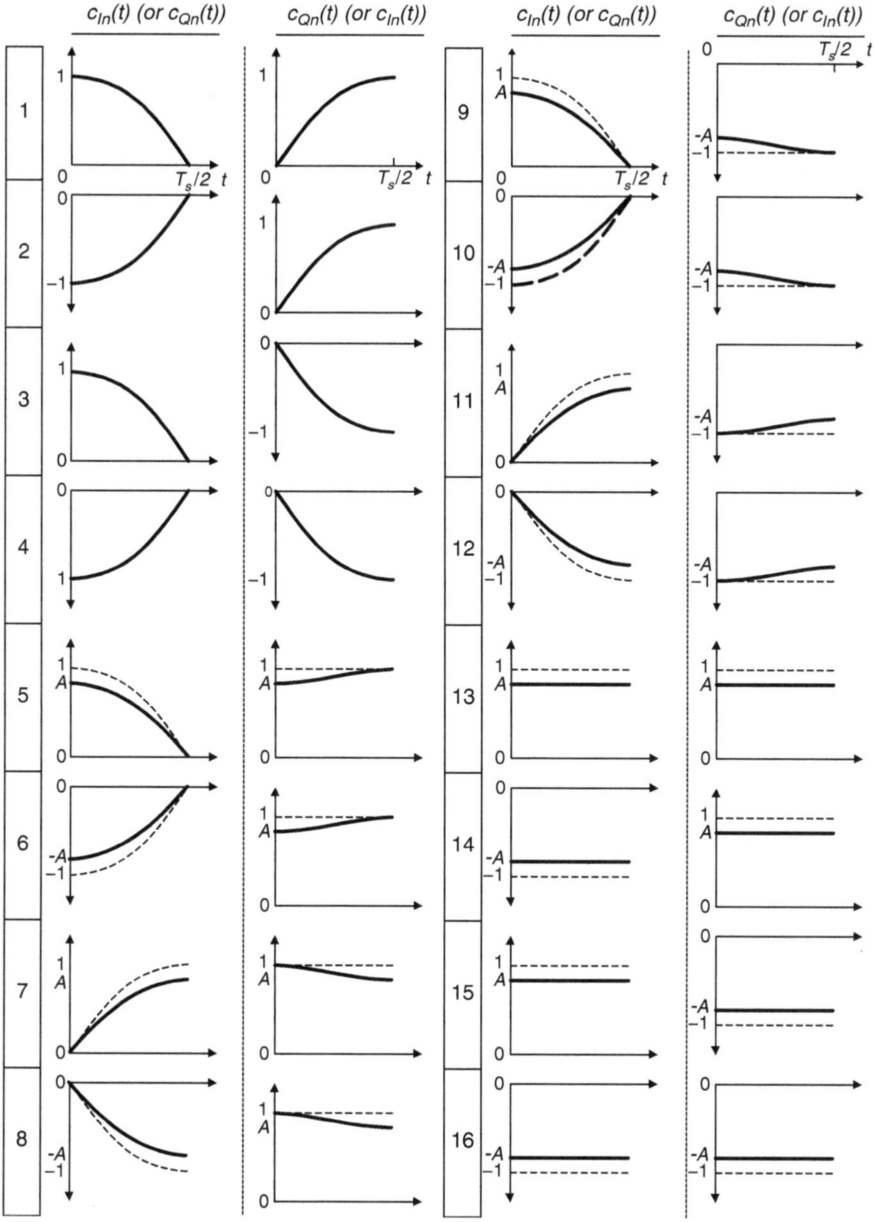

Fig. C.9 Sixteen combinations of FQPSK waveforms in half-symbol intervals

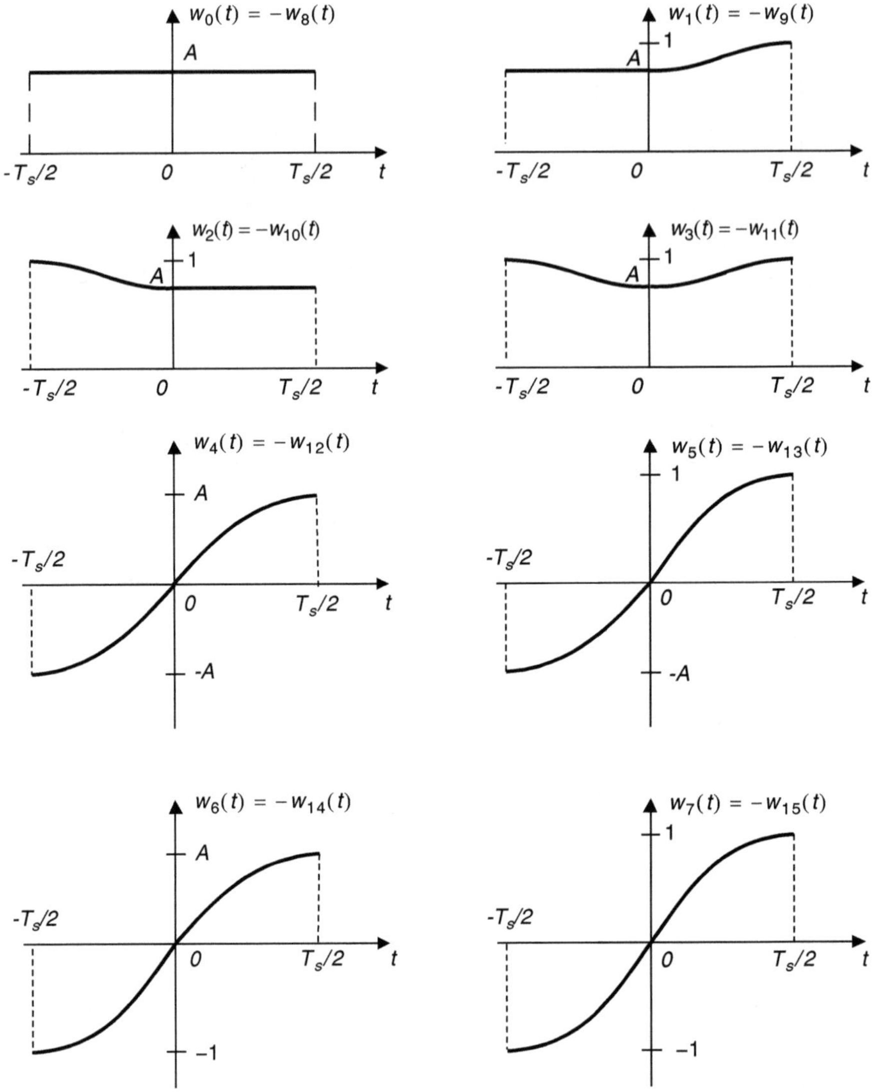

Fig. C.10 XPSK fundamental full symbol waveforms, where $A = 1/\sqrt{2}$ for nearly constant envelope. Redrawn from [13]

$$
\begin{aligned}
&w_0(t) = A, \quad -\frac{T_s}{2} \le t \le \frac{T_s}{2} \\
&w_1(t) = \begin{cases} A, & -\frac{T_s}{2} \le t \le 0 \\ 1-(1-A)\cos^2\frac{\pi t}{T_s}, & 0 \le t \le \frac{T_s}{2} \end{cases} \\
&w_2(t) = \begin{cases} 1-(1-A)\cos^2\frac{\pi t}{T_s}, & -\frac{T_s}{2} \le t \le 0 \\ A, & 0 \le t \le \frac{T_s}{2} \end{cases} \\
&w_3(t) = 1-(1-A)\cos^2\frac{\pi t}{T_s}, \quad -\frac{T_s}{2} \le t \le \frac{T_s}{2} \\
&w_4(t) = A\sin\frac{\pi t}{T_s}, \quad -\frac{T_s}{2} \le t \le \frac{T_s}{2} \\
&w_5(t) = \begin{cases} A\sin\frac{\pi t}{T_s}, & -\frac{T_s}{2} \le t \le 0 \\ \sin\frac{\pi t}{T_s}, & 0 \le t \le \frac{T_s}{2} \end{cases} \\
&w_6(t) = \begin{cases} \sin\frac{\pi t}{T_s}, & -\frac{T_s}{2} \le t \le 0 \\ A\sin\frac{\pi t}{T_s}, & 0 \le t \le \frac{T_s}{2} \end{cases} \\
&w_7(t) = \sin\frac{\pi t}{T_s}, \quad -\frac{T_s}{2} \le t \le \frac{T_s}{2} \\
&w_8(t) = -w_0(t),\ w_9(t) = -w_1(t),\ w_{10}(t) = -w_2(t),\ w_{11}(t) = -w_3(t) \\
&w_{12}(t) = -w_4(t),\ w_{13}(t) = -w_5(t),\ w_{14}(t) = -w_6(t),\ w_{15}(t) = -w_7(t)
\end{aligned}
\tag{C.16}
$$

The baseband waveform $c_{In}(t) = w_i(t)$ on the I channel in the nth symbol interval $[n-(1/2)]T_s \le t \le [n+(1/2)]T_s$ is not only dependent on its two successive symbol sequences, but also on its three successive symbol sequences on the Q channel, or the transition properties of the Q-channel symbols instead of their values. The mapping procedures are described in Table C.2. A similar mapping procedure for the baseband waveforms on the Q channel in the $nT_s \le t \le (n+1)T_s$ can be obtained analogously from Table C.2 and is omitted here.

A simple hardware implementation of the cross-correlated XPSK based on a look-up table method is illustrated in Fig. C.11. In the IJF-OQPSK-modulation case, the baseband waveforms (total four different waveforms) of the I channel in one symbol interval T_s are determined by the combinations of two successive input symbols, or one current symbol and one previous symbol. In the XPSK case, however, the I-channel baseband waveforms in the interval T_s are not only dependent on the two successive input symbols, but also on the three successive symbols of the Q channel, or transition properties of these three successive symbols.

Table C.2 Mapped baseband waveforms on the I channel in one symbol interval T_s

Output waveforms $c_{In}(t)$	I-CH input symbols $d_{I,n-1}d_{I,n}$	Q-CH input symbols $d_{Q,n-2}d_{Q,n-1}d_{Q,n}$
$w_0(t)$	11	−1−1−1 or 111 (no transition/no transition)
$w_1(t)$	11	−1−11 or 11−1 (no transition/transition)
$w_2(t)$	11	−111 or 1−1−1 (transition/no transition)
$w_3(t)$	11	−11−1 or 1−11 (transition/transition)
$w_4(t)$	−11	−1−1−1 or 111 (no transition/no transition)
$w_5(t)$	−11	−1−11 or 11−1 (no transition/transition)
$w_6(t)$	−11	−111 or 1−1−1 (transition/no transition)
$w_7(t)$	−11	−11−1 or 1−11 (transition/transition)
$w_8(t)$	−1−1	−1−1−1 or 111 (no transition/no transition)
$w_9(t)$	−1−1	−1−11 or 11−1 (no transition/transition)
$w_{10}(t)$	−1−1	−111 or 1−1−1 (transition/no transition)
$w_{11}(t)$	−1−1	−11−1 or 1−11 (transition/transition)
$w_{12}(t)$	1−1	−1−1−1 or 111 (no transition/no transition)
$w_{13}(t)$	1−1	−1−11 or 11−1 (no transition/transition)
$w_{14}(t)$	1−1	−111 or 1−1−1 (transition/no transition)
$w_{15}(t)$	1−1	−11−1 or 1−11 (transition/transition)

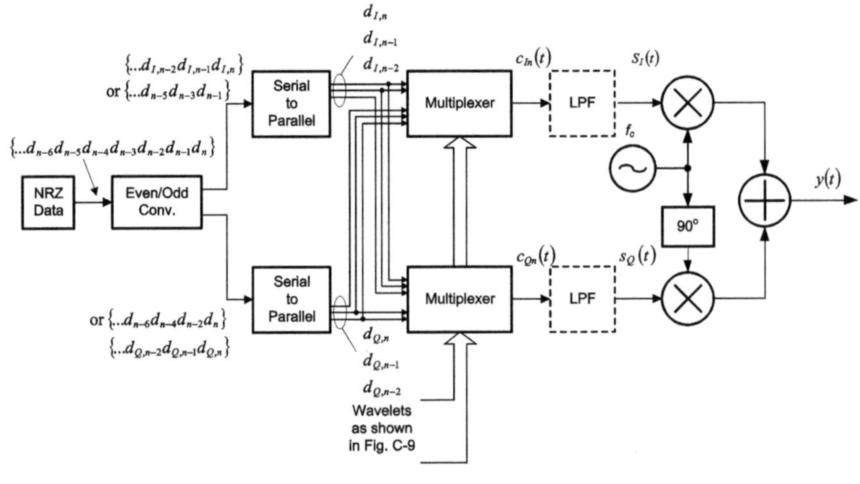

Fig. C.11 A block diagram of the cross-correlated XPSK based on a LUT method

Because there are four transition properties for three symbols, each of the four waveforms determined by the two I-channel symbols has four different waveforms, as shown in Table C.2. Thus, there are a total of 16 possible waveforms on the I channel or Q channel in the interval T_s.

The baseband signals of the cross-correlated XPSK are shown in Fig. 4.17. Compared with the baseband signals of IJF-OQPSK drawn with the dashed line,

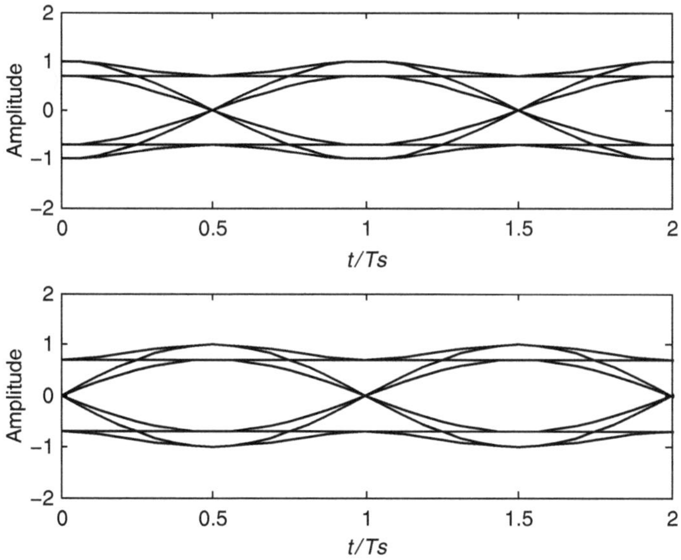

Fig. C.12 Eye diagrams of XPSK with $A = 1/\sqrt{2}$

the baseband signals of XPSK behave differently from that of IJF-OQPSK in order to achieve a constant envelope, such that the XPSK-modulated signal can avoid PSD side-lobe regrowth after passing through nonlinear channels.

The eye diagrams and constellation of XPSK with $A = 1/\sqrt{2}$ are shown in Figs. C.12 and C.13, respectively. In order to achieve nearly constant envelope, which is a necessary transmission condition without causing the PSD regrowth through a nonlinear channel, the ISI at the decision instants is intentionally introduced to XPSK. Fortunately, the ISI that is intentionally introduced in XPSK only slightly degrades the Bit Error Rate (BER), which will be shown in the following section.

It should be noted that the eye diagrams of the XPSK are very similar to that of *Tamed Frequency Modulation* (TFM) [14], which is a constant envelope modulation. Hence, the performance of the XPSK system is practically the same as that of TFM system. The XPSK signal, however, can be coherently demodulated by a conventional OQPSK demodulator, while the TFM demodulation processing is relatively complicated due to it property of frequency demodulation.

C.5 FQPSK-B

Although the XPSK modulation technique was first published in 1983, FQPSK modulation, however, did not achieve further significant spectrum improvement without significant BER degradation until 1996, when the baseband signals of the

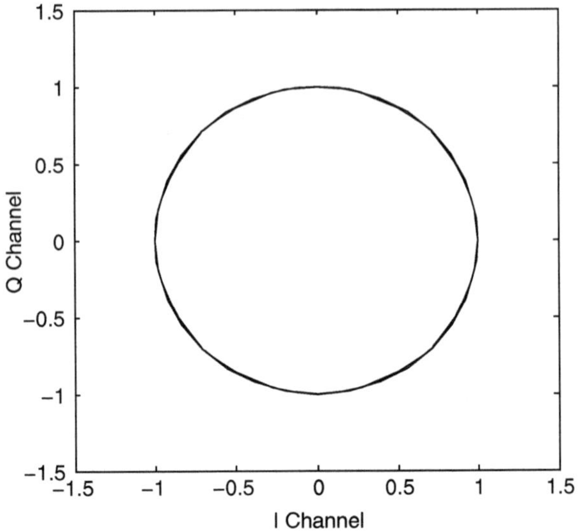

Fig. C.13 Constellation of XPSK with $A = 1/\sqrt{2}$

XPSK were filtered by Butterworth lowpass filters [15, 16], as illustrated in Fig. C.11. Due to such a filtering process, the filtered XPSK is called FQPSK-B. Now FQPSK simply stands for this advanced version. With such filtering the PSD's side-lobes of FQPSK-B (or filtered FQPSK) in a nonlinear channel roll off faster with a frequency increase when compared with the unfiltered FQPSK (or XPSK), while its envelope fluctuation slightly deviates from a nearly constant value. FQPSK-B, however, only suffers a BER degradation of 0.2 dB compared with the unfiltered FQPSK.

Figures C.14 and C.15 illustrate the power spectral densities of FQPSK-B and other modulations in either a linear or nonlinear channel, respectively. It is clear that the PSD of FQPSK-B is slightly affected by nonlinear amplification, but FQPSK-B still shows a significant spectral advantage when compared with filtered OQPSK, MSK, IJF-OQPSK, and XPSK (or unfiltered FQPSK) modulations. Even compared with GMSK, the PSD of FQPSK-B shows a spectral advantage over GMSK with $BT_b = 0.3$ up to –90 dB down.

C.6 BER Performance of FQPSK

Like the coherent detection for the OQPSK signal, the coherent detection for the FQPSK signal is preferable to non-coherent detection mainly due to a good BER performance. However, in some mobile channels, because of the frequency and phase offsets caused by multipath fading—such as Rayleigh fading, co-channel and adjacent channel interference, or other impairments—it is difficult, and sometimes impossible, to recover or track the carrier frequency and phase of the received

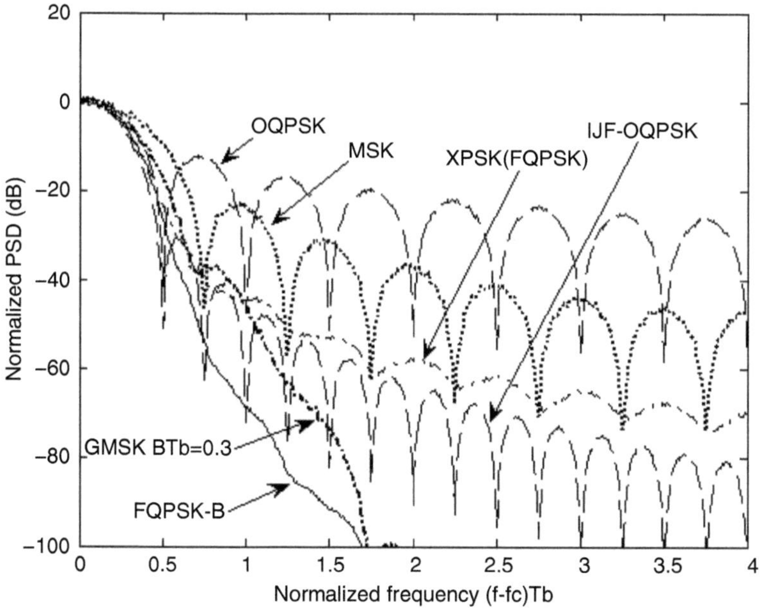

Fig. C.14 Power spectrum density of different modulations in a linear channel

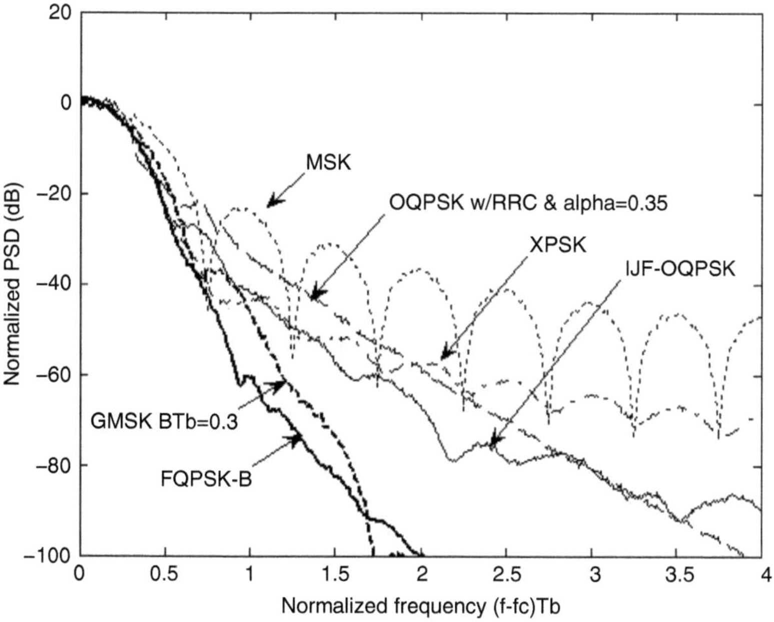

Fig. C.15 Power spectrum density of different modulations in a nonlinear channel

signal correctly, especially in the beginning of the reception. Under such situations, the receivers with such coherent detection suffer considerable performance loss. Even in some cases where the receivers finally synchronize their local oscillators with the carrier frequency and phase of the received signal, the receivers experience high burst errors and outages due to long acquisition times.

To solve these problems, non-coherent detection schemes such as differential detection [17] and limiter-discriminator detection [18] are the preferred countermeasures in the fading environment. Because of the robustness to both frequency and phase offsets provided by non-coherent reception systems, they have better performance in co-channel interference (CCI) and multipath fading, especially for fast fading with a large Doppler spread. For example, they can be used in the Bluetooth system, DECT system, and ZigBee system, where GFSK modulation is adopted, so that these systems have faster data recovery, lower cost, and lower implementation complexity. In addition, since they do not need the overhead to aid the carrier recovery, they can provide higher spectral efficiency and thus capacity than coherent systems. Therefore, non-coherent detection schemes are very attractive for systems that require low cost and low complexity.

In fact, the FQPSK-modulated signal was not available to be non-coherently detected at the receiver until 1999 [19]. Later, the discriminator detection for FQPSK and OQPSK was investigated in [20, 21]. To the author's best knowledge, it was the first time that this non-coherent detection scheme was reported for the OQPSK-type modulation signals, including FQPSK modulation. In this section, the simulation BER of FQPSK with coherent detection will be described due to its good BER performance. For differential detection and limiter-discriminator detection for FQPSK, the interested reader can refer to [20, 21].

Since FQPSK modulation is the same as OQPSK modulation, except for their different baseband waveforms, the coherent detection or demodulation used for OQPSK can also be used for FQPSK. As we have shown in Chap. 4, MSK and GMSK can also be treated as a quadrature phase modulation. Thus, some carrier recovery methods, such as the reverse modulation carrier recovery introduced in Chap. 4, can be used for FQPSK. A block diagram based on the reverse modulation carrier recovery shown in Fig. 4.27 can be used for coherent demodulation of FQPSK.

Usually, a pilot signal that allows the PLL to clock it first before the information-bearing signal is preferred. This pilot-aided carrier recovery scheme is very attractive in time-division multiple access (TDMA) system, such as the GSM system, in which data are transmitted in burst frames and fast carrier recovery and symbol timing synchronization are required. Each frame is further partitioned into assignable user time slots. In each slot, for example, alternating zero and one data pattern can be inserted prior to the information data for the pilot aided transmission. In the reverse-modulation–based carrier recovery, it is required for the PLL to lock its frequency and phase to the carrier frequency and phase of the received pilot signal first. Then, the received data after the pilot data are coherently detected. Meanwhile, the recovered data, in turn, are used to re-modulate the following received information-bearing modulated signal.

In the coherent demodulation of FQPSK shown in Fig. 4.27, a simple fourth-order Butterworth lowpass filters after the mixer is used to replace a signal correlator, or a so-called matched filter, in an optimum trellis-coded receiver for FQPSK [13]. In fact, in most practical applications, simple lowpass filters rather than correlators are preferred for their simplicity, especially in analog designs. Actually, FQPSK performance for coherent demodulation based on a Butterworth filter is competitive with that based on a signal correlator [13] due to its simple implementation and low cost.

Figure C.16 shows the recovered eye diagrams of FQPSK-B at the output of the fourth-order Butterworth lowpass filter. Due to the narrow bandwidth of the receiver channel selection filter, the Butterworth lowpass filter has large group delay variation within the bandwidth, and such group delay variation causes ISI. As a result, it degrades the system performance. Therefore, it is necessary for the receiver to have an allpass filter as a group delay equalizer to compensate for such group delay variation. It is obvious from Fig. C.16b that the compensated or equalized eyes have less ISI at the decision instants. Figure C.16c shows the experimental eye diagrams after the group delay equalizer.

Figure C.17 illustrates the BER curves of FQPSK/FQPSK-B with a Butterworth filter-based receiver. These results are obtained from MATLAB simulation. We observe that E_b/N_o required by the FQPSK-B (filtered FQPSK) receiver with group delay compensation at BER $= 10^{-4}$ is only about 0.2 dB more than the FQPSK (unfiltered FQPSK) receiver with group delay compensation, or only about 1.2-dB degradation compared to theoretical OQPSK performance. It is obvious that the group delay equalizer at the receiver LPF can improve BER performance by about 0.5 dB. If an optimum receiver is used for FQPSK, the BER performance of FQPSK-B with trellis-coded (Viterbi) is only about 0.6 dB inferior to the theoretical OQPSK performance [13] and 0.6 dB superior to that of FQPSK-B with Butterworth filter at the cost of increasing hardwire implementation.

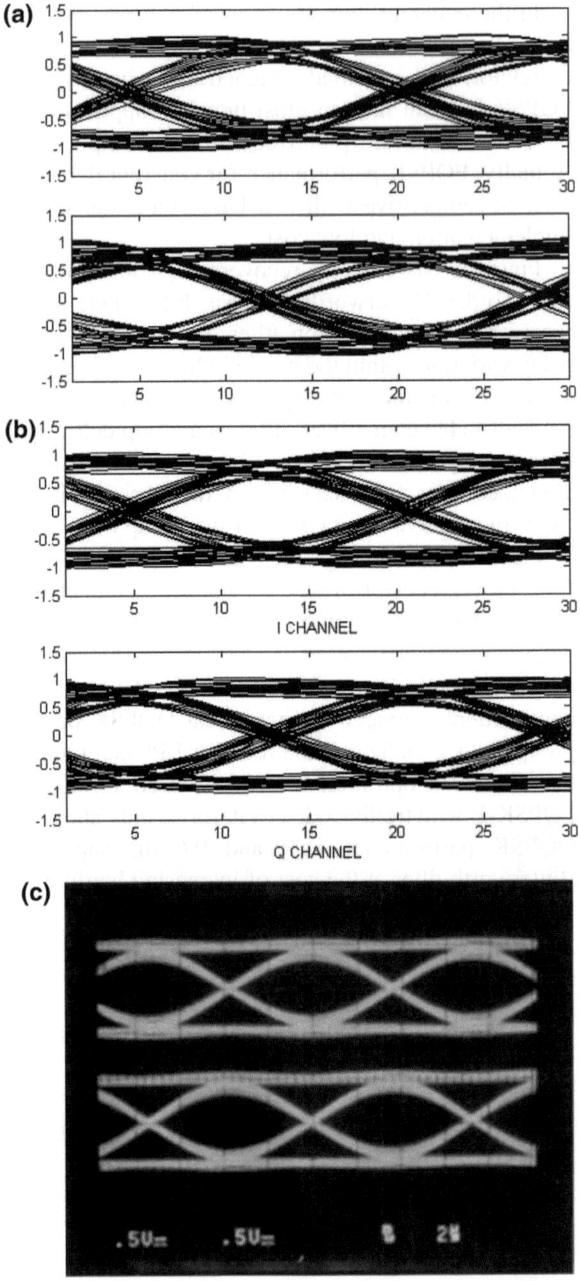

Fig. C.16 Received eye diagrams of FQPSK-B at receiver LPF output: (**a**) simulation before the second-order allpass filter, (**b**) simulation after the second-order allpass filter, and (**c**) hardware implementation after the second-order allpass filter at the bit rate of 270.833 kbps

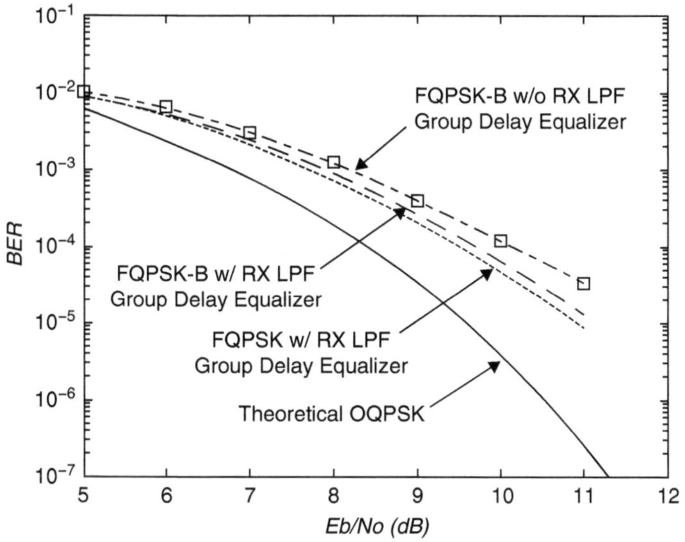

Fig. C.17 Bit error rate (BER) of FQPSK and FQPSK-B modulations in a nonlinear channel, where a second-order allpass filter is used for group delay compensation of receiver fourth-order Butterworth LPF with optimum $B_i T_b = 0.55$ at the receiver

Appendix D: Allpass Filter as Group Delay Equalizer

Due to its constant amplitude response, an allpass filter is widely used as a phase shifter to create the desired phase response over the specified frequency range. In digital communications, allpass filters are often employed as group delay compensators or equalizers to compensate for group delay variations caused by band-limited filtering at both transmitter and receiver in order to minimize ISI that degrades the bit error rate at the receiver. One of typical applications in the 3GPP WCDMA system is that an allpass filter is used at a receiver of an integrated RF transceiver as a group delay equalizer to compensate for group delay variations of the analog filters. In the following section, we introduce the first-order and second-order allpass as fundamental sections for constructing high-order allpass filters.

D.1 First-Order Allpass Filter

The transfer function of the first-order allpass filter is given by

$$H_a(s) = \frac{s - \sigma}{s + \sigma} \tag{D.1}$$

The amplitude is equal to constant 1 in all frequencies because the position σ of the zero is symmetrical to the position $-\sigma$ of the pole from the image axis. By using the normalized frequency $s_n = s/\omega_c$, (D.1) can be written as [22]

$$H_a(s_n) = \frac{s_n - \sigma_n}{s_n + \sigma_n} \tag{D.2}$$

where $\sigma_n = \sigma/\omega_c$ is the normalized zero or $-\sigma_n$ is the normalized pole. The transfer function above is also expressed as a polar format, or

$$H_a(s_n) = |H_a(j\omega_n)| e^{-j\theta_a(\omega_n)} \tag{D.3}$$

where the phase is

$$\theta_a(\omega_n) = -2\tan^{-1}\left(\frac{\omega_n}{\sigma_n}\right) \tag{D.4}$$

The group delay is obtained by taking the negative derivative of the phase $\theta_a(\omega_n)$ as given by

$$\mathrm{GD}_a(\omega_n) = -\frac{d\theta_a(\omega_n)}{d\omega_n} = \frac{2/\sigma_n}{1 + (\omega_n/\sigma_n)^2} \tag{D.5}$$

The group delay of the first-order allpass filter is plotted in Fig. D.1, where the cut-off frequency is normalized, and $\sigma_n = 1/(\mathrm{RC}) = 0.5, 1, \text{and } 2$, respectively.

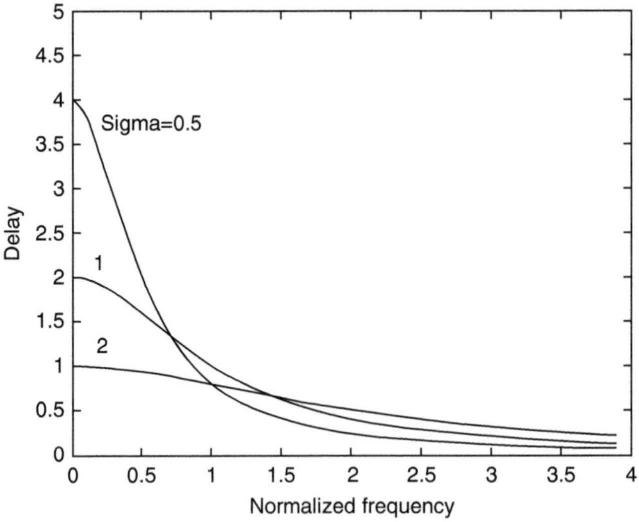

Fig. D.1 Group delay of the first-order allpass filter versus the normalized frequency ω_n for different values of σ_n

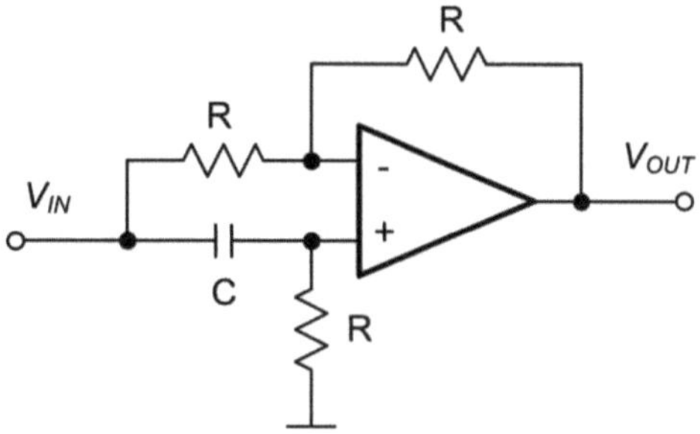

Fig. D.2 Active circuit with positive unit gain of the first-order allpass filter [22]

It can be seen from each curve that the group delay continuously decreases with frequency increase. Thus, the first-order allpass filter is suitable to compensating the filter's delay that has less delay at the low-frequency range and more delay at the range close to the cut-off frequency.

The maximum delay happens at $\omega_n = 0$

$$\mathrm{GD_{MAX}} = \mathrm{GD_a}(0) = \frac{2}{\sigma_n} \tag{D.6}$$

Either active circuits or passive circuits can realize the first-order allpass filter. Figure D.2 shows an active circuit of the first-order allpass filter with a positive gain. This circuit realizes $H_a(s)$ as in (D.1) with $\sigma = 1/(\mathrm{RC})$:

$$H_a(s) = \frac{s - 1/(\mathrm{RC})}{s + 1/(\mathrm{RC})} \tag{D.7}$$

Due to its simple circuit design, the first-order allpass filter is quite often used in the case where the group delay of the target filter has the minimum delay at $\omega_n = 0$, and then increases monotonically almost up to the normalized cut-off frequency, such as a Butterworth filter.

D.2 *Second-Order Allpass Filter*

A transfer function of the second-order allpass filter is given by

$$H_a(s) = \frac{s^2 - \frac{\omega_0}{Q}s + \omega_0^2}{s^2 + \frac{\omega_0}{Q}s + \omega_0^2} \tag{D.8}$$

where ω_0 is the pole frequency and Q is the quality factor. Compared with the first-order allpass filter, the second-order allpass filter has two adjustable parameters so that it has more shapes of the group delay. In general, it is easy to design the allpass filter starting from a normalized transfer function. Then, the actual transfer function can be obtained by de-normalizing the normalized transfer function through the actual cut-off frequency ω_c.

By using the normalized frequency, $s_n = s/\omega_c$, (D.8) can be written as

$$H_a(s_n) = \frac{s_n^2 - \frac{\widetilde{\omega}_0}{Q} s_n + \widetilde{\omega}_0^2}{s_n^2 + \frac{\widetilde{\omega}_0}{Q} s_n + \widetilde{\omega}_0^2} \tag{D.9}$$

where $\widetilde{\omega}_0 = \omega_0/\omega_c$ is the normalized pole frequency. The phase and group delay of the second-order allpass filter are given by

$$\theta_a(\omega_n) = -2\tan^{-1}\left(\frac{\frac{\omega_n \widetilde{\omega}_0}{Q}}{\widetilde{\omega}_0^2 - \omega_n^2}\right) \tag{D.10}$$

$$\mathrm{GD}_a(\omega_n) = -\frac{d\theta_a(\omega_n)}{d\omega_n} = \frac{\frac{2\widetilde{\omega}_0}{Q}\left(\widetilde{\omega}_0^2 + \omega_n^2\right)}{\left(\widetilde{\omega}_0^2 - \omega_n^2\right)^2 + \left(\frac{\omega_n \widetilde{\omega}_0}{Q}\right)^2} \tag{D.11}$$

The group delay response of the second-order allpass filter versus the factor Q at $\widetilde{\omega}_0 = 1$ is plotted in Fig. D.3. It is clear that the shape of the group delay is dependent on the factor Q. It was calculated that the group delay has a peak when $Q > 1/\sqrt{3} \approx 0.577$. Otherwise, the group delay decreases monotonously from the zero frequency, and has its maximum delay at the zero frequency. The group delay

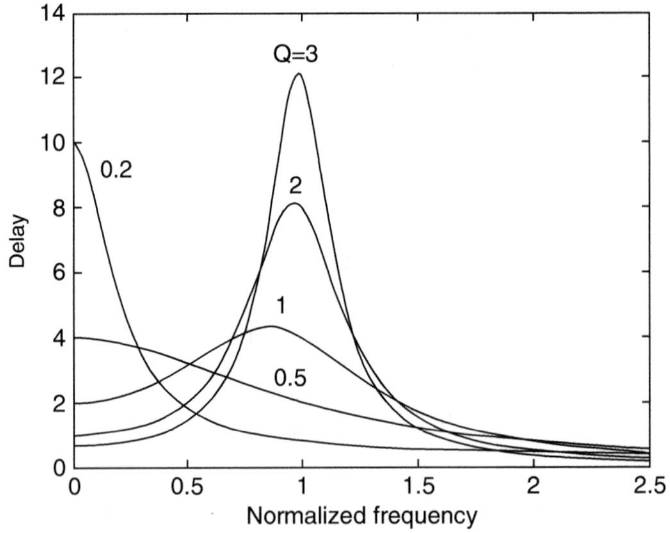

Fig. D.3 Group delay of the second-order allpass filter with different Q values at $\widetilde{\omega}_0 = 1$

with such a peak in the range from 0 to $\widetilde{\omega}_0 = 1$ makes the second-order allpass filter more flexible to compensate for the distorted delay with a shallow null in such a range, which cannot be compensated by the first-order allpass filter.

It can be seen from (D.11) that the delay at zero frequency is

$$\mathrm{GD_a}(0) = \frac{2}{Q\widetilde{\omega}_0} \tag{D.12}$$

When $Q > 1/\sqrt{3}$, the delay curve has the peak at about $\widetilde{\omega}_0$, and this peak is equal to

$$\mathrm{GD}_{a,\mathrm{MAX}}(\widetilde{\omega}_0) \approx \frac{4Q}{\widetilde{\omega}_0} \tag{D.13}$$

The group delays versus different values of $\widetilde{\omega}_0$ at $Q = 2$ are plotted in Fig. D.4. It is clear that the peak almost occurs at the pole frequency of $\widetilde{\omega}_0$. Thus, we can determine the peak position through $\widetilde{\omega}_0$.

Like the first-order allpass filter, the second-order allpass filter can be realized in either the active circuits or the passive circuits. Figure D.5 shows the second-order allpass filter constructed in the active circuits of *Thomas1*. Its transfer function is given by

$$H_\mathrm{a}(s) = -\frac{s^2 + \frac{R_5 - R_6}{R_3 R_5 C_2} s + \frac{R_6}{R_2 R_5 R_8 C_1 C_2}}{s^2 + \frac{1}{R_3 C_2} s + \frac{R_6}{R_2 R_5 R_8 C_1 C_2}} \tag{D.14}$$

For the realization of the allpass filter, the relationship between R_5 and R_6 is $R_6 = 2R_5$. Thus, (D.14) is rewritten as

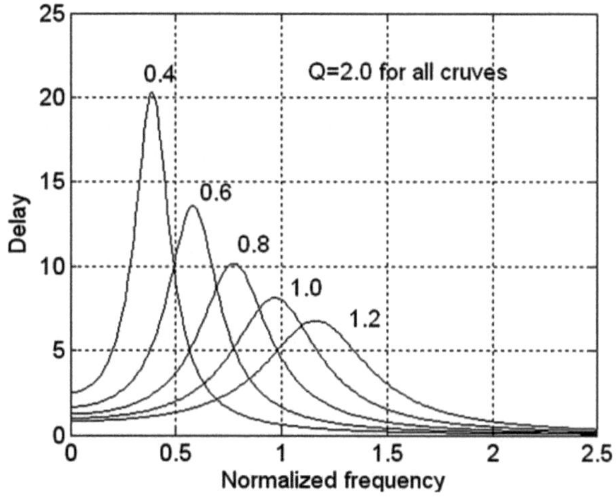

Fig. D.4 Group delay of the second-order allpass filter with different $\widetilde{\omega}_0$ values at $Q = 2$

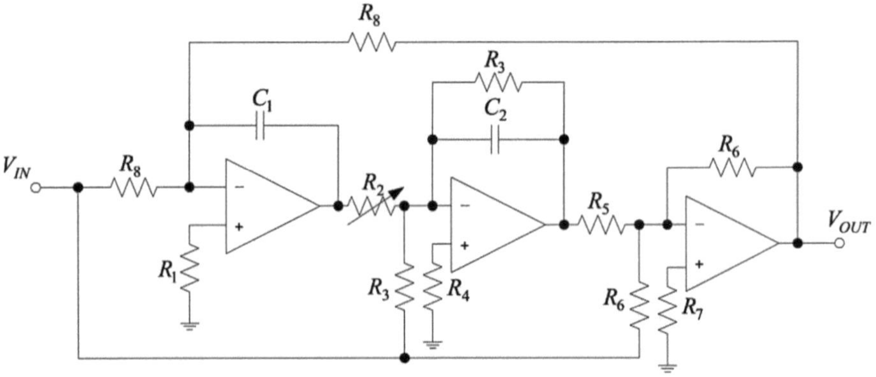

Fig. D.5 Active circuit of the second-order allpass filter

$$H_a(s) = -\frac{s^2 - \frac{1}{R_3C_2}s + \frac{R_6}{R_2R_5R_8C_1C_2}}{s^2 + \frac{1}{R_3C_2}s + \frac{R_6}{R_2R_5R_8C_1C_2}} \tag{D.15}$$

The minus sign in (D.15) is due to inverting amplification operation. This can be corrected by adding one more stage of inverting amplification.

Comparing (D.15) with the standard form of (D.9), we have the following appropriate parameters as

$$\tilde{\omega}_0^2 = \frac{R_6}{R_2R_5R_8C_1C_2}, \quad Q = R_3\sqrt{\frac{R_6C_2}{R_2R_5R_8C_1}} \tag{D.16}$$

In the following, we give an example of using the first-order and the second-order allpass filters to compensate for the group delay of a fourth-order Butterworth lowpass filter.

Design Example D.1 In digital communications, a bandlimited channel with a constant group delay or small group delay variation is preferable for minimizing ISI. Using both first-order and second-order allpass filters design a group delay equalizer to reduce the group delay variation of a fourth-order Butterworth lowpass filter with a cut-off frequency of 17.2 kHz. As introduced in Design Example 2.1, this analog filter was used to approximate a pulse-shaping root raised-cosine filter with $\alpha = 0.5$ for QPSK data transmission at a bit rate of 64 kbps.

Solution With a First-Order Allpass Filter We begin the design with a fourth-order Butterworth lowpass filter with a normalized frequency $\omega_n = \omega/\omega_c$, where $\omega_c = 2\pi f_c$ is the cut-off frequency. We use the subscript n here to distinguish the normalized frequency with the actual frequency. Thus, the normalized transfer function of the fourth-order Butterworth lowpass filter is

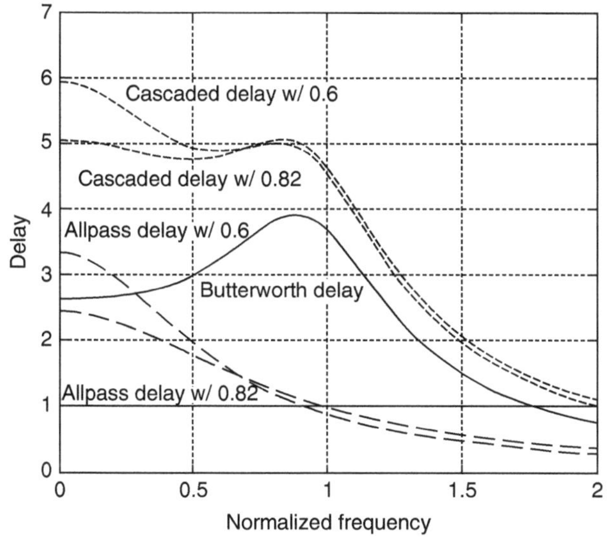

Fig. D.6 Group delay responses of the normalized fourth-order Butterworth lowpass filter and the first order allpass filter

$$H_L(s_n) = \frac{1}{\left(s_n^2 + 0.7654 s_n + 1\right)\left(s_n^2 + 1.848 s_n + 1\right)} \tag{D.17}$$

It is easy to calculate the group delay response of Butterworth lowpass filter by using a MATLAB calculation script. This group delay response is plotted in Fig. D.6. The group delay monotonically increases in the frequency range from 0 to 0.9 and variation is about 1.3 s within this frequency range.

The normalized transfer function of the first-order allpass filter is

$$H_a(s_n) = \frac{s_n - \sigma_n}{s_n + \sigma_n} \tag{D.18}$$

Its group delay is given in (D.5) and is rewritten here

$$\text{GD}_a(\omega_n) = 2\frac{\sigma_n}{\sigma_n^2 + \omega_n^2} \tag{D.19}$$

Figure D.6 shows the group delay of the first-order allpass filter with different σ values. Unlike the group delay shape of the Butterworth filter, the group delay of the first-order allpass filter with $\sigma_n = 1$ monotonically decreases in the same frequency range from 0 to 0.9 and variation is about 1 s.

Therefore, an appropriate compensation delay would be created with $\sigma_n < 1$. As a try, we first choose $\sigma_n = 0.6$ and plot the delay of the allpass filter in Fig. D.6, which is labeled "allpass delay *w*/0.6". The cascaded group delay is labeled "cascaded delay *w*/0.6". It is obvious that the allpass filter with $\sigma_n = 0.6$ adds too much delay to the Butterworth filter. Fortunately, it is relatively easy to find the optimal sigma value $\sigma_n = 0.82$ to achieve small delay variation with several trials due to only one parameter. Thus, the cascaded delay with $\sigma_n = 0.82$ gives the smallest delay variation of about $\Delta\text{GD}(\omega_n) = 0.3$ s, which is much smaller than the un-equalized delay variation of 1.3 s within the specified frequency range. After the delay equalization, the absolute delay increases about two times at the DC frequency, or from 2.6 to 5.0 s, but the absolute delay does not cause any problem in digital communications.

With $\sigma_n = 0.82$, the transfer function of the first-order allpass filter can be de-normalized by substituting $\sigma = \sigma_n \times \omega_c = 0.82 \times 2\pi \times 17,200 = 8.8618 \times 10^4$ into (D.18)

$$H_a(s) = \frac{s - 8.8618 \times 10^4}{s + 8.8618 \times 10^4} \tag{D.20}$$

Finally, the values of R and C are solved with $\sigma = 1/(RC)$, or $RC = 1/\sigma = 11.284\,\mu\text{s}$. If $C = 10$ nF is chosen, then the resistor is calculated to be equal to $R = 1.13$ kΩ.

Meanwhile, the transfer function of the Butterworth filter can be also de-normalized to the true transfer function by substituting the normalized frequency with two slightly different cut-off frequencies around the target cut-off frequency of 17.2 kHz, or $s_n = s/(2\pi \times 17,096)$ and $s_n = s/(2\pi \times 17,193)$, into two second-order sections in (D.17), respectively,

$$H_L(s) = \frac{(1.1539 \times 10^{10})(1.167 \times 10^{10})}{(s^2 + 8.2386 \times 10^4 s + 1.1539 \times 10^{10})(s^2 + 1.9724 \times 10^5 s + 1.167 \times 10^{10})} \tag{D.21}$$

If the lowpass filter $H_L(s)$ is implemented by cascading two *Sallen-Key* lowpass filters [22], its transfer function is expressed as

$$H_L(s) = \frac{\frac{1}{r_1^2 c_1 c_2}}{s^2 + \frac{2}{r_1 c_1} s + \frac{1}{r_1^2 c_1 c_2}} \times \frac{\frac{1}{r_2^2 c_3 c_4}}{s^2 + \frac{2}{r_2 c_1} s + \frac{1}{r_2^2 c_3 c_4}} \tag{D.22}$$

Parameters can be solved by comparing (D.21) and (D.22) as follows: $r_1 = 8.45$ kΩ, $c_1 = 1.2$ nF, $c_2 = 1.0$ nF, $r_2 = 3.57$ kΩ, $c_3 = 6.8$ nF, and $c_4 = 1.0$ nF.

Figure D.7 shows the group delay curves of two transfer functions that are expressed in (D.20) and (D.21) and their cascaded group delay curve in an actual frequency range. The actual delay variation is de-normalized by dividing the normalized group delay variation $\Delta\text{GD}(\omega_n) = 0.3$ by ω_c, or $\Delta\text{GD}(\omega) = 0.3/\omega_c = 0.3/(2\pi \times 17,200) = 2.776\,\mu\text{s}$ within the bandwidth, which can also be seen in Fig. D.7.

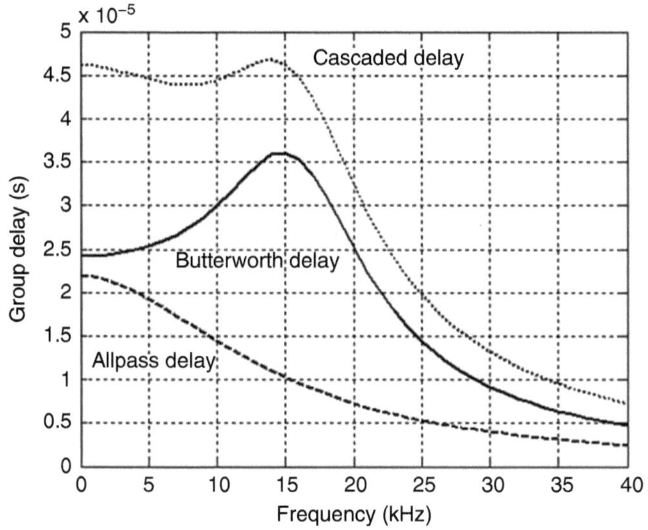

Fig. D.7 Group delay responses of the fourth-order Butterworth and first-order allpass filters in an actual frequency range

Fig. D.8 Group delay responses of the Butterworth lowpass and the second-order allpass filters for Example D.1

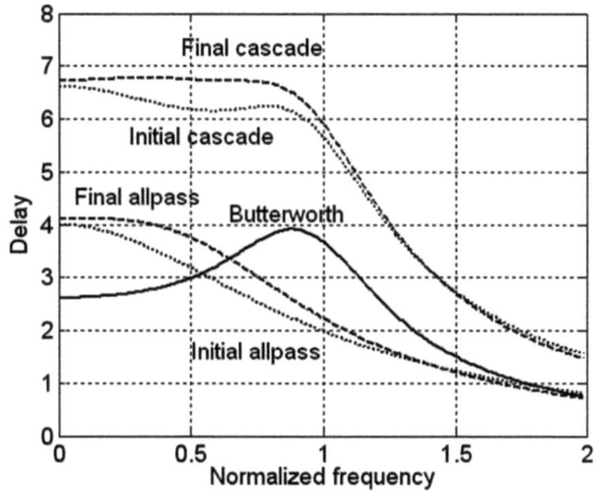

Solution With a Second-Order Allpass Filter First of all, we observe from Fig. D.6 that the group delay of the Butterworth filter increases monotonically up to the normalized frequency of 0.9. This means that the group delay of the second-order allpass filter should decrease monotonically in order to have the inverse characteristic of the group delay of the fourth-order Butterworth filter. From Fig. D.3, we can see that the group delay of the second-order allpass filter continuously decreases starting from zero frequency when $Q < 1/\sqrt{3} \approx 0.577$. We initially try to set $\widetilde{\omega}_0^2 = 1$ and $Q = 0.5$, and solve $\widetilde{\omega}_0/Q = 2$. Substituting these parameters into (D.11), we plot the group delay in Fig. D.8. From the initial cascade response, we can see that it is low at a frequency of around 0.5. To get more delay within such a range, we need to make the "initial allpass" delay flat around the normalized frequency of 0.5 by reducing $\widetilde{\omega}_0$ and increasing Q as well. With several further trials, the smallest delay variation curve labeled 'Final cascade' is obtained with $\widetilde{\omega}_0^2 = 0.71$ and $Q \approx 0.577$, and its peak-to-peak variation is about $\Delta\text{GD}(\omega_\text{n}) \approx 0.15\,\text{s}$, within the range from 0 to 0.8 rad/s, which is smaller by a half than $\Delta\text{GD}(\omega_\text{n}) \approx 0.3\,\text{s}$ in the case of the first-order allpass filter. Hence, the group delay variation with the second-order allpass filter is reduced to 0.15 from its original value of 1.3, or 8.5 times smaller than its original delay variation within the specified frequency range, respectively.

The normalized transfer function of the second-order allpass filter is given by substituting $\widetilde{\omega}_0 = \sqrt{0.71}$ and $Q \approx 0.577$ into (D.9)

$$H_\text{a}(s_\text{n}) = \frac{s_\text{n}^2 - 1.46 s_\text{n} + 0.71}{s_\text{n}^2 + 1.46 s_\text{n} + 0.71} \tag{D.23}$$

It is clearly shown that the second-order allpass filter with two adjustable parameters can achieve many different shapes, so that it is more flexible to compensate for different group delay responses than the first-order allpass filter. Figure D.9 shows the group delay responses of the Butterworth lowpass filter cascaded with the first-order allpass filter and the second-order allpass filter in the actual frequency range.

Next, the actual delay variation is $\Delta\text{GD}(\omega) = 0.15/\omega_\text{c} = 0.15/(2\pi \times 17{,}200)$ $= 1.388\,\mu\text{s}$ within the specified frequency range, which is also a half of $2.776\,\mu\text{s}$ in the case of the first-order allpass filter. The actual parameters of the second-order

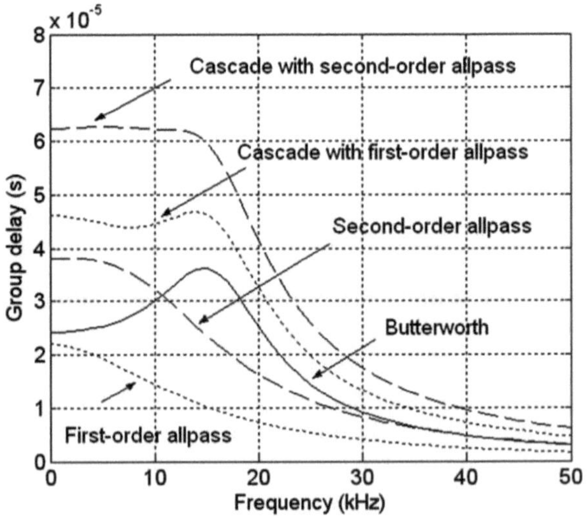

Fig. D.9 Group delay response of the fourth-order Butterworth lowpass filter cascaded with the first-order and second-order allpass filters in the actual frequency range for Example D.1

allpass filter are $\omega_0^2 = 0.71 \times (2\pi \times 17,200) = 8.292 \times 10^9$, and $\omega_0/Q = 1.578 \times 10^5$, and its transfer function is given by substituting these two parameters into (D.9):

$$H_a(s) = \frac{s^2 - 1.578 \times 10^5 s + 8.292 \times 10^9}{s^2 + 1.578 \times 10^5 s + 8.292 \times 10^9} \tag{D.24}$$

From (D.24), we can solve resistor and capacitor real values. Compared (D.24) with (D.15), we have the relationship $\omega_0/Q = 1/(R_3C_2) = 1.578 \times 10^5$. By choosing $C_2 = 1\,\text{nF}$ resistor is $R_3 = 1/\left(1.578 \times 10^5 \times 10^{-9}\right) = 6.34\,\text{k}\Omega$. Then, from the relationship $\omega_0^2 = R_6/(R_2R_5R_8C_1C_2)$ and with $R_6 = R_8 = 20\,\text{k}\Omega$, $R_5 = 10\,\text{k}\Omega$ and $C_1 = 1\,\text{nF}$, the resistor R_2 is given by $R_2 = 1/\left(R_5C_1C_2\omega_0^2\right) = 12.1\,\text{k}\Omega$. Figure D.10 shows the active implementation structure of the fourth-order Butterworth lowpass filter with a cut-off frequency of 17.2 kHz cascaded with the second-order allpass filter.

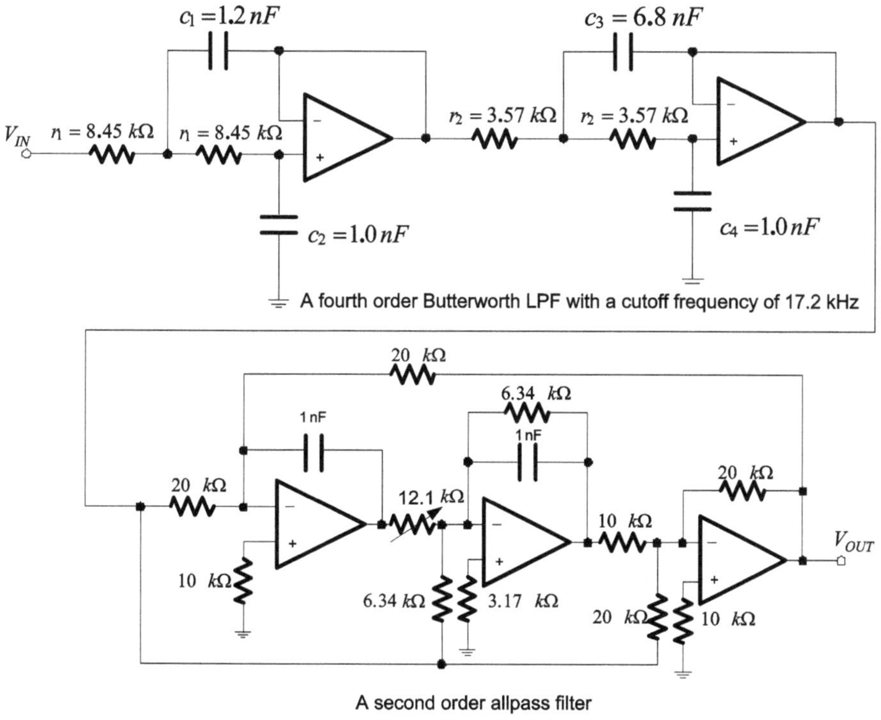

Fig. D.10 Active circuits of the fourth-order Butterworth lowpass filter cascaded with the second-order allpass filter

References

1. Kenney, J. S., & Leke, A. (1995, October). Power amplifier spectral regrowth for digital cellular and PCS applications. *Microwave Journal,* 74–92.
2. Ali-Ahmad, W. Y. (2004, April). Effective IM2 estimation for two-tone and WCDMA modulated blockers in zero-IF. In *RF Design* (pp. 32–40).
3. Razavi, B. (2003). *RF microelectronics*. Taiwan: Pearson Education Taiwan Ltd.
4. Le-Nook, T., & Feher, K. (1982). New modulation technique for low-cost power and bandwidth efficient satellite earth station. *IEEE Transactions on Communications, COM-30*(1), 275–283.
5. Le-Ngoc, T., & Fener, K. (1983). Performance of IJF-QOPSK modulation scheme in a complex interference environment. *IEEE Transactions on Communications COM-31*(1), 137–144.
6. Seo, J. S. (1983). *Superposed quadrature amplitude modulation (SQAM): A spectral and power efficient modulation technique.* M.A.Sc. thesis, University of Ottawa, Ottawa, Ont., Canada.
7. Seo, J. S., & Feher, K. (1985). SQAM: A new superpose*d QAM modem technique. Transactions on Communications, COM-33*(3), 296–300.
8. Kato, S., & Feher, K. (1983). XPSK: A new cross-correlated phase shift keying modulation technique. *IEEE Transactions on Communications, COM-31*(5), 701–707.
9. Range Commanders Council Telemetry Group, Range Commanders Council, White Sands Missile Range, New Mexico, *IRIG Standard 106-00:Telemetry Standards, 2000.*
10. Austin, M. C., & Chang, M. U. (1981). Quadrature overlapped raised-cosine modulation. *IEEE Transactions on Communications, COM-29*(3), 237–249.
11. Feher, K. (1983). *Digital communications: Satellite/earth station engineering*. Englewood Cliffs, NJ: Prentice-Hall.
12. Simon, M. K., & Yan, T. Y. (2000). Unfiltered Feher-patented quadrature phase shift-keying (FQPSK): Another interpretation and further enhancements: Parts 1, 2. *Applied Microwave & Wireless Magazine,* pp. 76–96/pp. 100–105, February/March 2000.
13. Simon, M. K. *Bandwidth-efficient digital modulation with application to deep-space communications.* JPL Publication 00-17, June 2001.
14. Jager, F. D., & Dekker, C. B. (1978). Tamed frequency modulation, a novel method to achieve spectrum economy in digital transmission. *IEEE Transactions Communications, COM-26,* 534–542.
15. Feher, K. et al., U.S. patents: 4,567,602; 4,339,724; 4,644,565; 5,784,402; 5,491,457. Canadian patents: 1,211,517; 1,130,871; 1,265,851.

© Springer International Publishing Switzerland 2017
W. Gao, *Energy and Bandwidth-Efficient Wireless Transmission*, Signals and Communication Technology, DOI 10.1007/978-3-319-44222-8

16. Hatamoto, C. (1998). *Improved FQPSK modulation technique*. MS thesis, University of California at Davis.
17. Simon, M. K., & Wang, C. C. (1984). Differential detection of Gaussian MSK in a mobile radio environment. *IEEE Transactions on Vehicular Technology, VT-33*(4), 307–320.
18. Pawula, R. F. (1981). On the theory of error rates for narrow-band digital FM. *IEEE Transactions on Communications, COM-29*(11), 1634–1643.
19. Park, H. C. (1999). *Differential detection techniques for spectrally efficient FQPSK signals*. Ph.D. dissertation, Dept of EIE, Seoul National University of Science and Technology, Seoul, Korea.
20. Lin, J. (2002). *Spectrum and RF Power Efficient Wireless Communication Systems*. Ph.D. Dissertation, Dept of ECE, University of California at Davis.
21. Lin, J., & Feher, K. (2003). Noncoherent limiter-discriminator detection of standardized FQPSK and OQPSK. In *IEEE Wireless Communications and Networking Conference* (WCNC) 2003, New Orleans, March 2003.
22. Schaumann, R., & Valkenburg, M. E. (2001). *Design of analog filters*. New York: Oxford University Press.
23. Part 11: Wireless LAN medium access control (MAC) and physical layer (PHY) specifications, IEEE Std. 802.11a, 1999.
24. Gao, W. (2013). Performance Enhancement of a WCDMA/HSDPA+ Receiver via Minimizing Error Vector Magnitude. IEEE International Test Conference (ITC'13) 2013, Anaheim, California, September 8–13, 2013.

Index

© Springer International Publishing Switzerland 2017

W. Gao, *Energy and Bandwidth-Efficient Wireless Transmission*, Signals and Communication Technology, DOI 10.1007/978-3-319-44222-8